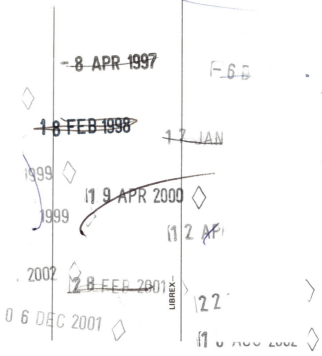

High Temperature Mechanical Behavior of Ceramic Composites

High Temperature Mechanical Behavior of Ceramic Composites

Edited by

Shanti V. Nair and Karl Jakus

Butterworth–Heinemann

Boston   Oxford   Melbourne   Singapore   Toronto   Munich   New Delhi   Tokyo

𝓡 A member of the Reed Elsevier group

Library of Congress Cataloging-in-Publication Data

High temperature mechanical behavior of ceramic composites / edited by
  Shanti V. Nair and Karl Jakus.
     p.   cm.
  Includes bibliographical references and index.
  ISBN 0-7506-9399-1 (acid-free paper)
  1. Ceramic–matrix composites—Mechanical properties. 2. Materials
at high temperatures. I. Nair, Shantikumar Vasudevan, 1953
II. Jakus, Karl.
TA418.9.C6H546 1995
620.1'4—dc20   95-9951
                                            CIP

British Library Cataloguing-in-Publication Data
A catalogue record for this book is available from the British Library

Butterworth–Heinemann
313 Washington Street
Newton, MA 02158-1626

10 9 8 7 6 5 4 3 2 1

Printed in the United States of America

Contents

PART **D**

Environmental Effects

PART **E**

Modeling

PART **F**

Elevated Temperature Mechanical Testing

Contributors

J. L. Bassani
Department of Mechanical
   Engineering and Applied
   Mechanics
University of Pennsylvania
Philadelphia, PA

D. C. Cranmer
Office of Science and Technology
   Policy
Washington, DC

S. F. Duffy
Department of Civil Engineering
NASA Research Associate
Cleveland State University
Cleveland, OH

A. G. Evans
Materials Department
College of Engineering
University of California
Santa Barbara, CA

E. R. Fuller, Jr.
US Department of Commerce
The National Institute of Standards
   and Technology
Gaithersburg, MD

J. P. Gyekenyesi
NASA Lewis Research Center
Cleveland, OH

J. W. Holmes
Ceramic Composites Research
   Laboratory
Department of Mechanical
   Engineering and Applied
   Mechanics
University of Michigan
Ann Arbor, MI

K. Jakus
Department of Mechanical
   Engineering
University of Massachusetts
Amherst, MA

A. S. Kobayashi
Department of Mechanical
   Engineering
University of Washington
Seattle, WA

T. J. Mackin
Materials Department
College of Engineering
University of California
Santa Barbara, CA

R. M. McMeeking
Department of Mechanical and
   Environmental Engineering
University of California
Santa Barbara, CA

S. V. Nair
Department of Mechanical
  Engineering
University of Massachusetts
Amherst, MA

S. R. Nutt
Department of Materials Science
University of Southern California
Los Angeles, CA

B. F. Sørensen
Materials Department
Risø National Laboratory
4000 Roskilde, Denmark

S. Suresh
Department of Engineering
Massachusetts Institute of Technology
Cambridge, MA

T. N. Tiegs
Metals and Ceramics Division
Oak Ridge National Laboratory
Oak Ridge, TN

S. M. Wiederhorn
US Department of Commerce
The National Institute of Standards
  and Technology
Gaithersburg, MD

Xin Wu
Ceramic Composites Research
  Laboratory
Department of Mechanical
  Engineering and Applied
  Mechanics
University of Michigan
Ann Arbor, MI

J.-M. Yang
Department of Materials Science and
  Engineering
University of California
Los Angeles, CA

F. W. Zok
Materials Department
College of Engineering
University of California
Santa Barbara, CA

Preface

The rapidly growing field of ceramic matrix composites has matured considerably over the last decade with the introduction of particulate-, whisker- and fiber-reinforced composites. During this period, advances in the processing of ceramic matrix composites have gone hand in hand with the development of sophisticated micromechanics models for their structural performance. Although ceramic composites have been primarily targeted for elevated temperature applications, the focus to date has been mainly on achieving structural integrity at ambient temperatures. Emphasis on the mechanics and mechanisms governing the elevated temperature structural performance of ceramic composites is relatively recent and fundamental understanding is still in a state of evolution.

This book provides an up-to-date, comprehensive coverage of the mechanical behavior of ceramic matrix composites at elevated temperature. The distinction between two important classes of behavior underlies the organization of this book. One is short-term behavior which includes such topics as strength, fracture toughness, R-curve behavior and dynamic fracture at elevated temperatures. Second is long-term behavior associated with creep and creep-fatigue processes. Topics related to long-term behavior include creep deformation, creep and creep-fatigue crack growth, delayed failure and composite lifetime. In addition to chapters on elevated temperature behavior, we have included chapters devoted to analytical modeling. These include modeling of creep deformation, micromechanics of elevated temperature crack bridging and crack growth, and reliability of ceramic composites at elevated temperature. The chapters on modeling will help delineate the differences between ambient and elevated temperature models of mechanical behavior. For example, crack bridging at elevated temperatures is thermally activated and time dependent, whereas bridging at ambient temperatures is time independent. Consequently, microstructural design based on ambient temperature models may not be suitable for elevated temperature applications. Finally, we have devoted a chapter exclusively to novel approaches and critical issues in the elevated temperature testing of ceramic composites.

In the presentation of the elevated temperature mechanical behavior of ceramic matrix composites, some degree of separation has also been made between fiber-reinforced and whisker- or particulate-reinforced composites. This has been necessary because of the way the field has evolved. The continuous fiber-reinforced composites area in many ways has evolved as a field in its own right, driven by developments in fiber processing technology.

Each chapter in the book has been written by an internationally recognized expert in the field of ceramic matrix composites. The chapters are organized to include a substantial review of the fundamentals. The authors have integrated material from a wide range of sources to provide a perspective for their own research contributions. The overview character of the chapters makes this book useful not only to researchers in the ceramics community but also to graduate students in advanced ceramics courses.

The editors are deeply grateful to the authors for agreeing to undertake the rather difficult task of writing comprehensive overviews in an evolving research area. We also thank the National Science Foundation for supporting our efforts through grants MSS-9201625 (Nair) and DMR-9012594 (Jakus). One of us (Nair) also acknowledges the University of Massachusetts at Amherst whose sabbatical support was crucial to the editing and completion of this book.

Shanti V. Nair
Karl Jakus

Overview

The Structural Performance of Ceramic Matrix Composites

A. G. Evans, F. W. Zok, and T. J. Mackin

Nomenclature

a_0	Length of unbridged matrix crack
a_i	Parameters found in the paper by Hutchinson and Jensen[1]— Table 1.2
a_m	Fracture mirror radius
a_N	Notch size
a_t	Transition flaw size
b	Plate dimension
b_i	Parameters found in the paper by Hutchinson and Jensen[1]— Table 1.2
c_i	Parameters found in the paper by Hutchinson and Jensen[1]— Table 1.2
d	Matrix crack spacing
d_s	Saturation crack spacing
f	Fiber volume fraction
f_l	Fiber volume fraction in loading direction
g	Function related to cracking of 90° plies
h	Fiber pull-out length
l	Sliding length
l_i	Debond length
l_s	Shear band length
m	Shape parameter for fiber strength distribution
m_m	Shape parameter for matrix flaw size distribution
n	Creep exponent
n_f	Creep exponent for fiber
n_m	Creep exponent for matrix
q	Residual stress in matrix in axial orientation
s_{ij}	Deviatoric stress
t	Time
t_b	Beam thickness
t_p	Ply thickness
u	Crack opening displacement (COD)
u_a	COD due to applied stress
u_b	COD due to bridging
v	Sliding displacement
w	Beam width
B	Creep rheology parameter $\dot{\varepsilon}_0/\sigma_0^n$
C_v	Specific heat at constant strain
E	Young's modulus for composite
E_0	Plane strain Young's modulus for composites
$\bar{E}$	Unloading modulus
E^*	Young's modulus of material with matrix cracks
E_f	Young's modulus of fiber
E_L	Ply modulus in longitudinal orientation
E_m	Young's modulus of matrix

E_s	Secant modulus	$\Delta\varepsilon_p$	Unloading strain differential
E_t	Tangent modulus	Δ^R	Displacement caused by matrix removal
E_T	Ply modulus in transverse orientation	ΔT	Change in temperature
G	Shear modulus	ε	Strain
$\mathcal{G}$	Energy release rate (ERR)	ε^*	Strain caused by relief of residual stress upon matrix cracking
$\mathcal{G}_{tip}$	Tip ERR		
$\mathcal{G}_{tip}^0$	Tip ERR at lower bound		
I_0	Moment of inertia	ε_0	Permanent strain
K	Stress intensity factor (SIF)	$\dot{\varepsilon}_0$	Reference strain rate for creep
K_b	SIF caused by bridging	ε_e	Elastic strain
K_m	Critical SIF for matrix	ε_s	Sliding strain
K_R	Crack growth resistance	ε_τ	Transient creep strain
K_{tip}	SIF at crack tip	κ	Beam curvature
L	Crack spacing in 90° plies	λ	Pull-out parameter
L_0	Reference length for fibers	μ	Friction coefficient
L_f	Fragment length	ν	Poisson's ratio
L_g	Gauge length	ξ	Fatigue exponent (of order 0.1)
N	Number of fatigue cycles	ρ	Density
N_s	Number of cycles at which sliding stress reaches steady state	σ	Stress
		σ^0	Stress on 0° plies
R	Fiber radius	σ_0	Creep reference stress
$\mathcal{R}$	R-ratio for fatigue ($\sigma_{max}/\sigma_{min}$)	σ_b	Bridging stress
$\mathcal{R}_c$	Radius of curvature	$\bar{\sigma}_b$	Peak, reference stress
S	Tensile strength of fiber	σ_e	Effective stress → $\sqrt{(3/2)s_{ij}s_{ij}}$
S_0	Scale factor for fiber strength	σ_f	Stress in fiber
S^*	UTS in presence of a flaw	σ_i	Debond stress
S_b	Dry bundle strength of fibers	σ_m	Stress in matrix
S_c	Characteristic fiber strength	σ_{mc}	Matrix cracking stress
S_g	UTS subject to global load sharing	σ_{rr}	Radial stress
		σ^R	Residual stress
S_p	Pull-out strength	σ_s	Saturation stress
S_{th}	Threshold stress for fatigue	σ_s^*	Peak stress for traction law
S_u	Ultimate tensile strength (UTS)	σ_τ	Lower bound stress for tunnel cracking
T	Temperature		
		σ^T	Misfit stress
α	Linear thermal coefficient of expansion (TCE)	τ	Interface sliding stress
		τ_0	Constant component of interface sliding stress
α_f	TCE of fiber	τ_f	Value of sliding stress after fatigue
α_m	TCE of matrix		
γ	Shear strain	τ_s	In-plane shear strength
γ_c	Shear ductility	τ_{ss}	Steady-state value of τ after fatigue
Γ	Fracture energy		
Γ_f	Fiber fracture energy	$\bar{\tau}_c$	Critical stress for interlaminar crack growth
Γ_i	Interface debond energy		
Γ_m	Matrix fracture energy	ϕ	Orientation of interlaminar cracks
Γ_R	Fracture resistance		
Γ_S	Steady-state fracture resistance	Ω	Misfit strain
Γ_T	Transverse fracture energy	Ω_0	Misfit strain at ambient temperature
δ_c	Characteristic length		
$\delta\varepsilon$	Hysteresis loop width		
$\Delta\varepsilon_0$	Reloading strain differential		

1.1 Introduction

1.1.1 Rationale

The strong interest in continuous fiber-reinforced ceramic matrix composites (CMCs) has arisen primarily because of their ability to retain good tensile strength in the presence of holes and notches.[2-4] This characteristic is important because CMC components generally need to be attached to other components. At these attachments (whether mechanical or bonded), stress concentrations arise, which dominate the design and reliability. Inelastic deformation at these sites is crucial. It alleviates the elastic stress concentration by locally redistributing stress.[5] Such inelasticity is present in CMCs.[6-9] In association with the inelastic deformation, various degradation processes occur which affect the useful life of the material. Several fatigue effects are involved:[10,11] cyclic, static and thermal. The most severe degradation appears to occur subject to out-of-phase thermomechanical fatigue (TMF). In addition, creep and creep rupture occur at high temperatures.[12]

All of the CMC properties that govern structural utility and life depend upon the constituent properties (fibers, matrix, interfaces), as well as the fiber architecture. Since the constituents are variables, optimization of the property profiles needed for design and lifing become prohibitively expensive if traditional empirical procedures are used. The philosophy of this article is based on the recognition that mechanism-based models are needed, which allow efficient interpolation between a well-conceived experimental matrix. The emphasis is on the creation of a framework which allows models to be inserted, as they are developed, and which can also be validated by carefully chosen experiments.

1.1.2 Objectives

The initial intent of this review is to address the *mechanisms* of stress redistribution upon monotonic and cyclic loading, as well as the *mechanics* needed to characterize the notch sensitivity.[5,13] This assessment is conducted primarily for composites with 2-D reinforcements. The basic *phenomena* that give rise to inelastic strains are matrix cracks and fiber failures subject to interfaces that debond and slide (Fig. 1.1).[14-16] These phenomena identify the essential constituent properties, which have the typical values indicated in Table 1.1.

Three underlying *mechanisms* are responsible for the nonlinearity.[17,18] (1) *Frictional dissipation* occurs at the fiber/matrix interfaces, whereupon the sliding resistance of debonded interfaces, τ, becomes a key parameter. *Control of τ is critical.* This behavior is dominated by the fiber coating, as well as the fiber morphology.[19,20] By varying τ, the prevalent damage mechanism and the resultant non-linearity can be dramatically modified. (2) The matrix cracks

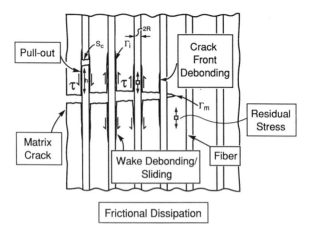

Fig. 1.1   The fundamental mechanisms that operate as a crack extends through the matrix.

Table 1.1   Constituent properties of CMCs and methods of measurement

Constituent property	Measurement methods	Typical range
Sliding stress, τ(MPa)	• Push-out force • Pull-out length, $\bar{h}$ • Saturation crack spacing, $\bar{l}_s$ • Hysteresis loop, $\delta\varepsilon$ • Unloading modulus, $\bar{E}_L$	1–200
Characteristic strength, S_c (GPa)	• Fracture mirrors • Pull-out length, $\bar{h}$	1.2–3.0
Misfit strain, Ω	• Bilayer distortion • Permanent strain, ε_p • Residual crack opening	0–2×10^{-3}
Matrix fracture energy, Γ_m (J m^{-2})	• Monolithic material • Saturation crack spacing, $\bar{l}_s$ • Matrix cracking stress, $\bar{\sigma}_{mc}$	5–50
Debond energy, Γ_i (J m^{-2})	• Permanent strain, ε_p • Residual crack opening, u_p	0–5

increase the *elastic compliance*.[21] (3) The matrix cracks also cause changes in the residual stress distribution, resulting in a *permanent strain*.[21]

The relative ability of these mechanisms to operate depends on the loading, as well as the fiber orientation. It is necessary to address and understand the mechanisms that operate for loadings which vary from tension along one fiber direction to shear at various orientations. For *tensile loading*,

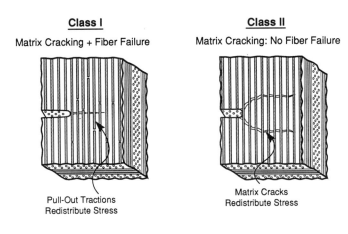

Fig. 1.2 Three prevalent damage mechanisms occurring around notches in CMCs. Each mechanism allows stress redistribution by a combination of matrix cracking and fiber pull-out.

several damange mechanisms have been found, involving matrix cracks combined with sliding interfaces (Fig. 1.2). These can be visualized by mechanism maps,[22] which then become an integral part of the testing and design activity. One damage mechanism involves Mode I cracks with simultaneous fiber failure, referred to as *Class I behavior*. Stress redistribution is provided by the tractions exerted on the crack by the failed fibers, as they *pull out*.[13,23–25] A second damage mechanism involves multiple matrix cracks, with minimal fiber failure, referred to as *Class II behavior* (Fig. 1.2). In this case, the plastic deformation caused by matrix cracks allows stress redistribution.[4,17] A schematic of a mechanism map based on these two damage mechanisms (Fig. 1.3) illustrates another important issue: the use of *nondimensional parameters* to interpolate over a range of constituent properties.† On the

†For ease of reference, all of the most important nondimensional parameters are listed in a separate table (Table 1.2).

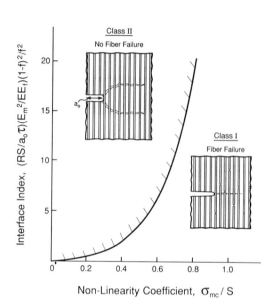

Fig. 1.3 A proposed mechanism map that distinguishes Class I and Class II behavior.

mechanism map, the ordinate is a nondimensional measure of sliding stress and the abscissa is a nondimensional *in situ* fiber strength. A third damage mechanism also exists (Fig. 1.2), referred to as *Class III*. It involves matrix shear damage prior to composite failure as a means for redistributing stress. A proposed mechanism map is presented in Fig. 1.4.[26]

A summary of tensile stress–strain curves obtained for a variety of 2-D composites (Fig. 1.5) highlights the most fundamental characteristic relevant to the application of CMCs. Among these four materials, the SiC/CAS (calcium aluminosilicate) system is found to be *notch insensitive* in tension,[4] even for quite large notches (~5 mm long). The other three materials exhibit varying degrees of *notch sensitivity*.[27] Moreover, the notch insensitivity in SiC/CAS arises *despite relatively small plastic strains*. These results delineate two issues that need resolution. (1) How much plastic strain is needed to impart notch insensitivity? (2) Is the ratio of the "yield" strength to ultimate tensile strength (UTS) an important factor in notch sensitivity? This review will address both questions.

The *shear behavior* also involves matrix cracking and fiber failure.[26] However, the ranking of the shear stress–strain curves between materials (Fig. 1.6) differs appreciably from that found for tension (Fig. 1.5). Preliminary

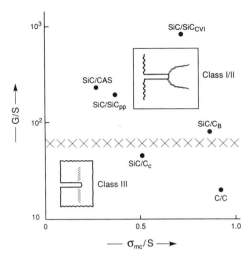

Fig. 1.4 A proposed mechanism map that distinguishes Class III behavior.

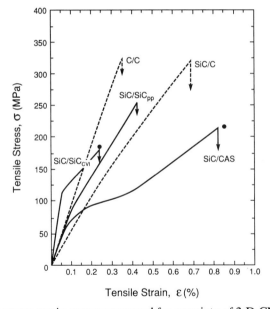

Tensile Strain, ε (%)

Fig. 1.5 Tensile stress–strain curves measured for a variety of 2-D CMCs.

efforts at understanding this difference and for providing a methodology to interpolate between shear and tension will be described.

Analyses of damage and failure have established that certain *constituent properties* are basic to composite performance (Table 1.1). These need to be *measured*, independently, and then used as characterizing parameters, analogous to the yield strength and fracture toughness in monolithic materials. The

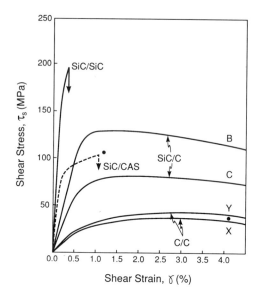

Fig. 1.6   Shear stress–strain curves measured for 2-D CMCs.

six major *independent* parameters are the interfacial sliding stress, τ, and debond energy, Γ_i, the *in situ* fiber properties, S_c and m, the fiber/matrix misfit strain, Ω, and the matrix fracture energy Γ_m, as well as the elastic properties, E, ν.[5] *Dependent* parameters that can often be used to infer the constituent properties include: the fiber pull-out length,[8,16] the fracture mirror radius on the fibers,[28] and the saturation crack spacing in the matrix.[29] Approaches for measuring the constituent properties in a *consistent, straightforward* manner will be emphasized and their relevance to composite behavior explored through models of damage and failure. Moreover, the expressions that relate composite behavior to constituent properties are often unwieldy, because a large number of parameters are involved. Consequently, throughout this article, the formulae used to represent CMC behaviors are the *simplest* capable of describing the major phenomena.†

In most composites with desirable tensile properties, linear elastic fracture mechanics (LEFM) criteria are violated.[30,31] Instead, various large-scale nonlinearities arise, associated with matrix damage and fiber pull-out. In consequence, alternate mechanics is needed to specify the relevant *material and loading parameters* and to establish *design rules*. Some progress toward this objective will be described and related to test data. This has been achieved

---

†The behaviors represented by these formulae are often applicable only to composites: the equivalent phenomenon being absent in monolithic ceramics. Consequently, the expressions should be restricted to composites with fiber volume fractions in the range of practical interest (f between 0.3 and 0.5). Extrapolation to small f would lead to erroneous interpretations, because mechanism changes usually occur.

using large-scale bridging mechanics (LSBM), combined with continuum damage mechanics (CDM).[13,23,25,32]

The preceding considerations dictate the ability of the material to survive thermal and mechanical loads imposed for short durations. In many cases, long-term survivability at elevated temperatures dictates the applicability of the material. Life models based on *degradation* mechanisms are needed to address this issue. For this purpose, generalized fatigue and creep models are required, especially in regions that contain matrix cracks. It is inevitable that such cracks exist in regions subject to strain concentrations and, indeed, are required to redistribute stress. In this situation, degradation of the interface and the fibers may occur as the matrix cracks open and close upon thermomechanical cycling, with access of the atmosphere being possible, through the matrix cracks. The rate of such degradation dictates the useful life.

1.1.3 Approach

To address the preceding issues, this chapter is organized in the following manner. Some of the basic thermomechanical characteristics of composites are first established, with emphasis on interfaces and interface properties, as well as residual stresses. Then, the fundamental response of *unidirectional* (1-D) materials, subject to *tensile loading*, is addressed, in accordance with several sub-topics: (1) mechanisms of non-linear deformation and failure; (2) constitutive laws that relate macroscopic performance to constituent properties; (3) the use of stress–strain measurements to determine constituent properties in a consistent, straightforward manner; and (4) the simulation of stress–strain curves. The discussion of 1-D materials is followed by the application of the same concepts to 2-D materials, subject to combinations of tensile and shear loading. At this stage, it is possible to address the *mechanisms of stress redistribution* around flaws, holes, attachments, and notches. In turn, these mechanisms suggest a *mechanics methodology* for relating strength to the size and shape of the flaws, attachment loading, etc.

Data regarding the effects of cyclic loading and creep on the life of brittle matrix composites are limited. The concepts to be developed thus draw upon knowledge and experience gained with other composite systems, such as metal matrix composites (MMCs) and polymer matrix composites (PMCs). The overall philosophy is depicted in Fig. 1.7.

1.2 Interfaces

1.2.1 Thermomechanical Representation

The thermomechanical properties of coatings at fiber–matrix interfaces are critically important. A consistent characterization approach is necessary and the most commonly adopted hypothesis is that there are two parameters (Fig. 1.8). One is associated with *fracture* and the other with *slip*.[1,33,136]

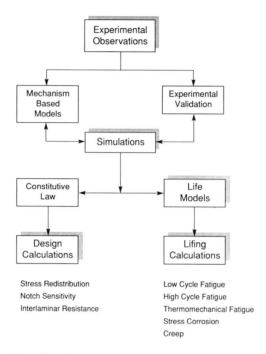

Fig. 1.7 The philosophy adopted for using models in the design and application of CMCs.

Fracture, or debonding, is considered to involve a debond energy, Γ_i.[34,35] Slip is expected to occur with a shear resistance, τ. A schematic representation (Fig. 1.9) illustrates the issues. *Debonding* must be a Mode II (shear) fracture phenomenon. In brittle systems, Mode II fracture typically occurs by the coalescence of microcracks within a material *layer*.[36] In some cases, this layer coincides with the coating itself, such that debonding involves a diffuse zone of microcrack damage (Fig. 1.9). In other cases, the layer is very thin and the debond has the appearance of a *single crack*. For both situations, it is believed that debond propagation can be represented by a debond energy, Γ_i, with an associated stress jump above and below the debond front.[1] Albeit, in several instances, Γ_i is essentially zero.[37] When a discrete debond crack exists, *frictional sliding* of the crack faces provides the shear resistance. Such sliding occurs in accordance with a friction law,[33,38,39]

$$\tau = \tau_0 - \mu\sigma_{rr} \qquad (1)$$

where μ is the Coulomb friction coefficient, σ_{rr} is the compression normal to the interface, and τ_0 is a term associated with fiber roughness. When the debond process occurs by diffuse microcracking in the coating, it is again assumed (without justification) that the interface has a constant shear resistance, τ_0.

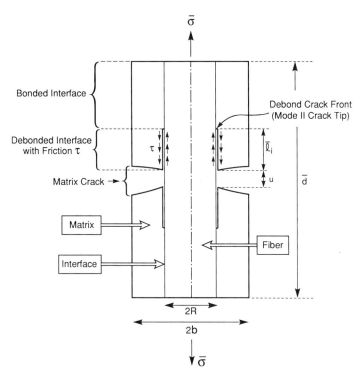

Fig. 1.8   A schematic indicating the sliding and debonding behavior envisaged in CMCs.

For debonding and sliding to occur, rather than brittle cracking through the fiber, the debond energy Γ_i must not exceed an upper bound, relative to the fiber fracture energy, Γ_f.[35] Calculations have suggested that the following inequality must be satisfied (Fig. 1.10)

$$\Gamma_i \lesssim (1/4)\Gamma_f \tag{2}$$

Noting that most ceramic fibers have a fracture energy, $\Gamma_f \approx 20\,\mathrm{J\,m^{-2}}$, Eqn. (2) indicates that the upper bound of the debond energy $\Gamma_i \approx 5\,\mathrm{J\,m^{-2}}$. This magnitude is broadly consistent with experience obtained on fiber coatings that impart requisite properties.[20,40–43]

1.2.2 Measurement Methods

Measurements of the sliding stress, τ, and the debond energy, Γ_i, have been obtained by a variety of approaches (Table 1.1). The most direct involve displacement measurements. These are conducted in two ways: (1) fiber push-through/push-in, by using a small-diameter indentor;[33,38,39] and (2) tensile loading in the presence of matrix cracks.[5,44,45] Indirect methods for obtaining τ

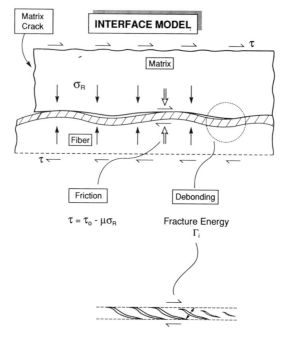

Fig. 1.9 The fiber sliding model.

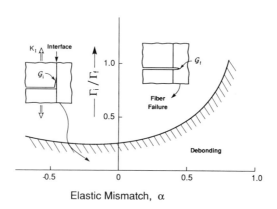

Fig. 1.10 A debond diagram for CMCs.

also exist. These include measurement of the saturation matrix crack spacing[29] and the fiber pull-out length.[16] The *direct* measurement methods require accurate determination of displacements, coupled with an analysis that allows rigorous deconvolution of load–displacement curves. The basic analyses used for this purpose are contained in papers by Hutchinson and Jensen,[1] Liang and Hutchinson,[47] Marshall[39] and Jero *et al.*[38] The fundamental features are

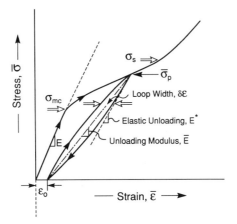

Fig. 1.11 A typical load–unload cycle showing the parameters that can be measured which relate to the interface properties.

illustrated by the behavior found upon tensile loading, subsequent to matrix cracking (Fig. 1.11). The hysteresis that occurs during an unload–reload cycle relates to the sliding stress, τ. Accurate values for τ can be obtained from *hysteresis measurements*.[5,18,45] Furthermore, these results are *relevant to the small sliding displacements*† that occur during matrix crack evolution in actual composites. The plastic strains contain *combined* information about τ, Ω and Γ_i. Consequently, if τ is already known, Γ_i can be evaluated from the plastic strains measured as a function of load, especially if Ω has been obtained from independent determinations.[5] The basic formulae that connect τ, Γ_i and Ω to the stress–strain behavior are presented in Section 1.5.

1.2.3 *Sliding Models*

The manipulations of interfaces needed to control τ can be appreciated by using a model to simulate the sliding behavior. A simplified sliding model has been developed (Fig. 1.10) which embodies the role of the pressure at contact points, due to the combined effects of a mismatch strain and roughness.[33,38,39] Coulomb friction is regarded as the *fundamental* friction law operating at *contacts*. Otherwise, the system is considered to be elastic. The variables in the analysis are (1) the amplitude and wavelength of the roughness; (2) the mismatch strain, Ω; (3) the Coulomb friction coefficient, μ; and (4) the elastic properties of the constituents. With these parameters as input, the sliding can be simulated for various loading situations. One set of simulations conducted for comparison with fiber push-out tests (Fig. 1.12) illustrates the relative

---

†Information about τ at larger sliding displacements is usually obtained from fiber push-through measurements.

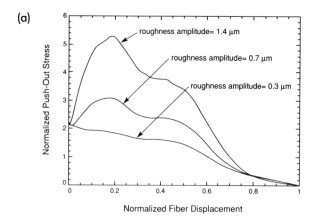

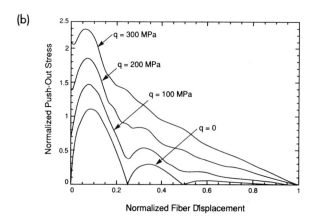

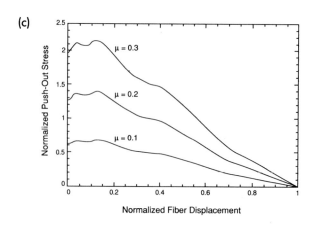

Fig. 1.12 Simulation of the effects of the key variables on the push-out behavior: (a) roughness, (b) residual stress, and (c) friction coefficient.

importance of each of the variables. For this set, the fiber roughness was characterized using a fractal method. The roughness within the section was selected at random, from the measured amplitude distribution, causing some differences in the push-out spectrum for each simulation. By using this simulation, *substantial* systematic changes in the sliding resistance have been predicted when the friction coefficient, the mismatch strain, and the roughness amplitude, are changed.† Generally, the mismatch strain and the roughness can be measured independently.[33] Consequently, the comparison between simulation and experiment actually provides an estimate of the *friction coefficient, μ*. If this is found to be within an acceptable range, the inferred μ is, thereafter, used to predict how τ can be expected to vary as either the misfit or the roughnesses are changed, if μ is fixed. This approach indicates that $\mu \approx 0.1$ for either C or BN coatings,[48] whereas $\mu \approx 0.5$ for oxide coatings.[49] Such values are compatible with macroscopic friction measurements made on bulk materials and thus appear to be reasonable. However, much additional testing is needed to validate the sliding model.

1.2.4 *Experimental Results*

Most of the experience with brittle matrix composites is on C, BN, or Mo fiber coatings.[19,20,41,50,51] Such coatings usually have a relatively low debond energy, Γ_i, and can provide a range of sliding stresses, τ (Table 1.1), as illustrated by comparison of three different carbon coatings on sapphire fibers in TiAl (Fig. 1.13a). A considerable range in τ has even been achieved with C coatings and values between 2 and 200 MPa have been found. Furthermore, this range is obtained even at comparable values of the misfit strain. The different values may relate to fiber roughness. Roughness effects are best illustrated by the sliding behavior of sapphire fibers in a glass matrix. During fiber manufacture, sinusoidal asperities are grown onto the surface of the sapphire fibers. The sinusoidal fiber surface roughness is manifest as a wavelength modulation in the sliding stress during push-out (Fig. 1.13b).[33] However, there must also be influences of the coating thickness and micro-structure. A model that includes an explicit influence of the coating has yet to be developed.

In most brittle matrix composites, the debond energy, Γ_i, has been found to be negligibly small ($\Gamma_i < 0.1 \, \text{J m}^{-2}$). Such systems include all of the glass ceramic matrix systems reinforced with Nicalon fibers, which have a C interphase formed by reaction during composite processing. Low values also seem to be obtained for SiC matrix composites with BN fiber coatings. The clear exception is SiC/SiC composites made by chemical vapor infiltration (CVI), which use a C *interphase*, introduced by chemical vapor deposition (CVD).[52] For such composites, the non-linear behavior indicates a debond energy $\Gamma_i \approx 5 \, \text{J m}^{-2}$ (Table 1.1). The interphase in this case debonds by a

---

†There are only minor effects of Poisson's ratio.

diffuse damage mechanism.[53] Moreover, it has been found that the coating behavior can be changed into one with $\Gamma_i \approx 0$, either by heat treatment of the composite (after CVI) or by chemical treatment of the fiber.[53] A basic understanding of these changes in Γ_i does not exist.

1.2.5 Environmental Influences

There are temperature and environmental effects on τ and Γ_i. There are also effects on τ of fiber displacement and cyclic sliding (Fig. 1.13c). These effects can critically influence composite performance. The basic effect of temperature on τ[54] concerns changes in the misfit strain and friction coefficient, evident from the simulations shown in Fig. 1.12. Environmental influences can be pronounced, especially in oxidizing atmospheres. The major effects arise either at high temperatures, or during fatigue.† When either C or Mo coatings are used, τ initially decreases upon either exposure or fatigue (Fig. 1.13d) because a gap is created between the fiber and matrix, caused by elimination of the coating, through volatile oxide formation.[8,20,55,56] This process occurs when the *local* temperature reaches ~800°C. The subsequent behavior depends on the fibers. When SiC fibers are used, further exposure causes SiO_2 formation[57] and this layer gradually fills the gap, leading to large values of τ. Eventually, a "strong" interface bond forms (with large Γ_i) that produces brittle behavior, without fiber pull-out. Conversely, oxide fibers in oxide matrices are inherently resistant to this embrittlement phenomenon[20,49] and are environmentally desirable, provided that the matrix does not sinter to the fibers.

1.3 Residual Stresses

The development of damage is sensitive to the residual stress caused by the misfit strain, Ω. Measurement of these stresses thus becomes an important aspect of the analysis and prediction of damage. These stresses arise at interlaminate and intralaminate levels. Within a laminate, the axial stress in the matrix is,[58]

$$q = (E_m/E_L)\sigma^T \tag{3}$$

where σ^T is the misfit stress, which is related to the misfit strain by (Table 1.2)

$$\sigma^T = (c_2/c_1)E_m\Omega \tag{4}$$

The average residual stress in a 0/90 laminate, with uniform laminate thickness, σ^R, depends on constituent properties in approximate accordance with[59,60]

$$\sigma^R \approx \frac{\Omega(1-f)E_L(1-E_m/E_L)}{(1+\nu_{LT})(1+E_L/E_T)} \tag{5}$$

---

†A consequence of internal heating associated with cyclic frictional sliding at the interfaces.

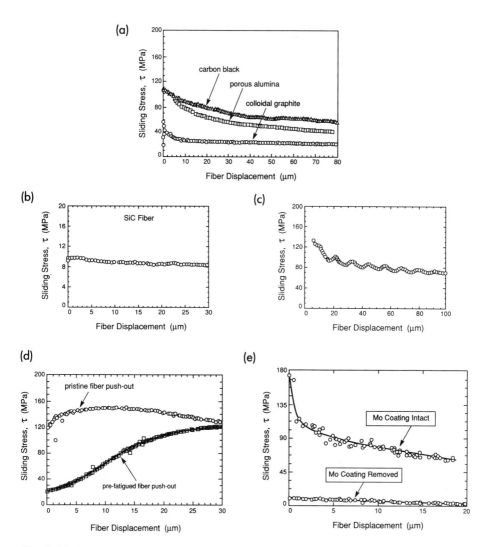

Fig. 1.13 Some typical fiber push-out measurements on metal, ceramic, and intermetallic composites: (a) $Al_2O_3/TiAl$ with C/Al_2O_3 double coatings, (b) SiC/glass, (c) Al_2O_3/glass showing effects of roughness, (d) SiC/Ti with C coating showing effect of fatigue and (e) Al_2O_3/Al_2O_3 with fugitive Mo coating.

Note that the residual stress $\sigma^R \to 0$ on the elastic properties becomes homogeneous ($E_f = E_m = E_L$). While connections between the residual stresses and constituent properties are rigorous, experimental determination is still necessary, because Ω is not readily predictable. In general, Ω includes terms associated with the thermal expansion difference, $\alpha_f - \alpha_m$, as well as volume changes that occur either upon crystallization or during phase transformations. For CVI systems, "intrinsic" stresses may also be present.

Table 1.2a   Inventory of nondimensional functions

Relative stiffness	$\xi \to fE_f/(1-f)E_m$
Sliding index	$\mathscr{T} \to \xi[\tau_0 E/\sigma E_f]^{1/2}$
Cyclic sliding index	$\Delta\mathscr{T} \to \xi[\tau_0 E/\Delta\sigma E_f]^{1/2}$
Loading index	$\mathscr{E} \to [2R\sigma/f\xi^2 a\tau_0]$
Cyclic loading indices	$\Delta\mathscr{E} \to [2R(\Delta\sigma)/f\xi^2 a\tau_0]$
	$\Delta\mathscr{E}_0 \to [2R(\Delta\sigma)/f\xi^2 a_0\tau_0]$
Bridging index	$\mathscr{E}_b \to [2R\sigma_b/f\xi^2 a\tau_0]$
Cyclic bridging indices	$\Delta\mathscr{E}_b \to [2R(\Delta\sigma_b)/f\xi^2 a\tau_0]$
	$\Delta\mathscr{E}_T \to [2RE_f(\alpha_f - \alpha_m)\Delta T/\xi^2 f\tau_0 a]$
Misfit index	$\Sigma_T \to \bar\sigma_T/\bar\sigma_p = (c_2/c_1)E_m\Omega/\bar\sigma_p$
Debond index	$\Sigma_i \to \bar\sigma_i/\bar\sigma_p = (1/c_1)\sqrt{E_m\Gamma_i/R\bar\sigma_p^2} - \Sigma_T$
Hysteresis index	$\mathscr{H} \to b_2(1-a_1f)^2 R\bar\sigma_p^2/4\mathscr{V}\tau E_m f^2$
Crack spacing index	$\mathscr{L} \to \Gamma_m(1-f)^2 E_f E_m/f\tau^2 E_L R$
Matrix cracking index	$M \to 6\tau\Gamma_m f^2 E_f/(1-f)E_m^2 RE_L$
Residual stress index	$\mathscr{Q} \to E_f f\Omega/E_L(1-\nu)$
Flaw index	$\mathscr{A} \to a_0 S^2/E_L\Gamma$
Flaw index for bridging	$\mathscr{A}_b \to [f/(1-f)]^2 (E_f E_L/E_m^2)(a_0\tau/RS_u)$
Flaw index for pull-out	$\mathscr{A}_p \to (a_0/\bar h)(S_p/E_L)$

Several experimental procedures can be used to measure the residual stresses. The three preferred methods involve diffraction (X-ray or neutron), beam deflection, and permanent strain determination. X-ray diffraction measurements have the limitation that the penetration depth is small, such that only near-surface information is obtained. Moreover, in composites, residual stresses are *redistributed* near surfaces.[47] Consequently, a full stress analysis is needed to relate the measured strains to either q or σ^R.

Beam deflection and permanent strain measurements have the advantage

Table 1.2b   Summary of Hutchinson–Jenson (H J) constants for Type II boundary conditions

$$a_1 = E_f/E$$

$$a_2 = \frac{(1-f)\, E_f\, [1 + E_f/E]}{[E_f + (1 - 2\nu)\, E]}$$

$$b_2 = \frac{(1 + \nu)\, E_m\, \{2(1 - \nu)^2\, E_f + (1 - 2\nu)\, [1 - \nu + f(1 + \nu)]\, (E_m - E_f)\}}{(1 - \nu)\, E_f\, [(1 + \nu)\, E_0 + (1 - \nu)\, E_m]}$$

$$b_3 = \frac{f(1 + \nu)\, \{(1 - f)\, (1 + \nu)\, (1 - 2\nu)\, (E_f - E_m) + 2(1 - \nu)^2\, E_m\}}{(1 - \nu)\, (1 - f)\, [(1 + \nu)\, E_0 + (1 - \nu)\, E_m]}$$

$$c_1 = \frac{(1 - fa_1)\, (b_2 + b_3)^{1/2}}{2f}$$

$$c_2 = \frac{a_2\, (b_2 + b_3)^{1/2}}{2}$$

$$c_1/c_2 = \frac{1 - a_1 f}{a_2 f}$$

with

$$E = fE_f + (1 - f)\, E_m$$

$$E_0 = (1 - f)\, E_f + fE_m$$

that they provide information *averaged* over the composite. The results thus relate directly to the misfit strain, Ω. An experimental approach having high reliability involves curvature measurements on beams made from 0/90 composites.[61] For such material, polishing to produce one 0° layer and one 90° layer results in elastic bending (Fig. 1.14).† The curvature κ is related to the residual stress by[61‡]

$$\sigma^R = E_L I_0 \kappa / t_b^2 w \tag{6}$$

where I_0 is the second moment of inertia, t_b is the beam thickness and w the beam width.

When only 1-D material is available, the preferred approach is to measure the displacement, Δ^R, that occurs when a section of matrix, length L_d, is

†Unless the material has a plain weave.
‡There is a typographical error in Beyerle *et al.*: the width w was omitted in their equation.

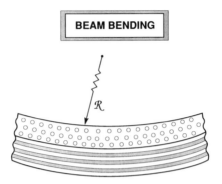

Fig. 1.14 A schematic of the beam bending effect used to evaluate the residual stress.

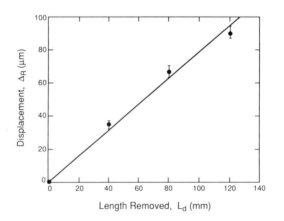

Fig. 1.15 Displacements caused by matrix dissolution as a function of length removed.

removed by dissolution (when possible). The residual stress in the matrix is then[62]

$$q = E_f f \Delta^R / (1 - f) L_d \qquad (7)$$

Typical results are plotted in Fig. 1.15.

Finally, the permanent strains that arise following tensile plastic deformation also relate to Ω. Measurement of these strains allows Ω to be assessed.[18] The relevant formulae are presented in Section 1.5.

Once the ambient misfit strain, Ω_0, has been inferred for the above measurements, the temperature dependence can be assessed from the thermal expansion mismatch, using,

$$\Omega = \Omega_0 - (\alpha_m - \alpha_f) \Delta T_R \qquad (8)$$

where ΔT_R is the temperature change from ambient.

Table 1.3 Important constituent properties for two typical CMCs: comparison between SiC/SiC and SiC/CAS

Property	Material	
	SiC/CAS	*SiC/SiC*
Matrix modulus, E_m (GPa)	100	400
Fiber modulus, E_f (GPa)	200	200
Sliding stress, τ (MPa)	15–20	100–150
Debond energy, Γ_i (J m^{-2})	~0.1	~2
Residual stress, q (MPa)	80–100	50–100
Fiber strength, S_C (GPa)	2.0–2.2	1.3–1.6
Shape parameter, m	3.3–3.8	4.2–4.7
Matrix fracture energy, Γ_m (J m^{-2})	20–25	5–10

Most *experimental results* are consistent with the misfit strain predicted from the thermal expansion difference, $\alpha_m - \alpha_f$, and the cooling range from the processing temperature. Examples for SiC/CAS and SiC/SiC are given in Table 1.3.

1.4 Fiber Properties

1.4.1 Load Sharing

The strength properties of fibers are *statistical* in nature. Consequently, it is necessary to apply principles of weakest link statistics, which define the properties of fibers *within a composite*. The initial decision to be made concerns the potential for interactions between failed fibers and matrix cracks. It has generally been assumed that matrix cracks and fiber failure are noninteracting and that global load sharing (GLS) conditions are obtained†.[16,63,64] In this case, the stress along a material plane that intersects a failed fiber is equally distributed among all of the intact fibers. Experience has indicated that these assumptions are essentially valid for a variety of CMCs.

Subject to the validity of GLS, several key results have been derived. Two characterizing parameters emerge:[65] a characteristic length

$$\delta_c^{m+1} = L_0(S_0 R/\tau)^m \qquad (9)$$

and a characteristic strength

$$S_c^{m+1} = S_0^m(L_0\tau/R) \qquad (10)$$

where m is the shape parameter, S_0 the scale parameter, L_0 the reference

---

†However, a criterion for GLS breakdown has yet to be devised.

length, and R the fiber radius. Various GLS results based on these parameters are described below.

When fibers do not interact, analysis begins by considering a fiber of length $2L$ divided into $2N$ elements, each of length δz. The probability that the fiber element will fail, when the stress is less than σ, is the area under the probability density curve[66,67]

$$\delta\phi(\sigma) = \frac{\delta z}{L_0} \int_0^\sigma g(S)\,dS \tag{11}$$

where $g(S)\,dS/L_0$ represents the number of flaws per unit length of fiber having a "strength" between S and $S+dS$. The local stress, σ, is a function of both the distance along the fiber, z, and the *reference* stress, $\bar{\sigma}_b$. The survival probability P_s for *all elements* in the fiber of length $2L$ is the product of the survival probabilities of each element,[68]

$$P_s(\bar{\sigma}_b, L) = \prod_{n=-N}^{N} [1 - \delta\phi(\bar{\sigma}_b, z)] \tag{12}$$

where $z = n\delta z$ and $L = N\delta z$. Furthermore, the probability Φ_S that the element at z will fail when the peak, reference stress is between $\bar{\sigma}_b$ and $\bar{\sigma}_b + \delta\bar{\sigma}_b$, but not when the stress is *less than* $\bar{\sigma}_b$, is the change in $\delta\phi$ when the stress is increased by $\delta\bar{\sigma}_b$ divided by the survival probability up to σ_b, given by[66,67,69]

$$\Phi_S(\bar{\sigma}_b, z) = [1 - \delta\phi(\bar{\sigma}_b, z)]^{-1} \left[\frac{\partial\delta\phi(\bar{\sigma}_b, z)}{\partial\bar{\sigma}_b} \right] d\bar{\sigma}_b \tag{13}$$

Denoting the probability density function for fiber failure by $\Phi(\bar{\sigma}_b, z)$, the probability that fracture occurs at a location z, when the peak stress is $\bar{\sigma}_b$, is governed by the probability that all elements *survive up to a peak stress* $\bar{\sigma}_b$, but that failure occurs, at z, when the stress reaches $\bar{\sigma}_b$.[69,70] It is given by the product of Eqn. (12) with Eqn. (13)

$$\Phi_S(\bar{\sigma}_b, z)\,\delta\bar{\sigma}_b\,\delta z = \frac{\Pi_{-N}^{N}[1 - \delta\phi(\bar{\sigma}_b, z)]}{[1 - \delta\phi(\bar{\sigma}_b, z)]} \left[\frac{\partial\delta\phi(\bar{\sigma}_b, z)}{\partial\bar{\sigma}_b} \right] d\bar{\sigma}_b \tag{14}$$

While the above results are quite general, it is convenient to use a power law to represent $g(S)$,

$$\int_0^\sigma g(S)\,dS = (\sigma/S_0)^m \tag{15}$$

Alternative representations of $g(S)$ are not warranted at the present level of development. Using this assumption, Eqn. (14) becomes[70]

$$\Phi(\bar{\sigma}_b,z) = \exp\left\{ -2 \int_0^L \left[\frac{\sigma(\bar{\sigma}_b,z)}{S_0} \right]^m \frac{dz}{L_0} \right\} \left(\frac{2}{L_0} \right) \frac{\partial}{\partial \bar{\sigma}_b} \left[\frac{\sigma(\bar{\sigma}_b,z)}{S_0} \right]^m \tag{16}$$

This basic result has been used to obtain solutions for several problems,[16,70,71] described below.

1.4.2 The Ultimate Tensile Strength

When multiple matrix cracking precedes failure of the fibers in the $0°$ bundles, the load along each matrix crack plane is borne entirely by the fibers. Nevertheless, the matrix has a crucial role, because stress transfer between the fibers and the matrix still occurs through the sliding resistance, τ. Consequently, some stress can be sustained by the failed fibers. This stress transfer process occurs over a distance related to the characteristic length, δ_c. As a result, the stresses on the intact fibers along any plane through the material are less than those experienced within a "dry" fiber bundle (in the absence of matrix). The transfer process also allows the stress in a failed fiber to be unaffected at distance $\gtrsim \delta_c$ from the fiber fracture site (Fig. 1.16). Consequently, composite failure requires that fiber bundle failure occurs *within* δ_c.[16] This phenomenon leads to a UTS *independent of gauge length*, L_g, provided that $L_g > \delta_c$.† The magnitude of the UTS can be computed by first evaluating the average stress on *all* fibers, failed plus intact, along an arbitrary plane through the material. Then, by differentiating with respect to the stress on the *intact* fibers, in order to obtain the maximum, the UTS becomes,

$$S_g = f_c S_c F(m) \tag{17a}$$

with

$$F(m) = [2/(m+1)]^{1/(m+1)}[(m+1)/(m+2)] \tag{17b}$$

It is of interest to compare this result to that found for a "dry" bundle. Then, the "fiber bundle" strength S_b, depends on the gauge length in accordance with,[72]

$$S_b = fS_0(L_0/L_g)^{1/m} e^{-1/m} \tag{17c}$$

In all cases, $S_g > S_b$.

As the load increases, the fibers fail systematically, resulting in a characteristic fiber fragment length. At composite failure, there can be multiple cracks within some fibers. The existence of many fiber fragments is still

---

†At small gauge lengths ($L_g < \delta_c$), the UTS becomes gauge-length dependent and exceeds S_u.[64]

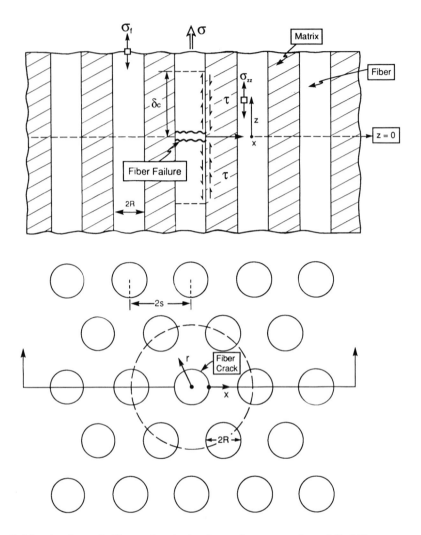

Fig. 1.16 A schematic illustrating the load transfer process from failed fibers.

compatible with a high ultimate tensile strength.† However, a diminished creep strength may ensue, as elaborated in Section 1.9.

The above results are applicable to *tensile loading*. When a bending moment is applied, the behavior is modified. In this case, the stress is redistributed by both matrix cracking and fiber failure. Predictions of the UTS in pure flexure (Fig. 1.17) indicate the salient phenomena.[64]

---

†A good analogy being the strength of a wire rope.

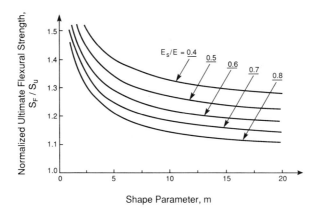

Fig. 1.17 Relationship between ultimate strengths measured in flexure and tension.

1.4.3 Fiber Pull-Out

In CMCs with good composite properties, fiber pull-out is evident on the tensile fracture surfaces.[73] Various measurements conducted on these surfaces provide valuable information. Regions with highly correlated fiber failures, with minimal pull-out, are indicative of *manufacturing flaws*. Such flaws often occur in regions where fiber coating problems existed. In zones where fiber failures are *uncorrelated*, the distribution of fiber pull-out lengths provides essential information. The pull-out lengths are related explicitly to the *stochastics* of fiber failure.[16,70] The basic realization is that, on average, fibers do not fail on the plane of the matrix crack, even though the stress in the fibers has its *maximum value* at this site. This unusual phenomenon relies exclusively on statistics, wherein the locations of fiber failure may be identified as a distribution function that depends on the shape parameter, m. Furthermore, the mean pull-out length, $\bar{h}$, has a connection with the charcteristic length, δ_c. Consequently, a functional dependence exists, dictated by the nondimensional parameters, $\tau\bar{h}/RS_c$ and m,

$$\bar{h}\tau/RS_c = \lambda(m) \tag{18}$$

There are two bounding solutions for the function λ (Fig. 1.18). Composite failure subject to *multiple matrix cracking* gives the *upper bound*. Failure in the presence of *a single crack* gives the *lower bound*.

Because of pull-out, a frictional *pull-out resistance* exists, which allows the material to sustain load, beyond the UTS. The associated "pull-out" strength S_p is an important property of the composite (Fig. 1.19). The strength, S_p, is given by[74]

$$S_p = 2\tau f\bar{h}/R$$

$$\equiv 2fS_c\lambda(m) \tag{19}$$

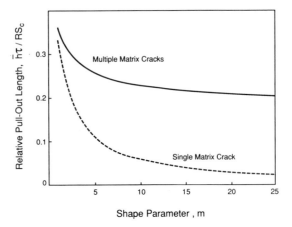

Fig. 1.18 Bounds on the relationship between the non-dimensional fiber pull-out length and the Weibull modulus.

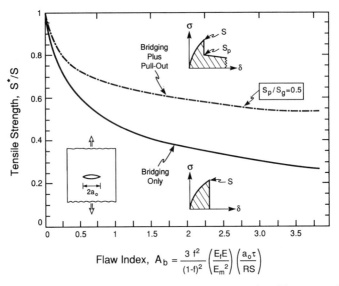

Fig. 1.19 The effect of unbridged regions, length $2a_0$, on the ultimate tensile strength.

1.4.4 Influence of Flaws

The preceding results are applicable provided that there are no unbridged segments along the matrix crack. *Unbridged regions* concentrate the stress in the adjacent fibers and weaken the composite.[13,75] Simple linear scaling considerations indicate that the diminished UTS depends on a nondimensional *flaw index* (Table 1.2a),

$$\mathscr{A} = a_0 S_g^2 / E_L \Gamma \tag{20}$$

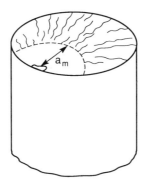

Fig. 1.20 A schematic indicating a fracture mirror and the dimension used to predict the *in situ* strength.

where Γ is the "toughness," reflected in the area under the stress–displacement curve for the bridging fibers, E is Young's modulus and $2a_0$ is the length of the unbridged segment. The flaw index, $\mathscr{A}$, can be specified, based on Γ, using LSBM. The dependence of the UTS, designated S^*, on the flaw index $\mathscr{A}$ can be determined from LSBM by numerical analysis (Fig. 1.19).[76] The results reveal that the ratio, S_p/S_g, is an important factor. Notably, relatively large values of the "pull-out" strength alleviate the strength degradation caused by unbridged cracks.

1.4.5 In Situ *Strength Measurements*

In general, composite consolidation degrades fiber properties and it becomes necessary to devise procedures that allow determination of S_c and m to be evaluated relevant to the fibers *within the composite*. This is a challenging problem. In some cases, it is possible to dissolve the matrix without further degrading the fibers and then measure the bundle strength.[77] This is not feasible with most CMCs of interest. The following two alternatives exist.

Some fibers exhibit fracture mirrors when they fail within a composite (e.g., Nicalon). A semi-empirical calibration has been developed that relates the mirror radius, a_m, to the *in situ* fiber tensile strength, S, given by (Fig. 1.20),

$$S \approx 3.5\sqrt{E_f \Gamma_f / a_m} \tag{21}$$

where Γ_f is the fracture energy of the fiber.[8,28,78] By measuring S on many fibers, and then plotting the cumulative distribution $G(S)$, both the shape parameter, m, and the characteristic *in situ* fiber strength, S_c, can be ascertained. Results of this type have been obtained for Nicalon fibers in a variety of different matrices (Fig. 1.21). This compilation indicates the sensitivity of the *in situ* strength to the composite processing approach. This

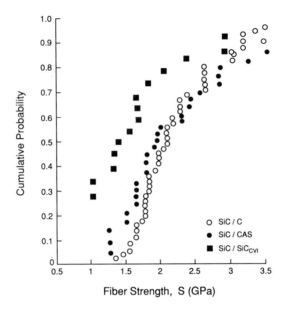

Fig. 1.21 *In situ* strength distributions measured for Nicalon fibers on three CMCs, using the fracture mirror approach.

fiber strength variation is also reflected in the range of UTS found among CMCs reinforced with these fibers (Fig. 1.4).

A problem in implementing the fracture mirror approach arises when a significant fraction of the fibers does not exhibit well-defined mirrors. Those fibers that do not have mirrors usually have a smooth fracture surface. It has thus been assumed that these are the weakest fibers in the distribution.[61,78] The order statistics used to determine $G(S)$ are adjusted accordingly. This assumption has not been validated.

The only alternative approaches for evaluating S_c, known to the authors, are based on pull-out and fragment length measurements.[46] Both quantities depend on S_c and m, as well as τ. Consequently, if τ is known, S_c can be determined. For example, m can be evaluated by fitting the distribution of fiber pull-out lengths to the calculated function. Then, S_c can be obtained for the mean value, $\bar{h}$, using Eqn. (12). This approach has not been extensively used and checked.

1.4.6 *Experimental Results*

Several studies have compared the multiple matrix cracking GLS prediction, S_g (Eqn. 17) with the UTS measured for either 1-D or 2-D CMCs. In most cases, the UTS is in the range (0.7–1) S_g, as indicated in Fig. 1.22. The two obvious discrepancies are the SiC/SiC$_{CVI}$ material and one of the SiC/C

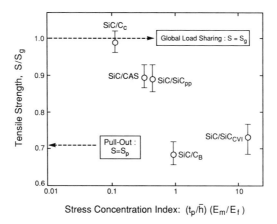

Stress Concentration Index: $(t_p/\bar{h})\,(E_m/E_f)$

Fig. 1.22 Comparison of the measured UTS with S_g predicted from GLS (Eqn. 17a) plotted against a stress concentration index. Note that $\bar{h}$ is inversely proportional to τ.

materials. In these cases, the GLS predictions overestimate the measured values. Moreover, τ is relatively large for both materials, as reflected in the magnitude of the stress concentration index (Fig. 1.22). Two factors have to be considered as these results are interpreted. (1) In some materials, the fraction of fibers that exhibit mirrors is not large enough to provide confidence in the inferred values of S_c and m. This issue is a particular concern for the SiC/SiC$_{CVI}$ material. (2) In other materials, manufacturing flaws are present that provide unbridged crack segments, which cause the UTS to be smaller than S_g (Section 1.4.4).

With the above provisos, it is surprising that the UTS measured for several 2-D CMCs is close to the GLS prediction. In these materials, cracks exist in the 90° plies at low stresses and these cracks should concentrate the stress on the neighboring fibers in the 0° plies. The UTS would thus be expected to follow the strength degradation diagram (Fig. 1.19). That this weakening does not occur remains to be explained. It probably reflects the influence on the strength degradation of elastic anisotropy, as well as pull-out (Fig. 1.19).

1.5 Matrix Cracking in Unidirectional Materials

The development of damage in the form of matrix cracks within 1-D CMCs subject to tensile loading has been traced by direct optical observations on specimens with carefully polished surfaces, and by acoustic emission detection,[7,9,62,79–81] as well as by ultrasonic velocity measurements.[82] Interrupted tests, in conjunction with sectioning and SEM observations, have also been used. Analysis of the matrix damage found in 1-D CMCs provides the basis upon which the behavior of 2-D and 3-D CMCs may be addressed. The *matrix cracks* are found to interact with predominantly intact fibers, subject to interfaces that debond and slide. This process commences at a lower bound stress, $\bar{\sigma}_{mc}$. The crack density increases with increase in stress above $\bar{\sigma}_{mc}$ and may eventually attain a saturation spacing, $\bar{d}_s$, at stress $\bar{\sigma}_s$. The details of crack evolution are governed by the distribution of matrix flaws. The matrix cracks reduce the unloading elastic modulus, $\bar{E}$, and also induce a permanent strain, ε_0 (Fig. 1.9). Relationships between $\bar{E}$, ε_p, and constituent properties provide the key connections between processing and macroscopic performance, via the properties of the constituents.

The deformations caused by matrix cracking, in conjunction with interface debonding and sliding, exhibit three regimes that depend on the magnitude of the debond stress, $\bar{\sigma}_i$. In turn, $\bar{\sigma}_i$ depends on the debond energy through the relationship[1]

$$\bar{\sigma}_i = (1/c_1)\sqrt{E_m \Gamma_i / R} - \sigma^T \tag{22a}$$

which has a useful nondimensional form

$$\Sigma_i = \bar{\sigma}_i / \sigma \tag{22b}$$

A mechanism map that identifies the three regimes is shown in Fig. 1.23.[18] When $\Sigma_i > 1$, debonding does not occur, whereupon matrix crack growth is an entirely elastic phenomenon. This condition is referred to as the no debond (ND) regime. When $\Sigma_i < 1/2$, small debond energy (SDE) behavior arises. The characteristic of this regime is that the reverse slip length at the interface, upon complete unloading, exceeds the debond length. In the SDE regime, Γ_i is typically small and does not affect certain properties, such as the hysteresis loop width. The term SDE is thus used, loosely, to represent the behavior expected when $\Gamma_i \approx 0$. An intermediate, large debond energy (LDE) regime also exists, when $1/2 \leq \Sigma_i \leq 1$. In this situation, reverse slip is impeded by the debond.

1.5.1 Basic Mechanics

The approach used to simulate Mode I cracking under monotonic loading is to define tractions σ_b acting on the crack faces, induced by the fibers (Fig. 1.1), and to determine their effect on the crack tip by using the J-integral,[31,58]

$$\mathscr{G}_{tip} = \mathscr{G} - \int_0^u \sigma_b \, du \tag{23}$$

where $\mathscr{G}$ is the energy release rate and u is the crack opening displacement. Cracking is considered to proceed when $\mathscr{G}_{tip}$ attains the pertinent fracture energy. Since the fibers are not failing, the crack growth criterion involves matrix cracking only. A lower bound is given by[58,83]

$$\mathscr{G}_{tip} = \Gamma_m(1-f) \tag{24}$$

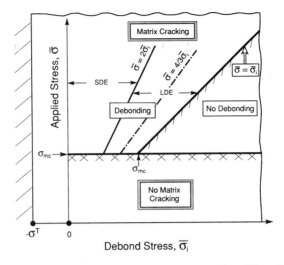

Fig. 1.23 A mechanism map representing the various modes of interface response.

with Γ_m being the matrix toughness. Upon crack extension, $\mathcal{G}$ becomes the crack growth resistance, Γ_R, whereupon

$$\Gamma_R = \Gamma_m(1-f) + \int_0^2 \sigma_b \, du \tag{25}$$

A traction law $\sigma_b(u)$ is now needed to predict Γ_R. A law based on frictional sliding along debonded interfaces has been used most extensively and appears to provide a reasonable description of many of the observed mechanical responses (Eqn. 1). The traction law also includes effects of the interface debond energy, Γ_i.[1] For many CMCs, Γ_i is small, as reflected in the magnitude of the debond stress, Σ_i.

For SDE, with a constant sliding stress, τ_0, the sliding distance, ℓ, in the absence of fiber failure, is related to the crack surface tractions, σ_b, by[14,31]

$$\ell = [RE_m(1-f)/2\tau_0 E_f f](\sigma_b + \sigma^T) \tag{26a}$$

For LDE, the corresponding solution is[1]

$$\ell = [RE_m(1-f)/2\tau_0 E_f f](\sigma_b - \bar{\sigma}_i) \tag{26b}$$

The sliding length is, in turn, related to the crack opening displacement. The corresponding traction laws are:[31,58] for SDE,

$$\sigma_b + \sigma^T = (2\xi\tau_0 E_L f u/R)^{1/2} \tag{27a}$$

and for LDE,

$$\sigma_b - \bar{\sigma}_i = (2\xi\tau_0 E_L f u/R)^{1/2} \tag{27b}$$

where ξ is defined in Table 1.2a. When fiber failure occurs, statistical considerations are needed to determine $\sigma_b(u)$.

The matrix fracture behavior can also be described by using stress intensity factors, K. This approach is more convenient than the J-integral in some cases: particularly for short cracks and for fatigue.[31,84] To apply this approach, it is first necessary to specify the contribution to the crack opening induced by the applied stress, as well as that provided by the bridging fibers. For a plane strain crack of length $2a$ in an infinite plate, the contribution due to the applied stress is[85]

$$u_\infty = (4/E_L)\sigma\sqrt{a^2 - x^2} \tag{28a}$$

and that caused by bridging is

$$u_b = -(4/E_L)\int_0^a \sigma_b(\hat{x})H d\hat{x} \tag{28b}$$

with H being a weight function. The net crack opening displacement is

$$u = u_\infty + u_b \tag{29}$$

The contribution to K from the bridging fibers is obtained using[85]

$$K_b = -2\sqrt{\frac{2}{\pi}}\int_0^a \frac{\sigma_b(x)\,dx}{\sqrt{a^2 - x^2}} \tag{30}$$

with σ_b given by Eqn. (26). The shielding associated with K_b leads to a tip stress intensity factor

$$K_{tip} = K + K_b \tag{31}$$

where K depends on the loading and specimen geometry.

A *criterion for matrix crack extension*, based on K_{tip}, is needed. For this purpose, to be consistent with the energy criterion (Eqn. (24)), the critical stress intensity factor is taken to be

$$K_{tip} = \sqrt{E\Gamma_m(1-f)} \tag{32}$$

Then, the two approaches (K and $\mathscr{G}$) lead to the same steady-state matrix cracking stress.

1.5.2 The Matrix Cracking Stress

The preceding basic results can be used to obtain solutions for matrix cracking.[14,29,31,58,83,86] Present understanding involves the following factors. Because the fibers are intact, a steady-state condition exists wherein the tractions on the fibers in the crack wake balance the applied stress. This special case may be addressed by integrating Eqn. (23) up to a limit $u = u_0$. This limit is obtained from Eqn. (27) by equating σ_b to σ. For SDE, this procedure gives[58]

$$\mathscr{G}_{tip}^0 = \frac{(\sigma + \sigma^T)^3 E_m^2 (1-f)^2 R}{6\tau_0 f^2 E_f E_L^2} \tag{33}$$

A *lower bound to the matrix cracking stress*, σ_{mc}, is then obtained by invoking Eqn. (24), such that†[58]

$$\bar{\sigma}_{mc} = E_L \left[\frac{6\tau \Gamma_m f^2 E_f}{(1-f) E_m^2 R E_L} \right]^{1/3} - \sigma^T$$

$$\equiv \sigma_{mc}^\dagger - \sigma^T \tag{34}$$

In some cases, *small* matrix cracks can form at stresses below $\bar{\sigma}_{mc}$.[7] These occur either within matrix-rich regions or around processing flaws. The nonlinear composite properties are usually dominated by fully developed matrix cracks that form at stresses above $\bar{\sigma}_{mc}$. However, these small flaws may provide access of the atmosphere to the interfaces and cause degradation.

Analogous results can be obtained using stress intensity factors.[31,84] For a small center crack in a tensile specimen,† Eqns. (30) and (32) give a steady-state result, at large crack lengths,[84]

$$K_{tip} = \frac{\sigma^* \sqrt{R}}{\sqrt{6\mathscr{T}}} \tag{35}$$

where $\mathscr{T}$ is a sliding index defined in Table 1.2a. When combined with the fracture criterion (Eqn. (32)), the matrix cracking stress, $\bar{\sigma}_{mc}$, is predicted to be the same as that given by Eqn. (34).

The K approach may also be used to define a transition crack length a_t, above which steady state applies. This transition length is given by[31,83]

$$a_t/R \approx E_m[\Gamma_m(1+\xi)^2(1-f)^4/\tau_0^2 f^4 E_f^2 R]^{1/3} \tag{36}$$

Namely, when the initial flaw size $a_i > a_t$, cracking occurs at $\sigma = \bar{\sigma}_{mc}$. Conversely, when the initial flaws are small, $a_i < a_t$, it has been shown that[84]

$$K_{tip} \approx K\left(1 - \frac{3.05}{\mathscr{E}} \sqrt{\mathscr{E} + 3.3} + \frac{5.5}{\mathscr{E}}\right) \tag{37}$$

where $\mathscr{E}$ is a loading index defined as (Table 1.2a),

$$\mathscr{E} = 2R(1-f)^2 E_m^2 \sigma/E_f E \tau_0 a f^2 (1-\nu^2) \tag{38}$$

This result for K_{tip}, when combined with Eqn. (32), gives a revised matrix cracking stress, which *exceeds* $\bar{\sigma}_{mc}$.

Analogous results can be derived for the LDE regime. In this case, Eqn.

$†K = \sigma\sqrt{\pi a}$.

(27b) may be used with Eqn. (23) to derive an energy release rate, which can be combined with the fracture criterion (Eqn. (24)) to predict σ_{mc}. The result is contained within the implicit formula[87]

$$\left(\frac{\sigma_{mc}}{\bar{\sigma}_i} - 1\right)^3 + 3\sqrt{\frac{E_m\Gamma_i}{R\bar{\sigma}_i}}\left(\frac{\sigma_{mc}}{\bar{\sigma}_1} - 1\right)^2 = \left(\frac{\sigma_{mc}^*}{\bar{\sigma}_1}\right)^3 \tag{39}$$

The trend in σ_{mc} with $\bar{\sigma}_i$ is plotted in Fig. 1.23.

1.5.3 Crack Evolution

The evolution of additional cracks at stresses above σ_{mc} is less well understood, because two factors are involved: *screening* and *statistics*.[29,80] When the sliding zones between neighboring cracks overlap, *screening* occurs and $\mathscr{G}_{tip}$ differs from $\mathscr{G}_{tip}^0$. The relationship is dictated by the location of the neighboring cracks. When a crack forms midway between two existing cracks with a separation $2d$, subject to SDE, $\mathscr{G}_{tip}$ is related to $\mathscr{G}_{tip}^0$ by[29]

$$\mathscr{G}_{tip}/\mathscr{G}_{tip}^0 = 4(d/2\ell)^3 \quad \text{(for } 0 \leqslant d/\ell \leqslant 1) \tag{40a}$$

and

$$\mathscr{G}_{tip}/\mathscr{G}_{tip}^0 = 1 - 4(1 - d/2\ell)^3 \quad \text{(for} 1 \leqslant d/\ell \leqslant 2) \tag{40b}$$

When d is sufficiently small, Eqn. (40a) applies and $\mathscr{G}_{tip}$ is independent of the stress. Once this occurs, $\mathscr{G}_{tip}$ cannot increase and is unable to satisfy again the matrix crack growth criterion (Eqn. (24)). This occurs with spacing, $\bar{d}_s$, at an associated stress $\bar{\sigma}_s$ (Fig. 1.11). This saturation spacing is given by

$$\bar{d}_s/R = \chi[\Gamma_m(1-f)^2 E_f E_m/f\tau_0^2 E_L R]^{1/3} \tag{41}$$

Note that this result is *independent* of the residual stress, because the terms containing $(\sigma_b + \sigma^T)$ in Eqns. (26) and (33) cancel when inserted into Eqn. (40a). The coefficient χ depends on the spatial aspects of crack evolution: periodic, random, etc. Simulations for spatial randomness indicate that $\chi = 1.6$.[29] Moreover, these same simulations indicate that the saturation stress should scale with $\bar{\sigma}_{mc}$, such that

$$\bar{\sigma}_s/\sigma_{mc} = 1.26 \tag{42}$$

This stress ratio depends *only* on the spatial characteristics of the cracks.

In addition to these screening effects, the actual *evolution of matrix cracks* at stresses above σ_{mc} is governed by *statistics* that relate to the size and spatial distribution of matrix flaws. If this distribution is known, the evolution can be predicted. Such statistical effects arise when the matrix flaws are smaller than the transition size, a_t, at which steady state commences (Eqn. (36)). In this case, a flaw size distribution must be combined with the short crack solution for K_{tip} (Eqn. (37)) in order to predict crack evolution. At the simplest level, this

(a)

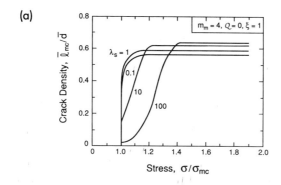

(b)

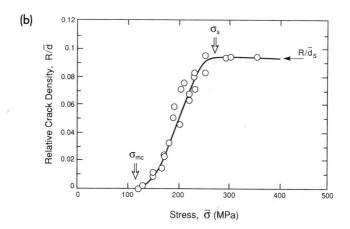

Fig. 1.24 (a) Simulation of crack evolution for various matrix flaw distributions characterized by λ_s when the shape parameter $\omega = 2$. (b) Evolution of matrix crack density with stress for unidirectional SiC/CAS.

has been done by assuming an exponential distribution for the matrix flaw size,[88]

$$p = \exp(L/L_*)(a_t/a)^\omega \tag{43}$$

where p is the fraction of flaws in a composite, length L, having size larger than a, ω is a shape parameter related to the Weibull modulus for the matrix ($\omega = m_m/2$), and L_* is a scale parameter

$$L_* = \lambda_s \ell_{mc} \tag{44}$$

with ℓ_{mc} being the slip distance at $\sigma = \sigma_{mc}$, and λ_s a flaw size coefficient. The condition $\lambda_s \leq 1$ corresponds to a high density of matrix flaws already large enough to be at steady state. Conversely, $\lambda_s > 1$ refers to a situation wherein most matrix flaws are smaller than the transition size, a_t.

Simulations can be performed in which the key variables are the shape parameter ω and the scale parameter λ_s. The simulated crack densities (Fig. 1.24a) indicate a sudden burst of cracking *at* $\sigma = \sigma_{mc}$, when $\lambda_s < 1$, followed by a gradual increase with continued elevation of the stress. The saturation stress is similar to that given by Eqn. (42). In contrast, when $\lambda_s \gg 1$, the cracks evolve more gradually with stress, reaching saturation at substantially higher levels of stress.† These simulated behaviors are qualitatively similar to those measured by experiment (Fig. 1.24b). Moreover, the values found for ω are in a reasonable range ($m_m = 2\omega \approx 4\text{–}8$). However, since ω and λ_s are not known *a priori*, in practice this approach becomes a fitting procedure rather than a predictive model. Despite this limitation, it has been found that a simple formula can be used to approximate crack evolution in most CMCs,[5] given by (Fig. 1.24b),

$$\bar{d} \approx \bar{d}_s \frac{(\bar{\sigma}_s/\bar{\sigma}_{mc} - 1)}{(\bar{\sigma}/\sigma_{mc} - 1)} \tag{45}$$

Analogous results can be derived for LDE, with the debond length given by Eqn. (26b) and the reference energy release rate by Eqn. (39). In this case, the saturation crack spacing is smaller than that given by Eqn. (42).

1.5.4 Constitutive Law

Analyses of the plastic strains caused by matrix cracks, combined with calculations of the compliance change, provide a constitutive law for the material. The important parameters are the permanent strain, ε_0 and the unloading modulus, $\bar{E}$. These quantities, in turn, depend on several constituent properties; the sliding stress, τ, the debond energy, Γ_i, and the misfit strain, Ω. The most important results are summarized below.

Matrix cracks *increase* the elastic compliance. Numerical calculations indicate that the *unloading elastic modulus*, E^*, is given by[21]

$$E_L/E^* - 1 = (R/\bar{d})\mathscr{B}(f, E_f/E_m) \tag{46}$$

where $\mathscr{B}$ is the function plotted in Fig. 1.25. The matrix cracks also cause a permanent strain associated with relief of the residual stress. This strain, ε^*, is related to the modulus and the misfit stress by (Fig. 1.26)[21]

$$\varepsilon^* \equiv \sigma^T(1/E^* - 1/E_L) \tag{47}$$

The preceding effects occur *without* interface sliding. The incidence of *sliding* leads to *plastic strains* that superpose onto ε^*. The magnitude of these strains depends on Σ_i (Fig. 1.23) and on the stress relative to the saturation stress, σ_s.

---

†Nevertheless, the saturation spacing remains insensitive to λ_s.[88]

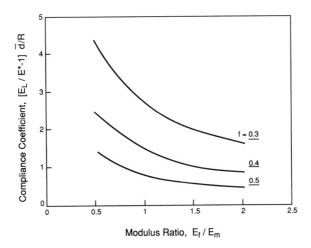

Fig. 1.25 Effects of modulus mismatch and fiber volume fraction on the elastic compliance.

1.5.4.1 Stresses below Saturation

Small Debond Energy.   For SDE, when $\sigma < \sigma_s$, the unloading modulus $\bar{E}$ depends on τ_0, but is *independent* of Γ_i and Ω. However, the permanent strain ε_0 depends on Γ_i and Ω, as well as τ_0. These differing dependencies of $\bar{E}$ and ε_0 on constituent properties have the following two implications. (1) To *simulate* the stress–strain curve, both ε_0 and $\bar{E}$ are required. Consequently, τ_0, Γ_i and Ω *must be known*. (2) The use of unloading and reloading to *evaluate the constituent properties* has the convenience that the hysteresis is dependent *only on* τ_0. Consequently, precise determination of τ_0 is possible. Moreover, with τ_0 known from the hysteresis, both Γ_i and Ω can be evaluated from the permanent strain. The principal SDE results are as follows.

The permanent strain is [5,9,18]

$$(\varepsilon_0 - \varepsilon^*)\mathscr{H}^{-1} = 4(1 - \Sigma_i)\Sigma_T + 1 - 2\Sigma_i^2 \tag{48}$$

where $\mathscr{H}$ is a hysteresis index (Table 1.2a)

$$\mathscr{H} = b_2(1 - a_i f)^2 R\bar{\sigma}^2/4\bar{d}\tau_0 E_m f^2 \tag{49}$$

and

$$\Sigma_T = \sigma^T/\bar{\sigma} \tag{50}$$

The maximum width of the hysteresis loop, at half maximum, $\delta\varepsilon_{1/2}$, is (Figs. 1.9 and 1.26)

$$\delta\varepsilon_{1/2} = \mathscr{H}/2 \tag{51}$$

The unloading strain is (Fig. 1.26),

$$\Delta\varepsilon_p = \mathscr{H} \tag{52}$$

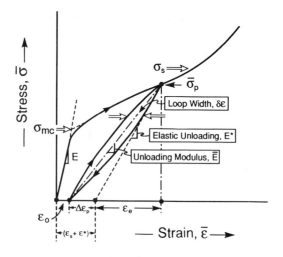

a) Debonding and Sliding Interface

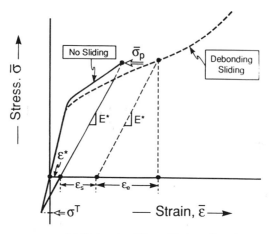

b) Behavior When Debonding Inhibited

Fig. 1.26 Basic parameters involved in the stress–strain behavior of CMCs. (a) Debonding and sliding subject to SDE, and (b) in the absence of debonding.

and the unloading modulus is

$$(\bar{E})^{-1} = (E^*)^{-1} + \mathcal{H}/\bar{\sigma} \tag{53}$$

Large Debond Energy. For LDE (Fig. 1.23), when $\sigma < \sigma_s$, the unloading modulus depends on both τ and Γ_i (Fig. 1.26). There are also linear segments to the unloading and reloading curves. These segments can be used to establish constructions that allow the constituent properties to be conveniently estab-lished. The principal results are as follows. The permanent strain is,[18]

$$(\varepsilon_0 - \varepsilon^*)\mathcal{H}^{-1} = 2(1 - \Sigma_i)(1 - \Sigma_i + 2\Sigma_T) \tag{54}$$

and the unloading modulus is

$$(\bar{E})^{-1} = (E^*)^{-1} + 4\Sigma_i)\mathcal{H}/\bar{\sigma} \tag{55}$$

In this case, the hysteresis loop width depends on the magnitude of Σ_i. For intermediate values, $1/2 \leqslant \Sigma_i \leqslant 3/4$

$$\delta\varepsilon_{1/2} = \mathcal{H}[1/2 - (1 - 2\Sigma_i)^2] \tag{56}$$

whereas, for $3/4 \leqslant \Sigma_i \leqslant 1$

$$\delta\varepsilon_{1/2} = 4\mathcal{H}(1 - \Sigma_i)^2 \tag{57}$$

1.5.4.2 Stresses above Saturation

At stress, $\sigma > \sigma_s$, the crack density remains essentially constant. It has thus been assumed that there is no additional stress transfer between the fibers and the matrix. In this case, the tangent modulus is given by:[14]

$$E_t \equiv d\sigma/d\varepsilon = fE_f \tag{58}$$

In practice, the tangent modulus is usually found to be smaller than predicted by Eqn. (58). Two factors are involved: changes in the sliding stress and fiber failure. At high fiber stresses, the Poisson condition of the fiber reduces the radial stress, σ_{rr}. Consequently, whenever the sliding stress can be represented by Eqn. (1), τ decreases as the stress increases. The associated tangent modulus at fixed crack spacing is[1]

$$d\bar{\sigma}/d\bar{\varepsilon} = b_1 E_m \bar{d}/a_3 b_2 R[1 + \vartheta + \exp(-\vartheta)] \tag{59}$$

where $\vartheta = 2\mu b_1 d/R$, with μ being the friction coefficient.

As the UTS is approached, significant fiber failures occur, which further reduce the tangent modulus. The basic stress–strain relationship is,[64]

$$\bar{\sigma} = fE_f\bar{\varepsilon}\left\{1 + \sum_{n \geqslant 1} \frac{(-1)^n}{2n!}\left[\frac{2 + n(m+1)}{1 + n(m+1)}\right](E_f\bar{\varepsilon}/S_c)^{n(m+1)}\right\} \tag{60}$$

1.5.5 Simulations

The preceding constitutive laws may be used to simulate stress–strain curves for comparison with experiments. In order to conduct the simulations, the constituent properties, τ, Γ_i and Ω are first assembled into the non-dimensional parameters $\mathcal{H}$, Σ_i, and Σ_T. For this purpose, it is necessary to have independent knowledge of $\bar{d}(\bar{\sigma})$. When this does not exist, an estimation procedure is needed, based on Eqn. (45), through evaluation of $\bar{d}_s$, $\bar{\sigma}_{mc}$ and $\bar{\sigma}_s$. The first step is to use Eqn. (41) to evaluate the saturation crack spacing $\bar{d}_s$

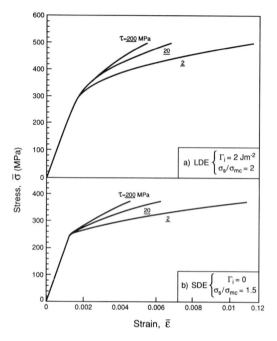

Fig. 1.27 Simulated stress–strain curves for 1-D CMCs indicating the relative importance of constituent properties.

from the constituent properties. One limitation of this procedure concerns the accuracy with which χ and Γ_m are known. An alternative option exists when crack spacing data are available for another CMC with the *same matrix*. Then, Eqn. (41) can be used to *scale* $\bar{d}_s$ in accordance with,

$$\bar{d}_s^3 \sim E_f R^2 / \tau_0^2 E_L \tag{61}$$

It is also possible to estimate $\bar{\sigma}_{mc}$ from the constituent properties, by using Eqn. (34). Then Eqn. (42) is used to estimate σ_s.

When $d(\bar{\sigma})$ has been established in this manner, stress–strain curves can be simulated for 1-D materials. Based on this approach, simulations have been used to conduct sensitivity studies of the effects of constituent properties on the inelastic strain. Examples (Fig. 1.27) indicate the spectrum of possibilities for CMCs.

1.5.6 Experiments

Relatively complete matrix cracking and *inelastic strain* measurements have been made on two unidirectional CMCs:[5,18,48,89] SiC/CAS, as well as SiC/SiC (produced by CVI). The stress–strain curves for these two materials (Fig. 1.28) indicate a contrast in inelastic strain capability, which demand

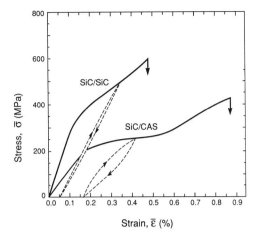

Fig. 1.28 Stress–strain curves and typical hysteresis measurements obtained on SiC/CAS and SiC/SiC unidirectional composites.

interpretation. Some typical hysteresis measurements for these materials (Fig. 1.28) reveal major differences, which must reflect differences in constituent properties. There are also considerable differences in the evolution of matrix cracks (Fig. 1.24). An analysis of the hysteresis loops (Fig. 1.29) and the permanent strain (Fig. 1.30), as well as other characteristics, indicate the substantial differences in interface properties summarized in Table 1.3. These differences arise despite the fact that the fibers are the same and that the fiber coatings are C in both cases.†

The constituent properties from Table 1.3 can, in turn, be used to simulate the stress–strain curves (Fig. 1.31). The agreement with measurements affirms the simulation capability whenever the constituent properties have been obtained from *completely independent tests* (Table 1.1). This has been done for the SiC/CAS material, but not yet for SiC/SiC. While the limited comparison between simulation and experiment is encouraging, an unresolved problem concerns the predictability of the saturation stress, σ_s. In most cases, *ab initio* determination cannot be expected, because the flaw parameters for the matrix (ω, λ_s) are processing sensitive. Reliance must therefore be placed on experimental measurements, which are rationalized, *post facto*. Further research is needed to establish whether formalisms can be generated from the theoretical results which provide useful bounds on σ_s. A related issue concerns the necessity for matrix crack density information. Again, additional insight is needed to establish meaningful bounds. Meanwhile, experimental methods that provide crack density information in an

---

†Analysis of the coating structure by TEM provides a rationale for specifying the differing interface responses in accordance with the basic model (Fig. 1.9).

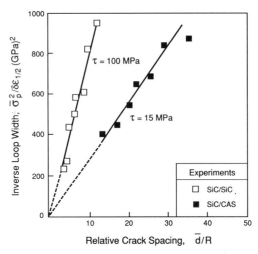

Fig. 1.29 Analysis of hysteresis loop results for unidirectional SiC/CAS and SiC/SiC.

efficient, straightforward manner require development. One possibility involves measurements of the acoustic velocity, ν, which can be conducted continuously, during testing.[82] These measurements relate to changes in the elastic modulus E^* as matrix cracks develop ($E^* = \rho_0 \nu^2$). This modulus can be related to the crack spacing, through a model (Eqn. (46)).

1.6 Matrix Cracking in 2-D Materials

General loadings of 2-D CMCs involve mixtures of tension and shear. For design purposes, it is necessary to have models and experiments that combine these loadings. Matrix cracking and fiber failure are the basic phenomena that dictate all of the non-linearities. However, there are important differences between tension and shear. The behavior subject to tensile loading has been widely investigated.[42,55,77,90–92] The behavior in shear is only appreciated at an elementary level.[26] Furthermore, the intermediate behaviors have had even less study.[90,93] Nevertheless, the basic concept is clear. It is required that matrix cracking, as well as fiber failure, phenomena be incorporated into the models in a consistent manner, such that *interpolation* approaches can be devised and implemented, which interrelate the tensile and shear properties.

1.6.1 Tensile Properties

General comparison between the tensile stress–strain [$\sigma(\varepsilon)$], curves for 1-D and 2-D materials (Fig. 1.32) provides important perspective. It is found that $\sigma(\varepsilon)$ for 2-D materials is quite closely matched by simply scaling down the 1-D curves by 1/2. The behavior of 2-D materials must, therefore, be

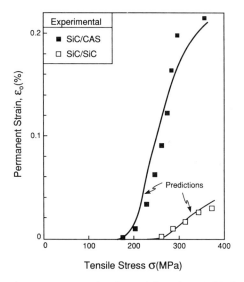

Fig. 1.30 Analysis of permanent strains for unidirectional SiC/CAS and SiC/SiC.

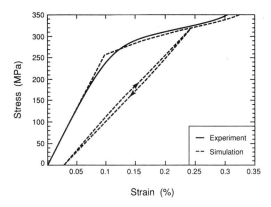

Fig. 1.31 Simulated stress–strain curve for unidirectional SiC/SiC and comparison with experiment.

dominated by the 0° plies,† because these plies provide a fiber volume fraction in the loading direction about half that present in 1-D material.[5]

The most significant 2-D effects occur at the *initial deviation from linearity*. At this stage, matrix cracks that form either in matrix-rich regions or in 90° plies evolve at lower stresses than cracks in 1-D materials. The associated nonlinearities are usually slight and do not normally contribute substantially to the overall nonlinear response of the material. However, these

---

†Furthermore, since some of the 2-D materials are woven, the 1/2 scaling infers that the curvatures introduced by weaving have minimal effect on the stress–strain behaviors.

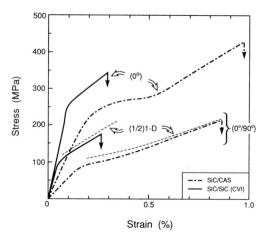

Fig. 1.32 A comparison of stress–strain curves measured for 1-D and 2-D CMCs. The dotted lines labeled 1/2 (1-D) represent the behavior expected in 2-D materials when the 90° plies carry zero load.

cracks have important implications for oxidation embrittlement and creep rupture, and require analysis. Matrix cracking in the 90° plies often proceeds by a tunneling mechanism (Fig. 1.33). Tunnel cracking occurs subject to a lower bound stress σ_τ,[94,95] given by

$$\sigma_\tau = \sigma_\tau^0 - \sigma^R(E_L + E_T)/2E_T \tag{62}$$

with

$$\sigma_\tau^0 = (E\Gamma_R/t_P)^{1/2} g(f, E_f/E_m) \tag{63}$$

The function g depends on whether the transverse fibers either remain in contact with the matrix upon loading, or separate. The relative unloading modulus associated with such tunnel cracks, $\overline{E}/E$, depends primarily on the crack density, $t_p/\overline{L}$, with $\overline{L}$ being the mean crack spacing in the 90° plies.[94,96] At large crack densities, a limiting value is reached, given by,

$$\overline{E}/E_0 = E_L/(E_L + E_T) \tag{64}$$

where

$$E_0 = E_L(1 + E_2/E_T)/2(E_L/E_T - \nu_L^2) \tag{65}$$

The corresponding permanent strain is

$$\varepsilon_0 = (1/\overline{E} - 1/E_0)\sigma^R(E_L + E_T)/2E_L \tag{66}$$

Examples of the overall stress–strain response are summarized in Fig. 1.34. In practice, the stresses at which these cracks evolve may be larger, because the formation of cracks, at stresses above σ_τ, depends on the availability of flaws in the 90° plies.

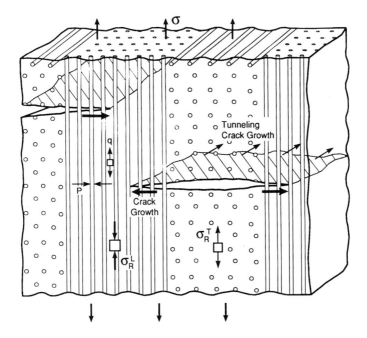

Fig. 1.33 The matrix crack growth mechanisms that operate in 2-D CMCs.

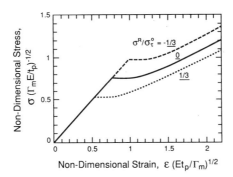

Fig. 1.34 Simulated stress–strain response for a 2-D CMC subject to tunnel cracking.

Lateral extension of these tunnel cracks into the matrix of the 0° plies (Fig. 1.33) results in behavior similar to that found in 1-D material. Moreover, if the stress σ^0 acting on the 0° plies is known, the 1-D solutions may be used directly to predict the plastic strain. Otherwise, this stress must be estimated.[94] For a typical 0/90 system, σ^0 must range between $\bar{\sigma}$ and $2\bar{\sigma}$, depending upon the extent of matrix cracking in the 90° plies and upon E_T/E_L. Preliminary analysis has been conducted using, $\sigma^0 = 2\bar{\sigma}$, as implied by the comparison

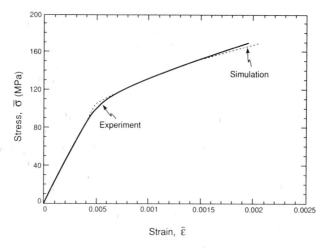

Fig. 1.35 A simulated stress–strain curve for a 2-D SiC/SiC in which the plastic strains are dominated by the 0° plies. The experimental results for a SiC/SiC composite are shown for comparison.

between 1-D and 2-D stress–strain curves (Fig. 1.32). Additional modeling on this topic is in progress.

Using this simplified approach, simulations of stress–strain curves have been conducted.[89,97] These curves have been compared with experimental measurements for several 2-D CMCs. One result is summarized in Fig. 1.35. It is apparent that the simulations lead to somewhat larger flow strengths than the experiments, especially at small inelastic strains. To address this discrepancy, further modeling is in progress, which attempts to couple the behavior of the tunnel cracks with the matrix cracks in the 0° plies.

1.6.2 Shear Properties

The matrix cracking that occurs in 2-D CMCs, subject to shear loading, depends on the loading orientation and the properties of the matrix. Two dominant loading orientations are of interest: in-plane shear along one fiber orientation and out-of-plane (or interlaminar) shear. The key difference between these loading orientations concerns the potential for interaction between the matrix cracks and the fibers (Fig. 1.36). For the out-of-plane case, matrix cracks evolve without significant interaction with the fibers. Conversely, for in-plane loading, the matrix crack *must interact* with the fibers. These interactions impede matrix crack development. Consequently, *the in-plane shear strength always exceeds the interlaminar shear strength*.

1.6.2.1 In-Plane Shear

Experiments that probe the in-plane shear properties have been' performed by using Iosipescu test specimens.[26] A summary of experimental results

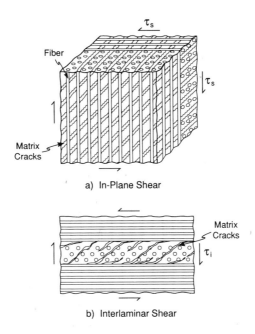

a) In-Plane Shear

b) Interlaminar Shear

Fig. 1.36 Schematic indicating the two modes of shear damage: (a) in-plane, and (b) interlaminar.

(Fig. 1.6) indicates that the matrix has a major influence on the shear flow strength τ_s and the shear ductility, γ_c. Moreover, it has been found that the shear flow strengths can be ranked using a parameter, $\mathscr{W}$, derived from the matrix cracking stress in the absence of interface sliding,[58] given by (Fig. 1.37),

$$\mathscr{W} = \sqrt{\Gamma_m/RG} \qquad (67)$$

The property of principal importance within $\mathscr{W}$ is the shear modulus, G, which reflects the *increase in compliance* caused by the matrix cracks. However, it remains to develop a model that gives a complete relationship between the composite strength and the constituent properties.

The shear ductility also appears to be influenced by the shear modulus, but in the opposite sense: high modulus matrices result in low ductility. This behavior has been rationalized in terms of the effect of matrix modulus on the bending deformation experienced by fibers between matrix cracks.[26] As yet, there have been no calculations that address this phenomenon.

1.6.2.2 Interlaminar Shear

The matrix cracks that form upon interlaminar shear loading and provide the plastic strains are material dependent. The simplest case, depicted in Fig. 1.36b, involves multiple tunnel cracks that extend across the layer and orient

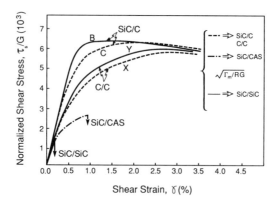

Fig. 1.37 Normalized in-plane shear stress–strain curves with the non-dimensional parameter $\mathcal{W}$ indicated.

normal to the maximum tensile stress *within* the layer.[93] In other cases, the matrix cracks are confined primarily to the matrix-only layers between plies.[61] A general understanding of these different behaviors does not yet exist.

When the interlaminar cracks form by tunneling, the solutions have a direct analogy within the transverse cracking results described above.[98] In shear loading, the tunnel cracks evolve and orient such that a Mode II crack develops, as sketched on Fig. 1.36. There is a critical shear stress, $\bar{\tau}_c$, at which interlaminar shear failure occurs, given by

$$\bar{\tau}_c \approx 1.5\sqrt{G\Gamma_R/t_p} \tag{68}$$

where t_p is now the thickness of the material layer that governs cracking. There must also be effects of residual stress, but these have yet to be included in the model. The form of the critical shear stress relation (Eqn. (68)) is the same as that for transverse tunnel cracking (Eqn. (62)), verifying that these two phenomena are interrelated. The elastic properties dictate whether $\bar{\tau}_c$ or σ_τ is the larger: usually $\bar{\tau}_c < \sigma_\tau$ because $G \ll E$.

1.6.3 Transverse Tensile Properties

CMCs with 2-D fiber architecture are susceptible to interlaminar cracking in various component configurations (Fig. 1.38). In such cases, as the crack extends through the component, conditions range from Mode I to Mode II. Tests and analyses are needed that relate to these issues. Most experience has been gained from polymer matrix composites (PMCs).[99] The major issue is the manner whereby the interlaminar (transverse) cracks interact with the fibers. In principle, it is possible to conduct tests in which the cracks do not interact. In practice, such interactions always occur in CMCs, as the crack front meanders and crosses over inclined fibers.[100,101] These interactions dominate

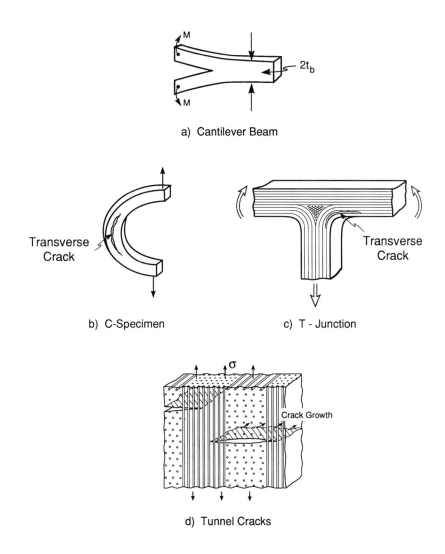

a) Cantilever Beam

Transverse
Crack

Transverse
Crack

b) C-Specimen

c) T - Junction

d) Tunnel Cracks

Fig. 1.38   Schematic of the various modes of transverse cracking in CMCs.

the measured fracture loads in conventional double cantilever beam (DCB) specimens, as well as in flexure specimens.[102,103] Some typical results for the transverse fracture energy (Fig. 1.39), indicate the large values (compared with $\Gamma_m \approx 20\,\mathrm{J\,m^{-2}}$) induced by these interactions.

Analysis indicates that large-scale bridging (LSB) is involved and the bridging behavior can be explicitly ascertained from the measured curves.[103] For the particular case of a DCB specimen (Fig. 1.38a), the J-integral is explicitly defined in terms of the bending moment, M, *and* the traction law.[13]

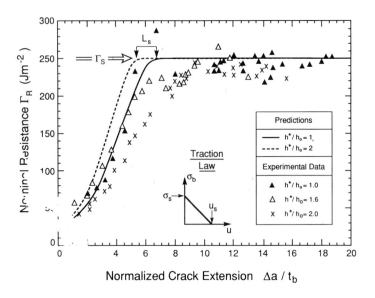

Fig. 1.39 The transverse fracture resistance of a SiC/CAS material. Also shown are the traction law assumed for inclined bridging fibers upon transverse cracking and the predicted resistance curves.

For example, the steady-state resistance, Γ_s, for a linear softening traction law, is

$$\Gamma_s \equiv 12M_s^2/W_L t_b^3$$
$$= \sigma_s u_s/2 + \Gamma_m \tag{69}$$

and the zone length at steady state is

$$L_s = (Eu_s/3\sigma_s)^{1/4} t_b^{3/4} \tag{70}$$

where $2t_b$ is the DCB beam thickness, M_s is the bending moment, with the quantities Γ_s and L_s defined on Fig. 1.39. Experimental measurements made with DCB specimens can be used to evaluate the parameters, σ_s and u_s, by simply fitting the data to Eqns. (69) and (70). This information can then be used to *predict* Γ_s and Γ_R for other configurations.

An example is given for SiC/CAS composites in Fig. 1.39. Experimental results for this material[100] give $u_s \approx 100\ \mu m$ and $\sigma_s \approx 10$ MPa. One application of these results is the prediction of the tunnel cracking found in 0/90 laminates (Eqn. (62)). The analysis of tunnel cracking[94] has established that for typical laminate thicknesses, the crack opening displacements are small ($<1\ \mu m$). For such small displacements, there is a *negligible influence of the fibers*. Consequently, $\Gamma_R \approx \Gamma_m(1-f)$. Other applications to C-specimens and T-junction are in progress.

An obvious limitation of the procedure is the uncertainty about the manner whereby the matrix crack interacts with the fibers in other geometries and hence, the universality of σ_s and u_s. This is a topic for further research.

1.7 Stress Redistribution

1.7.1 Background

CMCs usually have substantially lower notch sensitivity than monolithic brittle materials and, in several cases, exhibit notch insensitive behavior.[3,4] This desirable characteristic of CMCs arises because the material may *redistribute stresses* around strain concentration sites. Notch effects appear to depend on the class of behavior. Moreover, a different *mechanics* is required for each class, because the stress redistribution mechanisms class operates over different physical scales. Class I behavior involves stress redistribution by fiber bridging/pull-out, which occurs along the crack plane.[13,104] Large-Scale Bridging Mechanics (LSBM) is preferred for such materials. Class II behavior allows stress redistribution by large-scale matrix cracking[4] and Continuum Damage Mechanics (CDM) is regarded as most appropriate. Class III behavior involves material responses similar to those found in metals, and a comparable mechanics might be used:[27] either LEFM for small-scale yielding or non-linear fracture mechanics for large-scale yielding. Since a unified mechanics has not yet been identified, it is necessary to use *mechanism maps* that distinguish the various classes (Figs. 1.3 and 1.4).

1.7.2 Mechanism Transitions

The transition between Class I and Class II behaviors involves considerations of both matrix crack growth and fiber failure. One hypothesis for the transition may be analyzed using LSBM. Such analysis allows the condition for fiber failure at the end of an unbridged crack segment to be solved simultaneously with the energy release rate of the matrix front. The latter is equated to the matrix fracture energy.[75] By using this solution to specify that fiber failure occurs *before the matrix crack extends into steady state*, Class I behavior is presumed to ensue. Conversely, Class II behavior is envisaged when steady-state matrix cracking occurs prior to fiber failure. The resulting mechanism map involves two indices (Table 1.2),

$$\mathscr{S} = (RS/a_0\tau)(E_m^2/E_L E_f)[(1-f)/f]^2$$
$$\equiv 3\mathscr{S}_b \tag{71}$$

and

$$\mathscr{U} = \sigma_{mc}/S \tag{72}$$

With $\mathscr{S}$ and $\mathscr{U}$ as coordinates, a mechanism map may be constructed that distinguishes Class I and Class II behavior (Fig. 1.3). While this map has qualitative features consistent with experience, the experiments required for

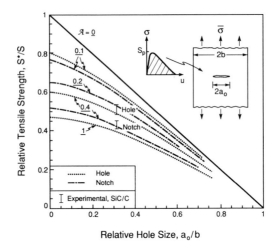

Fig. 1.40 Effects of relative notch size on the UTS. Also shown are experimental data for a SiC/C material.

validation have not been completed. In practice, the mechanism transition in CMCs probably involves additional considerations.

The incidence of Class III behavior is found at relatively small magnitudes of the ratio of shear strength, τ_s, to tensile strength S. When τ_s/S is small, a shear band develops at the notch front and extends normal to the notch plane. Furthermore, since τ_s is related to G, the parameter G/S is selected as the ordinate of a mechanism map. Experimental results suggest that Class III behavior arises when $G/S \gtrsim 50$ (Fig. 1.4).

1.7.3 Mechanics Methodology

1.7.3.1 Class I Materials

The Class I mechanism, when dominant, has features compatible with LSBM.[104-106] These mechanics may be used to characterize effects of notches, holes and manufacturing flaws on tensile properties, whenever a single matrix crack is prevalent. For cases wherein the flaw or notch is small compared with specimen dimensions, the tensile strength may be plotted as functions of *both* flaw indices: $\mathscr{A}_b$ and $\mathscr{A}_p$ (Fig. 1.18). For the former, the results are sensitive to the ratio of the pull-out strength S_p to the UTS.[76] These results should be used whenever the unnotched tensile properties are compatible with global load sharing. Conversely, $\mathscr{A}_p$ should be used as the notch index when the unnotched properties appear to be pull-out dominated.

When the notch and hole have dimensions that are significant fraction of the plate width $(a_0/b > 0)$, *net section* effects must be included.[76] Some results (Fig. 1.40) illustrate the behavior for different values of the notch sensitivity index, $\mathscr{A}$. Experimental validation has not been undertaken, although partial

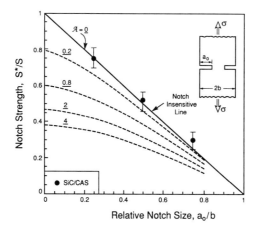

Fig. 1.41 Experimental results for SiC/CAS indicating notch insensitivity.

results for one material (SiC/C$_B$) are compatible with LSBM,[27] as shown for data obtained with center notches and holes (Fig. 1.40). The promising feature is that LSBM explains the difference between notches and holes (upon requiring that $\mathcal{A} \approx 0.8$).

1.7.3.2 Class II Materials

The non-linear stress–strain behavior governed by matrix cracking (expressed through $\bar{E}$, Eqn. (55), and ε_0, Eqn. (54)) provides a basis for a CDM approach that may be used to predict the effects of notches and holes.[32] Such developments are in progress.† In practice, several Class II CMCs have been shown to exhibit notch insensitive behavior, at notch sizes up to 5 mm.[3,4] The notch insensitivity is manifest in the effect of the relative notch size, a_0/b, on the ratio of the UTS measured in the presence of notches (designated S^*), to the strength in the absence of notches (designated S). Results for SiC/CAS are illustrated in Fig. 1.41. In this material, the nonlinearity provided by the matrix cracks allows sufficient stress redistribution that the stress concentration is *eliminated*. This occurs despite the low ductility (<1%). A CDM procedure capable of predicting this behavior will be available in the near future, using the stress–strain simulation capability based on constituent properties (Figs. 1.25, 1.34, and 1.35).

1.7.3.3 Class III Materials

Class III behavior has been found in several C matrix composites.[27] In these materials, the shear bands can be imaged using an X-ray dye penetrant method. Based on such images, the extent of the shear deformation zone, ℓ_s, is

---

†An important factor that dictates whether continuum or discrete methods are used concerns the ratio of the matrix crack spacing to the radius of curvature of the notch.

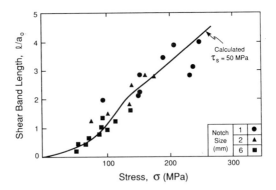

Fig. 1.42 Relationship between shear band length and stress for a C/C composite. The calculated curve is also shown.

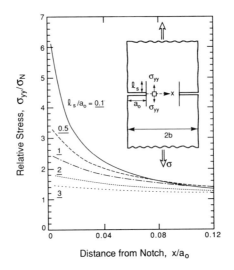

Fig. 1.43 Effect of shear bands on the stress ahead of a notch.

found to be predictable from measured shear strengths, τ_s (Fig. 1.6), in approximate accordance with (Fig. 1.42)

$$\ell_s/a_0 \approx \sigma/\tau_s - 1 \qquad (73)$$

Calculations have indicated that this shear zone diminishes the stress ahead of the notch (Fig. 1.43), analogous to the effect of a plastic zone in metals. For C/C materials, it has been found that the shear band lengths are small enough that LEFM is able to characterize the experimental data over a range of notch lengths, such that, $K_{Ic} \approx 16 \, \text{MPa}\sqrt{\text{m}}$ (Fig. 1.44). However, conditions must exist where LEFM is violated. For example, when $\ell_s/a_0 \gtrsim 4$, the stress concentration is essentially eliminated (Fig. 1.43) and the material must then

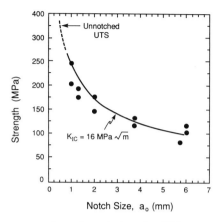

Fig. 1.44 LEFM representation of notch data for C/C.

become notch insensitive. Further work is needed to identify parameters that bound the applicability of LEFM, as well as establish the requirements for notch insensitivity.

1.7.4 Measurements

Notch sensitivity data (Figs. 1.40, 1.41, and 1.44) provide an explicit measure of stress redistribution. However, further understanding requires techniques that probe the stress and strain around notches, as CMCs are loaded to failure. Many of the methods have been developed and used for the same purpose on PMCs.[107,108] These techniques can measure both strain and stress distributions.

Strain distributions are measured with high spatial resolution by using Moiré interferometry. In this method, the fringe spacings relate to the in-plane displacements which, in turn, govern the strains. There has been only limited use of this technique for CMCs.[109] Preliminary measurements suggest that the inelastic deformations that arise from shear bands result in strains similar to *elastic* strains. Based on such similarity, it may be speculated that the reduced stress concentrations may relate explicitly to the lower stresses that arise upon inelastic deformation at fixed strain (Fig. 1.45). Further exploration of this simple concept is in progress.

Since strain measurements appear to have minimal sensitivity to the stress redistribution mechanisms operative in CMCs, a technique that measures the *stress* distribution is preferred. One such method involves measurement of thermoelastic emission. This method relies on the *temperature rise* ΔT that occurs when an element of the composite is subject to a *hydrostatic stress* $\Delta \sigma_{kk}$ under adiabatic conditions. The fundamental adiabatic relationship for a homogeneous solid is,[110]

$$\Delta \dot{\sigma}_{kk} = (C_v \rho_0 / \alpha T_0) \Delta \dot{T} \tag{74}$$

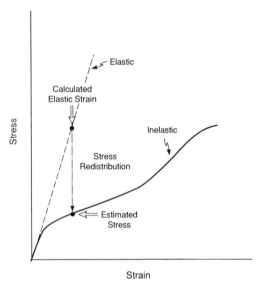

Fig. 1.45 A schematic indicating an approximate method for obtaining the stress by using the strain obtained from elastic calculations.

where C_v is the specific heat at constant strain, and ρ_0 is the density. One experimental implementation of this concept is a technique referred to as Stress Pattern Analysis by Thermoelastic Emission (SPATE).[110] It involves the use of high sensitivity infrared detectors, which measure the temperature in a lock-in mode, as a cyclic stress is applied to the material. This feature essentially eliminates background problems and has good signal-to-noise characteristics. SPATE measurements are conventionally performed at small stress amplitudes, which elicit "elastic" behavior in the material. Experimental results[137] for a Class II material (SiC/CAS) have confirmed that the stress concentration can be eliminated by matrix cracks (Fig. 1.46). In addition, results for a Class III material (C/C) have provided a direct measure of the stress redistribution caused by shear bands (Fig. 1.47).

Another method for strain measurement uses fluorescence spectroscopy.[111] This method has particular applicability to oxides, especially Al_2O_3 (either as fiber or matrix). The technique has the special advantage that strains can be measured in individual fibers, such that stress changes caused by matrix cracks can be measured. Such measurements permit the material to be probed at the spatial resolution needed to understand mechanisms in detail.

1.8 Fatigue

1.8.1 Basic Phenomena

Upon cyclic loading, matrix cracking and fiber failure occur in brittle matrix composites,[11,56,77,112] in accordance with the same three classes found

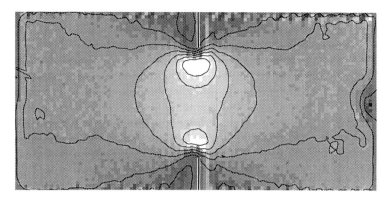

Fig. 1.46 A SPATE image obtained from a SiC/CAS composite compared with that for a monolithic material.

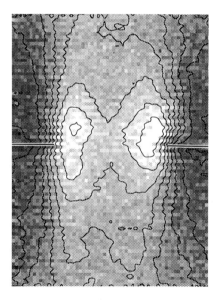

Fig. 1.47 A SPATE image obtained from a C/C composite.

for monotonic loading (Fig. 1.2). The preceding matrix cracking and fiber failure models still apply, except that some additional factors need to be introduced.[113] The experimental results needed to establish the specific fatigue mechanisms that operate in CMCs are sparse. However, similar mechanisms operate in metal (MMC) and polymer (PMC) matrix composites. Observations, modeling, and measurements performed on these materials provide insights that facilitate and hasten an understanding of the cyclic behaviors of brittle matrix composites.

Among the new features that enter when cyclic loading is used are *degradation mechanisms* and, in some cases, revised *crack growth criteria*. The macroscopic characteristics associated with the degradation mechanisms are fatigue life (σ-N) curves and changes in compliance. In addition, the hysteresis loops change as fatigue proceeds. Analyses of compliance and hysteresis changes, as well as differences in fiber pull-out, indicate that the interface sliding stress changes as fatigue proceeds. *A cyclic sliding function τ_f (N) thus becomes a new constituent property.*[113] In some cases, at high temperature and upon thermomechanical fatigue, a particularly low fatigue threshold stress (compared with the UTS) implies fiber strength degradation. Consequently, a cyclic fiber strength function, $S_f(N)$, may also be needed to predict fatigue life.

Several possible matrix *crack growth criteria* are applicable to fatigue. These relate to the conditions at the crack front. When the matrix itself is susceptible to cyclic fatigue, the Paris law relates crack growth in the matrix to the stress intensity *range* at the crack front, ΔK_{tip}, by

$$da/dN = \beta(\Delta K_{tip}/E)^{n_f} \tag{75}$$

where N is the number of cycles, n_f is a power-law exponent and β is a material-dependent coefficient. In some cases, n_f is sufficiently large that matrix crack growth is dominated by the peak value of either K_{tip} or $\mathcal{G}_{tip}$. Then, the same criterion used for monotonic loading (Eqn. (24)) may be preferred. Finally, when the dominant mechanism involves stress corrosion, crack growth can be described in terms of $\mathcal{G}_{tip}$ through the commonly used power law[114]

$$\frac{da}{dt} = \dot{a}_0 \left(\frac{\mathcal{G}_{tip}}{\mathcal{G}_m} \right)^{\eta} \tag{76}$$

where $\dot{a}_0$ is a reference velocity, η is the power-law exponent, and $\mathcal{G}_m$ is the matrix toughness, taken to be $\Gamma_m(1-f)$.

1.8.2 Matrix Crack Growth

When the interfaces are "weak," fibers can remain intact in the crack wake and cyclic frictional dissipation resists fatigue crack growth.[84] The latter has been extensively demonstrated on Ti matrix composites reinforced with SiC fibers.[115,116] The essential features of the "weak" interface behavior are as follows: intact, sliding fibers acting in the crack wake shield the crack tip, such that the stress intensity range at the crack tip, ΔK_{tip}, is less than that expected for the applied loads, ΔK. Using this approach, a simple transformation converts the monotonic crack growth parameters into cyclic parameters that can be used to interpret and simulate fatigue growth of each matrix crack. The key transformation is based on the relationship between interface sliding during loading and unloading, which relates the monotonic result to the cyclic equivalent through[84]

$$(\tfrac{1}{2})\Delta\sigma_b(x/a, \Delta\sigma) = \sigma_b(x/a, \Delta\sigma/2) \tag{77}$$

where $\Delta\sigma$ is the range in the applied stress. Notably, the amplitude of the *change* in fiber traction $\Delta\sigma_b$ caused by a change in applied stress, $\Delta\sigma$, is twice the fiber traction σ_b which would arise in the monotonic loading of a previously unopened crack, caused by an applied stress equal to half the stress change. This *result is fundamental to all subsequent developments*.[84]

The stress intensity factor for bridging fibers subject to cyclic conditions is

$$\Delta K_b(\Delta\sigma) = -2\sqrt{\frac{a}{\pi}}\int_0^a \frac{\Delta\sigma_b(x,\Delta\sigma)}{\sqrt{a^2-x^2}}\,dx \tag{78}$$

which, with the use of Eqn. (77), becomes

$$\Delta K_b(\Delta\sigma) = 2K_b^{max}(\Delta\sigma/2) \tag{79}$$

where the superscript *max* refers to the maximum values of the parameters achieved in the loading cycle and thus, K_b^{max} is the bridging contribution that would arise when the crack is loaded by an applied stress equal to $\Delta\sigma/2$. Furthermore, since ΔK is linear, Eqn. (77) is also valid for the tip stress intensity factor:

$$\Delta K_{tip} = 2K_{tip}(\Delta\sigma/2) \tag{80}$$

When the fibers remain intact, a cyclic *steady state* (ΔK independent of crack length) is obtained when the cracks are long, given by the condition $\Delta\mathscr{E} \leqslant 4$,[84] where $\Delta\mathscr{E}$ is defined in Table 1.2a. The result is†

$$\Delta K_{tip} = \Delta\sigma\sqrt{R}(\sqrt{12}\Delta\mathscr{T})^{-1} \tag{81}$$

where $\Delta\mathscr{T}$ is defined in Table 1.2a.

The corresponding crack growth rate is determined by using a crack growth criterion. When a Paris law applies, Eqns. (75) and (81) give[84]

$$\frac{da}{dN} = \beta\left(\frac{\Delta\sigma\sqrt{R}}{\sqrt{6}\Delta\mathscr{T}E_m}\right)^{n_f} \tag{82}$$

When the matrix does not fatigue, such that Eqn. (24) represents the crack growth criterion, *fatigue crack growth after the first cycle is only possible, provided that τ reduces upon cycling*, as elaborated below.

When short cracks are of relevance ($\Delta\mathscr{E} > 4$),

$$\Delta K_{tip} = \Delta\sigma\sqrt{\pi a}\left(1 - \frac{4.31}{\Delta\mathscr{E}}\sqrt{\Delta\mathscr{E} + 6.6} + \frac{11}{\Delta\mathscr{E}}\right) \tag{83}$$

Consequently, at fixed $\Delta\sigma$, ΔK_{tip} increases as the crack extends, and the Paris law matrix crack growth accelerates. However, the bridged matrix fatigue crack always grows at a *slower rate* than an unbridged crack of the same length.

---

†For cyclic loading, the residual stress q does not affect ΔK_{tip}.

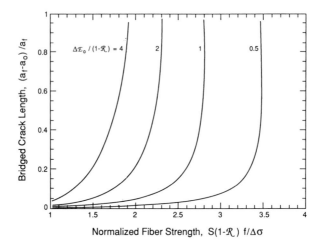

Fig. 1.48 The length of matrix crack a_f at first fiber failure as a function of fiber strength for a range of stress amplitudes, $\Delta\mathscr{E}$.

Consequently, the composite always has *superior* crack growth resistance relative to the monolith.

To incorporate the effects of fiber breaking into the fatigue crack growth model, a deterministic criterion has been used:[117] the statistical characteristics of fiber failure have yet to be incorporated. To conduct the calculation, once the fibers begin to fail, the unbridged crack length has been continuously adjusted to maintain a stress at the unbridged crack tip equal to the fiber strength. These conditions lead to the determination of the crack length, a_f, when the first fibers fail, as a function of the fiber strength and the maximum applied load (Fig. 1.48). Note that when either the fiber strength is high or the applied estress is low, no corresponding value of a_f can be identified and the fibers do not fail.

After the first fiber failure, fibers continue to break as the crack grows. Continuing fiber failure creates an unbridged segment larger than the original notch size. However, only the current unbridged length $2a_u$ and the current total crack length $2a$ are relevant (Fig. 1.49).[117]

If the fibers are relatively weak and break close to the crack tip ($a_\theta/a \rightarrow 1$), the bridging zone is always a small fraction of the crack length. In this case, there is minimal shielding. If the fibers are moderately strong, the fibers remain intact at first, but when the first fibers fail, subsequent failure occurs quite rapidly as the crack grows. The unbridged crack length then increases more rapidly than the total crack length and the ΔK_{tip} also increases as the crack grows. When the fibers are even stronger, first fiber failure is delayed, but once such failure occurs, many fibers fail simultaneously and the unbridged length increases rapidly. This causes a sudden increase in the crack growth rate. Finally, when the fiber strength exceeds a critical value, the fibers

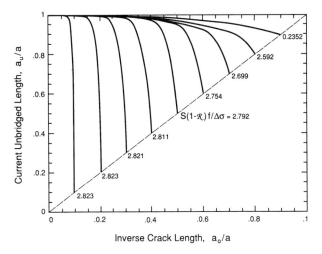

Fig. 1.49 Fiber breaking rate as manifest in the current unbridged matrix crack length $2a_u$ as a function of total crack length $2a$.

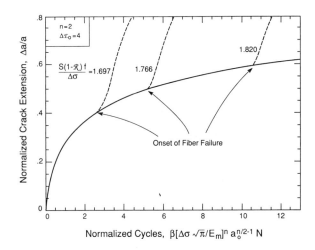

Fig. 1.50 Predicted matrix crack growth curves.

never break and the fatigue crack growth rate always diminishes as the crack grows. The sensitivity of these behaviors to fiber strength is quite marked (Fig. 1.49), with the different types of behavior occurring over a narrow range of fiber strength. Some typical crack growth curves predicted using this approach are plotted in Fig. 1.50. Finite geometry effects associated with LSB also exist.[117]

The results of Fig. 1.48 can be used to develop a criterion for a "threshold" stress range, $\Delta\sigma_t$, below which fiber failure does not occur for *any*

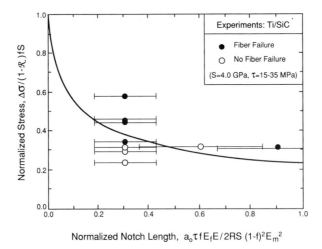

Fig. 1.51   The threshold stress diagram. Also shown are experimental results for Ti matrix composites.

crack length. Within such a regime, the crack growth rate approaches the steady-state value given by Eqn. (82), with all fibers in the crack wake remaining intact. The variation in the "threshold" stress range with fiber strength is plotted in Fig. 1.51.

A notable feature of the predictions pertains to the role of the stress ratio $\mathcal{R}_s$ in composite behavior. Prior to fiber failure, the crack growth rate is independent of $\mathcal{R}_s$ (except for its effect on the fatigue properties of the matrix itself). However, $\mathcal{R}_s$ has a strong influence on the transition to fiber failure, as manifest in its effect on the maximum stress. It thus plays a dominant role in the fatigue lifetime.

In many cases, CMCs are subject to multiple matrix cracking upon cyclic and static loading, which leads to reductions in the unloading modulus $\bar{E}$, as well as changes in the hysteresis. The basic mechanics is essentially the same as that described for monotonic loading, except that the matrix crack growth criterion must be changed.

1.8.3 Thermomechanical Fatigue

The basic matrix crack growth model can be extended to include thermomechanical fatigue (TMF). This can be achieved by means of another transformation wherein all of the stress range terms in Eqns. (77) to (83) are replaced, as follows,[118]

$$\Delta\sigma \Rightarrow \Delta t = \Delta\sigma + fE_f(\alpha_f - \alpha_m)\Delta T$$
$$\Delta\sigma_b \Rightarrow \Delta t_b = \Delta\sigma_b + fE_f(\alpha_f - \alpha_m)\Delta T \qquad (84)$$

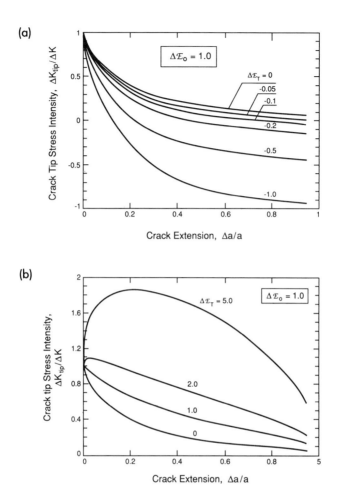

Fig. 1.52 Effects of TMF on the tip stress intensity factor: (a) in-phase, and (b) out-of-phase.

where ΔT represents the temperature cycle and $\Delta \sigma$ the stress cycle. With these transformations, it is possible to represent the crack growth using two nondimensional parameters, $\Delta \mathcal{E}_0$ and $\Delta \mathcal{E}_T$ (Table 1.2a) that specify the stress cycling and the temperature cycling, respectively. It is immediately apparent that matrix crack growth and fiber failure are expected to be quite different for out-of-phase and in-phase TMF. The salient predictions are presented for both cases.

For materials in which $\alpha_m > \alpha_f$, in-phase TMF causes ΔT to be *less* than that expected for stress cycling alone and vice versa. These effects are apparent from trends in the stress intensity range, ΔK_{tip} (Fig. 1.52), calculated for cases wherein fiber failure does not occur. A key result is that, whereas ΔK_{tip} always reduces upon initial crack extension either for stress cycling alone or for in-phase TMF, it can *increase* for out-of-phase TMF. Furthermore, for an extreme ratio of $\Delta \mathscr{E}_T$ to $\Delta \mathscr{E}_0$, ΔK_{tip} can *exceed* that for the monolithic matrix without fibers. This result implies that the crack growth rate also *exceeds* that for the monolith (at the equivalent ΔK). Then, the composite has crack growth resistance *inferior* to the monolithic matrix. The implications for the choice of fiber and the allowable temperature range ΔT are immediate.

When fiber failure effects are introduced, in-phase and out-of-phase cycling result in behaviors *that oppose* those associated with matrix crack growth. Namely, the crack size, a_f, at which fiber failure commences is *smaller* for in-phase loading than for out-of-phase loading (Fig. 1.53). Consequently, in order to ensure a threshold, the material is required to operate under conditions of fiber integrity. Then, in-phase TMF represents the more severe problem.

1.8.4 *Experimental Results*

Experimental measurements performed on CMCs and Ti MMCs reflect features associated with the cyclic degradation of the sliding stresses and fiber strength, and also provide a critique of crack growth criteria. These features are manifest in phenomena ranging from the growth characteristics of individual cracks to changes in modulus, and in fatigue life curves. The salient cyclic and static fatigue characteristics are illustrated using various experimental results.

The growth of individual cracks has been investigated on Ti MMCs, but not on CMCs. The crack growth trends found in Ti MMCs are in broad agreement with the predictions of the matrix crack growth models (Fig. 1.54), upon using a Paris law applicable to the matrix (Eqn. (75)). The results indicate that sliding stress τ decreases upon cycling, because of "wear" mechanisms operating within the fiber coating.[119] The reduction in τ occurs after a relatively small number of cycles (<1000) and thereafter, remains at an essentially constant value. It is also evident for these materials that the fiber strength is *not degraded by cyclic sliding of the interface*, even after $>10^5$ cycles.

Tensile fatigue testing of CMCs has been conducted under conditions which produce multiple cracking. There are consequent changes in modulus and hysteresis loop width, which relate to the fatigue life. Such results do not provide a critical test of the matrix crack growth criterion, but clearly illustrate the influence of cycling on the interface sliding stress and the fiber strength. Reductions in unloading modulus $\bar{E}$ are found at fixed *stress amplitude* (Fig. 1.55).[11,120] In some cases, there is also a small subsequent increase. The modulus changes have been analyzed, such that constituent properties during fatigue may be obtained. For example, measurements made for SiC/CAS (at frequencies $<10\,\text{Hz}$) have been correlated with the crack density (Fig. 1.56). Comparisons with model predictions indicate a substantial reduction in sliding

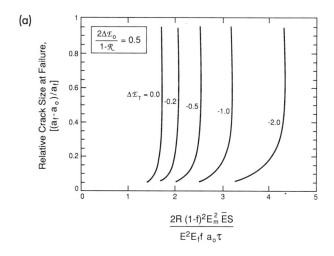

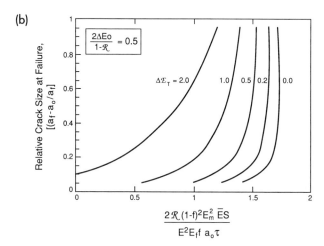

Fig. 1.53 Effects of TMF on the crack size at which fiber failure occurs: (a) in-phase, and (b) out-of-phase.

stress, from $\tau \approx 15\,MPa$ for the pristine composite[121] to $\tau \approx 5\,MPa$. At higher frequencies ($\gtrsim 50\,Hz$), frictional heating also occurs, accompanied by a *larger reduction* in τ.[122,131] The hypothesis is that the frictional heating causes the C fiber coating to be eliminated. Such behavior would be consistent with that found upon isothermal heat treatment.[93]

The occurrence of cyclic fatigue failure at peak stresses substantially lower than the UTS (Fig. 1.57) has been found at high temperatures and, especially, for TMF. Such results suggest that the fiber strength systematically diminishes for certain cyclic thermomechanical loadings. There are three primary mechanisms of fiber weakening: abrasion, oxidation, and stress

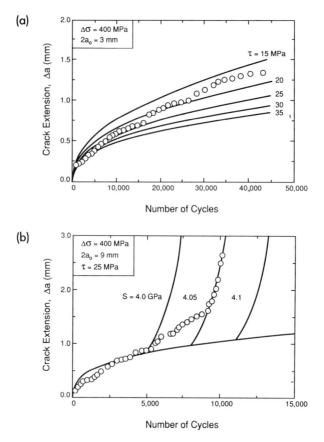

Fig. 1.54   Comparison between experimental crack growth results and predictions for unidirectional SiC/Ti composites, showing effects of stress and notch length on fiber failure.

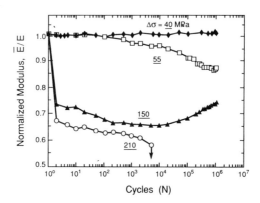

Fig. 1.55   Modulus reduction found upon fatigue in a glass matrix composite.[11]

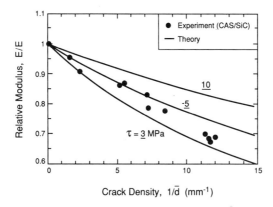

Fig. 1.56 Influence of cyclic loading on modulus reduction as a function of crack density for a unidirectional CAS/SiC composite indicating that τ has been decreased by fatigue.

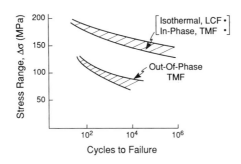

Fig. 1.57 Isothermal fatigue and TMF data for a glass matrix composite.[3]

corrosion. These mechanisms might be distinguished in the following manner. The strength degradation caused by stress corrosion occurs abruptly, following *time* accumulated at *peak load*.[114] Abrasion occurs systematically with cyclic sliding at the interfaces (Fig. 1.58) and should be enhanced by out-of-phase TMF, which accentuates the sliding displacement. Oxidation is strictly time and temperature dependent. The strong effect of out-of-phase TMF on the fatigue life at high temperature[3] suggests that fiber degradation by abrasion is an important mechanism, perhaps accentuated by oxide formation at higher temperatures. Much additional study is required on this topic.

In some CMCs, modulus changes and rupture occur at *constant stress*.[120] Substantial matrix crack growth has been found at stresses below that required to produce cracks in short duration, monotonic tensile tests. Furthermore, the crack densities following extended periods under load ($\sim 10^6$ s) are *higher* than those obtained in the short duration tests. The development of cracks with time

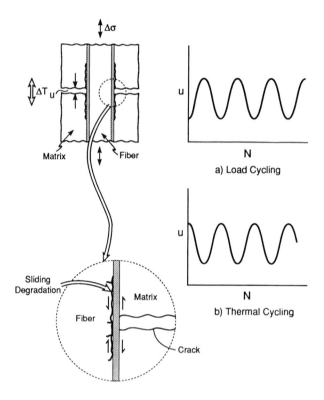

Fig. 1.58 Mechanism of fiber degradation by fatigue, coupled with oxidation.

and stress (Fig. 1.59) has been considered to involve stress corrosion of the matrix. The behavior is consistent with a revised matrix crack growth criterion (Eqn. (76)), without any changes in the sliding stress. Fiber weakening may also be occurring by stress corrosion.

1.9 Creep

1.9.1 Basic Behavior

The creep behavior and relationships with constituent properties are critically influenced by fiber failure, matrix cracks, and interface debonding. Some of the basic stress–time characteristics are sketched in Figs. 1.60 and 1.61. When the fibers and matrix are intact and the interfaces are bonded, the creep deformations of the composite and the constituent properties are related in a straightforward manner.[118,123] When one constituent is elastic (fiber or matrix) and the other creeps, the *longitudinal* creep strain is *transient* and stops

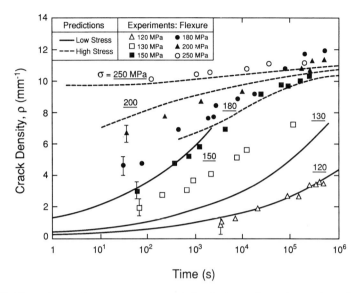

Fig. 1.59 Experimental measurements and simulations of matrix crack evolution in a SiC/CAS composite caused by stress corrosion at constant stress.

when all of the strain is transferred onto the elastic material (Figs. 1.60 and 1.61).[118] The creep law needed to describe this behavior is

$$\dot{\varepsilon}_{ij} = \frac{1}{2G}\dot{s}_{ij} + \frac{1}{9K}\delta_{ij}\dot{\sigma}_{kk} + \frac{3}{2}B\sigma_e^{n-1}s_{ij} + \alpha\delta_{ij}\dot{T} \tag{85}$$

where $\dot{\varepsilon}$ is the strain rate, $\dot{\sigma}$ is the stress rate, δ_{ij} is the Kronecker delta, n is the creep index, s_{ij} is the deviatoric stress, and the effective stress, σ_e, is defined by

$$\sigma_e = \sqrt{\tfrac{3}{2}s_{ij}s_{ij}} \tag{86}$$

and B is the rheology parameter for steady-state creep,

$$B = \dot{\varepsilon}_0/\sigma_0^n \tag{87}$$

with σ_0 being the reference stress, and $\dot{\varepsilon}_0$ the reference strain rate. If the fibers are elastic and the matrix creeps, the stress in the matrix, σ_m, evolves at constant applied stress as $(n \neq 1)$,[118,123]

$$\sigma_m(t) = \left\{ \frac{(n-1)fE_fE_m\,Bt}{E_L} + \frac{1}{[\sigma_m(0)^{n-1}]} \right\}^{1-n} \tag{88}$$

where $\sigma_m(0)$ is the matrix stress at time, $t = 0$. When the matrix stress,

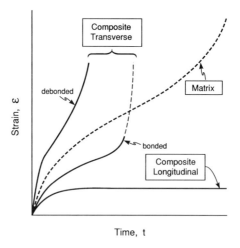

Fig. 1.60 Schematic indicating creep anisotropy in unidirectional CMCs.

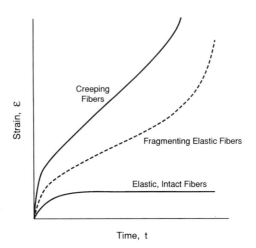

Fig. 1.61 Schematic indicating effects of intact and creeping fibers, as well as fiber failure on the longitudinal creep.

$\sigma_m \to 0$, the stress on the fibers increases to, $\sigma_f = \sigma/(1-f)$, such that the transient strain ε_t is

$$\varepsilon_t = \sigma/E_f(1-f) \tag{89}$$

Similar results apply when the fibers creep, but the matrix is elastic.

When both the fiber and the matrix creep, steady state develops in the

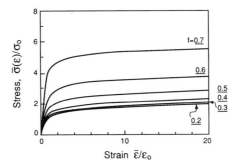

Fig. 1.62 Transverse strength of a unidirectional composite with a power-law hardening matrix.

composite following an initial transient (Fig. 1.61). The evolution of the matrix stress occurs according to[124]

$$\left[\frac{E}{fE_mE_f}\right]\dot{\sigma}_m = B_m\sigma_m^{n_m} - B_f\left[\frac{\sigma - (1-f)\sigma_m}{f}\right]^{n_f} \tag{90}$$

where n_m and n_f are the creep indices for the matrix and fibers, respectively. When a *steady state* is reached $(\dot{\sigma}_m = 0)\,\sigma_m$ and σ_f are related by,

$$[\sigma_m^{n_m}(B_m/B_f)]^{1/n_f} + \frac{(1-f)}{f}\,\sigma_m = \frac{\sigma}{f} \tag{91}$$

and

$$\sigma_m(1-f) + \sigma_f f = \sigma \tag{92}$$

These formulae can be solved for specific n_m and n_f to obtain σ_m and σ_f. With the stresses known, the composite creep rate can be readily obtained.

Transverse creep with well-bonded fibers is usually *matrix dominated*. Solutions which have been generated for bonded rigid fibers thus have utility. All such solutions indicate that the creep attains steady state, with a creep rate *lower* than that for the matrix alone (Fig. 1.60). Moreover, strengthening solutions derived for transverse deformation with a power-law hardening matrix (Fig. 1.62) also apply to a power-law creeping matrix, in steady state.† The reduction in creep rate depends on the power-law exponent for the matrix and the spatial arrangement of the fibers. For a composite with a square arrangement of fibers, and a matrix subject to diffusional creep $(n_m = 1)$, since there is no creep in the fiber direction (z),[118]

$$\dot{\varepsilon}_{yy} = -\dot{\varepsilon}_{xx} = (\sigma_{yy} - \sigma_{xx})k_1(f) \tag{93}$$

†With the strains becoming the strain rates.

with

$$k_1(f) = (3/4)[(1-f)/(1+2f)] \tag{94}$$

In essence, k_1 gives the *reduction* in creep rate upon incorporating the bonded fibers. For nonlinear matrices, the equivalent results have the form

$$\dot{\varepsilon}_{xx} = -\dot{\varepsilon}_{yy} = B_m(\sigma_{xx} - \sigma_{yy})^{n_m-1}(\sigma_{xx} - \sigma_{yy})k_n(f) \tag{95}$$

where k_n is the function of the fiber volume fraction and spatial arrangement. For example, when $n_m = 5$ and a square fiber array is used,

$$k_n = 0.42[(1-f)/(1+f^2)]^5 \tag{96}$$

1.9.2 Effect of Fiber Failures

When stresses are applied along the fiber axis in a system with a creeping matrix, the time-dependent stress elevation on the fibers may cause some fiber failures. Following fiber failure, *sliding* would initiate at the interface, accompanied by further creep in the matrix. The time constant for this process is much longer than that for the initial transient, described above, and can be analyzed as a separate creep problem.[118] While the process is complicated, several factors are important. If the stress on the fibers reaches their strength, S, the composite will fail. Moreover, the relevant S is probably that with a *small* τ, associated with creep sliding of the interface. In this limit, composite failure is possible at all stresses above the "dry bundle" strength, S_b (Eqn. 17c). Conversely, the composite *cannot* rupture at stress below S_b, *unless the fibers are degraded by creep.* The dry bundle strength thus represents a "threshold." At stresses below S_b, creep must be transient.

At higher stresses, the fibers will fracture and may fragment. Then, steady-state creep is possible (Fig. 1.60), proceeding in accordance with a creep law devised for a material with aligned rigid reinforcements of *finite aspect ratio.* This behavior is represented by the Mileiko[125] model. The solution for a nonsliding interface is[118,126]

$$\dot{\varepsilon} = B_m \sigma^{n_m}(R/L_f)^{n_m+1}\mathcal{L}(n_m, f) \tag{97}$$

where L_f is the fragment length and

$$\mathcal{L}(n_m, f) = 2^{n_m+1}\sqrt{3}\left[\frac{\sqrt{3}(2n_m+1)}{2n_m f}\right]^{n_m}\frac{(1-f)^{(n_m-1)/2}}{(n_m-1)} \tag{98}$$

However, the fragment length *decreases* as the stress increases. This occurs in accordance with the scaling,[46]

$$L_f/R \sim (S_c/\sigma)^m \tag{99}$$

Consequently, steady-state creep rate should occur with a large power-law exponent, $n_m + m + mn_m$.[118] Such behavior has been reported in composites

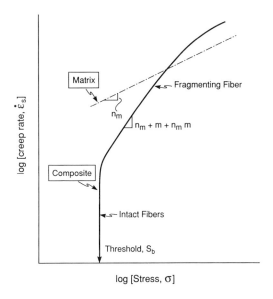

Fig. 1.63 A sketch indicating the longitudinal creep threshold and the behavior above the threshold.

with discontinuous fibers.[127] The overall behavior is sketched in Fig. 1.63. In practice, because of the large stress exponent at stress above S_b, *adequate creep performance can only be ensured at stresses below S_b.*

1.9.3 Interface Debonding

While there are no solutions known to the authors for transverse creep with debonding interfaces, the analogy (noted above) between power-law deformation and steady-state creep provides insight. Calculations of transverse deformation with, and without, interface bonding (Fig. 1.64) indicate a major strength degradation when debonding occurs.[128,129] Furthermore, the composite behavior approaches that for a body containing *cylindrical* holes. Creep results for porous bodies[130] may thus provide rough estimates of the transverse creep strength when the interfaces debond.

1.9.4 Matrix Cracking

In some CMCs, the fibers creep more readily than the matrix. Such materials include SiC/SiC and SiC/C. In this case, fiber creep and matrix cracking appear to proceed in a synergetic manner that accelerates the creep and causes premature creep rupture. The basic phenomenon is as follows. Creep in the fiber increases the stress on the matrix, as described above. Above a threshold, the stress on the matrix then exceeds σ_τ (Eqn. (62)),

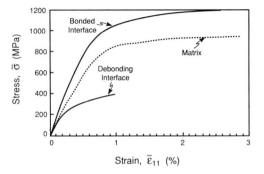

Fig. 1.64 Comparison of transvere behavior with, and without, interface debonding.

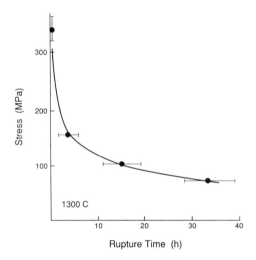

Fig. 1.65 Creep rupture data for a SiC/C composite which is susceptible to fiber creep and matrix cracking.

causing multiple matrix cracks to form in the 90° plies. These cracks gradually extend into the 0° plies, because creep of the fibers relaxes the bridging traction. As a result, the stress at these locations is borne entirely by the fibers, which creep, without impediment, leading to rupture of the composite (Fig. 1.65). The rupture ductility of polycrystalline ceramic fibers is typically quite low because of void formation along the grain boundaries. Consequently, matrix cracking often leads to creep rupture with *creep brittle* characteristics. A creep analogy to the tunneling stress, σ_τ (Eqn. (64)) is thus representative of a *threshold stress*. At stresses above σ_τ, matrix cracks eventually extend across the composite and the composite fails by fiber rupture.

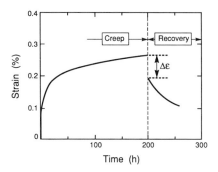

Fig. 1.66 Creep recovery effects in a SiC/Si$_3$N$_4$ with "elastic" fibers and a creeping matrix.[132]

1.9.5 Strain Recovery

Since creep in composites redistributes stresses between matrix and fiber, strain recovery *must occur when the loads are removed*.[131] This behavior is well established for a system with one elastic constituent and one viscoplastic constituent, in accordance with standard Kelvin concepts. Notably, the elastic stretch in one constituent is gradually relaxed when the load is removed. The specifics depend, of course, on the nature of the viscoplasticity. A simple example illustrates the salient phenomena. A composite with elastic fibers and a creeping matrix, loaded along the fiber direction, has been crept until the stress in the matrix is essentially zero (Fig. 1.66). The load is then removed. The instantaneous elastic shrinkage $\Delta\varepsilon$ must satisfy

$$\Delta\varepsilon = \frac{\sigma_m}{E_m} = \frac{\Delta\sigma_f}{E_f} \tag{100}$$

The stresses after elastic unloading are, thus,

$$\sigma_m = -\frac{f\sigma E_m}{(1-f)E_L} \tag{101}$$

$$\sigma_f = \sigma E_m/E_L \tag{102}$$

Thereafter, holding at temperature causes σ_m to relax according to Eqn. (88), with $\sigma_m(0)$ given by Eqn. (101).

1.9.6 Experimental Results

Experimental data for a range of different composites are used to illustrate some of the features described above, and to anticipate trends. The longitudinal behaviors found when the fibers are elastic are addressed first.

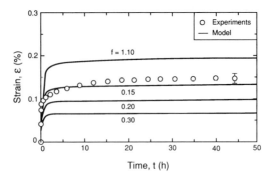

Fig. 1.67 Transient longitudinal creep in a TiAl matrix composite reinforced with sapphire fibers.[133]

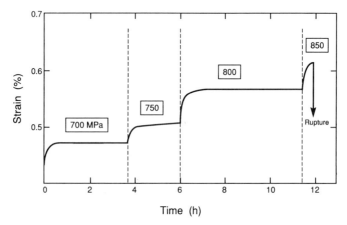

Fig. 1.68 Transient creep and rupture data obtained for a SiC/Ti composite subject to incremental loading.[138]

Results obtained on TiAl reinforced with sapphire fibers (Fig. 1.67) establish the existence of transient creep in the longitudinal orientation when the fibers are elastic and intact, but the matrix is subject to creep.[133] At higher loads, when some fibers fail, creep can continue and rupture may occur, as demonstrated by data obtained on a Ti matrix composite reinforced with SiC fibers (Fig. 1.68).[134] Removal of the load after creep results in reverse deformation, as demonstrated for a SiC/Si₃N₄ composite (Fig. 1.66). Upon using a creep index applicable to monolithic Si_3N_4 ($n = 2$), the stress in the matrix relaxes in the manner

$$\sigma_m = \left[\frac{fE_f E_m Bt}{E_L} - \frac{(1-f)E_L}{f\sigma E_m} \right]^{-1}$$

(103)

Note that B has units $(\text{stress})^{-2}$.

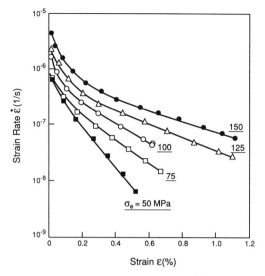

Fig. 1.69 Longitudinal creep of a SiC/CAS composite.[12]

The inverse situation may also be important in some CMCs, wherein the fibers creep but the matrix is *elastic*.[12,135] Typical examples include SiC/SiC and SiC/C composites, which have SiC fibers with fine grain size (such as Nicalon). In these materials, matrix cracks are created upon loading above a threshold stress, σ_T. When these cracks exist, fiber creep results in continuous deformation and creep rupture (Fig. 1.65). However, if the stress is below the threshold, creep will occur in a transient manner.

When both the matrix and fibers creep, and there are no matrix cracks, continued deformation of the composite proceeds in the longitudinal orientation.[12] Results obtained on SiC/CAS (Fig. 1.69) verify that creep continues. However, interpretation is complicated by microstructural changes occurring in the fibers, which lead to creep hardening. The deformation is, thus, entirely primary in nature. These results identify microstructural stability as an important fiber selection criterion.

1.10 Challenges and Opportunities

Reasonable progress has been made in understanding inelastic strain mechanisms, although the continued development of models, and experimental validation, is still necessary. It is now possible to appreciate how stress redistribution occurs and to characterize the notch sensitivity. The analysis of the degradation mechanism is much less mature.

There are several challenges and opportunities that arise. With regard to the short duration performance, it is necessary to develop simple constitutive laws that can be used with finite element codes in order to calculate stresses around attachments, holes, etc. Mechanism-based models of the inelastic strain are preferred for this purpose. However, there is insufficient basic understanding about the inelastic strains that occur upon *shear loading* and their dependence on constituent properties. Basic inelastic strain models with matrix cracks inclined to the fibers are needed to address this deficiency.

Degradation mechanisms that operate upon cyclic loading in the presence of matrix cracks require concerted study. Interface changes and fiber degradation are both possible. Moreover, there may be detrimental synergistic interaction with the environment. The models developed for MMCs indicate that the retention of fiber strength upon cyclic loading is particularly important, because this strength governs the fatigue threshold. Mechanism and models that predict fiber strength degradation are critically important.

References

1. J. W. Hutchinson and H. Jensen, *Mech. of Mtls.*, **9**, 139 (1990).
2. K. M. Prewo, *J. Mater. Sci.*, **17**, 3549 (1982).
3. S. Mall, D. E. Bullock, and J. J. Pernot, *Composites*, in press.
4. C. Cady, T. J. Mackin, and A. G. Evans, *J. Am. Ceram. Soc.*, to be published.
5. A. G. Evans, J. M. Domergue, and E. Vagaggini, *J. Am. Ceram. Soc.*, **77**, 1425 (1994).
6. V. C. Nardonne and K. M. Prewo, *J. Mater. Sci.*, **23**, 168 (1988).
7. R. Y. Kim and N. Pagano, *J. Am. Ceram. Soc.*, **74**, 1082–90 (1991).
8. H. C. Cao, E. Bischoff, O. Sbaizero, M. Rühle, and A. G. Evans, *J. Am. Ceram. Soc.*, **73**[6], 1691–99 (1990).
9. A. W. Pryce and P. Smith, *J. Mater. Sci.*, **27**, 2695–2704 (1992).
10. K. M. Prewo, *J. Mater Sci.*, **22**, 2595–2701 (1987).
11. L. P. Zawada, L. M. Butkus, and G. A. Hartman, *J. Am. Ceram. Soc.*, **74**, 2851–2858 (1991).
12. C. Weber and A. G. Evans, *J. Am. Ceram. Soc.*, **77**, 1745 (1994).
13. G. Bao and Z. Suo, *Appl. Mech. Rev.*, **45**[8], 355–366 (1992).
14. J. Aveston, G. A. Cooper, and A. Kelly, in *The Properties of Fiber Composites*, IPC, UK, 1971, pp. 15–26.
15. A. G. Evans and D. G. Marshall, *Acta Metall.*, **37**, 2567–2583 (1989).
16. W. A. Curtin, *J. Am. Ceram. Soc.*, **74**, 2837 (1991).
17. A. G. Evans and F. W. Zok, in *Topics in Fracture and Fatigue*, ed. A. S. Argon, Springer-Verlag, New York, NY, 1992, pp. 271–308.
18. E. Vagaggini and A. G. Evans, *J. Am. Ceram. Soc.*, in press.
19. A. G. Evans, F. W. Zok, and J. Davis, *Composites Science and Technology*, **42**, 3–24 (1991).
20. J. B. Davis, J. P. A. Löfvander, A. G. Evans, E. Bischoff, and M. L. Emiliani, *J. Am. Ceram. Soc.*, **76**, 1249 (1994).
21. M. Y. He, B. X. Wu, A. G. Evans, and J. W. Hutchinson, *Mech. of Mtls.*, in press.
22. A. G. Evans, *Mat. Sci. Eng.*, **A143**, 63 (1991).
23. B. N. Cox and D. B. Marshall, *Fatigue and Fracture of Eng. Mtls.*, **14**, 847 (1991).

24. B. N. Cox, *Acta Metall. Mater.*, **39**[6], 1189–1201 (1991).
25. B. N. Cox and C. S. Lo, *Acta Metall. Mater.*, **40**, 69 (1992).
26. P. A. Brøndsted, F. E. Heredia, and A. G. Evans, *J. Am. Ceram. Soc.*, to be published.
27. F. E. Heredia, S. M. Spearing, M. Y. He, T. J. Mackin, P. A. Brønsted, A. G. Evans, and P. Mosher, *J. Am. Ceram. Soc.*, in press.
28. J. F. Jamet, D. Lewis, and E. Y. Luh, *Ceram. Eng. Sci. Proc.*, **5**, 625 (1984).
29. F. W. Zok and S. M. Spearing, *Acta Metall. Mater.*, **40**, 2033 (1992).
30. D. B. Marshall and B. N. Cox, *Mech. of Mtls.*, **7**, 127 (1986).
31. D. B. Marshall, B. N. Cox, and A. G. Evans, *Acta Metall.*, **33**, 2103–2121 (1985).
32. D. Hayhurst, F. A. Leckie, and A. G. Evans, *Proc. Roy. Soc.*, *London*, **A434**, 369 (1991).
33. T. Mackin, P. Warren, and A. G. Evans, *Acta Metall. Mater.*, **40**[6], 1251–1257 (1992).
34. A. G. Evans, *J. Am. Ceram. Soc.*, **73**[2], 187–206 (1990).
35. M. Y. He and J. W. Hutchinson, *Intl. J. Solids Structures*, **25**[9], 1-53–1-67 (1989).
36. N. A. Fleck, *Proc. Roy. Soc.*, **A432**, 55–76 (1991).
37. D. B. Marshall and W. C. Oliver, *J. Am. Ceram. Soc.*, **70**, 542–548 (1987).
38. P. D. Jero, R. J. Kerans, and T. A. Parthasarathy, *J. Am. Ceram. Soc.*, **74**, 2793 (1991).
39. D. B. Marshall, *Acta Metall. Mater.*, **40**, 427–441 (1992).
40. R. W. Rice, "BN Coating of Ceramic Fibers for Ceramic Fiber Composites," U.S. Patent 4,642,271, February 10, 1987; assigned to the United States of America as represented by the Secretary of the Navy.
41. R. W. Rice, J. R. Spann, D. Lewis, and W. Coblenz, *Ceram. Eng. Sci. Proc.*, **5**[7–8], 614–624 (1984).
42. D. C. Cranmer, *Am. Ceram. Soc. Bull.*, **68**[2], 415 (1989).
43. J. J. Brennan and K. M. Prewo, *J. Mater. Sci.*, **17**, 2371–83 (1982).
44. D. B. Marshall and A. G. Evans, *J. Am. Ceram. Soc.*, **68**, 225–231 (1985).
45. T. J. Kotil, J. W. Holmes, and M. Comninou, *J. Am. Ceram. Soc.*, **73**, 1879 (1990).
46. W. A. Curtin, *J. Mater. Sci.*, **26**, 5239–5253 (1991).
47. C. Liang and J. W. Hutchinson, *Mech. of Mtls.*, **14**, 207–221 (1993).
48. J. Lamon, P. Raballiat, and A. G. Evans, *J. Am. Ceram. Soc.*, in press.
49. T. Mackin, J. Yang, C. Levi, and A. G. Evans, *Mat. Sci. Eng.*, **A161**, 285 (1993).
50. J. J. Brennan, in *Tailoring of Multiphase Ceramics*, eds. R. E. Tressler *et al.*, Plenum Press, New York, NY, 1986, **20**, 549.
51. B. Bender, O. Shadwell, C. Bulik, L. Incorvati, and D. Lewis III, *J. Am. Ceram. Soc. Bull.*, **65**[2], 363–369 (1986).
52. R. Naslain, International Conference on Composite Materials, CWRU, in press.
53. X. Bourrat, unpublished research at LCTS, Bordeaux, France, 1993.
54. E. Y. Luh and A. G. Evans, *J. Am. Ceram. Soc.*, **70**[7], 466–469 (1987).
55. J. W. Holmes and S. F. Shuler, *J. Mater. Sci. Lett.*, **9**, 1290–1291 (1990).
56. J. W. Holmes and C. Cho, *J. Am. Ceram. Soc.*, **75**[4], 929–938 (1992).
57. E. Bischoff, M. Rühle, O. Sbaizero, and A. G. Evans, *J. Am. Ceram. Soc.*, **72**[5], 741–745 (1989).
58. B. Budiansky, J. W. Hutchinson, and A. G. Evans, *J. Mech. Phys. Solids*, **34**, 167–189 (1986).
59. K. K. Chawla, *Composite Materials Science and Engineering*, Springer-Verlag, New York, NY, 1987.

60. F. W. Zok and A. G. Evans, to be published.
61. D. Beyerle, S. M. Spearing, and A. G. Evans, *J. Am. Ceram. Soc.*, **75**[12], 3321–3330 (1992).
62. D. Beyerle, S. M. Spearing, F. Zok, and A. G. Evans, *J. Am. Ceram. Soc.*, **75**[10], 2719–2725 (1992).
63. L. Phoenix and R. Raj, *Acta Metall. Mater.*, **40**, 2813–2828 (1992).
64. F. Hild, J. M. Domergue, F. A. Leckie, and A. G. Evans, *Int. J. Solids Structures*, to be published.
65. R. B. Henstenburg and S. L. Phoenix, *Polym. Comp.*, **10**[5], 389–406 (1989).
66. J. R. Matthews, W. J. Shack, and F. A. McClintock, *J. Am. Ceram. Soc.*, **59**, 304 (1976).
67. A. Freudenthal, in *Fracture*, ed. H. Liebowitz, Academic Press, New York, NY, 1967.
68. H. E. Daniels, *Proc. Roy. Soc.*, **A183**, 405 (1945).
69. H. L. Oh and I. Finnie, *Intl. J. Frac.*, **6**, 287 (1970).
70. M. D. Thouless and A. G. Evans, *Acta Metall.*, **36**, 517 (1988).
71. M. Sutcu, *Acta Metall.*, **37**[2], 651–661 (1989).
72. H. T. Corten, in *Modern Composite Materials*, eds. L. J. Broutman and R. H. Krock, Addison-Wesley, Reading, MaA, 1967, p. 27.
73. M. D. Thouless, O. Sbaizero, L. S. Sigl, and A. G. Evans, *J. Am. Ceram. Soc.*, **72**[4], 525–532 (1989).
74. D. C. Phillips, *J. Mater. Sci.*, **9**[11], 1874 (1974).
75. L. Cui and B. Budiansky, to be published.
76. Z. Suo, S. Ho, and X. Gong, *J. Matl. Engr. Tech.*, **115**, 319–326 (1993).
77. K. M. Prewo, *J. Mater. Sci.*, **21**, 3590–3600 (1986).
78. A. J. Eckel and R. C. Bradt, *J. Am. Ceram. Soc.*, **72**, 435 (1989).
79. R. Y. Kim, *Ceram. Eng. Sci. Proc.*, **13**, 281–300 (1992).
80. C. Cho, J. W. Holmes and J. R. Barber, *J. Am. Ceram. Soc.*, **75**[2], 316–324 (1992).
81. R. Y. Kim and A. P. Katz, *Ceram. Eng. Sci. Proc.*, **9**, 853–860 (1988).
82. S. Baste, R. El Guerjouma, and B. Andoin, *Mech. of Mtls.*, **14**, 15–32 (1992).
83. L. N. McCartney, *Proc. Roy. Soc.*, **A409**, 329–350 (1987).
84. R. M. McMeeking and A. G. Evans, *Mech. of Mtls.*, **9**, 217–227 (1990).
85. H. Tada, P. C. Paris, and G. R. Irwin, *The Stress Analysis of Cracks Handbook*, Del Research Corp., St. Louis, MO, 1985.
86. R. Singh, *J. Am. Ceram. Soc.*, **72**, 1764 (1989).
87. A. G. Evans and J. M. Domergue, *J. Am. Ceram. Soc.*, in press.
88. S. M. Spearing and F. W. Zok, *J. Eng. Mtls. Tech.*, **115**, 314–318 (1993).
89. J. M. Domergue, A. G. Evans, and D. Roach, *J. Am. Ceram. Soc.*, to be published.
90. B. Harris, R. A. Habib, and R. G. Cooke, *Proc. Roy. Soc., Series A*, **437**, 109–131 (1992).
91. R. F. Cooper and K. Chyung, *J. Mater. Sci.*, **22**, 126 (1987).
92. K. M. Prewo and J. J. Brennan, *J. Mater. Sci.*, **17**[4], 1201–1206 (1982).
93. O. Sbaizero and A. G. Evans, *J. Am. Ceram. Soc.*, **69**[6], 481 (1986).
94. C. Xia, R. R. Carr, and J. W. Hutchinson, "Harvard Univ. Report Mech-202," *Acta Metall. Mater.*, in press.
95. J. W. Hutchinson and Z. Suo, *Appl. Mech. Rev.*, **29**, 63–191 (1992).
96. N. Laws and G. Dvorak, *J. Composite Mtls.*, **22**, 900 (1988).
97. X. Aubard, Thèse de Doctorat de l'Université de Paris, France, November 1992.

98. C. Xia and J. W. Hutchinson, "Harvard Univ. Report Mech-208," *Intl. J. Solids Structures*, to be published.
99. H. Chai, *Composites*, **15**[4], 277–290 (1984).
100. S. M. Spearing and A. G. Evans, *Acta Metall.*, **40**[9], 2191–2199 (1992).
101. D. A. W. Kaute, H. R. Shercliff, and M. F. Ashby, *Acta Metall. Mater.*, to be published.
102. R. Bordia, B. J. Dalgleish, P. G. Charalambides, and A. G. Evans, *J. Am. Ceram. Soc.*, **74**[11], 2776–2780 (1991).
103. G. Bao, B. Fan, and A. G. Evans, *Mech. of Mtls.*, **13**, 59–66 (1992).
104. F. W. Zok, O. Sbaizero, C. Hom, and A. G. Evans, *J. Am. Ceram. Soc.*, **74**[1], 187–193 (1991).
105. J. Bowling and G. W. Groves, *J. Mater. Sci.*, **14**, 43 (1979).
106. F. W. Zok and C. L. Hom, *Acta Metall. Mater.*, **38**, 1895 (1990).
107. W. W. Stinchcomb and C. E. Bakis, in *Fatigue of Composite Materials*, ed. K. L. Reifsnider, Elsevier Science Publishers B.V., Amsterdam, The Netherlands, 1991, pp. 105–180.
108. H. R. Bakis, H. R. Yih, W. W. Stinchcomb, and K. L. Reifsnider, ASTM STP 1012, ASTM, Philadelphia, PA, 1989, pp. 66–83.
109. M. C. Shaw, D. B. Marshall, B. J. Dalgleish, M. Dadkah, M. Y. He, and A. G. Evans, *Acta Metall. Mater.*, in press (1994).
110. N. Harwood and W. M. Cummings, *Thermoelastic Stress Analysis*, Adam Hilger IOP Publishing, Philadelphia, PA, 1991.
111. S. E. Molis and D. R. Clarke, *J. Am. Ceram. Soc.*, **73**, 3189 (1990).
112. C. Q. Rousseau, in *Thermal and Mechanical Behavior of Metal Matrix and Ceramic Matrix Composites*, eds. J. M. Kennedy, H. H. Moeller, W. S. and Johnson, ASTM STP 1080, American Society for Testing and Materials, Philadelphia, PA, 1990, pp. 240–252.
113. D. Rouby and P. Reynaud, *Comp. Sci. Tech.*, **48**, 109–118 (1993).
114. S. M. Wiederhorn, *J. Am. Ceram. Soc.*, **50**, 45 (1967).
115. D. Walls, G. Bao, and F. Zok, *Scripta Metall. Mater.*, **25**, 911 (1991).
116. M. Sensmeier and K. Wright, in *Proceedings TMS Fall Meeting*, ed. P. K. Law, and M. N. Gungor, 1989, p. 441.
117. G. Bao and R. McMeeking, *Acta Metall. Mater.*, **42**, 2415 (1994).
118. R. M. McMeeking, *Int. J. Solids Structures*, in press.
119. D. Walls, G. Bao, and F. W. Zok, *Acta Metall. Mater.*, in press.
120. S. M. Spearing, F. W. Zok, and A. G. Evans, *J. Am. Ceram. Soc.*, to be published.
121. T. Mackin and F. W. Zok, *J. Am. Ceram. Soc.*, **75**, 3169–3171 (1993).
122. J. W. Holmes, *J. Am. Ceram. Soc.*, **74**[7], 639–645 (1991).
123. M. McLean, *Composites Science and Technology*, **23**, 37–52 (1985).
124. D. McLean, *J. Mater. Sci.*, **7**, 98–104 (1972).
125. S. T. Mileiko, *J. Mater. Sci.*, **5**, 254–261 (1970).
126. A. Kelly and K. N. Street, *Proc. Roy. Soc., London*, **A328**, 283–293 (1972).
127. T. G. Nieh, *Metallurgical Transactions A*, **15A**, 139–146 (1984).
128. S. Gunawadena, S. Jansson, and F. E. Leckie, *Acta Metall. Mater.*, in press.
129. S. Jansson and F. A. Leckie, *J. Mech. Phys. Sol.*, **40**, 593–612 (1992).
130. P. Sofranis and R. M. McMeeking, *J. Appl. Mech.*, **59**, 588 (1992).
131. J. W. Holmes, *J. Mater. Sci.*, **26**, 1808–1814 (1991).
132. J. W. Holmes, Y. H. Park, and J. W. Jones, *J. Am. Ceram. Soc.*, submitted.
133. C. Weber, J. Y. Yang, J. P. A. Löfvander, C. G. Levi, and A. G. Evans *Acta Metall. Mater.*, in press.
134. C. Weber, S. J. Connell, Z. Z. Du, and F. W. Zok, submitted to *Acta Metall. Mater.* (1994).

135. F. Abbe, J. Vicens, and J. L. Chermant, *J. Mater. Sci. Lett.*, **8**, 1026–1028 (1989).
136. R. J. Kerans and T. A. Parthasarathy, *J. Am. Ceram. Soc.*, **74**, 1585 (1991).
137. T. J. Mackin, A. G. Evans, M. Y. He, and T. E. Purcell, *J. Ceram. Soc.*, in press (1994).
138. C. Weber and F. W. Zok, unpublished work (1993).

Short-Term Behavior

CHAPTER 2

Strength and Toughness of Ceramic Composites at Elevated Temperatures

J.-M. Yang and T. N. Tiegs

2.1 Introduction

Ceramic matrix composites reinforced with fibers and whiskers have recently been under intensive development for high temperature structural applications. They possess the attractive properties of the ceramics such as high melting temperature, high strength and stiffness especially at high temperature, low density and excellent environmental resistance, combined with improved toughness, strength, and mechanical reliability. These unique properties make them indispensable for applications in high temperature, high stress and aggressive environments such as turbine engines and heat-conversion systems. However, before these materials can be committed to critical high temperature applications, a thorough understanding of their mechanical behavior and fracture processes at room and elevated temperature is necessary. During past years, considerable progress has been made toward understanding the complex relationship between the properties of the constituents (fiber or whisker, matrix, interface) and mechanical properties of ceramic composites are now known and validated.[1-4] However, high temperature mechanical properties of these composites have not yet been extensively studied. Only limited experimental data have been reported in the literature regarding the strength and toughness of the composites at elevated temperatures. This chapter provides a brief review of the strength and toughness of both the whisker- and fiber-reinforced ceramic matrix composites at elevated temperatures.

2.2 Whisker-Reinforced Ceramics

Whisker-reinforced ceramic composites have come into prominence for structural applications because of their potential for high strength and fracture toughness which can be retained at elevated temperatures. Originally applied

87

Table 2.1 Room temperature fracture strength and toughness comparisons for selected ceramic matrices showing changes with the addition of SiC whiskers

Matrix	Whisker content (vol. %)	Fracture strength (MPa)	Fracture toughness (MPa $\sqrt{m}$)	Reference
Alumina	None	400	3–4	15
Alumina	20 vol.%	650	7.5–8.3	15
Mullite	None	220	2.2–2.3	23
Mullite	20 vol.%	420	4.7	23
Silicon nitride	None	660	5.5	9
	20 vol.%	770	6.9	9

to the reinforcement of metal matrices, such as aluminum, the first applications of whisker reinforcement to ceramics was in the alumina and mullite matrix systems.[5,6]

Whiskers are usually defined as short, discontinuous, rod- or needle-shaped single crystal fibers in the size range of 0.1–3 μm in diameter and 5–200 μm in length. They typically have very high tensile strengths, up to 20 GPa, and elastic moduli, up to 550 GPa.[7] SiC whiskers were first commercially available in the early 1960s.[8] Since that time, interest in whiskers has increased due to composite development and the commercial success of some products. Currently, SiC whiskers have the most commercial value for use and thus generate the greatest interest.

In addition to the initial work in the alumina and mullite matrix systems previously mentioned, SiC whiskers have also been used to reinforce other ceramic matrices such as silicon nitride,[9–13] glass,[14,15] magnesia–alumina spinel,[16] cordierite,[17] zirconia,[18] alumina/zirconia,[18,19] mullite/zirconia,[18–21] and boron carbide.[22] A summary of the effect of SiC whisker additions on the mechanical properties of various ceramics is given in Table 2.1. As shown, the addition of whiskers increases the fracture toughness of the ceramics in all cases as compared to the same monolithic materials. In many instances, improvements in the flexural strengths were also observed. Also important is the fact that these improvements over the monolithic materials are retained at elevated temperatures in many cases.

2.2.1 Whisker Characterization

SiC whiskers are known to vary in their physical and chemical characteristics from manufacturer to manufacturer and in some cases from batch to batch. While the differences can be quite minor, they can make major impacts in the performance of ceramic matrix composites both at room and elevated temperature.

Table 2.2 Summary of whisker characteristics

	American matrix	ARCO/ACMC	Tokai carbon	Tateho
Av. diameter	1.3	0.6	0.5	0.4
Approx. mean aspect ratio	—	30:1	25:1	50:1
Impurities (ppm)				
B	>1000	3	—	3
Ca	400	400	20	400
Fe	200	200	50	200
K	100	100	3	20
Mg	100	100	3	30
Mn	3	300	5	30
Na	100	100	50	5
O	2.9%	1.3%	N.D.	1.0%

A summary of the typical whisker characteristics for several manufacturers is given in Table 2.2. The differences in impurity contents are dependent on the precursors, catalysts, and growth techniques used in the manufacture of the whiskers. For example, American Matrix (AMI) used boric acid as a catalyst, while ARCO/ACMC and Tateho both start with rice hulls as precursor materials. Consequently, the AMI whiskers have significant amounts of associated boron, and the calcium in the ARCO/ACMC and Tateho whiskers comes from the rice hulls. The impurities are present mainly as oxides and their distribution occurs both on external surfaces and within internal inclusions. The oxygen contents are predominantly attributable to silica and depend mainly on post-growth oxidation treatments given that the whiskers remove residual carbon. Post-oxidation treatments, such as HF leaching, can be used to lower the surface oxygen contents. The SiC-whisker surface chemistry is also important in ceramic matrix composites because it influences the nature of the interface between the matrix and the whisker. In alumina/SiC composites, it was determined that the presence of excess surface carbon on the whiskers resulted in a weak interfacial bond and high toughness and strength materials. Where the interfacial bond is too strong, no debonding or crack bridging occurs and the composites have relatively low fracture toughness.

2.2.2 Toughening Behavior of Whisker-Reinforced Composites

Research into the toughening behavior responsible in the composite materials shows that crack–whisker interaction resulting in crack bridging, whisker pull-out and crack deflection are the major toughening mechanisms.

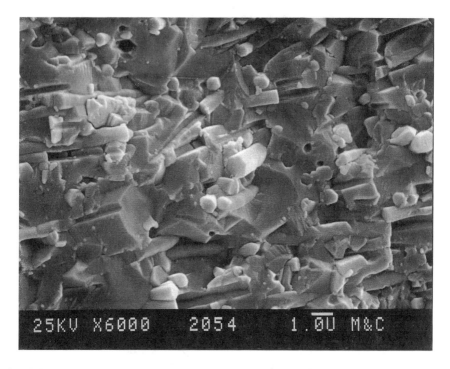

Fig. 2.1 Fracture surface of an alumina/20 vol.% SiC whisker composite showing microscopically rough surface. Presence of whiskers readily evident due to debonding along the matrix–whisker interface.

For this mechanism to operate, debonding along the crack–whisker interface (often associated with crack deflection) must occur during crack propagation and allow the whiskers to bridge the crack in its wake. Examination of the fracture surfaces of whisker-reinforced composites reveals that they are microscopically rough with whiskers readily evident (Fig. 2.1).

Micromechanical modeling and available experimental evidence indicates that the composite toughness, K_{Ic} (composite), can be described as the sum of the matrix toughness, K_{Ic} (matrix), and a contribution due to whisker toughening, ΔK_{Ic} (whisker reinforcement).[1,23] In other words,

$$K_{Ic} \text{ (composite)} = K_{Ic} \text{ (matrix)} + \Delta K_{Ic} \text{ (whisker reinforcement)} \quad (1)$$

Becher *et al.*[23] derived a relationship for the increase in fracture toughness due to whisker reinforcement:

$$\Delta K_{Ic}(W.R.) = \sigma_f \sqrt{\frac{V_f r}{B(1 - \nu^2)} \frac{E_c}{E_w} \frac{\gamma_m}{\gamma_i}} \quad (2)$$

where $\Delta K_{Ic}(W.R.)$ is the increase in fracture toughness due to whisker reinforcement, σ_f is the fracture strength of whiskers, V_f is the volume fraction of whiskers, r is whisker radius, ν is Poisson's ratio for whiskers, E is Young's modulus for composite (c) and whiskers (w), γ is fracture energy for matrix (m) and matrix–whisker interface (i), and B is a constant that depends on the bridging stress profile (approximately 6 for alumina/SiC whisker composites).

Evans[1] used the following to relate whisker reinforcement to toughening:

$$\Delta K_{Ic}(W.R.) \sim fdS^2/E + 4L_if(d/R)/(1-f) \tag{3}$$

where $\Delta K_{Ic}(W.R.)$ is the increase in fracture toughness due to whisker reinforcement, f is the volume fraction of whiskers, d is whisker debond length, S is strength of whiskers, E is Young's modulus of composite, and R is whisker radius.

In both derivations of toughening behavior, increases in toughness for whisker-reinforced composites are dependent on the following parameters: (1) whisker strength, (2) volume fraction of whiskers, (3) elastic modulus of the composite and whisker, (4) whisker diameter, and (5) interfacial fracture energies.

2.2.3 Alumina Matrix Composites

The high temperature, short-term properties of alumina and alumina/zirconia composites have been examined in a number of studies.[18,19,24–32] Typical fracture strength and toughness at elevated temperatures are summarized in Figs. 2.2 and 2.3.

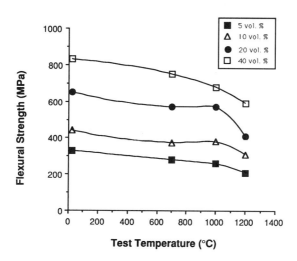

Fig. 2.2 Flexural strength of alumina/SiC whisker composite at elevated temperatures and different whisker volume contents.

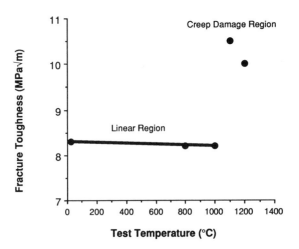

Fig. 2.3 Elevated fracture toughness of alumina/20 vol.% SiC whisker composite showing a linear response up to temperatures of 1000°C. At temperatures ≥1100°C, creep damage results in apparent increases in toughness.

The studies show that the fracture behavior remains linear up to temperatures of ~1100°C with little variation in flexural strength or fracture toughness. However, at temperatures of ≥1200°C, the composites show considerable strength degradation even with short time exposures at stresses of 1/2 or 2/3 of the fast fracture strength. Becher and Tiegs observed that at 2/3 of the fast fracture strength, failure occurred in <10 h.[27] This marked strength degradation is associated with creep, which is covered in depth in Chapter 4 of this book. At temperatures >1100°C, apparent increases in toughness are also associated with creep crack nucleation and growth.

Thermal shock testing of an alumina/20 vol.% SiC whisker composite showed no decrease in flexural strength with temperature transients up to 900°C.[33] Monolithic alumina, on the other hand, shows significant decreases in flexural strength with temperature changes of >400°C. The improvement is a result of interaction between the SiC whiskers and thermal-shock induced cracks in the matrix, which prevents coalescence of the cracks into critical flaws.

2.2.4 Mullite Matrix Composites

Whereas the ambient mechanical properties have been studied extensively, the high temperature properties of mullite and mullite/zirconia matrix composites have been examined by only a few studies.[18,30,32,34]

The flexural strength of a mullite/20 vol.% SiC whisker composite as a function of temperature is presented in Fig. 2.4. As shown, the significant increase in the strength of the composite is retained up to temperatures of at

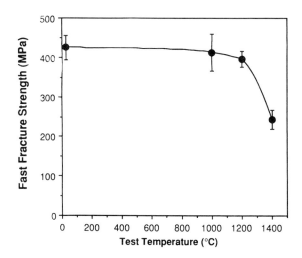

Fig. 2.4   Fracture strength of mullite/20 vol.% SiC whiskers at elevated temperature.

least 1200°C. The drop in strength at 1400°C is attributable to a degradation or softening of the intergranular glassy phase that is present in the composites. TEM examination showed pockets of silica-rich glass, high in calcium, in selected areas surrounding some of the SiC whiskers. The source of the intergranular glassy phase is believed to be due to excess silica in the mullite powder used in that study and the calcium being introduced with the whiskers. Similar glassy phases have been observed in other studies using the same mullite powder.[35] Previous results on mullite without glassy grain boundary phases have shown no strength degradation up to temperatures of 1500°C.[36] It would be anticipated that elimination or crystallization of the glassy phases in the present material would improve the high temperature properties.

Shaw and Faber[30] reported that the toughness of mullite/20 vol.% SiC whisker composites with either ARCO/ACMC or Tateho whiskers remained relatively constant up to temperatures of 1100°C. At temperatures ≥900°C they also observed considerable whisker pull-out lengths up to 100 μm. However, the number of pull-outs was always <0.01% of the total whisker volume and only observed with whiskers that were perpendicular to the crack plane.

2.2.5 Silicon Nitride Matrix Composites

The addition of SiC whiskers to silicon nitride matrices resulted in only moderate increases in the fracture toughness a compared to monolithic materials. Fracture strengths of composite materials showed increases in some cases and slight decreases in others. Summaries of the variations in toughness and strength are given in Tables 2.3 and 2.4. The moderate increases in

Table 2.3 Summary of short-term fracture toughness of silicon nitride/SiC whisker composites at elevated temperatures

Additive content	Whisker content and type	Densification method	Fracture toughness (MPa√m)				Reference
			25°C	1000°C	1200°C	1375–1400°C	
4% Y_2O_3		Reaction-bond/HIP	4.7	4.1	3.7	4.1	33
4% Y_2O_3	30 vol.% Tateho	Reaction-bond/HIP	4.6		3.9	4.1	33
4% Y_2O_3	30 vol.% AMI	Reaction-bond/HIP	5.8	5.2	5.6	4.7	33
4% Y_2O_3	30 vol.% AMI[a]	Reaction-bond/HIP	4.9		6.0	4.9	33
GN-10[b]	None	HIP	5.67 ± 0.13				35
GN-10[b]	20 vol.% AMI	HIP	5.99 ± 0.45		5.48 ± 0.11		35
					5.74 ± 0.23		35
GN-10[b]	30 vol.% AMI	HIP	6.40 ± 0.40		5.95 ± 0.42		35
GN-10[a,b]	20 vol.% AMI	HIP	6.11 ± 0.20		5.48 ± 0.18		35
GN-10[a,b]	30 vol.% AMI	HIP	6.02 ± 0.46		5.42 ± 0.30		35
None	None	HIP	3.1	2.7	2.6		37
None	30 vol.% Tokai	HIP	4.1	4.0	3.8		37

[a]Sintering aid content increased to adjust for whisker addition.
[b]Proprietary sintering aid formulation of Garrett Ceramic Components, Torrance, CA.

Table 2.4 Summary of short-term flexural strength at elevated temperatures

Additive content	Whisker content and type	Densification method	Flexural strength (MPa)				Reference
			25°C	1000°C	1200°C	1375–1400°C	
6% Y_2O_3 2% Al_2O_3	—	Hot-press	981 ± 44			361 ± 3	32
6% Y_2O_3 2% Al_2O_3	30 vol.% ACMC	Hot-press	994 ± 73			350 ± 3	32
6% Y_2O_3 2% Al_2O_3[a]	30 vol.% ACMC	Hot-press	1083 ± 113			369 ± 12	32
GN-10[b]	None	HIP	890 ± 124			514 ± 22	35
GN-10[b]	20 vol.% AMI	HIP	708 ± 83			381 ± 20	35
GN-10[b]	30 vol.% AMI	HIP	626 ± 49			362 ± 34	35
GN-10[a,b]	20 vol.% AMI	HIP	756 ± 77			393 ± 26	35
GN-10[a,b]	30 vol.% AMI	HIP	648 ± 110			419 ± 57	35
6% Y_2O_3 2% Al_2O_3	—	HIP	1043 ± 50		618 ± 27	407 ± 62	34
6% Y_2O_3 2% Al_2O_3	30 vol.% ACMC	HIP	1195 ± 71		616 ± 105	386 ± 58	34
None	None	HIP	~500	~500	~410	~260	37
None	30 vol.% Tokai	HIP	~690	~425	~360	~280	37

[a]Sintering aid content increased to adjust for whisker addition.
[b]Proprietary sintering aid formulation of Garrett Ceramic Components, Torrance, CA.

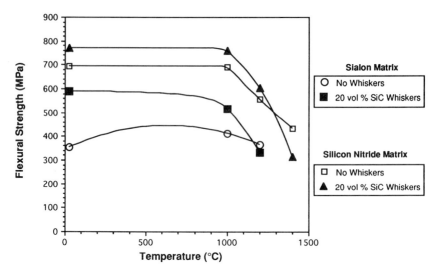

Fig. 2.5   High temperature flexural strength of sialon ($Si_{4.5}Al_{1.5}O_{1.5}N_{6.5}$–2%3$Y_2O_3$–5$Al_2O_3$) and silicon nitride ($Si_3N_4$–6% Y_2O_3–2% Al_2O_3) materials with and without SiC whiskers. Whiskers are from American Matrix.

toughness were less than originally expected and are attributable to the fact that the whiskers inhibit acicular growth of the β-Si_3N_4 grains that normally occurs in conventional fabrication of silicon nitride. This results in a lower matrix toughness and so the contribution from the whisker reinforcement is diminished. The differences in strength are mainly a result of the degree and quality of whisker dispersion in the composites.

The high temperature properties have been examined in a number of studies.[37–44] Typical high temperature behavior is summarized in Figs. 2.5 and 2.6. The flexural strengths shown in Fig. 2.5 are relatively stable up to temperatures of 1000°C. However, at higher temperatures the strengths decrease considerably. In fact, for temperatures of 1200°C for the sialon and 1400°C for the silicon nitride, the mean strengths for the monolithic materials are greater than the composites. It is believed that these decreases are due to the impurities associated with the whiskers and their effect on the composition of the intergranular phases. Many of the impurities listed in Table 2.2 are located on the whisker surfaces and can be incorporated into the liquid phase during densification. All of the species listed would have a detrimental effect on the refractory nature of the intergranular phases.

The effect of different whisker types from various manufacturers on the high temperature flexural strength is shown in Fig. 2.6. As before, the strengths remain relatively constant up to 1000°C and decrease at higher temperatures. The decrease is most dramatic for the composite with whiskers from American Matrix. This is related to the relatively high impurity content of

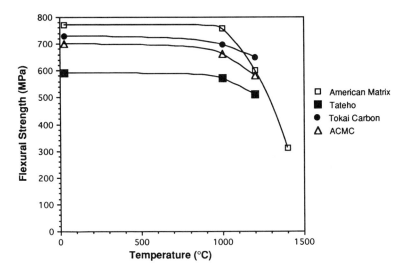

Fig. 2.6 High temperature flexural strength of silicon nitride (Si_3N_4–6% Y_2O_3– 2% Al_2O_3) materials with various types of SiC whiskers. Whisker content was 20 vol.% for all composites. ACMC is Advanced Composites Materials Corp.

these whiskers and the fact that much of the boron (as B_2O_3) is available to the liquid phases. The other whisker types examined all have less associated impurities and the compositions of the intergranular phases are more refractory.

2.2.6 Zirconia, Cordierite, and Spinel Matrix Composites

As mentioned previously, the main body of research on whisker-reinforced composites was concerned with alumina, mullite, and silicon nitride matrix materials. None the less, selected work examined zirconia, cordierite, and spinel as matrix materials.[16–18] The high temperature strength behavior reported for these composites is summarized in Table 2.5. As shown, the zirconia matrix composites exhibited decreases in room temperature strength with the addition of SiC whiskers. However, the retained strength at 1000°C, was significantly improved for the whisker composites over the monolithic. Claussen and co-workers attributed this behavior to loss of transformation toughening at elevated temperatures for the zirconia monolith, whereas the whisker-reinforcement contribution did not decrease at the higher temperature.[17,18]

The cordierite and spinel matrix composites both showed strength improvements over the monolithics that were retained up to 1000°C. Measurements on the spinel matrix composite showed a significant decrease at 1200°C.

Table 2.5 Summary of elevated temperature mechanical properties for zirconia, cordierite, and spinel matrix composites

Matrix	Whisker content (vol.%)	Fracture strength (MPa) 25°C	Fracture strength (MPa) 1000°C	Fracture strength (MPa) 1200°C	Fracture toughness (MPa $\sqrt{m}$)	Reference
Zirconia	None	1150	160		6.8	13
Zirconia	30 vol.%	590	380		11.0	13
Cordierite	None	180	170		2.2	14
Cordierite	20 vol.%	260	245		3.7	14
Spinel	None	320	300	275	—	12
Spinel	30 vol.%	415	370	275	—	12

This is probably due to the introduction of impurities from the whiskers similar to the effects observed with the silicon nitride matrix materials discussed above.

2.2.7 Applications Dependent on Short-Term Behavior

At the present time, whisker-reinforced alumina is being fabricated and marketed extensively as a cutting tool for high-nickel alloys.[45-47] Cutting rates up to ten times higher than conventional tools have been obtained. Improvements in machining of cast iron and steel products have also been observed. Because of the excellent wear resistance of whisker-reinforced ceramic composites, short-term applications include wear parts, such as pump seal rings, grit blast nozzles, aluminum can tooling and dies for metal extrusion and wire pulling. While many of these applications would be considered in the realm of long-term behavior, many involve short-term exposure to elevated temperatures followed by thermal cycling to lower temperatures.

2.2.8 Future Directions for Advanced High Toughness Whisker-Reinforced Composites

As shown in the discussion on toughening mechanisms, whisker-reinforced composite toughness is dependent on the following parameters: (1) whisker strength, (2) volume fraction of whiskers, (3) elastic modulus of the composite and whisker, (4) whisker diameter, and (5) interfacial fracture energies. Many of these parameters are essentially fixed in a narrow range, either by matrix selection (E), whisker selection (E_w, v), or processing considerations (V_f). Thus to increase composite toughness by whisker reinforcement significantly, increases in whisker strength and diameter are necessary.

Calculated whisker strengths for commercially available whiskers were in the range of 5–10 GPa.[48] Strength limiting flaws included voids, accumulations of smaller inclusions in core regions, and excessive surface roughness. Estimates indicate that if high strength SiC whiskers were made by eliminating strength-limiting flaws, alumina or silicon nitride matrix composites with toughnesses $(K_{Ic}) > 15$ MPa$\sqrt{m}$ may be realized.

2.3 Fiber-Reinforced Ceramic Matrix Composites

2.3.1 Stress–Strain Characteristics

The tensile stress–strain curves of several fiber-reinforced ceramic matrix composites versus temperature are shown in Fig. 2.7.[49–52] Most of the composites exhibit a non-linear stress–strain behavior in tension analogous to elastic–plastic behavior of metallic alloys at both room and elevated temperatures. The first deviation from linearity is due to the development of a single crack that passes completely through the matrix, but remains bridged by intact fibers. A detailed fracture mechanics analysis has been developed to relate the matrix cracking stress to microstructural properties by Marshall *et al.*[53] This analysis is based upon the fact that the bridging fibers resist the opening of a matrix crack by frictional forces at the fiber–matrix interface. The lower-bound, steady-state matrix cracking stress, σ_0, was derived as

$$\sigma_0 = [6E_f f^2 \tau \Gamma_m E^2/(1-f)E_m^2 R]^{1/3} - E\sigma_r/E_m \tag{4}$$

where f is the fiber volume fraction; E, E_f and E_m are the modulus of the composite, fiber and matrix, respectively, R is the fiber diameter, τ is the interfacial sliding stress, Γ_m is the matrix toughness, and σ_r is residual stress. Also, to allow crack bridging by the fibers, debonding at the fiber–matrix interface must occur at the matrix crack front. An analysis conducted by He and Hutchinson indicates that in systems in which fiber and matrix have similar elastic moduli, debonding will occur provided that the interfacial fracture energy is less than one quarter of the fiber fracture energy $(\Gamma_i/\Gamma_f < 1/4)$.[54] However, this criterion has not yet been experimentally validated.

As the stress is raised above the first matrix cracking stress, further cracking occurs resulting in a periodic crack array. Cao *et al.*[55] have shown that the magnitude of the saturation crack spacing, d, is governed by the sliding stress, τ, such that τ and d are related by

$$\tau = 1.34[(1-f)^2 E_f E_m R^2/fEd^3]^{1/2} \tag{5}$$

Following the multiple matrix cracking, the composite can continue to carry the load until fiber failure. The ultimate tensile strengths of several continuous fiber-reinforced ceramic composites versus temperatures are shown in Fig.

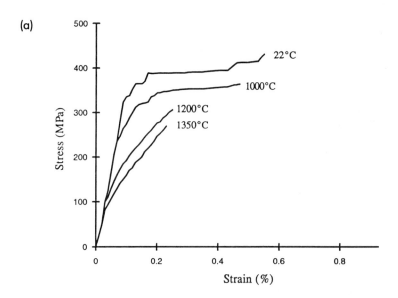

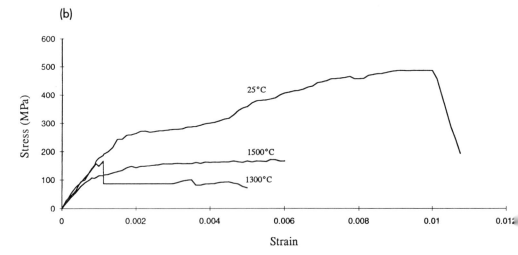

Fig. 2.7 The tensile stress–strain curves versus temperatures for (a) unidirectional SCS-6 fiber-reinforced hot-pressed Si_3N_4, (b) unidirectional SCS-6 fiber-reinforced reaction-bonded Si_3N_4, (c) 2-D Nicalon fabric-reinforced CVI-SiC, and (d) 3-D braided Nicalon fabric-reinforced CVI-SiC.[49–52]

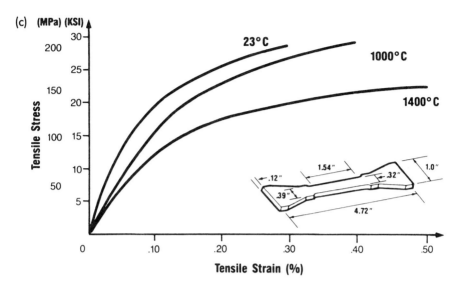

(c)

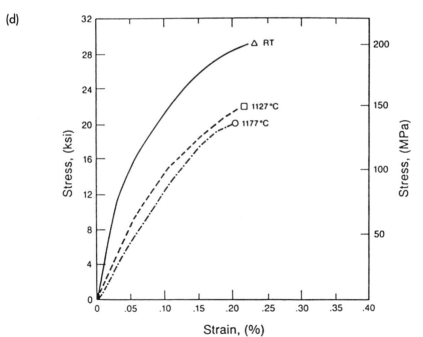

(d)

Fig. 2.7–*contd.*

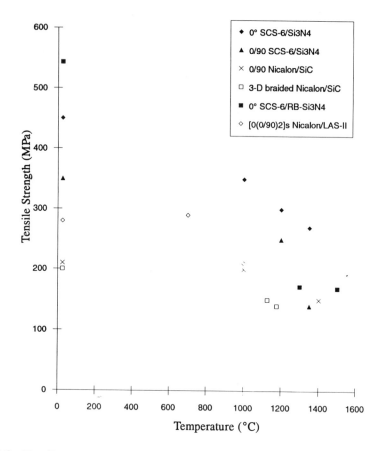

Fig. 2.8 Tensile strength versus temperature for several fiber-reinforced ceramic composites.[49–52,56,57]

2.8.[49–52,56,57] It is obvious that the tensile strength and the first matrix cracking stress of these composites decreases significantly as temperature increases.

The typical load–deflection curves for an unidirectional SCS-6 fiber-reinforced Zircon composite as a function of temperature are shown in Fig. 2.9.[58] The load–displacement curve at room temperature shows an initial linear elastic behavior, followed by a sudden load drop at the onset of first matrix cracking. Beyond this, composite also retained its load-bearing capability to higher loads, the matrix cracking density increasing until the ultimate strength is reached. After the maximum load, there is a gradual decrease in load as more of the intact fibers begin to fail and pull out from the matrix. The stress–strain curve varies considerably as temperature is increased. Composites tested at 1160°C and 1315°C also display an initial elastic behavior; however, the maximum load-carrying capability of the composite is very close to the first matrix cracking stress. The flexural strength of several fiber-reinforced ceramic

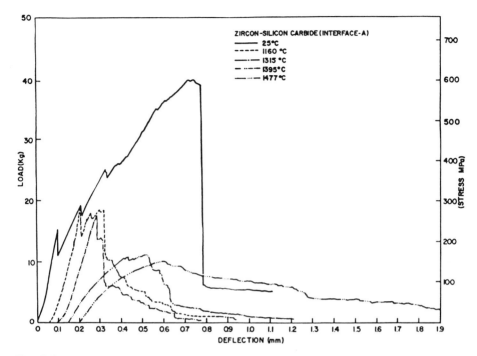

Fig. 2.9 The load-displacement curves for unidirectional SCS-6/Zircon composites versus temperature.[58]

composites versus temperature is given in Fig. 2.10.[52,58–60] The three-point or four-point flexural strength also decreases rapidly as the temperature increases, especially in air. In a non-oxidizing atmosphere, strength can be retained to temperatures at which creep of the matrix causes loss of strength or fiber strength loss.

2.3.2 Strength Degradation Mechanisms

The strength degradation at elevated temperatures is primarily due to fiber strength degradation, interface degradation due to chemical reactions, or environmental attack and matrix microcracking. Currently the major practical temperature limitation on ceramic matrix composites is due to the lack of available ceramic fibers with good properties above 1000°C. Above this temperature range, degradation or creep of existing ceramic fibers becomes excessive. Table 2.6 lists the chemical compositions, densities, physical and mechanical properties of several commercially available ceramic fibers.[61–66] The tensile strength versus temperatures and the creep behavior of these fibers are shown in Figs. 2.11–2.13.[61,62,64] The AVCO SCS-6 fiber, produced by chemical vapor deposition (CVD) of SiC onto a pyrolytic carbon core, and double coated with carbon graded to silicon carbide, shows the greatest

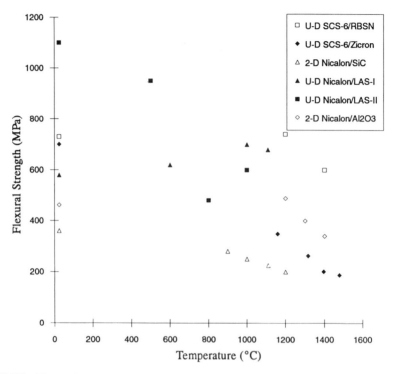

Fig. 2.10 Flexural strength versus temperature for several fiber-reinforced ceramic composites.[52,58–60]

strength retention at high temperature. However, it is limited to use in flat laminates as a result of its large diameter (143 μm). All the other fibers listed in Table 2.1 show substantial decrease in strength at 1000–1200°C. Creep is also observed at high temperature; however, creep strain is an order of magnitude less than that observed for Nicalon at temperatures below 1400°C.[67]

The Nicalon fiber, a polycarbosilane-derived SiC, is manufactured by Nippon Carbon.[68] It is not a stoichiometric SiC, but contains a significant amount of free carbon, excess silicon and oxygen, leading to compositional and microstructural changes at elevated temperatures, including accelerated grain growth and creep. Nicalon has been shown to be comprised of amorphous or microcrystalline SiC embedded in an amorphous matrix of silica which also contains agglomerations of free carbon. Degradation at elevated temperatures is attributed to the evolution of CO and SiO, accompanied by β-SiC grain growth. The Tyranno fiber is produced from a titanium-doped (1.5–4.0 w/o) polycarbosilane using the same process as that used in the preparation of Nicalon.[69] It is amorphous due to the presence of titanium which inhibits crystallization, with properties similar to those of Nicalon, though loss of strength above 1000°C is less abrupt. It also contains higher oxygen (17%) and

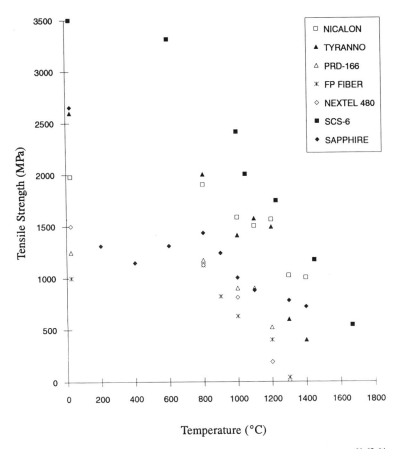

Fig. 2.11 Tensile strength versus temperature for several ceramic fibers.[61,62,64]

more free carbon as compared with ceramic-grade Nicalon. Above 1400°C, the free carbon would be expected to react with Si-O bonds, evolving CO and accounting for the decrease in tensile strength. Two Si-C-N-O fibers are also under development for high temperature applications: HPZ and MPDZ.[63,70] The HPZ fiber is processed from hydriodopolysilazane precursor and pyrolyzed in an oxygen-free atmosphere. The fiber is amorphous silicon carbonitride which contains about 10 w/o carbon. The MPDZ fiber is processed from methylpolydisilazane precursor and has higher carbon content than the HPZ fiber. The high temperature strength and creep resistance of the HPZ fiber are still inferior to those of the Nicalon fiber as shown in Figs. 2.12 and 2.13.

A number of oxide fibers with diameters ranging from 10 to 150 μm are commercially available. The use of oxide fibers is most likely limited to composites with oxide matrices, to avoid chemical reactions at the interface and interdiffusion between the fiber and matrix. The Nextel series of fibers

Table 2.6 The chemical compositions, densities, physical and mechanical properties of several commercially available ceramic fibers

Manufacturer	Fiber type	Typical composition (wt.%)	Tensile strength (GPa)	Strain to failure (%)	Young's modulus (GPa)	Specific gravity	Diameter (μm)
Nippon Carbon	Nicalon	65 SiC 15 C 20 SiO$_2$	2.7	1.4	185	2.55	15
Ube Chemicals	Tyranno	Si, C, O Ti < 5	3	1.5	200	2.4	9
Dow Corning/Celanese	MPDZ	47 Si 30 C 15 N 8 O	1.9	1.1	180	2.3	12
Dow Corning/Celanese	HPZ	59 Si 10 C 28 N 3 O	2.2	1.5	150	2.35	10

Manufacturer	Fiber	Composition					
Dow Corning/Celanese	MPS	69 Si 30 C 1 O	1.2	1.6	190	2.65	11
Du Pont de Nemours	Fiber FP	>99 α-Al_2O_3	1.40	0.4	380	3.9	20
Du Pont de Nemours	PRD-166	80 α-Al_2O_3 20 ZrO_2	2.07	0.6	380	4.2	20
Sumitomo Chemicals	Alf	85 Al_2O_3 15 SiO_2	2	1.1	180	3.2	18
ICI	Safimax	96 δ-Al_2O_3 4 SiO_2	2.0	0.7	300	3.3	3
3M	Nextel 312	62 Al_2O_3 24 SiO_2 14 B_2O_3	1.75	1.1	154	2.7	11
	Nextel 440	70 Al_2O_3 28 SiO_2 2 B_2O_3	2.1	1.1	189	3.05	11
	Nextel 480	70 Al_2O_3 28 SiO_2 2 B_2O_3	2.3	1.0	224	3.05	11
Textron	SCS-6	SiC	4.0	0.9	406	3.0	143
Saphikon	Sapphire	Al_2O_3	3.5	0.7	524	3.9	75–150

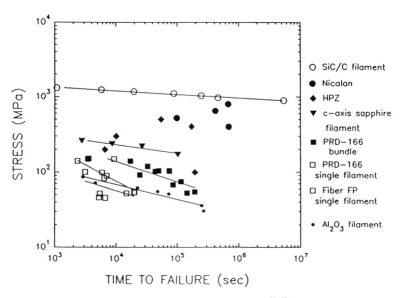

Fig. 2.12   Stress rupture plot for several ceramic fibers.[61,62] All data obtained at 1200°C except *c*-axis sapphire (1700°C).

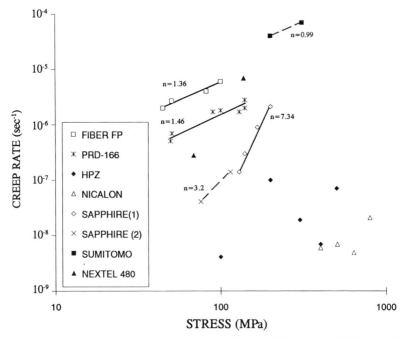

Fig. 2.13   Creep rate versus stress plot for several ceramic fibers and bulk sapphire (*c*-axis).[61] All data were obtained at 1200°C except sapphire, which were obtained at 1700°C.

manufactured by 3M are sol–gel derived having a microcrystalline mullite structure. The Nextel fibers retain at least 75% of their tensile properties up to 1000°C. However, the Nextel fibers creep at 1000°C and above. The small-grain polycrystalline α-Al_2O_3 fiber produced by Du Pont (FP) retains its strength well up to 1000°C. However, it exhibits grain growth and creep at temperatures above 1000°C. The PRD-166, also produced by Du Pont, is an α-Al_2O_3 fiber having approximately 20 w/o of Y_2O_3 partially stabilized ZrO_2 as a second phase.[64] The ZrO_2 addition inhibits grain growth and improves both room temperature strength and strength retention after exposure to elevated temperatures. The single crystal Al_2O_3 fiber produced by the melt-pulling technique has a number of advantages compared to the polycrystalline ceramic fibers. These advantages are microstructural stability at high temperature (grain growth is not an issue), high elastic modulus retention at high temperature, and good creep resistance compared to the polycrystalline ceramics. As shown in Fig. 2.11, the strength of the sapphire fiber is quite high at room temperature, but then drops precipitously to an initial minimum at about 400°C, increases slightly and then falls off above 800°C.[71] The mechanisms which lead to the rapid tensile strength loss of sapphire fiber at low temperatures have not yet been determined.

Interfacial degradation resulting from chemical reactions, oxidation or thermal expansion-induced residual stresses is another critical factor responsible for the degradation of ceramic matrix composites at elevated temperatures. Although the as-fabricated composites may exhibit limited interfacial bonding with acceptable ambient temperature properties, exposure at elevated temperature and oxidizing environment causes diffusion, coupled with the ingress of O_2, N_2, etc., from the environment, resulting in chemical bonding across the interface. Such reaction may result in a high interfacial shear strength, leading to brittle failure or degradation of fiber strength. It has been shown that the interfacial shear resistance in a Nicalon/LAS (lithium aluminosilicate) composite increased from 2 MPa at room temperature to approximately 40 MPa at 1000°C.[72] The low interfacial shear strength in the as-processed materials is due to the presence of a carbon-rich interlayer formed during composite fabrication. However, extended heat-treatment in air leads to the formation of a continuous SiO_2 layer between the fiber and matrix. The fracture mechanics analysis of matrix cracking indicates that the increase in interfacial frictional stress would increase the matrix cracking stress above the fiber bundle stress. The corresponding failure behavior of the composite changes from fiber-controlled failure (damage-tolerant) at room temperature to matrix-controlled (catastrophic) failure at high temperature. The effect of fiber strength and interfacial shear strength on the fracture behavior of uniaxial fiber-reinforced ceramic composites has been summarized in a fracture mechanism map as shown in Fig. 2.14.[72]

Residual stress resulting from the thermal expansion mismatch between the fiber and matrix is another factor which will affect the interfacial properties, matrix cracking stress and the mechanical properties of ceramic

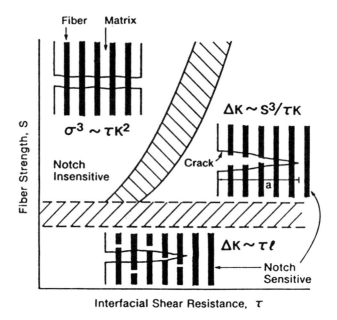

Fig. 2.14 Fracture mechanism map for uniaxially fiber-reinforced ceramic composites under tensile loading.[72]

composites. It has been shown that the interfacial shear strength of the SCS-6 fiber-reinforced reaction-bonded silicon nitride composite increases approximately twofold with increasing temperature—between 5 and 18 MPa at room temperature, compared to between 12 and 32 MPa at 1300°C.[73] Since the thermal expansion of the fiber is higher than that of the matrix, the fiber–matrix interface is subjected to residual tension at ambient temperatures. As the temperature is increased, this tensile stress will be gradually relieved which results in an increased mechanical interlocking between the fiber and matrix. The increasing interfacial properties will change the crack growth behavior from crack deflection along the fiber–matrix interface at ambient temperature to crack growth perpendicular to the fibers at elevated temperatures.[74] A similar study conducted by Abbe and Chermont showed that the frictional stress of a Nicalon/SiC composite decreases with increasing temperatures as shown in Fig. 2.15.[75] The weakening of interface can be due to a relief of thermal residual stress, leading to a smaller tangential compressive stress at the interface since the thermal expansion coefficient of the fiber is lower than that of the matrix. The influence of residual stress on the matrix cracking stress and the fracture resistance of ceramic composites has been summarized by Marshall and Evans.[76]

Matrix microcracking may also impose a significant limitation on the engineering use of ceramic composites through its effect on strength at

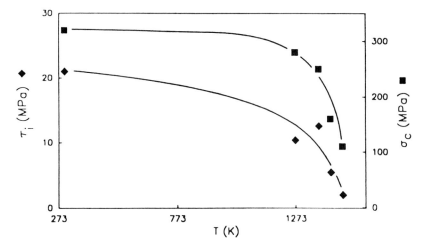

Fig. 2.15 The interfacial frictional stress versus temperature for a Nicalon/SiC composite.[75]

moderate to high temperatures in oxidizing atmospheres. This is because the cracking provides an easy path for inward progression of oxygen to the interface, leading to oxidative embrittlement. It has been shown that the rapid loss of strength of Nicalon/LAS-II composite at 700–800°C was due to a mechanism involving atmospheric attack of the carbon-rich interface between fibers and matrix promoted by the matrix microcracks.[56] In the SCS-6 fiber-reinforced ceramic composite, the C-rich layer on the fiber surface is also susceptible to oxidative attack once the matrix cracking occurs. Since the matrix cracking stress decreases substantially at high temperature, it thus is important to either improve the matrix cracking stress at a level close to the ultimate strength or to develop a stable interface which is resistant to environmental attack.

2.3.3 *Fracture Behavior and Toughening Mechanisms*

Numerous investigators have demonstrated that the fracture toughness of polycrystalline ceramics at ambient temperature can be significantly improved by reinforcing them with continuous fibers. A number of different toughening mechanisms have also been identified as schematically shown in Fig. 2.16.[1] These include crack deflection, fiber bridging, microcracking and fiber pull-out. A rising R-curve behavior at ambient temperature (an increase of fracture toughness with crack extension) has also been reported for several fiber-reinforced ceramic matrix composites.[77–80] This is primarily due to the fiber bridging and matrix microcracking in the crack wake. Since fibers do not fail

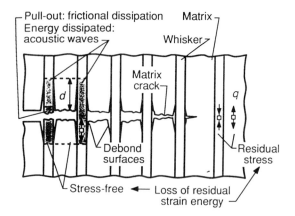

Fig. 2.16 Schematic indicating the various contributions to the steady-state toughness.[1]

upon matrix cracking, the composite can sustain additional load as the crack extends. Therefore, the bridging effect causes the resistance to increase appreciably, resulting in stable crack extension. The magnitudes of the interface sliding stress, the fiber bundle strength, and the residual stress in the matrix have been shown as the key parameters that govern the fracture characteristics of a unidirectional fiber-reinforced ceramic composite with notch under monotonic loading.[80]

Fundamental understanding of the details of the failure modes and toughening mechanisms of these composites at elevated temperature is still very limited. The fracture toughness of several fiber-reinforced ceramic composites versus temperature is shown in Fig. 2.17.[77–81] The toughness values are obtained using a single-edge notched-beam specimen under four-point bending. The fracture toughnesses of the C/SiC and Nicalon/SiC composites are between 30 to 35 MPa m$^{1/2}$ and are retained up to 1400°C in inert atmosphere. However, in an oxidizing atmosphere, the fracture toughness also decreases rapidly as temperature increases. A recent study conducted by Nair and Wang indicated that the R-curve effect of a woven Nicalon fabric-reinforced/SiC composite at ambient temperature was substantially reduced at 1200°C in air as shown in Fig. 2.18.[82] The magnitude of the fracture toughness at elevated temperatures is in the range of 12–18 MPa m$^{1/2}$ which is still higher than that of a monolithic SiC. Microstructural examination showed that growth of major crack was associated with significant degree of delamination cracking, resulting in a substantial degree of crack branching. Little fiber bridging at the crack wake was observed as a result of degradation of the Nicalon fibers at high temperatures. Crack branching and microcracking appear to be the possible toughening mechanisms of this composite at elevated temperature.

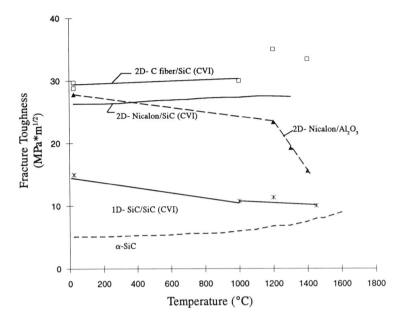

Fig. 2.17 The fracture toughness versus temperature for several fiber-reinforced ceramic composites.[77–81]

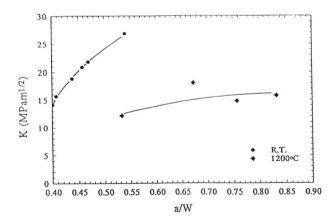

Fig. 2.18 The R-curve behavior of a woven Nicalon/SiC composite at ambient and elevated temperatures.[82]

2.3.4 Prospects of Developing Fiber-Reinforced Ceramics for High Temperature Applications

As stated above, fiber-reinforced ceramic composites with satisfactory strength and toughness at elevated temperatures have so far not been achieved. The primary reasons for this appear to be structural degradation or oxidation of the fibers and matrix at high temperatures and incompatibility of the constituents in oxidizing atmospheres at high temperature. Therefore, continuing research is needed to address the following critical issues.

(1) The need to develop fibers with better microstructural stability at elevated temperatures and ability to retain their properties between 1000–2000°C. The requirements of fiber properties for strong and tough ceramic composites have been discussed by DiCarlo.[83] A small diameter, stoichiometric SiC fiber fabricated by either CVD or polymer pyrolysis, and a microstructurally stable, creep-resistant oxide fiber appear to be the most promising reinforcements.

(2) The need to tailor a stable interface either by coating or *in situ* reaction. The strong dependence of ceramic composite properties on the properties of the interface generally requires consideration of fiber coating and/or reaction layers for high temperature applications. Evans shows that the most promising approach appears to be the use of dual coating: the inner coating satisfies the debonding and sliding requirements, while the outer coating provides protection against the matrix during processing.[1] The principal challenge is to identify an inner coating that has the requisite mechanical properties while also being thermodynamically stable in air at elevated temperatures.

(3) The need to improve the matrix cracking stress to a level close to the ultimate strength. The matrix cracking stress can be improved by increasing the interfacial bond strength, fiber volume fraction, E_f/E_m ratio or by decreasing the fiber diameter. Another promising approach is to improve the fracture strength of the matrix by incorporating the nanophase ceramic particles. Nihara has shown that the addition of nano-size ceramic dispersoids within the matrix grain or at the grain boundaries has resulted in substantial improvement of mechanical properties of Al_2O_3 and Si_3N_4 at both room and elevated temperatures.[84] Therefore, the ceramic nanocomposites may serve as potential matrix materials with excellent matrix cracking resistance.

Acknowledgments

J.-M. Yang acknowledges the support from the National Science Foundation (MSM 8809790 and 9057030). The assistance of Dr. S. M. Jeng in preparing this manuscript is also gratefully acknowledged. T. N. Tiegs acknowledges the support from the U.S. Department of Energy, Assistant Secretary for Conservation and Renewable Energy, Office of Transportation

Technologies, as part of the Ceramic Technology for Advanced Heat Engines Project of the Advanced Materials Development Program, under contract DE-AC05-84OR21400 with Martin Marietta Energy Systems, Inc.

References

1. A. G. Evans, "Perspective on the Development of High Toughness Ceramics," *J. Am. Ceram. Soc.*, **73**[2], 187–206 (1990).
2. A. G. Evans and D. B. Marshall, "The Mechanical Behavior of Ceramic Matrix Composites," *Acta Metall.*, **37**[10], 2567–2583 (1989).
3. D. B. Marshall and J. E. Ritter, "Reliability of Advanced Structural Ceramics and Ceramic Matrix Composites—A Review," *Am. Ceram. Soc. Bull.*, **66**[2], 309 (1987).
4. P. F. Becher, "Microstructural Design of Toughened Ceramics", *J. Am. Ceram. Soc.*, **74**[2], 255–269 (1991).
5. P. F. Becher and G. C. Wei, "Toughening Behavior in SiC-Whisker-Reinforced Alumina," *J. Am. Ceram. Soc.*, **67**[12], C-267–C-269 (1984).
6. G. C. Wei and P. F. Becher, "Development of SiC-Whisker-Reinforced Ceramics," *Am. Ceram. Soc. Bull.*, **64**[2], 298–304 (1985).
7. J. J. Petrovic, J. V. Milewski, D. L. Rohr, and F. D. Gac, "Tensile Mechanical Properties of SiC Whiskers," *J. Mater. Sci.*, **20**, 1167–77 (1985).
8. J. V. Milewski, "Whiskers," in *Handbook of Reinforcements for Plastics*, eds. J. V. Milewski and H. S. Katz, Van Nostrand Reinhold Co., New York, NY, 1987, p. 205.
9. S. T. Buljan, J. G. Baldoni, and M. L. Huckabee, "Si₃N₄-SiC Composites," *Am. Ceram. Bull.* **66**[2], 347–52 (1987).
10. R. Hayami, K. Ueno, I. Kondou, N. Tamari, and Y. Toibana, "Si₃N₄-SiC Whisker Composite Material," in *Tailoring Multiphase and Composite Ceramics*, eds. R. E. Tressler, G. L. Messing, C. G. Pontano, and R. E. Newnham, Materials Science Research Series, Plenum Press, New York, NY, 1986, pp. 663–674.
11. R. Lundberg, L. Kahlman, R. Pompe, and R. Carlsson, "SiC-Whisker-Reinforced Si₃N₄ Composites," *Am. Ceram. Soc. Bull.*, **66**[2], 330–333 (1987).
12. P. D. Shalek, J. J. Petrovic, G. F. Hurley, and F. D. Fac, "Hot-Pressed SiC Whisker/Si₃N₄ Matrix Composites," *Am. Ceram. Soc. Bull.*, **65**[2], 351–56 (1986).
13. T. N. Tiegs, *et al.*, "Dispersion-Toughened Composites," in Ceram. Tech. for Adv. Heat Eng. Proj. Semiann. Prog. Rep. for April through Sept. 1988, ORNL/TM-11116, Oak Ridge National Laboratory, Oak Ridge, TN, 1989, pp. 92–97.
14. F. D. Gac, J. J. Petrovic, J. V. Milewski, and P. D.Shalek, "Performance of Commercial and Research Grade SiC Whiskers in a Borosilicate Glass Matrix," *Ceram. Eng. Sci. Proc.*, **7**[7–8], 978–982 (1986).
15. K. P. Gadjaree and K. Chyung, "Silicon Carbide Whisker Reinforced Glass and Glass–Ceramic Composites," *Am. Ceram. Soc. Bull.*, **65**[2], 370–376 (1986).
16. P. C. Panda and E. R. Seydel, "Near-Net-Shape Forming of Magnesia-Alumina Spinel/Silicon Carbide Fiber Composites," *Am. Ceram. Soc. Bull.*, **65**[2], 338–41 (1986).
17. N. Claussen and G. Petzow, "Whisker-Reinforced Zirconia-Toughened Ceramics", in *Tailoring Multiphase and Composite Ceramics*, eds. R. E.

Tressler, G. L. Messing, C. G. Pontano, and R. E. Newnham, Materials Science Research Series, Plenum Press, New York, NY, 1986, pp. 649–662.

18. N., Claussen, K. L. Weisskopf, and M. Ruhle, "Mechanical Properties of SiC-Whisker-Reinforced TZP," in *Fracture Mechanics of Ceramics, Vol. 7*, eds. R. C. Bradt, A. G. Evans, D. P. H. Hasselman, and F. F. Lange, Plenum Press, New York, NY, 1986, pp. 75–86.

19. P. F. Becher, T. N. Tiegs, J. C. Ogle, and W. H. Warwick, "Toughening of Ceramics by Whisker Reinforcement," in *Fracture Mechanics of Ceramics, Vol. 7*, eds. R. C. Bradt, A. G. Evans, D. P. H. Hasselman, and F. F. Lange, Plenum Press, New York, NY, 1986, pp. 61–73.

20. P. F. Becher and T. N. Tiegs, "Toughening Behavior Involving Multiple Mechanisms: Whisker Reinforcement and Zirconia Toughening," *J. Am. Ceram.*, **70**[9], 651–654 (1987).

21. R. Ruh, K. S. Mazdiyasni, and M. G. Mendiratta, "Mechanical and Microstructure Characterization of Mullite and Mullite-SiC-Whisker and ZrO_2-Toughened-Mullite-SiC-Whisker Composites," *J. Am. Ceram. Soc.*, **71**[6], 503–512 (1988).

22. G. C. Wei (U.S. Dept. of Energy), "Silicon Carbide Whisker Reinforced Ceramic Composites and Method for Making Same," U.S. Patent 4,543,345, September 24, 1985.

23. P. F. Becher, C. H. Hsueh, P. Angelini, and T. N. Tiegs, "Toughening Behavior in Whisker Reinforced Ceramic Matrix Composites," *J. Am. Ceram. Soc.*, **71**[12], 1050–1061 (1988).

24. P. F. Becher and T. N. Tiegs, "Temperature Dependence of Strengthening by Whisker Reinforcement: SiC Whisker-Reinforced Alumina in Air," *Adv. Ceram. Mater.*, **3**[2], 148–153 (1988).

25. T. N. Tiegs, "Tailoring of Properties of SiC Whisker-Oxide Matrix Composites," in *Proc. 3rd Internatl. Symp. Ceram. Mater. & Components for Engines*, American Ceramic Society, Westerville, OH, 1989, pp. 937–949.

26. T. N. Tiegs and P. F. Becher, "Whisker Reinforced Ceramic Composites," in *Tailoring Multiphase and Composite Ceramics*, eds. R. E. Tressler, G. L. Messing, C. G. Pontano, and R. E. Newnham, Materials Science Research Series, Plenum Press, New York, NY, 1986, pp. 639–647.

27. P. F. Becher and T. N. Tiegs, "Elevated-Temperature-Delayed Failure of Alumina Reinforced with 20 vol% Silicon Carbide Whiskers," *J. Am. Ceram. Soc.*, **73**[1], 91–96 (1990).

28. K. W. White and L. Guazzone, "Elevated-Temperature Toughening Mechanisms in a SiC_w/Al_2O_3 Composite," *J. Am. Ceram. Soc.*, **74**[9], 2280–2285 (1991).

29. M. G. Jenkins, A. S. Kobayashi, K. W. White, and R. C. Brandt, "Elevated Temperature Fracture Resistance of a SiC Whisker Reinforced/ Polycrystalline Al_2O_3 Matrix Composite," *Eng. Fract. Mech.*, **30**[4], 505–515 (1988).

30. M. C. Shaw and K. T. Faber, "Temperature-Dependent Toughening in Whisker-Reinforced Ceramics," in *Ceramic Microstructures '86: Role of Interfaces*, Vol. 21, Materials Science Research Series, Plenum Press, New York, NY, 1987, pp. 929–938.

31. S. Inoue, K. Niihara, T. Uchiyama, and T. Hirai, "Al_2O_3/SiC (Whisker)/ ZrO_2 Ceramic Composites," in *Proc. 2nd Internat. Symp. on Ceramic Materials and Components for Engines*, eds. W. Bunk and H. Hausner, Verlag Deutsche Keramische Gesellschaft, Bad Honnef, Germany, 1986, pp. 609–617.

32. T. N. Tiegs and P. F. Becher, "Development of Alumina- and Mullite-SiC

Whisker Composites: High Temperature Properties," in *Proc. 24th Auto. Tech. Dev. Contractors' Coord. Meeting*, Vol. P-197, Society of Automotive Engineers, Warrendale, PA, 1987, pp. 279–283.

33. T. N. Tiegs and P. F. Becher, "Thermal Shock Behavior of an Alumina-SiC Whisker Composite", *Am. Ceram. Soc. Comm.*, **70**[5], C-109-C-111 (1987).
34. T. N. Tiegs, P. F. Becher, and P. Angelini, "Microstructures and Properties of SiC Whisker-Reinforced Mullite Composites," in *Mullite and Mullite Matrix Composites, Ceramic Transactions, Vol. 6*, eds. S. Somiya, R. F. Davis, and J. A. Pask, American Ceramic Society, Westerville, OH, 1990, pp. 463–472.
35. R. D. Nixon, S. Chevacharoenkul, R. F. Davis, and T. N. Tiegs, "Creep of Hot-Pressed SiC Whisker-Reinforced Mullite", in *Mullite and Mullite Matrix Composites, Ceramic Transactions, Vol. 6*, eds. S. Somiya, R. F. Davis, and J. A. Pask, American Ceramic Society, Westerville, OH, 1990, pp. 579–603.
36. T. Mah and K. S. Mazdiyasni, "Mechanical Properties of Mullite", *J. Am. Ceram. Soc.*, **66**[1], 699–703 (1983).
37. S. T. Buljan, *et al.*, "Ceramic Matrix Composites," in Ceram. Tech. for Adv. Heat Eng. Proj. Semiann. Prog. Rep. for Oct. 1988–Mar. 1989, ORNL/TM-11239, Oak Ridge National Laboratory, Oak Ridge, TN, 1989, pp. 55–67.
38. N. D. Corbin, *et al.*, "Material Development in the Si_3N_4/SiC(w) System Using Encapsulated HIPing," in Ceram. Tech. for Adv. Heat Eng. Proj. Semiann. Prog. Rep. for Oct. 1988–Mar. 1989, ORNL/TM-11239, Oak Ridge National Laboratory, Oak Ridge, TN, pp. 68–89.
39. S. T. Buljan, *et al.*, "Silicon Nitride-Metal Carbide Composites," in Ceram. Tech. for Adv. Heat Eng. Proj. Semiann. Prog. Rep. for Oct. 1989–Mar. 1990, ORNL/TM-11586, 1990, pp. 132–149.
40. H. Yeh, *et al.*, "SiC Whisker Toughened Silicon Nitride," in Ceram. Tech. for Adv. Heat Eng. Proj. Semiann, Prog. Rep. for Oct. 1990–Mar. 1991, ORNL/TM-11859, Oak Ridge National Laboratory, Oak Ridge, TN, 1991, pp. 160–170.
41. G. Pezzotti, *et al.*, "Processing and Mechanical Properties of Dense Si_3N_4-SiC-Whisker Composites without Sintering Aids," *J. Am. Ceram. Soc.*, **72**[8], 1461–1464 (1989).
42. G. Pezzotti, I. Tanaka, and T. Okamoto, "Si_3N_4/SiC-Whisker Composites without Sintering Aids: III, High-Temperature Behavior," *J. Am. Ceram. Soc.*, **74**[2], 326–332 (1991).
43. S. R. Choi, *et al.*, "Dynamic FatigueProperty of Silicon Carbide Whisker-Reinforced Silicon Nitride," *Ceram. Eng. Sci. Proc.*, **12**[7–8], 1524–1536 (1991).
44. B. J. Hockey, S. M. Wiederhorn, W. Liu, J. G. Baldoni,and S. T. Buljan, "Tensile Creep of SiC Whisker-Reinforced Silicon Nitride," in *Proc. 27th Auto. Tech. Dev. Contractors' Coord. Meeting*, Vol. P-230, Society of Automotive Engineers, Warrendale, PA, 1990, pp. 251–257.
45. K. H. Smith, "Ceramic Composite Offers Speed, Feed Gains," *Mach. Tool Blue Book*, **81**[1], 71–72 (1986).
46. S. F. Wayne and S. T. Buljan, "The Role of Thermal Shock on Tool Life of Selected Ceramic Cutting Tool Materials," *J. Am. Ceram. Soc.*, **72**[5], 754–760 (1989).
47. C. R. Blanchard and R. A. Page, "Effect of Silicon Carbide Whisker and Titanium Carbide Particulate Additions on the Friction and Wear Behavior of Silicon Nitride," *J. Am. Ceram. Soc.*, **73**[11], 3442–3452 (1990).
48. T. N. Tiegs, L. F. Allard, P. F. Becher, and M. K. Ferber, "Identification and Development of Optimum Silicon Carbide Whiskers For Silicon Nitride

Matrix Composites," in *Proc. 27th Auto. Tech. Dev. Contractors' Coord. Meeting*, Vol. P-230, Society of Automotive Engineers, Inc., Warrendale, PA, 1990, pp. 167–172.

49. D. A. Jablonski and R. B. Bhatt, "High-temperature Tensile Properties of Fiber-Reinforced Reaction Bonded Silicon Nitride," *J. Comp. Tech. & Res.*, **12**[3], 139–146 (1990).

50. C. V. Burkland and J.-M. Yang, "Chemical Vapor Infiltration of Fiber-Reinforced SiC Matrix Composites", *SAMPE Journal*, **25**[5], 29–33 (1989).

51. Textron Specialty Materials Product Data, Lowell, MA,1990.

52. Du Pont Ceramic Matrix Composites Engineering Data, Wilmington, DE, 1988.

53. D. B. Marshall, B. N. Cox, and A. G. Evans, "The Mechanics of Matrix Cracking in Brittle-Matrix Fiber Composites," *Acta Metall.*, **33**, 2013–2021 (1985).

54. M. He and J. W. Hutchinson, "Crack Deflection at an Interface Between Dissimilar Elastic Materials," *Int. J. Solids Struct.*, **25**, 1053 (1989).

55. H. C. Cao, E. Bischoff, O. Sbaizero, M. Ruhle, A. G. Evans, D. B. Marshall, and J. J. Brennan, "Effects of Interfaces on the Properties of Fiber-Reinforced Ceramics," *J. Am. Ceram. Soc.*, **73**, 1691 (1990).

56. K. M. Prewo, "Silicon Carbide Fiber-Reinforced Glass-Ceramic Composite Tensile Behavior at Elevated Temperature," *J. Mat. Sci.*, **24**, 1373 (1989).

57. K. M. Prewo, "Tensile and Flexural Strength of SiC Fiber-Reinforced Glass-Ceramic Composite", *J. Mat. Sci.*, **21**, 3590 (1986).

58. R. N. Singh, "High-Temperature Mechanical Properties of a Uniaxially Reinforced Zicon-Silicon Carbide Composite," *J. Am. Ceram. Soc.*, **73**[8], 2399–2406 (1990).

59. R. T. Bhatt, "The Properties of Silicon Carbide Fiber-Reinforced Silicon Nitride Composites", in *Whisker- and Fiber-Toughened Ceramics*," eds. R. A. Bradley, D. E. Clark, D. C. Larsen, and J. O. Stiegler, ASM, Materials Park, PA, 1988, p. 199.

60. M. Gomina, P. Fourvel, and M.-H. Rouillon, "High Temperature Mechanical Behavior of an Uncoated SiC–SiC Composite Material", *J. Mat. Sci.*, **26**, 1891 (1991).

61. R. E. Tressler and D. J. Pysher, "Mechanical Behavior of High Strength Ceramic Fibers at High Temperatures," in *Advanced Structural Inorganic Composites*, ed. P. Vincenzini, Elsevier Science Publishers, Amsterdam, 1991, pp. 3–18.

62. D. J. Pysher, K. C. Goretta, R. S. Hodder, and R. E. Tressler, "Strengths of Ceramic Fibers at Elevated Temperatures," *J. Am. Ceram. Soc.*, **72**[2], 284 (1989).

63. T. I. Mah, M. G. Mendiratta, A. P. Katz, and K. S. Mazdiyasni, "Recent Development in Fiber–Reinforced High Temperature Composite Materials," *Ceram. Bull.*, **66**[2], 304 (1987).

64. A. R. Bunsell, "Ceramic Fibers for Reinforcement," in *Ceramic Matrix Composites*, ed. R. Warren, Chapman & Hall, New York, 1991, pp. 12–34.

65. K. Okamura, "Ceramic Fibers From Polymer Precursors," *Composites*, **18**[2] 107 (1987).

66. F. Hurwitz, "Ceramic Fiber-Reinforced Ceramic Matrix Composites," in *International Encyclopedia of Composites*, Vol. 1, VCH Publishers, New York, NY, 1990, p. 297.

67. J. A. DiCarlo, "Creep of CVD SiC Fibers," *J.Mat. Sci.*, **21**, 217 (1986).

68. S. Yajima, K. Okamura, J. Hayashi, and M. Omori, *J. Am. Ceram. Soc.*, **59**, 324 (1976).

69. T. Yamamura, T. Harashima, M. Shibuya, and Y. Iwai, "Development of Continuous Si-Ti-C-O Fiber With High Mechanical Strength and Heat Resistance," 6th World Congress on High Tech. Ceramics (CIMTEC), Milan, Italy, 1986.
70. G. E. Legrow, T. F. Lim, J. Lipowitz, and R. S. Reaoch, "Ceramic Fibers From Hydridopolysilazane," *Ceram. Bull.*, **66**[2], 363 (1987).
71. Sapphikon Engineering Data, Milford, NH, 1992.
72. E. Y. Luh and A. G. Evans, "High-Temperature Mechanical Properties of a Ceramic Matrix Composite," *J. Am. Ceram. Soc.*, **70**[7], 466–469 (1987).
73. G. Morscher, P. Pirouz, and A. H. Heuer, "Temperature Dependence of Interfacial Shear Strength in SiC-Fiber-Reinforced Reaction-Bonded Silicon Nitride," *J. Am. Ceram. Soc.*, **73**[3], 713–720 (1990).
74. S. V. Nair, T.-J. Gwo, N. M. Narbut, J. G. Kohl, and G. J. Sundberg, "Mechanical Behavior of a Continuous-SiC-Fiber-Reinforced RBSN-Matrix Composite," *J. Am. Ceram. Soc.*, **74**[10], 2551–2558 (1991).
75. F. Abbe and J.-L. Chermant, "Fiber-Matrix Bond Strength Characterization of Silicon Carbide-Silicon Carbide Materials," *J. Am. Ceram. Soc.*, **73**[8], 2573–2575 (1990).
76. D. B. Marshall and A. G. Evans, "The Influence of Residual Stress on the Fracture Toughness of Composites," *Mat. Forum*, **11**, 304 (1988).
77. M. Bouquet, J. M. Birbis, and J. M. Quenisset, "Toughness Assessment of Ceramic Matrix Composites", *Comp. Sci. Tech.*, **37**, 223 (1990).
78. P. J. Lamicq, G. A. Bernhart, M. M. Dauchier, and J. G. Mace, "SiC/SiC Composite Ceramics," *Ceram. Bull.*, **65**[2], 336 (1986).
79. L. Heraud and P. Spriet, "High Toughness S-SiC and SiC-SiC Composites in Heat Engines," in *Whisker- and Fiber-Toughened Ceramics*, eds. R. A. Bradley, D. E. Clark, D. C. Larsen, and J. O. Stiegler, ASM, Materials Park, PA, 1988, p. 217.
80. A. G. Evans, "The Mechanical Properties of Reinforced Ceramic, Metal and Intermetallic Matrix Composites," *Mat. Sci. Eng.*, **A143**, 63–76 (1991).
81. C. A. Anderson, P. Barron-Antolin, A. S. Fareed, and G. H. Schiroky, "Properties of Fiber-Reinforced Lanxide Alumina Matrix Composites," in *Whisker- and Fiber-Toughened Ceramics*, eds. R. A. Bradley, D. E. Clark, D. C. Larsen, and J. O. Stiegler, ASM, Materials Park, PA, 1988, p. 209.
82. S. V. Nair and Y.-L. Wang, "Failure Behavior of a 2-D Woven SiC Fiber/SiC Matrix Composite at Ambient and Elevated Temperatures," *Ceram. Eng. Sci. Proc.*, **13**[7–8], 433 (1992).
83. J. A. DiCarlo, "Fibers for Structurally Reliable Metal and Ceramic Composites," *J. Metals*, **37**[6], 44 (1985).
84. K. Nihara, "New Design Concept of Structural Ceramics—Ceramic Nanocomposites," *The Centen. Memor. Iss. of the Ceram. Soc. Japan*, **99**[10], 974 (1991).

CHAPTER **3**

Dynamic and Impact Fractures of Ceramic Composites at Elevated Temperature

A. S. Kobayashi

3.1 Introduction

Although A. A. Griffith[1,2] developed the crack stability criterion for brittle material, i.e., glass, based on the balance of the released and dissipated energy rates in the 1920s, fracture mechanics, as is known today, did not come into being until the late 1940s when G. R. Irwin[3] and E. Orowan[4] utilized Griffith's criterion to describe fracture of metals. After a flurry of research activities for the past three decades, fracture mechanics has now evolved into a matured discipline. It is not only used routinely in post-mortem studies of failed structural components but has also been incorporated into structural codes for failure prevention or for fail-safe/safe-fail designs. These developments are amply documented in numerous textbooks and reference books and in at least two international journals which are exclusively devoted to fracture mechanics.

The above refers primarily to linear elastic fracture mechanics (LEFM) studies related to the onset of brittle fracture as well as fatigue. The states of the sciences of ductile and dynamic fracture, which only became of serious concern in the early 1960s and 1970s, respectively, have not reached a similar level of maturity despite the immense research efforts expended on these topics in recent years. Yet to be resolved in the former is a viable ductile fracture criterion in view of the recently uncovered uncertainties regarding the J-integral as a crack tip parameter.[5,6] As for the latter, a reliable dynamic crack propagation criterion is yet to be established, as will become apparent in subsequent sections of this chapter.

In spite of its developing state, dynamic fracture must be dealt with in designing ceramic structural components, such as heat engine components, ceramic armor and ceramic cutting tools. Shattering, which is characteristic of a ceramic component under dynamic loading, involves dynamic fracture initiation and dynamic crack propagation. In particular, the absence of dynamic crack arrest, which is commonly observed in dynamic fracture of metals, is the underlying cause of shattering in many ceramics. Dynamic fracture mechanics encompasses these three phenomena of crack initiation under dynamic loading, rapid crack propagation, and arrest of a rapidly propagating crack. While early papers on dynamic fracture mechanics date back to the 1950s,[7-10] serious studies on dynamic fracture mechanics started in the 1970s with the need to predict the extent of rapid crack propagation in a nuclear power pressure vessel subjected to emergency core cooling, as well as to determine the effectiveness of crack arresters in a large marine structure. As a result of such concerted efforts, much is known on the dynamic responses of rapidly propagating cracks in relatively ductile metals and polymers. Unfortunately, the same cannot be said about ceramics and ceramic composites due to the community's preoccupation in static fracture initiation and in its seemingly endless and costly attempts to elevate the extremely low fracture toughness of ceramics. Needless to say, the study of dynamic fracture of ceramics and ceramic composites at elevated temperature is non-existent, except for the limited efforts by the author and his colleagues. With no breakthrough in sight, an alternative to the above is to design safe-fail ceramic components based on an inevitable fracture, or to promote the use of ceramic components as one-time energy absorbers, in which case fracture should be enhanced. The basic mechanics in either of these two applications is dynamic fracture mechanics, of which little is known by those involved in ceramic mechanics research today.

In the following sections, a cursory review of dynamic fracture mechanics will be given. Procedures for impact and dynamic fracture characterization of ceramics and ceramic composites will then be described, and the properties peculiar to ceramics and ceramic composites will be discussed.

3.2 Historical Review

The early papers on dynamic fracture mechanics were simple extensions of Griffith's instability criterion for predicting the onset of crack propagation. Mott,[7] Roberts and Wells[9] and Berry[10] added varying forms of an estimated kinetic energy rate term to Griffith's balance of energy rate equation to account for the global kinetics associated with a moving crack. Unfortunately, these approaches did not consider the actual crack tip state of stress which governs both the static and the dynamic fracture processes.

In contrast, the moving Griffith's crack, which was derived by Yoffe[8] during this early period, did provide a crack velocity-independent stress intensity factor in a crack velocity-dependent crack tip stress field. Using her solution, Yoffe predicted that the crack will kink to an angle of about 63° when the crack velocity reaches about 60% of the shear wave velocity, thus leading to a crack branching criterion which is governed by a critical crack velocity. The crack branching criterion is particularly appropriate since shattering of ceramics and ceramic composites consists of many multiples of crack branching. While Yoffe's solution was a historical first, the anomaly of her modeling resulted in an infinite energy release rate as the crack velocity approached the Rayleigh wave velocity. Subsequent solutions by Broberg[11] and Baker[12] for a constant velocity crack initiating from zero and finite crack lengths, respectively showed that the energy release rate approached zero as the crack velocity approached the Raleigh wave velocity. The corresponding crack tip stress fields were characterized by crack velocity-dependent stress intensity factors and stress distributions. Unlike Yoffe's paper, crack branching was not discussed in either of these two papers.

Early views on crack arrest considered the arrest phenomenon to be an inverse of the onset of crack propagation, namely that a propagating crack would arrest when the instantaneous static stress intensity factor $K_I < K_{Ic}$ where K_{Ic} is the fracture toughness.[13] Many tests and research programs were conducted to verify or discredit this postulate with, at times, raging controversies on the physical significance of a dynamic crack arrest stress intensity factor. Ample experimental,[14,15] numerical,[16] and some theoretical[17] analyses now suggest that dynamic arrest stress intensity factor is a separate material property and that static analysis is not sufficient for predicting the arrest of a propagating crack.

3.3 Fundamental Equations in Dynamic Fracture

Ceramics and the matrix in ceramic composites exhibit cleavage fracture at room temperature as well as at elevated temperatures. This is fortunate since most of the theoretical developments in dynamic fracture are confined to linear elastic fracture mechanics which is then applicable to fracture of the ceramic matrix. However, the additional complexities of crack deflection and fiber–matrix interface cracking, as well as fiber/whisker/particulate pull-outs, are at this time yet to be addressed.

Available theoretical solutions in dynamic fracture are few, and limited to finite or semi-infinite cracks in an infinite solid for Mode I, self-similar crack extension. Despite the above limitations, short of conducting detailed numerical analysis of the crack tip state of stress, these solutions must be used to deduce the characteristics of the crack tip state of stress, as well as to extract the dynamic stress intensity factor for elastodynamic fracture mechanics. In the following sections, a brief description of available theoretical solutions is presented.

3.3.1 Stationary Crack Impacted by a Tension Wave

The dynamic stress intensity factor, K_{Id}, of a stationary semi-infinite crack, which is impacted by a square plane tension wave, σ_0, of duration t, in

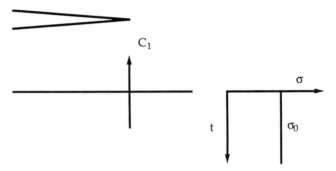

Fig. 3.1   Plane stress wave impacting a crack.

an infinite solid as shown in Fig. 3.1, was given by Freund[18] as:

$$K_I(t,0) = \frac{2\sigma_0}{(1-\nu)}\left(\frac{C_1(1-2\nu)}{\pi}\right)^{1/2} t^{1/2} \tag{1}$$

For a ramp tensile pulse loading, K_{Id} is a simple superposition of the discrete stress values as:

$$K_{Id} = \frac{2}{(1-\nu)}\left(\frac{C_1(1-2\nu)}{\pi}\right)^{1/2} \sum_{i=1}^{n} \Delta\sigma_i \Delta t_i^{1/2} \tag{2}$$

where C_1 is the dilatational stress wave velocity and ν is the Poisson's ratio.

Note that the stress intensity factors do not involve a characteristic dimension nor a characteristic time where t represents the time varying boundary condition. If the crack starts to propagate rapidly after an incubation time of t^*, then the above stress intensity factors should be modified with a scalar function of the crack velocity.

3.3.2 Crack Propagating at Constant Velocity

The crack tip state for a crack propagating at constant velocity in a two-dimensional, isotropic, homogeneous elastic material has been derived by Freund[19] and Nishioka and Atluri[20] in addition to those cited in the previous section. Probably the most general of these derivations is the asymptotic crack tip stress and displacement fields given as infinite series in Ref. 20. Obviously, the singular first order term in the infinite series with mixed mode dynamic stress intensity factors of K_I^{dyn} is the most significant crack tip term in the stress field. In addition, the second order term has been shown to govern crack kinking and branching angle.[21,22] The first and second order terms of the crack tip stress and displacement fields of a crack which is propagating at a constant velocity in an isotropic homogeneous elastic continuum are represented in terms of the coordinate system shown in Fig. 3.2 as:

$$
\sigma_x = \frac{K_I^{dyn} B_I(V)}{\sqrt{2\pi}} \left\{ (1 + 2\beta_1^2 - \beta_2^2) \frac{\cos\dfrac{\theta_1}{2}}{r_1^{1/2}} - \frac{4\beta_1\beta_2}{(1+\beta_2^2)} \frac{\cos\dfrac{\theta_2}{2}}{r_2^{1/2}} \right\} + \sigma_{ox}^{dyn}(\beta_1^2 - \beta_2^2)
$$

$$
+ \frac{K_{II}^{dyn} B_{II}(V)}{\sqrt{2\pi}} \left\{ -(1 + 2\beta_1^2 - \beta_2^2) \frac{\sin\dfrac{\theta_1}{2}}{r_1^{1/2}} + (1+\beta_2^2) \frac{\sin\dfrac{\theta_2}{2}}{r_2^{1/2}} \right\} \tag{3a}
$$

$$
\sigma_y = \frac{K_I^{dyn} B_I(V)}{\sqrt{2\pi}} \left\{ -(1 + \beta_2^2) \frac{\cos\dfrac{\theta_1}{2}}{r_1^{1/2}} + \frac{4\beta_1\beta_2}{(1+\beta_2^2)} \frac{\cos\dfrac{\theta_2}{2}}{r_2^{1/2}} \right\}
$$

$$
+ \frac{K_{II}^{dyn} B_{II}(V)}{\sqrt{2\pi}} \left\{ (1 + \beta_2^2) \frac{\sin\dfrac{\theta_1}{2}}{r_1^{1/2}} - (1+\beta_2^2) \frac{\sin\dfrac{\theta_2}{2}}{r_2^{1/2}} \right\} \tag{3b}
$$

$$
\sigma_{xy} = \frac{K_I^{dyn} B_I(V)}{\sqrt{2\pi}} \left\{ 2\beta_1 \frac{\sin\dfrac{\theta_1}{2}}{r_1^{1/2}} - 2\beta_2 \frac{\sin\dfrac{\theta_2}{2}}{r_2^{1/2}} \right\}
$$

$$
+ \frac{K_{II}^{dyn} B_{II}(V)}{\sqrt{2\pi}} \left\{ 2\beta_1 \frac{\cos\dfrac{\theta_1}{2}}{r_1^{1/2}} - \frac{1+\beta_2^2}{2\beta_2} \frac{\cos\dfrac{\theta_2}{2}}{r_1^{1/2}} \right\} \tag{3c}
$$

The corresponding plane strain displacements are:

$$
u = \frac{K_I^{dyn} B_{II}(V)}{\mu} \sqrt{\frac{2}{\pi}} \left\{ r_1^{1/2} \cos\frac{\theta_1}{2} - \frac{2\beta_1\beta_2}{1+\beta_2} r_2^{1/2} \cos\frac{\theta_2}{2} \right\}
$$

$$
- \frac{K_{II}^{dyn} B_{II}(V)}{\mu} \sqrt{\frac{2}{\pi}} \left\{ r_1^{1/2} \sin\frac{\theta_1}{2} - \frac{1+\beta_2^2}{2} r_2^{1/2} \sin\frac{\theta_2}{2} \right\}
$$

$$
+ \frac{1}{2\mu} \sigma_{ox}^{dyn} B_I(V) \left\{ r_1 \cos\theta_1 - \frac{1+\beta_2^2}{2} r_2 \cos\theta_2 \right\} \tag{4a}
$$

$$v = \frac{K_I^{dyn} B_I(V)}{\mu} \sqrt{\frac{2}{\pi}} \left\{ -\beta_1 r_1^{1/2} \sin\frac{\theta_1}{2} + \frac{2\beta_1}{1+\beta_2^2} r_2^{1/2} \sin\frac{\theta_2}{2} \right\}$$

$$+ \frac{K_{II}^{dyn} B_{II}(V)}{\mu} \sqrt{\frac{2}{\pi}} \left\{ -\beta_1 r_1^{1/2} \cos\frac{\theta_1}{2} - \frac{1+\beta_2^2}{2\beta_2} r_2^{1/2} \cos\frac{\theta_2}{2} \right\}$$

$$+ \frac{1}{2\mu} \sigma_{ox}^{dyn} \cdot B_I(V) \left\{ -\beta_1 r_1 \sin\theta_1 + \frac{1+\beta_2^2}{2\beta_2} r_2 \sin\theta_2 \right\} \tag{4b}$$

where

$$\beta_1^2 = 1 - \frac{V^2}{C_1^2} \qquad\qquad \beta_2^2 = 1 - \frac{V^2}{C_2^2}$$

$$r_j e^{i\theta_j} = x + i\beta_j y$$

$$B_I(V) = \frac{1+\beta_2^2}{4\beta_1\beta_2 - (1+\beta_2^2)^2} \qquad B_{II}(V) = \frac{2\beta_2}{4\beta_1\beta_2 - (1+\beta_2^2)^2} \tag{5}$$

where x and y are the orthogonal coordinates, and r and θ are the polar coordinates both with their origin at the moving crack tip, as shown in Fig. 3.2. C_1 and C_2 are the dilatational and distortional stress wave velocities, respectively, V is the crack velocity, and μ is the shear modulus.

Note that for the asymptotic equations of Eqns. (2) and (3) to be valid, $r < d/10$ where d is the governing characteristic length, and is normally the crack length or the remaining ligament, whichever is the smaller, of a fracture specimen. Also, the above asymptotic equations are not valid for an orthotropic elastic continuum, such as a ceramic fiber/ceramic matrix composite. While the *static* crack tip state for an orthotropic elastic continuum has been derived, to the author's knowledge, no dynamic counterpart is available to date. Nevertheless, the above crack tip state should be applicable to particulate/whisker-filled ceramic matrix composites which macroscopically behave like an isotropic homogeneous continuum.

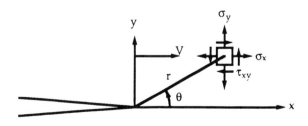

Fig. 3.2 Constant velocity crack.

3.3.3 Dynamic Crack Curving and Branching

The physical significance of the remote stress component, σ_{ox}^{dyn}, in linear elastodynamic fracture mechanics is its influence on crack curving and crack branching. The mechanics of elastic crack curving, as well as crack branching under Mode I crack tip state, was studied by the author and his colleagues.[21,22] The dynamic crack curving criterion postulates that the state of stress ahead of a crack tip dictates the direction of crack propagation. The crack curving criterion thus assumes that when the circumferential stress within a prescribed crack tip region attains a maximum value off the axis of a self-similar crack extension, crack curving will occur. When this maximum circumferential stress exists at a characteristic crack tip distance, r_0, then the propagating crack will kink at a given angular orientation, θ_0, from its axis. For a crack propagating at a constant velocity, this value is

$$r_0 = \frac{9}{128\pi}\left(\frac{K_1}{\sigma_{ox}}V_0(\nu, C_1, C_2)\right)^2 \tag{6a}$$

where

$$F_o(V, C_1, C_2) = B_1(V)\left\{-(1+\beta_2^2)(2-3\beta_1^2) - \frac{4\beta_1\beta_2}{1+\beta_2^2}(14+3\beta_2^2)\right.$$

$$\left. - 16\beta_1(\beta_1 - \beta_2) + 16(1+\beta_1^2)\right\} \tag{6b}$$

The above elastic crack curving criterion requires that $r_0 < r_c$ for the crack to curve away from its axis, where r_c is a material constant which specifies the characteristic crack tip region in which the off-axis microcracks enlarge and connect to the main crack tip, as shown schematically in Fig. 3.3. The angular deviation of the crack from its original direction of self-similar crack extension is given in Ref. 21.

In the presence of a large driving force, i.e., a large K_I^{dyn}, the crack will bifurcate in order to shed the excess driving force, thus resulting in crack branching where the crack branching angle is governed by the crack curving criterion. This is a necessary condition for crack branching which requires, as a sufficiency condition, the above-mentioned crack curving criterion. This crack branching criterion was used successfully to correlate the predicted and measured crack branching angle and the estimated crack branching stress intensity factor, K_{Ib}.[22]

Again, the above crack kinking and branching criteria are limited to isotropic homogeneous material, which for all practical purposes will include particulate/whisker-filled ceramic matrix composites. No equivalent criterion exists for orthotropic/inhomogeneous material. Limited experimental results show that self-similar crack extension is a rare phenomenon in fracture of fiber-reinforced ceramic matrix composites and thus the kinking and branching criterion, if developed, must necessarily be a three-dimensional one.

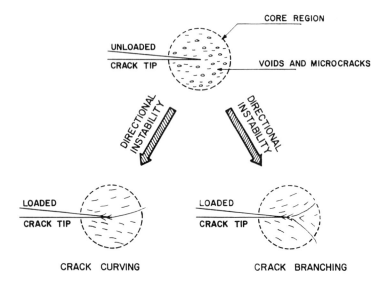

Fig. 3.3 Crack curving and branching mechanism.

3.4 Impact Fracture of Ceramics and Ceramic Composites

The recent proliferation of papers on impact studies of ceramics at room temperature is motivated by armament research.[23-25] Papers on impact studies of ceramics, which are destined for use in heat engines, are also limited to room temperature testing[26] and are an indication of the paucity of suitable elevated temperature testing procedures. These impact studies can be classified into three testing procedures, namely the plate and the bar impact testing, the split Hopkinson bar testing, and the drop-weight testing procedures. These three procedures are described briefly in the following sections.

3.4.1 Plate Impact Testing

The plate impact test provides high stress and high strain rate loading of a plate specimen under well-characterized loading conditions.[27,28] As shown in Fig. 3.4, a typical testing system consists of a flyer plate which impacts the specimen plate at high velocity and generates a one-dimensional plane wave in the center of the specimen. The star-shaped flyer plate, shown in Fig. 3.4, was developed to reduce the spurious influences of the diffracted stress waves from its edges.[28] The momentum trap behind the specimen provides a soft recovery of the specimen. Upon impact, the transmitted compression wave reflects as a tension wave at the gap between the specimen and the momentum trap as shown by the Lagrangian t–x diagram in Fig. 3.5. The critical section, which is

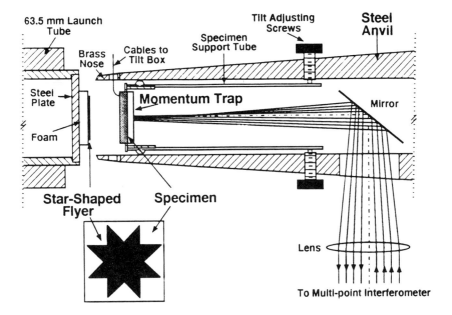

Fig. 3.4 Plate impact tester.[29]

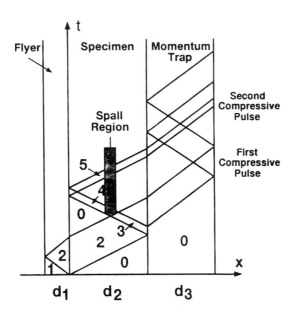

Fig. 3.5 Lagrangian *t–x* diagram for soft recovery in a plate impact tester.[29]

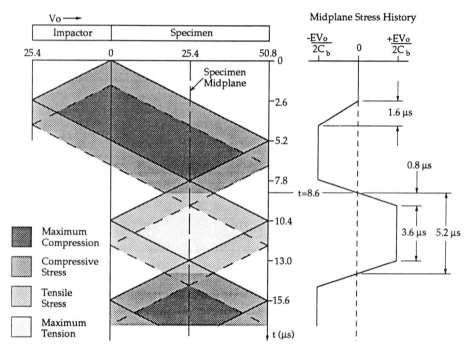

Fig. 3.6 Lagrangian diagram for bar impact experiment.

identified as the spall region in Fig. 3.5, is parallel to the plate surface. The velocity–time history, which is measured by an interferometer along the back-side of the momentum trap, enables the evaluation of inelastic responses in compression and tension. By impacting the flyer plate and specimen plate at a skewed angle, the resultant longitudinal and transverse waves together generate different stress histories at different sections of the specimen and require detailed analysis using the interferometry data.[29]

3.4.2 Bar Impact Testing

The bar impact test, which is a variation of the plate impact test, produces a one-dimensional compressive square stress pulse which neglects the effects of lateral inertia caused by Poisson's effect. This testing procedure was developed in the laboratory of the author and his colleague,[30] and is described here in some detail.

The impactor is made of the same material and its length is half of the specimen length, as shown in Fig. 3.6, and it produces a stress pulse with a length equal to the specimen length. The incoming compressive wave, which reflects off the free end as a tensile wave, is canceled by the oncoming compressive wave. The result is a shrinking compressive wave centered in the middle of the specimen. After the incoming and reflected stress waves cancel

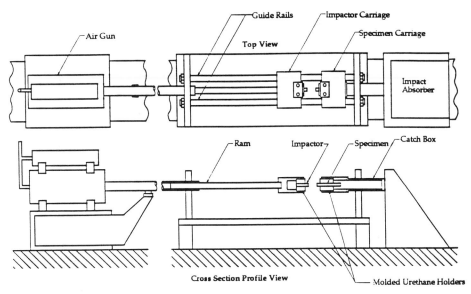

Fig. 3.7 Impact apparatus and air gun.

each other, a tensile component suddenly appears at the center of the specimen. This square tensile wave expands outward to both ends until the entire specimen is in a state of tension. The maximum stress amplitude of the pulse is given as[31]

$$\sigma = \frac{EV_0}{2C_b} \quad \text{and} \quad C_b^2 = \frac{E}{\rho} \tag{7}$$

where C_b is the bar wave velocity, E is the elastic modulus, ρ is the density, and V_0 is the impactor velocity.

Experience with metals shows that spall damage can occur at the location of maximum stress and is a function of the tensile stress amplitude and pulse duration.[32] Based on the Lagrangian diagram, the region of maximum damage due to the stress pulse is located in the middle region of the specimen. For example, plate impact experiments on MgO crystals have generated micro-cracking near the midplane[33] and plate impact experiments on Cu-SiO$_2$ crystals have produced microvoid formation near the midplane.[34]

The impact apparatus, shown in Fig. 3.7, consists of an air gun, pneumatic controls, and the impact system. The impact carriage, which contains the impactor, is propelled down the guide rails towards the specimen carriage. A critical requirement of a bar impact experiment is that the two impacting faces must meet with perfect flatness. Thus, the two mating surfaces must be flat and perpendicular to the adjacent sides. Urethane molds, which hold the specimen and impactor, were molded directly in the carriages with a single ground-steel bar as an alignment guide for this purpose.

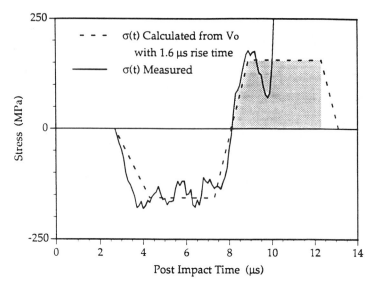

Fig. 3.8   Typical measured stress and ideal stress.

The impact velocity, V_0, was measured by a low-power laser light which strikes a mirror mounted on the impactor carriage. The mirror, with six black lines spaced 10 mm apart, reflects the laser beam into a photodiode which records the passage of the black lines and hence the translation of the impact carriage. The impact velocity, which is a function of the barrel pressure, P; the barrel cross-sectional area, A; the impactor carriage travel, d; and its mass, m, was thus computed. A calibration curve was then established by a straight-line fit of the plot of V_0^2 versus P.

A strain gauge of 1.6 mm gauge length was mounted 21.6 mm from the impact face in order to monitor the transient strain wave in the specimen. A smaller gauge could not be used since it did not dissipate enough heat due to the poor heat conductivity of the ceramic specimens.

The longitudinal stress, which was computed from the measured strain, was correlated with the Lagrangian diagram at an impact velocity of 12.2 m/s, as shown in Fig. 3.8. The wave form is basically trapezoidal as predicted by the elementary bar theory, with higher frequency oscillations which are attributed to lateral inertia. The waveforms compare closely to experimental results for long rod impact obtained by Miklowitz.[35] Figure 3.9 shows the predicted and experimentally determined stresses plotted against the impact velocity. The predicted stress was calculated using Eq. (7), the measured impact velocity, and the wave speed. The measured stress was computed from the measured strain.

Alumina (Coors AD-85) were impacted at velocities up to 14.0 m/s. Those specimens impacted above 12.2 m/s failed by complete spall during the second cycle of tension, and possibly due to cumulative damage. Specimens

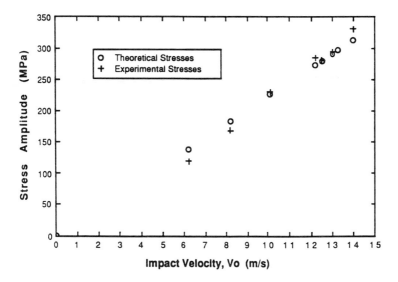

Fig. 3.9 Measured and predicted stress levels.

which did not fail by complete spall were tested for fracture toughness, K_{Ic}, by the procedure presented in Refs. 36–38. The fracture toughness is plotted as a function of impact velocity in Fig. 3.10. The K_{Ic} was found to be independent of impact velocity and close to those values given by the manufacturer.

The fracture faces of the spalled specimens and quasi-static fractured specimens were examined under the SEM. The spall-failed specimens exhibited extensive crack branching along the fracture surface while the quasi-statically fractured specimens showed none. The fracture surfaces were qualitatively examined for intergranular and transgranular fractures. In both cases, only a small percentage (~5–10%) of the grains failed by transgranular fracture and the remaining surface was intergranular failure. The spall fracture surface exhibited substantial microcracking at the grain boundaries.

3.4.3 Split Hopkinson Bar Tester

The split Hopkinson bar tester, which is used extensively for impact testing of metals, is also used to test ceramics and ceramic composites in compression[39] and in tension.[40,41] Figure 3.11 shows a schematic of the traditional split Hopkinson bar tester, which was used together with a furnace to study compression failure of ceramic composites in air at temperatures ranging from 23°C to 1100°C.[39] The stress waves in the input and output bars were monitored with strain gauges and provided information on the stress wave, which traversed back and forth through the specimen, as well as on the inelastic energy absorbed by the specimen. This system was used to test cylindrical specimens fabricated from pyroceramic matrix reinforced with about

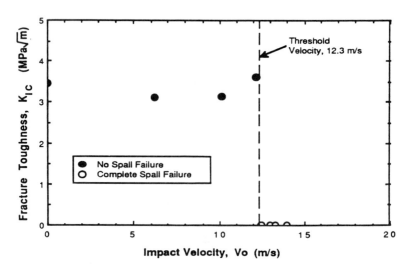

Fig. 3.10 Fracture toughness as a function of impact velocity.

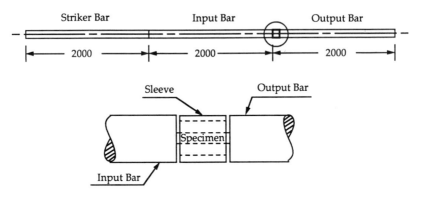

Fig. 3.11 Split Hopkinson bar tester.[39]

46% vol. of SiC fibers laid up in 0° (unidirectional) and 0/90°. Figure 3.12 shows that while the compressive strength varied with temperature for this material, it remained essentially constant up to a strain rate of approximately 10^2 1/s. In contrast, the compressive strength of the matrix, i.e., pyroceram, increased with strain rate at 1100°C as shown in Fig. 3.13.

An ingenious split Hopkinson bar tester, which imparts tension directly to the specimen without the precompression wave, was developed by Costin and Duffy.[40] The compressive wave developed by the explosive charge is

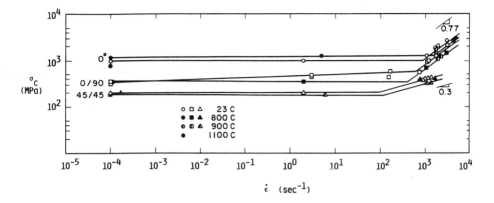

Fig. 3.12 Compressive strength versus strain rate for SiC fiber-reinforced pyroceram.[39]

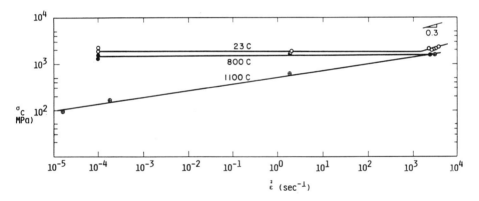

Fig. 3.13 Compressive strength versus strain rate for pyroceram.[39]

shaped into a tensile pulse through multiple reflections at the detonation end and then propagates down the steel bar. Duffy, Suresh and their colleagues[41,42] determined the dynamic fracture initiation fracture toughness, K_{Id}, of pre-cracked Al_2O_3, Si_3N_4, SiC, and SiC_w/Al_2O_3 bar specimens using this split Hopkinson bar tester, as shown in Fig. 3.14. Tables 3.1 and 3.2 show K_{Id} values at room temperature and the K_{Id} for Al_2O_3 at elevated temperature. By adding a pretorque to the bar specimen, this test setup was also used to measure the fracture toughness under combined Modes I and III fracture at room temperature.[43]

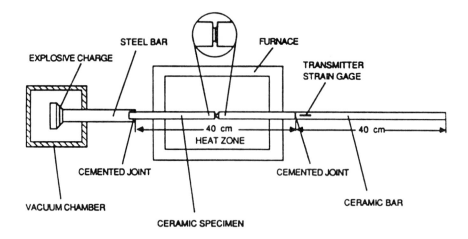

Fig. 3.14 Schematic diagram of the experimental setup for the elevated-temperature dynamic fracture test.[42]

Table 3.1 Room temperature dynamic and quasi-static fracture initiation toughness values of the ceramic materials tested in the present study

Material	K_{Id} ($MPa\,m^{1/2}$)	K_{Ic} ($MPa\,m^{1/2}$)	K_{Id}/K_{Ic}
Al_2O_3	3.5	2.7	1.3
Si_3N_4	3.8	2.8	1.4
SiC	3.3	3.1^b	1.1
Al_2O_3-SiC whisker	8.0	5.9	1.4
$Al_2O_3{}^{a,17}$	5.7	4.3	1.3
Al_2O_3-SiC whisker^{a,17}	7.9	6.6	1.2

^{a}Based on drop-weight impact testing.
^{b}Reported by the supplier.

Table 3.2 Experimentally measured dynamic and quasi-static fracture initiation toughness values of Al_2O_3 as a function of temperature

Temperature (°C)	K_{Id} ($MPA\,m^{1/2}$)	K_{Ic} ($MPA\,m^{1/2}$)	K_{Id}/K_{Ic}
20	3.5	2.7	1.3
900	3.4	2.2	1.5
1100	3.1	2.2	1.4
1300	2.0	1.4	1.4

3.5 Dynamic Fracture of Ceramics and Ceramic Composites

As mentioned previously, impact failures of ceramics and ceramic matrix composites (CMCs) is a complex phenomenon involving a multitude of simultaneous microcrack generation, growth and coalescence into macrocracks which, in turn, grow, branch and coalesce. Intact fibers in a CMC will not always arrest a propagating crack in the brittle ceramic matrix where the propagating crack is known to tunnel around the fibers with little resistance. Macroscopically, a CMC should have a significantly larger dynamic crack initiation toughness, K_{Id}, and a dynamic crack arrest stress intensity factor, K_{Ia}, in order to arrest a running crack. Once the laws governing a propagating single crack are known, a statistical or a fractal analysis of the many branched cracks and the laws governing fiber fracture/pull-out can be used to predict the overall dynamic response of an impacted CMC. The ongoing impact studies,[23–25] unfortunately, will not provide such understanding of these mechanisms. Dynamic fracture mechanics studies of ceramics and CMCs are virtually non-existent except for one paper,[44] in addition to those of the author and his colleagues[45–50] A brief review of the limited results obtained to date is given here.

Dynamic fracture testing of ceramics and CMCs differs from that of structural metals and polymers in that the available test specimens are often as small as $6 \times 6 \times 40$ mm bars, dictated by the high fabrication cost and the availability of blank material. In terms of experimental techniques, the popular optical methods for determining the dynamic stress intensity factor at room temperature, such as photoelasticity[51] and caustics,[52] cannot be used due to the opaqueness and the low Poisson's ratio, respectively, of ceramics. As for crack length measurement, the traditional photographic and crack gauge techniques for monitoring rapid crack propagation are unusable at elevated temperature and, at best, inaccurate due to the small crack opening displacement in these brittle materials. In addition, under dynamic loading at elevated temperature, the impact load, which must be measured outside of the furnace, could be vastly different from the applied load at the impact point of the heated specimen. These physical constraints require the development of a new dynamic fracture testing procedure, which up to now was not available. The hybrid experimental-numerical procedure which circumvented some of the above difficulties, was thus developed[50] to characterize the dynamic fracture responses of ceramics and CMCs at room and elevated temperatures. Details of this procedure have been reported elsewhere[48,49] and thus only a brief description is given here.

The loading system, as shown in Fig. 3.15, consists of a drop-weight tower, which is mounted integrally with the furnace, with a maximum operating temperature of 1500°C. The impact load was measured by a load transducer at the top end of the push rod and outside of the furnace. An additional load transducer was placed between the bottom end of the push rod

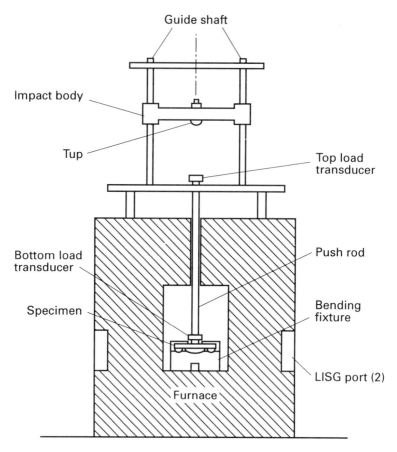

Fig. 3.15 Schematic diagram of drop-weight impact and furnace with LIDG ports.

and the fracture specimen for room temperature testing. The crack tip loading rate was of the order of $10^5\ \mathrm{MPa}\sqrt{m/s}$ which is comparable with other impact tests of ceramics.[45]

The specimens used in this study were single-edge-notched, three-point bend, alumina specimens of $9.1 \times 6.4 \times 76.2\ \mathrm{mm}$, TiB_{2p}/SiC and SiC_w/Al_2O_3 ceramic composite specimens of $6.4 \times 6.4 \times 76.2\ \mathrm{mm}$ in size. Two platinum tabs were positioned adjacent to the crack tip and indented by a Vicker's diamond indenter to generate optical interference fringes when illuminated with a coherent light source at room and elevated temperatures. Each indentation was located at 0.5 mm behind the crack tip. The relative displacement between the two indentations was measured by the laser interferometric displacement gauge (LIDG) system.[53,54] Rigid body motion of the specimen was effectively removed from the data analysis by the difference in the two symmetric LIDG data.

An implicit dynamic finite element code was executed in its generation mode[55] to compute the dynamic initiation, K_{Id}, and the dynamic propagation stress intensity factor (dynamic SIF) K_I^{dyn}. This analysis, which incorporated the push rod in the finite element model, is relatively cumbersome since the specimen vibrates at its natural frequency upon impact. As a result, the three contacts at the two end supports and at the push rod are made and broken intermittently during the fracture process. These variable boundary conditions must be incorporated into the generation analysis[55] of the fracturing specimen. The input data consisted of the crack extension history, which was determined by the LIDG data, and the measured impact load. The dynamic initiation fracture toughness, K_{Id}, was computed by using a calibrated crack opening displacement (COD) procedure. Finally, the dynamic SIF was determined from the numerically computed dynamic energy release rate using Freund's relation.[56]

The accuracy of the test procedure was validated by room temperature testings of two to three fracture specimens of each material. Additional load data, which were obtained through the bottom load transducer, were used to check the accuracy of the finite element modeling of the load train. A KRAK† gauge at the remaining ligament of the prenotched bar was used to check the master curve which related the crack extension history with the COD data at room temperature. Details of this validation analysis are described in Ref. 57.

The above procedure was used to determine the K_{Id} and the dynamic SIF of alumina, TiB_{2p}/SiC and SiC_w/Al_2O_3 CMCs impacted at room temperature and at 1000–1400°C. Figure 3.16 shows the resultant crack velocity versus dynamic SIF relation for Al_2O_3 where little differences are noted between the data of room temperature and 1000°C. If the cluster of data at the left end did not exist, then the well-known gamma shape curve, which has been observed in metals and polymers,[51,52,58,59] could have been obtained. Figure 3.16, however, shows that the crack continues to propagate slowly, i.e., at speeds ranging from 10 to 40 m/s under a dynamic SIF considerably less than the fracture toughness, K_{Ic}, and is consistent with previous findings.[45,46] The K_{Id} values of this material were found to be 5.7 MPa$\sqrt{m}$ at room temperature and 5.1 MPa$\sqrt{m}$ at 1000°C. Also shown in this figure are the dynamic SIF versus crack velocity relation for statically loaded specimens at room temperature.

Figure 3.17 shows the crack velocity versus dynamic SIF relations of TiB_{2p}/SiC CMC impacted at room temperature and 1200°C. The crack velocity under impact loading is relatively constant during the entire crack propagation history. Also shown are the dynamic SIF versus crack velocity relation for statically loaded specimens at room temperature.

Figure 3.18 shows the resultant crack velocity versus dynamic SIF relation of SiC_w/Al_2O_3 CMC impacted at room temperature and 1200°C. Also shown is the dynamic SIF versus crack velocity relation for statically loaded

†TTI Division, Hartrun Corporation, St. Augustine, FL 32084.

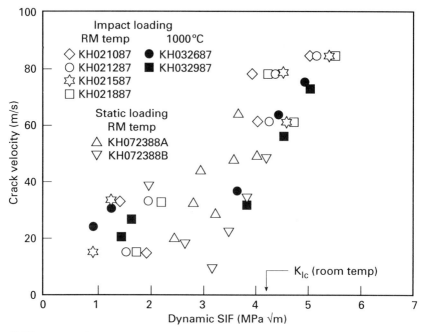

Fig. 3.16 Dynamic SIF versus crack velocity relation for alumina.

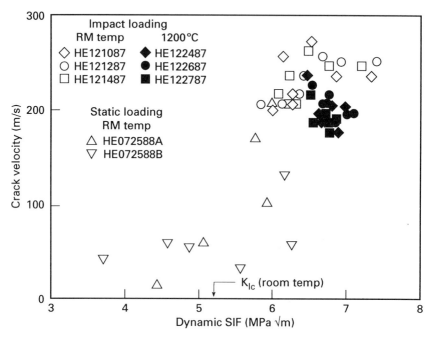

Fig. 3.17 Dynamic SIF versus crack velocity relation, TiB$_2$-particulate/SiC-matrix composite.

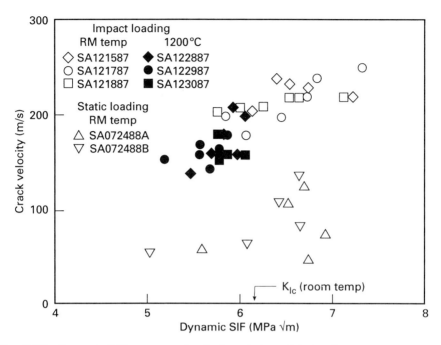

Fig. 3.18 Dynamic SIF versus crack velocity relation, SiC_w/Al_2O_3-matrix composite.

specimens at room temperature. While the trend of a decreasing crack velocity with decreasing dynamic SIF is observed, the available data does not indicate the existence or lack of existence of a dynamic crack arrest SIF. The K_{Id} values at room temperature and 1200°C were 8.3 MPa$\sqrt{m}$ and 7.6 MPa$\sqrt{m}$, respectively. Little difference is noted between the data of room temperature and 1200°C.

As mentioned previously, the crack velocity versus dynamic SIF relations for metals[51,52] and polymers[58,59] represent a gamma curve with distinct crack arrest SIFs. The results reported here, as well as those of Ref. 60, show unequivocally that such dynamic crack arrest SIF does not exist in the ceramics and the CMCs studied by the author and his colleagues. Crack arrest, however, has been observed in chevron-notched, three-point bend specimens, which were machined from the same SiC_w/Al_2O_3 ceramic composites and which were loaded under an extremely small displacement rate of 0.01 mm/min.[61] The run-arrest events in this test were characterized by small crack jumps of about 0.8 mm, which initiated at the sharp crack tip in the chevron-notched specimens. In contrast, the results reported here were generated from blunt machined cracks with larger stored energy prior to crack propagation. Once the excess driving force had been dissipated during rapid crack propagation under static loading and the crack had entered the region of dynamic SIF, which is lower than K_{Ic}, crack arrest was to be expected. Figures 3.16–3.18 show that such was not the case.

The lack of crack arrest was then attributed to the difference in the fracture morphologies of extremely slow and rapid crack extensions. This postulate was tested by extensive fractography analysis of the statically and impact loaded Al_2O_3 specimens. While intergranular fracture was the dominant failure mode in both specimens, some transgranular fracture was observed in all regions of the fracture surface. The percentage areas of transgranular fracture decreased from an average of 16% during the dynamic crack initiation phase to an average of 10% at slower crack propagation in the impacted specimen. For the statically loaded specimen, the percentage of transgranular areas decreased from 5 to 2%. The higher percentage areas of transgranular fracture during the initiation phase can be attributed to the higher crack velocity and the higher dynamic SIF due to the overdriving force generated by the blunt crack tip. This fractography analysis also showed that rapid crack propagation is always accompanied by transgranular fracture regardless of the magnitude of the driving force, i.e., the dynamic SIF, and the crack velocity. Unfortunately, no comparison could be made with the fracture morphology associated with stable crack growth since our test results consisted of only rapid crack propagation events even in the statically loaded specimen. Figure 3.19 shows the percentage area of transgranular fracture versus dynamic SIF. The percentage area in excess of 10% at the lowest dynamic SIF of about 1.5 MPa$\sqrt{m}$ was obtained from the impacted specimen. This data suggests that the continuous input of work during the fracture process generated a higher percentage area of transgranular failure with little chance of crack arrest.

The above finding regarding transgranular fracture is consistent with that of Nose et al.[63] who reported not only transgranular failure of the Al_2O_3 matrix but also SiC whisker failure under rapid crack propagation, i.e., pop-in, in the SiC_w/Al_2O_3 CMC. In contrast, the fracture morphology for stable crack growth in Ref. 62 showed the dominance of intergranular failure accompanied by SiC whisker pull-outs. The SIF for the former, i.e., pop-in fracture or rapid crack jump, remained a constant 6 MPa$\sqrt{m}$ while the SIF continued to increase to about 9 MPa$\sqrt{m}$ with increased stable crack extension. In a previous paper,[64] Nose et al. reported similar findings of a lower SIF of about 4 MPa$\sqrt{m}$ associated with pop-in fracture and a higher SIF varying from 4 to 5.5 MPa$\sqrt{m}$ associated with subcritical crack growth in Al_2O_3.

The results of Nose et al. suggest that the added transgranular failure among the dominant intergranular failures of Al_2O_3 resulted in the lower fracture resistance for a rapidly propagating crack. The dominant intergranular failure suggests a higher fracture resistance due to crack deflection[66] and is consistent with the analysis of Okada and Sines.[66] Transgranular failure, on the other hand, will reduce the effectiveness of crack deflection and thus the crack is driven by a lower driving force. The failure energy of a single crystal ceramic,[67] i.e., energy required for transgranular fracture, is generally higher than that of a polycrystalline ceramic, thus suggesting that transgranular failure

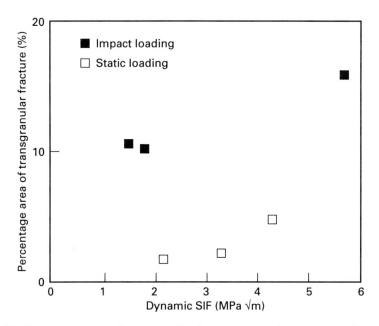

Fig. 3.19 Percentage area of transgranular fracture versus dynamic stress intensity factor for alumina.

requires more energy than intergranular failure. Transgranular failure thus provides a larger driving force but also a competing higher resistance.

Our results on alumina showed that under rapid crack propagation, transgranular fracture does occur both at high as well as low dynamic SIFs associated with the corresponding high and low crack velocities, respectively. Despite Davidge's observation,[67] the kinematic constraint of a rapidly extending flat crack front must have enforced a locally moderate transgranular failure and driven the crack at a lower SIF, thus reducing the chance for crack arrest even at a dynamic SIF less than K_{Ic}. A low percentage area of transgranular failures, i.e., 2%, thus continued to drive the crack at a subcritical dynamic SIF.

The above limited data, which are confined to those of the author and his colleagues, regarding dynamic fracture of ceramics and CMCs have highlighted the two conflicting influences of transgranular fracture in ceramics and CMCs. A similar effect has been observed by the author in his studies on dynamic fracture of concrete where rapid crack propagation induces transaggregate fracture. An understanding of the micromechanics which induce the above mentioned transgranular fracture could possibly lead to an engineered ceramics matrix in which dynamic crack arrest could be possible in the matrix portion of the CMC.

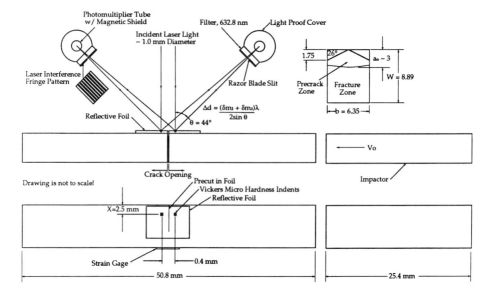

Fig. 3.20   Specimen geometry and LIDG technique.

3.5.1 Bar Impact Testing

The bar impact apparatus, which was described previously, can be used for dynamic fracture testing by initiating rapid crack propagation at a precrack at the center test section. The instantaneous crack length is monitored indirectly by a LIDG[53] which indirectly measures the transient COD. Figure 3.20 shows the specimen configuration used in the dynamic fracture tests with the LIDG arrangement. Figure 3.21 shows a typical LIDG record. The latter required a sharp precrack, which was generated by the single-edge precrack beam (SEPB) method.[64] Details of the test setup and the data reduction procedure, which is based on Eqn. (4), are given in Ref. 30.

The procedure was used to obtain the crack length histories of the ceramic specimens impacted at velocities of approximately 5.8 and 10 m/s for Al_2O_3 and SiC_w/Al_2O_3. The crack length histories are presented in Fig. 3.22 where the crack velocity increased with increasing impact velocity, V_0, but otherwise did not differ between Al_2O_3 and SiC_w/Al_2O_3. The dynamic stress intensity factors, K_I^{dyn}, for the two materials are shown in Fig. 3.23. The fracture toughness increased with increasing impact velocities for each material. The K_I^{dyn} data was plotted as a function of the non-dimensional crack length, $\alpha = a/W$. Initially, K_I^{dyn} increased rapidly and then increased more slowly as the crack approached the traction-free lateral boundary. Figure 3.23 shows that the crack velocities varied with impact velocities but not with the driving force, i.e., the dynamic stress intensity factors.

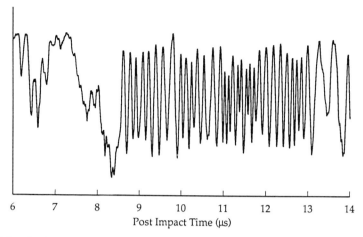

Fig. 3.21 Typical signal representing LIDG fringe motion.

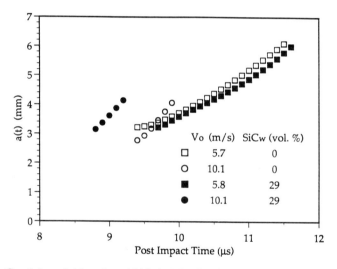

Fig. 3.22 Crack length histories of SiC$_w$/Al$_2$O$_3$ CMC.

The dynamic initiation fracture toughness, K_{Id}, was calculated by the technique described in Ref. 30, which is based on Eqn. (2), and the results are summarized in Table 3.3. The technique for determining K_{Id} required the stress history at the crack tip to be correctly measured and the time at the initiation of crack propagation to be correctly determined from the LIDG signals. The initiation dynamic fracture toughness, K_{Id}, was in the range of previously published values for similar materials.[41,42,60,61] The specimens impacted at 10 m/s showed macroscopic crack branching at the crack length, α_b, marked in Fig. 3.23. The crack branching toughness,[22] K_{Ib}, which is

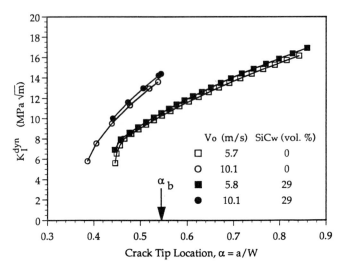

Fig. 3.23 Dynamic stress intensity factor of SiC_w/Al_2O_3 CMC.

defined by $K_{Ib} = K_I^{dyn}$ at $\alpha = \alpha_b$, is also listed in Table 3.3. This K_{Ib} is a necessary but not a sufficient condition for crack branching.[22] Crack branching did not occur when the sufficiency condition, as represented by Eqn. (6), was not satisfied despite the fact that K_I^{dyn} exceeded K_{Ib}.

3.5.2 Fracture Surface Morphology

A scanning electron microscope was used to correlate the fracture surface morphology of the Al_2O_3 and the 29% versus SiC_w/Al_2O_3 CMC with K_I^{dyn} which increased threefold over K_{Id} in both materials at the highest impact velocity. The fracture surfaces exhibited as many as four distinct regions. The precrack region was a relatively flat and uniformly textured surface. This was

Table 3.3 Dynamic fracture characterization of Al_2O_3 and SiC_w/Al_2O_3

Material	E (GPa)	ν	ρ (g/cm³)	V_0 (m/s)	V (m/s)	K_{Ic}^a (MPa √m)	K_{Id} (MPa √m)	K_{Ib} (MPa √m)
$Al_2O_3^b$	221	0.22	3.41	0–14	—	3–4	—	—
$Al_2O_3^c$				5.7	1546		5.9	—
	387	0.22	3.97	10.1	2465	4	5.8	13.6
29 vol.%				5.8	1536		6.5	—
$SiC_w/Al_2O_3^b$	408	0.23	3.73	10.1	2383	7	9.8	14.2

[a] From static fracture toughness measurements.

[b] Coors AD-85. Coors Ceramic Company, Golden, CO.

[c] Hot pressed processing, 99.9% theoretical density. Advanced Composite Materials Corp., 1525 S. Buncombe Road, Greer, SC 29651 USA.

Table 3.4 Fracture surface features correlated with dynamic stress intensity factor

Surface feature	K_I^{dyn} for Al_2O_3 ($MPa \sqrt{m}$)	
	$V_0 = 5.7 \, m/s$	$V_0 = 10.1 \, m/s$
Mirror	5.9–8.6	5.8–8.1
Mist	8.6–11.7	8.1–11.5
Hackle	>11.7	11.5–13.6
Crack branching	No branching	>13.6

followed by the "mirror," the "mist" and the "hackle" regions. The latter consisted of small localized crack branches or branching attempts which were difficult to observe with the SEM or the optical microscope, but were more apparent with the naked eye. These regions were correlated with K_I^{dyn} and data are given in Table 3.4. All materials impacted at 10 m/s exhibited a macro-branched region (4th region). The specimens impacted at 5.8 m/s did not display the macroscopic crack branch region.

The microscopic feature of the fracture surfaces generated by stable crack growth and rapid crack propagation were then compared. Both fracture surfaces of the Al_2O_3 exhibited a combination of intergranular and transgranular fracture surfaces where the intergranular fracture was dominant. The dynamic fracture surface, however, displayed a larger percentage of transgranular fracture surface. Similar results were observed in the 29% versus SiC_w/Al_2O_3 where the fracture surface exhibited both fiber pull-out and trans-fiber fracture, with increased trans-fiber fracture in the dynamic fracture region. The apparent lack of increased fiber pull-out, and hence the lack in increased crack bridging toughening mechanism, would help explain the nearly equal dynamic fracture toughness in the Al_2O_3 and SiC_w/Al_2O_3.

3.6 Summary

Much of the text in this chapter was devoted to the review of the basic elastodynamic stress wave and the crack tip stress field equations, and of testing procedures which utilize these theoretical relations. Conspicuously missing are the actual impact and fracture data, particularly at elevated temperature. The paucity of data is symptomatic of the undeveloped state of this field where few laboratories throughout the world have the resources to undertake dynamic and impact characterization of ceramic matrix composites at operating temperatures in excess of 1000°C. The lack of resources is further compounded by the high cost and limited supply of CMCs from which the test specimens are fabricated. Thus, much of the actual test data was limited to room temperature testing of Al_2O_3 and SiC_w/Al_2O_3, although all testing procedures described in this chapter can potentially be used at elevated temperature. Another review of the state of science in dynamic and impact fracture of ceramic composites in the future could be blessed with a wealth of experimental data on high temperature ceramic matrix composites if the community recognizes the importance of such information, since much of the needed testing and data reduction methodologies are in place.

Acknowledgment

The author gratefully acknowledges the financial support of the Office of Naval Research, ONR Contract N00014-87-K-0326, through which many of the results reported in this chapter were generated.

References

1. A. A. Griffith, "The Phenomena of Rupture and Flow of Solids," *Philosophical Transactions, Royal Society of London*, **A221**, 163–197 (1921).
2. A. A. Griffith, "The Theory of Rupture," in *Proceedings of 1st International Congress on Applied Mechanics*, eds. C. B. Biezeno and J. M. Burgers, Waltman, Delft, The Netherlands, 1925, pp. 55–63.
3. G. R. Irwin, "Fracture Dynamics," *Fracturing of Metals*, ASM, Materials Park, PA, 1948, pp. 147–166.
4. E. Orowan, "Energy Criteria of Fracture," *Welding Journal*, **34**, 1575–1605 (1955).
5. M. S. Dadkhah and A. S. Kobayashi, "Further Studies in the HRR Field of a Moving Crack, An Experimental Analysis," *Journal of Plasticity*, **6**, 635–650 (1990).
6. M. S. Dadkhah, A. S. Kobayashi, and W. L. Morris, "Crack Tip Displacement Fields and J-R Curves of Four Aluminum Alloys," in *Fracture Mechanics, Twenty-Second Symposium*, Vol. 2, eds. S. N. Atluri, J. C. Newman, Jr., I. S. Raju, and J. S. Epstein, ASTM STP 1131, ASTM, Philadelphia, PA, 1992, pp. 135–153.
7. N. F. Mott, "Fracture of Metals: Theoretical Considerations," *Engineering*, **165**, 16–18 (1948).
8. E. H. Yoffe, "The Moving Griffith Crack," *Philosophical Magazine*, **42**, 739–750 (1951).
9. D. K. Roberts and A. A. Wells, "The Velocity of Brittle Fracture," *Engineering*, **178**, 820–821 (1954).
10. J. P. Berry, "Some Kinetic Considerations of the Griffith Criterion for Fracture," *Journal of Mechanics and Physics of Solids*, **8**, 194–216 (1960).
11. K. G. Broberg, "The Propagation of a Brittle Crack," *Arkiv for Fysik*, **18**, 159–192 (1960).
12. B. R. Baker, "Dynamic Stresses Created by a Moving Crack," *ASME Journal of Applied Mechanics*, **29**, 449–458 (1962).
13. G. R. Irwin, "Basic Concepts for Dynamic Fracture Testing," *ASME Journal of Basic Engineering*, **91**[3], 519–524 (1969).
14. J. G. Kalthoff, J. Beinert, S. Winkler, and W. Klemm, "Experimental Analysis of Dynamic Effects in Different Crack Arrest Test Specimens," in *Crack Arrest Methodologies and Applications*, eds. G. T. Hahn and M. E. Kanninen, ASTM STP 711, ASTM, Philadelphia, PA, 1980, pp. 109–127.
15. L. Dahlberg, F. Nilsson, and B. Brickstad, "Influence of Specimen Geometry on Crack Propagation and Arrest Toughness," in *Crack Arrest Methodologies*

and Applications, eds. G. T. Hahn and M. E. Kanninen, ASTM STP 711, ASTM, Philadelphia, PA, 1980, pp. 89–108.

16. T. Nishioka and S. N. Atluri, "Numerical Analysis of Dynamic Crack Propagation: Generation and Prediction Studies," *Engineering Fracture Mechanics*, **16**[3], 303–332 (1982).

17. F. Nilsson, "A Suddenly Stopping Crack in an Infinite Strip Under Tearing Action," in *Fast Fracture and Crack Arrest*, eds. G. T. Hahn and M. E. Kanninen, ASTM STP 627, ASTM, Philadelphia, PA, 1977, pp. 77–91.

18. L. B. Freund, "Crack Propagation in an Elastic Solid Subjected to General Loading IV-Stress Wave Loading," *Journal of Mechanics and Physics of Solids*, **21**, 47–61 (1973).

19. L. B. Freund "Crack Propagation in an Elastic Solid Subjected to General Loading—I. Constant Rate of Extension," *Journal of Mechanics and Physics of Solids*, **20**, 129–140 (1972).

20. T. Nishioka and S. N. Atluri, "Path-Independent Integrals, Energy Release Rates, and General Solutions of Near-Tip Fields in Mixed-Mode Dynamic Fracture Mechanics," *Engineering Fracture Mechanics*, **18**, 1–22 (1983).

21. M. Ramulu and A. S. Kobayashi, "Dynamic Crack Curving—A Photoelastic Evaluation," *Experimental Mechanics*, **23**, 1–9 (1983).

22. M. Ramulu, A. S. Kobayashi, and B.S.-J. Kang, "Dynamic Crack Branching—A Photoelastic Evaluation," in *Fracture Mechanics: Fifteenth Symposium*, ed. R. J. Sanford, ASTM STP 833, ASTM, Philadelphia, PA, 1984, pp. 130–148.

23. International Conference on Mechanical Behavior of Materials under Dynamic Loading, *Journal de Physique*, Colloque **C5**, Supplément au No. 8, Tome 46 (Aout 1985).

24. C. Y. Chiem, H.-D. Kunze, and L. W. Meyer (eds.), *Impact Loading and Dynamic Behavior of Materials*, Vol. 1, DGM Informationgesellschaft mbH, Verlag, Oberurset, Germany, 1988.

25. International Conference on Mechanical Behavior of Materials under Dynamic Loading, *Journal de Physique*, Colloque **C3**, Supplément au No. 9, Tome 49 (September 1988).

26. V. J. Tenery (ed.), *Third International Symposium on Ceramic Materials and Components for Engines*, American Ceramic Society, Columbus, OH, 1989.

27. R. J. Clifton and R. W. Klopp, "Pressure-Shear Plate Impact Testing," in *Metal Handbook*, Vol. 8, 9th edn., American Society for Metals, Metals Park, OH, 1985, pp. 230–239.

28. G. F. Raiser, R. J. Clifton, and M. Ortiz, "A Soft-Recovery Plate Impact Experiment for Studying Microcracking in Ceramics," *Mechanics of Materials*, **10**, 1–43 (1990).

29. H. D. Epinosa and R. J. Clifton, "Place Impact Experiments for Investigating Inelastic Deformation and Damage of Advanced Materials," in *Experiments in Micromechanics of Failure Resistant Materials*, AMD Vol. 130, ed. K.-S. Kim, ASME, New York, 1991, pp. 37–56.

30. L. R. Deobald and A. S. Kobayashi "A Bar Impact Tester for Dynamic Fracture Testing of Ceramics and Ceramic Composites," *Experimental Mechanics*, **32**, 109–116 (1992).

31. J. A. Zukas, T. Nicholas, S. F. Hallock, L. B. Greszczuk, and D. R. Curran, *Impact Dynamics*, John Wiley and Sons, New York, 1982, pp. 8–17.

32. M. A. Meyers and C. T. Aimone, "Dynamic Fracture (Spalling) of Metals," *Progress in Material Science*, **28**, 1–96 (1983).

33. K. S. Kim and R. J. Clifton, "Dislocation Motion in MgO Crystals Under Plate Impact," *Journal of Material Science*, **19**, 1428–1438 (1984).

34. M. Taya, I. W. Hall, and H. S. Yoon, "Void Growth in Single Crystal

Cu-SiO$_2$ During High Strain-Rate Deformation," *Acta Metall.*, **33**[12], 2143–2153 (1985).

35. J. Miklowitz, *The Theory of Elastic Waves and Waveguides*, North-Holland, Publishing Co., Amsterdam, The Netherlands, 1978, pp. 42–60, 383–394.
36. M. G. Jenkins, "Ceramic Crack Growth Resistance Determination Utilizing Laser Interferometry," PhD Dissertation, University of Washington, Seattle, WA, 1987.
37. L. Chuck, E. R. Fuller, and S. W. Freiman, "Chevron-Notch Bend Testing in Glass: Some Experimental Problems," in *Chevron-Notched Specimens: Testing and Stress Analysis*, eds. J. H. Underwood, S. W. Freiman, and F. R. Baratta, ASTM STP 855, ASTM, Philadelphia, PA, 1984, pp. 167–175.
38. D. Munz, R. T. Bubsey, and J. L. Shannon, Jr., "Fracture Toughness Determination of Al$_2$O$_3$ using Four-Point-Bend Specimens with Straight-Through and Chevron Notches," *Journal of the American Ceramic Society*, **63**[5–6], 300–305 (1980).
39. J. Lankford, "Dynamic Compressive Fracture in Fiber-Reinforced Ceramic Matrix Composites," *Material Science and Engineering*, **A107**, 261–268 (1989).
40. L. S. Costin and J. Duffy, "The Effect of Loading Rate and Temperature on the Initiation of Fracture in a Mild, Rate Sensitive Steel," *J. Engr. Material Technol.*, **101**, 258–264 (1979).
41. J. Duffy, S. Suresh, K. Cho, and E. Bopp, "A Method for Dynamic Fracture Initiation Testing of Ceramics," *Trans. ASME*, **110**[4], 325–331 (1989).
42. S. Suresh, T. Nakamura, Y. Yeshurun, K.-H. Yang, and J. Duffy, "Tensile Fracture Toughness of Ceramic Materials: Effects of Dynamic Loading and Elevated Temperatures," *Journal of the American Ceramic Society*, **73**[8], 2457–2466 (1990).
43. S. Suresh and E. K. Tschegg, "Combined Mode I–Mode III Fracture of Fatigue Precracked Alumina," *Journal of the American Ceramic Society*, **70**[10] 726–733 (1987).
44. S. T. Gonczy and D. L. Johnson, "Impact Fracture of Ceramics at High Temperature," in *Fracture Mechanics of Ceramics*, Vol. 3, eds. R. C. Bradt, D. P. H. Hasselman, and F. F. Lange, Plenum Press, New York, NY, 1978, pp. 495–506.
45. A. S. Kobayashi, A. F. Emery, and B. M. Liaw, "Dynamic Toughness of Glass," in *Fracture Mechanics of Ceramics*, Vol. 6, eds. R. C. Bradt, A. G. Evans, D. P. H. Hasselman, and F. F. Lange, Plenum Press, New York, NY, 1983, pp. 47–62.
46. A. S. Kobayashi, A. F. Emery, and B. M. Liaw "Dynamic Fracture Toughness of Reaction Bonded Silicon Nitride," *Journal of the American Ceramic Society*, **66**[2], 151–155 (1983).
47. B. M. Liaw, A. S. Kobayashi, and A. F. Emery, "Effect of Loading Rates on Dynamic Fracture of Reaction Bonded Silicon Nitride," in *Fracture Mechanics: Seventeenth Volume*, eds. J. H. Underwood, R. Chait, C. W. Smith, D. P. Wilhem, W. A. Andrews, and J. C. Newman, ASTM STP 905, ASTM, Philadelphia, PA, 1986, pp. 95–107.
48. K.-H. Yang, A. S. Kobayashi, and A. F. Emery, "Dynamic Fracture Characterization of Ceramic Matrix Composites," *Journal de Physique*, Colloque **C3**, Supplément au No. 9, C3-223-C3-230 (1988).
49. K.-H. Yang, A. S. Kobayashi, and A. F. Emery, "Effects of Loading Rates and Temperature on Dynamic Fracture of Ceramics and Ceramic Matrix Composites," in *Ceramic Materials and Components for Engines*, eds. V. J. Tennery and M. K. Ferber, American Ceramic Society, Columbus, OH, 1989, pp. 766–775.

50. A. S. Kobayashi and K.-H. Yang "A Hybrid Technique for High-Temperature Dynamic Fracture Analysis," in *Applications of Advanced Measurement Techniques*, British Society for Strain Measurements, Whittles Publishing Caithess, U.K., 1988, pp. 109–120.
51. T. Kobayashi and J. W. Dally, "Dynamic Photoelastic Determination of the Å-K Relation for 4340 Alloy Steel," in *Crack Arrest Methodology and Applications*, eds. G. T. Hahn and M. F. Kanninen, ASTM STP 711, ASTM, Philadelphia, PA, 1980, pp. 189–210.
52. J. F. Kalthoff, J. Beinert, S. Winkler, and W. Klemm, "Experimental Analysis of Dynamic Effects in Different Crack Arrest Test Specimens," in *Crack Arrest Methodology and Applications*, eds. G. T. Hahn and M. F. Kanninen, ASTM STP 711, ASTM, Philadelphia, PA, 1980, pp. 109–127.
53. W. N. Sharpe, Jr., "Interferometric Surface Strain Measurement," *Int. J. of Nondestructive Testing*, **3**, 59–76 (1979).
54. M. G. Jenkins, A. S. Kobayashi, M. Sakai, K. W. White, and R. C. Bradt, "Fracture Toughness Testing of Ceramics using a Laser Interferometric Strain Gage," *Bull. J. Amer. Cer. Soc.*, **66**[12], 1734–1738 (1987).
55. A. S. Kobayashi, "Dynamic Fracture Analysis by Dynamic Finite Element Method-Generation and Propagation Analysis," in *Nonlinear and Dynamic Fracture Mechanics*, eds. N. Perrone and S. N. Atluri, ASME, New York, 1979, pp. 19–36.
56. L. B. Freund, "Crack Propagation in an Elastic Solid Subjected to General Loading: II, Non-uniform Rate of Extension," *J. of the Mechanics and Physics of Solids*, **20**, 141–152 (1972).
57. K.-H. Yang and A. S. Kobayashi, "An Experimental-Numerical Procedure for High Temperature Dynamic Fracture Analysis," in *Computational Mechanics '88*, eds. S. N. Atluri and G. Yagawa, Springer-Verlag, Berlin, 1988, pp. 48.v.1–8.
58. J. W. Dally, "Dynamic Photoelastic Studies of Fracture," *Experimental Mechanics*, **19**, 349–367 (1979).
59. A. S. Kobayashi, M. Ramulu, M. S. Dadkhah, K.-H. Yang, and B. S.-J. Kang, "Dynamic Fracture Toughness," *Intl. J. of Fracture*, **30**, 275–285 (1980).
60. K.-H. Yang and A. S. Kobayashi, "Dynamic Fracture Responses of Alumina and Two Ceramic Composites," *Journal of the American Ceramic Society*, **73**[8], 2309–2315 (1990).
61. Y. Takagi and A. S. Kobayashi, "Further Studies on Dynamic Fracture Response of Alumina and SiC$_w$/Al$_2$O$_3$ Composite," in *Symposium on Elevated Temperature Crack Growth*, ASME, Vol. 18, eds. S. Mall and T. Nicholas, ASME, New York, 1990, pp. 145–148.
62. M. G. Jenkins, A. S. Kobayashi, K. W. White, and R. C. Bradt, "Crack Initiation and Arrest in SiC Whisker/Al$_2$O$_3$ Matrix Ceramic/Ceramic Composites," *Journal of the American Ceramic Society*, **70**[6], 393–395 (1987).
63. T. Nose, M. Ueki, T. Fujii, and H. Kubo, "Toughening Behavior in SiC Whisker Reinforced Al$_2$O$_3$ Ceramics," Paper presented at the 1st International Ceramic Science & Technology Congress, Oct. 31–Nov. 3, 1989, Anaheim, CA.
64. T. Nose and T. Fujii, "Evaluation of Fracture Toughness for Ceramic Materials by a Single-Edge-Precracked-Beam Method," *Journal of the American Ceramic Society*, **71**[5], 328–333 (1988).
65. K. T. Faber and A. G. Evans, "Crack Deflection Processes—I. Theory," *Acta Metall.*, **31**[4], 565–576 (1983).
66. T. Okada and G. Sines, "Prediction of Delayed Fracture from Crack Coalescence-Alumina," in *Fracture Mechanics of Ceramics*, Vol. 7. eds. R. C.

Bradt, A. G. Evans, D. P. H. Hasselman, and F. F. Lange, Plenum Press, New York, NY, 1986, pp. 297–310.

67. R. W. Davidge, "Effects of Microstructure on Mechanical Properties of Ceramics," in *Fracture Mechanics of Ceramics*, Vol. 2, eds. R. C. Bradt, D. P. H. Hasselman, and F. F. Lange, Plenum Press, New York, NY, 1974, pp. 447–468.

Long-Term Behavior

CHAPTER 4

Creep Deformation of Particulate-Reinforced Ceramic Matrix Composites

S. M. Wiederhorn and E. R. Fuller, Jr.

4.1 Introduction

Current interest in ceramic matrix composites (CMCs) for high temperature applications has its origins in the potential value of these materials for heat engines. Metallic alloys in these applications have a temperature limit of $\approx 1100°C$.[1] Ceramics and ceramic composites, by contrast, have a potential application temperature of 1400–1500°C. If engines could be made to operate in this temperature regime, substantial increases in operating efficiencies would be obtained, resulting in substantial savings in fuel, and/or increases in performance.

One of the main impediments to using ceramic composites in such applications is the occurrence of creep rupture at high temperature.[2,3] Failure occurs either by crack growth from a dominant flaw, or by the nucleation, growth and coalescence of cavities. Mechanisms that control crack growth at elevated temperatures are quite different from those at low temperature. Creep processes in the vicinity of the major flaw, or at the tips of cracks that grow from these flaws, are important. These creep processes control the rate of crack growth and the final rupture time. Likewise, creep processes play a crucial role in the nucleation and growth of cavities and in their coalescence, leading to final rupture.

Particulate ceramic matrix composites are two-phase ceramics, in which hard refractory particles or fibers are embedded in a ceramic matrix which is usually less creep resistant. Creep begins when the matrix begins to flow: hence, the more refractory the matrix, the more creep resistant the composite. However, creep resistance of the composite is determined by both the creep behavior of the matrix and the concentration of particles in the matrix. At low particle concentrations, creep is determined primarily by the rheological

155

properties of the matrix.[4] As the particle concentration is increased, creep resistance also increases, primarily as a consequence of particle interference with the flow process. Above a critical concentration of particles, the mode of creep changes to one in which creep is controlled by interparticle contact.[5] Processes that occur at contact sites now dominate the creep process. In addition, as particles slide over one another, cracks and cavities are generated. This damage gradually accumulates forming the macroscopic cracks that result in failure.[6] Thus, creep behavior controls failure at high temperatures.

In this chapter, we discuss the creep behavior of ceramic matrix composites in terms of their microstructure. The role of interparticle interactions during creep is emphasized. The relationship between creep rate and lifetime is also discussed.

4.2 Types of Ceramic Matrix Composites

Particulate ceramic matrix composites are made by reinforcing normal ceramics with chopped fibers, platelets, particles or whiskers. These composites are usually made by normal ceramic processing techniques.[7-13] Powders of the ceramic matrix are either dry or wet mixed with the particles or fibers. The powders are cast or pressed in a die and then hot pressed or HIPed to final form. HIPing or hot pressing assists the sintering process to obtain high concentrations of reinforcement, e.g., up to 40 vol.% for whisker or chopped fiber composites. Hot pressing is used to produce simple parts such as commercial cutting tools and bearings.[14] HIPing is used to produce more complicated parts for both experimental and industrial purposes.[14,15]

Less-conventional processing techniques are also used to make ceramic matrix composites. Siliconized silicon carbide, for example, is made by liquid infiltration.[16,17] A compact of SiC particles is formed and then presintered, or reaction bonded. Liquid silicon is then infiltrated into the structure. Many different microstructures of siliconized silicon carbide can be made in this manner. The volume fraction of SiC particles can be as high as 90 vol.%. Bimodal structures have also been made by this technique. These materials are used for radiant heaters and heat exchangers.[17,19]

Although silicon nitrides are usually considered to be monolithic ceramics, their microstructures under certain conditions can be similar to those of composite materials. They are made by either sintering, hot pressing, or HIPing, during which a polymorphic transformation occurs converting α-Si_3N_4 to β-Si_3N_4.[20,21] During this transformation needle-like Si_3N_4 grains form, increasing resistance to both fracture and creep. These elongated grains are embedded in a matrix of more-or-less equi-axed Si_3N_4 grains; all of the grains are bonded by an amorphous phase.[22-24] Research efforts to enhance and control this needle-forming process are currently in progress at several companies and research laboratories. Needle growth is enhanced by using glass-forming sintering aids.[25,26] Values of fracture toughness as high as 11 MPa$\sqrt{m}$ have been obtained in this manner. Similar techniques are being used to grow elongated grains in α-SiC[27]). These *in situ* growth techniques offer the possibility of fully dense composites with a higher volume fraction of "whiskers" within the ceramic matrix, because there are no whiskers in the solid to interfere with material flow during hot pressing or sintering.

4.3 Phenomenology of Creep

In this section we review experimental observations on the creep of ceramic matrix composites. Observations that apply to all ceramic matrix composites are discussed. Creep curves obtained on ceramic matrix composites are compared with curves obtained on metals and metallic alloys. The role of a second phase in increasing the creep resistance of composites is emphasized. Finally, a discussion of creep asymmetry is presented, wherein creep occurs more easily in tension than in compression.

4.3.1 General Observations

A typical tensile creep curve for a particulate reinforced ceramic matrix composite, siliconized silicon carbide (Si/SiC),[28] is shown in Fig. 4.1. In comparison to the behavior of metals and metallic alloys, tertiary creep is suppressed in this material. There is only a slight upward curvature of the creep curve prior to failure. In many other ceramic matrix composites, tertiary

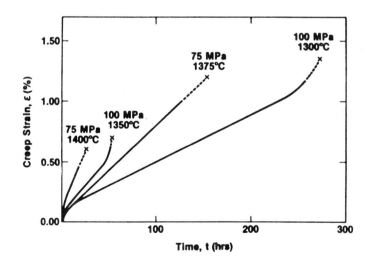

Fig. 4.1   Tensile creep curves for siliconized silicon carbide (Carborundum KX01). Over most of the data range, these data can be represented by a constant creep rate; there is a short primary creep stage, and almost no tertiary creep. The rupture strain decreases with increasing creep rate. The strain to failure, $\approx 1.5\%$, indicates brittle behavior even at low rates of creep deformation. Figure from Ref. 28.

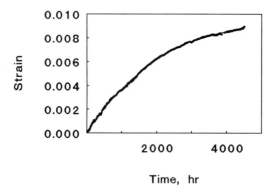

Fig. 4.2   Creep data on a commercial vintage of silicon nitride sintered with 4 w/o yttria. These data were taken from an early version of NT154. Later versions were heat treated to reduce the amount of primary creep. Primary creep occurs over the entire test period. Figure from Ref. 33.

creep is missing entirely; creep curves consist only of primary and secondary stages. This has been observed for some grades of silicon nitride,[29–34] for vitreous-bonded aluminum oxide[29] and for siliconized silicon carbide.[35] For some ceramic composites, creep consists entirely of a primary creep stage;[32–34] see Fig. 4.2.

Some other aspects of creep in ceramic matrix composites are also shown in Fig. 4.1. The failure strain in these materials is generally small, typically less than 1–2%. The strain at failure is also a function of the minimum strain rate: the lower the minimum strain rate, the greater the strain to failure. This is easily seen in Fig. 4.1 where the failure strain of this Si/SiC is much greater for lower creep rates. This effect is illustrated more quantitatively in Fig. 4.3 for the same material.[36] As can be seen, the failure strain varies from 0.5 to 1.5%, as the minimum strain rate varies from $\approx 10^{-7}\,s^{-1}$ to $\approx 10^{-8}\,s^{-1}$. This same type of behavior is obtained for other ceramic matrix composites.

4.3.2 Creep Resistance of Composites

A commonly used particulate composite at high temperatures is siliconized silicon carbide. Made by liquid-phase infiltration of a presintered silicon carbide powder compact,[16,17] the silicon carbide content ranges from approximately 60 vol.% to over 90 vol.%, depending on processing conditions. Higher concentrations of SiC are obtained by adding more carbon to the compact prior to infiltration. This method of manufacture is a relatively inexpensive way to produce composites for high temperature use. Examples of the microstructure obtained for this type of material are shown in Fig. 4.4. By adding silicon carbide to silicon, the room temperature toughness of the material in Fig. 4.4a

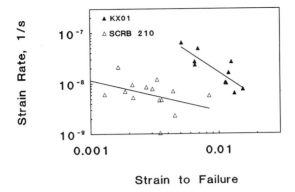

Fig. 4.3 Relation between strain-to-failure and minimum creep rate for two commercial grades of siliconized silicon carbide: Carborundum KX01 and COORS SCRB 210. In both cases, the strain-to-failure increases as the creep rate decreases.

is raised from $\approx 0.9\,\text{MPa}\sqrt{m}$ for silicon[37] to $\approx 4\,\text{MPa}\sqrt{m}$ for the composite.[38] At the same time, the creep rate at 1300°C and an applied shear stress of 100 MPa is decreased from $\approx 10^{-2}\,\text{s}^{-1}$ for silicon[39] to $\approx 10^{-8}\,\text{s}^{-1}$ for the composite,[35] an improvement of six orders of magnitude.

A second example of improved creep resistance through particle reinforcement is found in glass-bonded aluminum oxide. Although these materials are not intended for high temperature use, their structure is similar to that of particulate composites which are used at elevated temperatures. For example, a vitreous-bonded aluminum oxide containing approximately 8 vol.% glass has a creep rate of $10^{-7}\,\text{s}^{-1}$ at 1050°C when measured in flexure at an initial applied stress of 60 MPa.[40]† If we assume the effective viscosity of the composite in $\eta = \sigma/\dot{\varepsilon}$, then $\eta = 6 \times 10^{14}\,\text{Pa s}$ for this material. The viscosity of the grain boundary glass was estimated by a semi-empirical method described by Urbain *et al.*[41] The chemical composition of the grain-boundary glass, needed in this estimate, was analyzed by energy dispersive X-ray analysis.[40] At 1050°C, the glass viscosity was approximately $3 \times 10^4\,\text{Pa s}$. Thus, by adding alumina particles to this glass, an effective increase in creep resistance of ≈ 10 orders of magnitude is obtained. The improvement in the creep resistance of siliconized silicon carbide and vitreous-bonded alumina is typical of what can be achieved for particulate composites when the concentration of particles is high. The improved creep behavior of these materials is a direct consequence of interparticle contact.

Improvement in the creep resistance of whisker-reinforced composites is modest compared to that obtained in particulate-reinforced composites. Creep

---

†Creep measurements were made on the as-received material before the grain-boundary glass began to devitrify significantly. The stress given here is the initial applied stress in the outer fiber of the flexure bar. This outer-fiber stress relaxes as the bar bends and the neutral axis shifts.

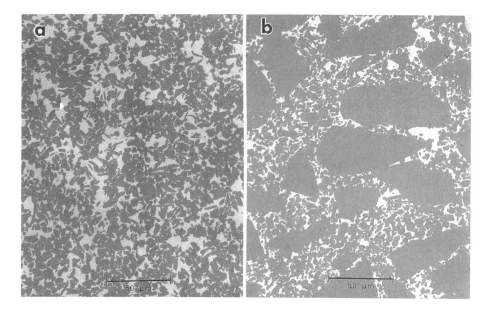

Fig. 4.4 Examples of the microstructure of siliconized silicon carbide. (a) Carborundum KX01 contains approximately 33 vol.% Si, (b) COORS SCRB 210 contains approximately 18 vol.% Si.

studies have been conducted on at least four types of whisker-reinforced ceramic matrix composites: $MoSi_2$,[42–44] Al_2O_3,[45–49] mullite $(Al_2O_3 \cdot SiO_2)$[50] and Si_3N_4,[50–54] all reinforced with SiC whiskers, SiC_w. The behavior of these materials is illustrated by flexural creep data collected by Lin and Becher[46] on an alumina composite containing 20 wt.% SiC_w: see Fig. 4.5. These data suggest an increase in the effective creep resistance of ≈2 orders of magnitude over the unreinforced Al_2O_3. Other authors studying the creep behavior of alumina reinforced by SiC_w have found similar increases in creep resistance, ≈1 to 2 orders of magnitude. In studies on mullite[50] and on $MoSi_2$ reinforced with SiC whiskers,[42–44] a similar enhancement in creep resistance is obtained.

In contrast to the findings just discussed, SiC whisker reinforcement is not effective in improving creep resistance of silicon nitride.[53] This lack of effect is a consequence of the fact that Si_3N_4 normally contains elongated grains, which improve toughness and increase creep resistance. When SiC whiskers are added to Si_3N_4, growth of elongated Si_3N_4 grains is suppressed, so that they are no longer present in the composite. Whiskers take the place of the elongated grains; hence, there is no net improvement in creep resistance. The importance of the elongated grains to the creep resistance of Si_3N_4 has been demonstrated by Buljan et al.[55] When either SiC whiskers or elongated grains are present, creep resistance of the composite is about 30 times that of Si_3N_4 containing only equi-axed grains.

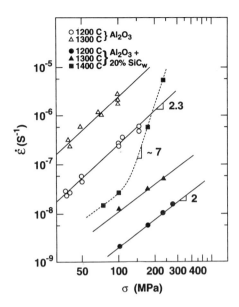

Fig. 4.5 Comparison of flexural creep of whisker-reinforced Al_2O_3 with that of whisker-free aluminum oxide. A sharp increase in the stress exponent is observed at the higher stresses for the whisker-reinforced materials tested at 1400°C. Figure from Ref. 46.

4.3.3 Creep Asymmetry

An important feature of the creep behavior of composites is creep asymmetry, wherein a composite creeps faster in tension than in compression. Creep asymmetry was first observed by Morrell and Ashbee[56] in a lithium zinc silicate glass-ceramic, which has the microstructure of a particulate-reinforced composite. Approximately two to six times as much stress was required in compression to achieve the same creep rate as that in tension: see Fig. 4.6. This creep asymmetry is often accompanied by a change in stress exponent with applied stress. The slope of the data plotted in Fig. 4.6 increases as a function of applied stress, indicating an increase in the stress exponent of the creep behavior with applied stress. This behavior was confirmed by Wang and Raj[57] on commercial glass ceramics. Again, the stress in compression to achieve a given creep rate was two times that required in tension.

Results on other composite materials are similar to those obtained by Morrell and Ashbee.[56] Creep asymmetry has been demonstrated for two grades of siliconized silicon carbide,[35,60,61] SiC whisker-reinforced silicon nitride,[53] HIPed silicon nitride,[29] and vitreous-bonded aluminum oxide.[29] Again, stresses required to achieve the same creep rate were at least a factor of two greater in compression than in tension. In two grades of siliconized silicon carbide,[35,58–61] the stress exponent changed from ≈ 4 at creep rates below

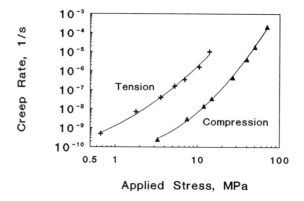

Fig. 4.6 Temperature-compensated creep plot for tension and compression on a glass-ceramic. Data are compensated by an Arrhenius correction to 700°C. Data from Ref. 56 were replotted. The curves are least square fits of quadratic polynomials to the data. An increase in the stress exponent of the creep data is clearly indicated.

$10^{-8}\,s^{-1}$ to ≈ 10 at higher creep rates: see Fig. 4.7. In the other materials, stress exponents ranged from 1 to 2 in compression, and from 5 to 6 in tension.[29,50,54,53] The lower value obtained in compression is probably attributable to diffusional creep, whereas the higher value in tension is probably a consequence of cavitation.

The most common explanation for asymmetric creep in ceramic composites is the formation of cavities during the creep process. Cavities usually form in tension, not in compression.† Cavity formation in tension has been observed in SiC/Si,[35,58,59] SiAlON,[62] Si_3N_4,[30] SiC_w/Si_3N_4[50,53] and vitreous-bonded Al_2O_3.[30] Due to the irregular nature of the microstructure of composites, stresses build up locally at contact points during creep. These contacts act as pinning points and thus slow down the creep process. In tension, these stresses can be relieved by cavitation, which releases the pinning point and accelerates the creep process. In compression, however, these stresses are not relieved and a higher stress is required to achieve the same creep rate.

Although cavitation is believed to be the most important contribution to creep asymmetry, there are other contributing factors. In Fig. 4.7, for example, data from the lower portion of the tensile curve labeled KX01 were taken from specimens in which few cavities were observed. Therefore, cavitation cannot be used as the sole explanation for the difference in behavior in tensile and compressive creep. Other factors seem to be important in establishing creep

†Lange *et al.*[63] report cavitation in silicon nitride tested in compression. The amount of cavitation depended on both composition and compressive stress. For a given strain, cavitation was low for low applied stresses. Some compositions exhibited very little cavitation regardless of stress. It is probable that the results reported here were a consequence of the combined effect of the low stresses normally used in flexure bars, and the use of cavitation-resistant compositions in our studies.

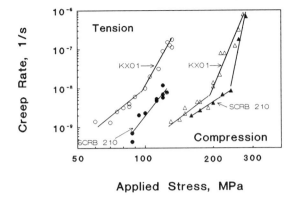

Fig. 4.7 Asymmetric creep in siliconized silicon carbide. Two materials were tested: Carborundum KX01 and COORS SCRB 210. Tests were conducted at 1300°C. Data were from Refs. 36 and 60.

asymmetry. These factors involve mechanisms that control the rheology of particulate systems, i.e., dilation of the network of particles and matrix flow between the particles.

4.4 Rheology of Two-Phase Systems

Creep deformation of composite materials is a subset of the science of flow of two-phase materials. For most ceramic matrix composites, the matrix phase is less refractory than the reinforcement phase so that it softens and flows at high temperatures, whereas the refractory phase remains rigid. The temperature of softening determines the temperature at which the composite begins to creep. The matrix phase may be Newtonian (i.e., the stress exponent is unity), as for glass or fine-grain aluminum oxide, or it may be non-Newtonian, as for silicon. At the same time, the reinforcing particles or whiskers usually behave elastically during deformation. Thus, ceramic matrix composites may be considered to be dispersions or suspensions of rigid particles in a flowing matrix.

In this section, some results on the rheology of polymeric suspensions are discussed. Within the range of application, the viscosity of the polymers ($\approx 10^3$ Pa s) is very low compared to that of the matrix of ceramic composites ($\approx 10^{13}$ Pa s). Nevertheless, the deformation process for both types of materials is similar, and so the results of one field should be transferable to the other. In both cases an abrupt increase in viscosity occurs at high densities of particles. Much of the review in this section is taken from articles published in the rheology literature.[4]

The first study of the rheology of suspensions was made by Einstein[65] who analyzed the flow of dilute suspensions of spherical particles. The relative

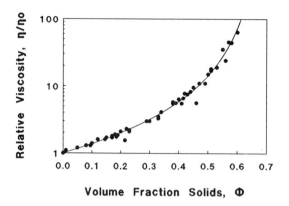

Fig. 4.8 Effect of volume fraction rigid spheres on the relative viscosity of the suspension. In these studies, the particle diameter ranged from 0.1 to 440 μm. Data from Ref. 66.

viscosity, η/η_0, was found to be independent of particle size and linearly proportional to the volume fraction, ϕ, of particles:

$$\eta/\eta_0 = 1 + 2.5\phi \tag{1}$$

where η_0 is the viscosity of the fluid, and η is the viscosity of the suspension.

Studies of the effect of particle size on viscosity suggest that Eqn. (1) is obeyed in the limit as ϕ goes to zero. As the volume fraction of particles increases, the relative viscosity increases at a faster rate than linear rate, approaching infinity as the packing density of the suspension approaches that of densely packed solid particles: Fig. 4.8.[66] Regardless of particle size, all data in Fig. 4.8 scatter about a common curve. An empirical fit for this type of curve has been discussed by Kitano et al.:[67]

$$\eta/\eta_0 = [1 - (\phi/\phi_0)]^{-2} \tag{2}$$

where ϕ_0 is an empirical constant having the value of 0.68 for a suspension of spherical particles. The constant ϕ_0 behaves as a critical volume fraction or packing threshold, above which the relative viscosity increases without limit.†

Kitano et al.[67] also conducted a study on suspensions of elongated particles: Fig. 4.9. Aspect ratios ranged from 1 (for spheres) to 27 (for carbon fibers). The results of this study are similar to those obtained for spherical particles. However, the larger the aspect ratio, the lower the volume fraction of solid at which the packing threshold is reached. The viscosity of the suspensions follow Eqn. (2), with a constant ϕ_0 that depends on the aspect ratio: see Table 4.1.

---

†As ϕ/ϕ_0 approaches 1, stresses between particles within the suspension become high enough that other modes of deformation take over. Thus, η/η_0 remains finite. As ϕ approaches 0, $\eta/\eta_0 \approx 1 + 2.94\phi$ for spheres, which differs slightly from the Einstein relation.

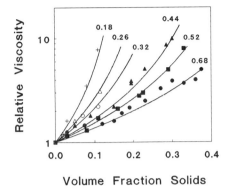

Fig. 4.9 Effect of particle asperity on the relative viscosity of molten polymer suspensions. Particles studied were as follows: •, glass spheres; ■, natural calcium carbonate; ▲, precipitated calcium carbonate; ○, glass fibers—aspect ratio = 18; △ carbon fiber—aspect ratio = 23; + carbon fibers—aspect ratio = 27. Data are from Ref. 67.

More recently, Yoon and Chen[68] developed a theory to predict the deformation behavior of particulate composites. Their theory treats the case of rigid particles embedded in a nonNewtonian matrix. The relative deformation rate, $\dot{\varepsilon}/\dot{\varepsilon}_0$, is related to the volume fraction of particles, ϕ, the creep stress exponent of the matrix, n, and the stress concentration factor, k, of the inclusion in the matrix:

$$\dot{\varepsilon}/\dot{\varepsilon}_0 = (1 - \phi)^q \qquad (3)$$

where $q = 1 + (k - 1)n$. Experimental data collected on a commercial grade of zirconia (2Y-TZP) reinforced with mullite was used to support their theory. As in treatments of the flow of Newtonian liquids containing rigid particles, creep resistance increases with particle concentration and with particle asperity ratio.

Table 4.1 Critical volume fraction, ϕ_0, for various aspect ratios of particles, after Refs. 4 and 57

Aspect ratio	ϕ_0
1	0.68 (for smooth spheres) 0.44 to 0.52 (for rough crystals)
8	0.44
18	0.32
23	0.26
27	0.18

However, this theory does not predict a concentration threshold at which the creep resistance increases abruptly.

The volume fraction ϕ_0 in Eqn. (2) defines a percolation limit for creep.[69,70] Above this threshold, most particles within the slurry are part of an extended cluster that dominates the flow process. Two-dimensional experiments on slurries demonstrate that, at the threshold, 60–80% of the particles within the slurry form part of the cluster.[71,72] When this happens, particle motion within the slurry is controlled by the cluster, and further deformation of the slurry is dominated by deformation of the cluster itself.

Comparing the type of information obtained on suspensions with that obtained on composites gives useful insight into the types of mechanisms that control creep of ceramic matrix composites. The very large increase in creep resistance of dense particulate composites, i.e., more than 65 vol.% particles, suggests that the particle packing density is above the percolation threshold. Creep of particulate composites is, therefore, controlled by direct interparticle contract, as modified by the presence of relatively inviscid matrices. Mechanisms that control such super-threshold creep are discussed in Section 4.5.

The much smaller improvement of creep resistance for whisker-reinforced composites is typical of the subthreshold behavior of particulate suspensions reported in the rheology literature. Here the whisker volume concentrations range from $\phi = 0\%$ to $\phi = 30\%$. In this regime, creep is controlled primarily by flow of the matrix and the volume fraction of particles in the matrix. These results suggest that creep behavior could be improved by increasing the volume fraction, ϕ, of particles, to values that exceed the threshold limit. To further test this analogy between composites and suspensions, creep data are needed on composites as a function of the volume fraction of the reinforcement phase. However, since whisker-reinforced composites are typically processed by "flow-type" processes (with rigid whiskers), higher volume fractions are difficult to obtain. Super-threshold fiber concentrations may be possible when the "whisker" phase is grown *in situ*, as in *in situ* toughened silicon nitride[25,26] and silicon carbide.[27] Then, the fiber concentration is limited not by flow constraints but by constraints of fiber growth, and higher concentrations of fiber may be achievable.

4.5 Super-Threshold Deformation

In most two-phase ceramics, particle concentrations exceed the rheological threshold discussed above. Therefore, deformation is controlled not by fluid flow, but by direct interactions between the particles that make up the solid. Above the packing threshold, many particles are either in direct contact, or are sufficiently close that particle interactions are important. Processes that occur close to the particle surface dominate the deformation behavior. These processes include solution–precipitation, matrix flow (or matrix percolation) between the particles, and cavitation.

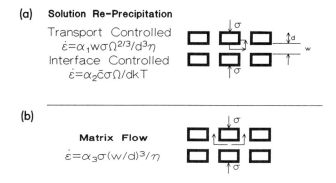

Fig. 4.10 Models of two-phase creep: (a) solution-precipitation and (b) matrix flow. In the equations given above: α is a dimensionless constant (different for each equation); Ω is the molar volume of the grains; k is Boltzmann's constant; η is the viscosity of the bonding phases; and $\bar{c}$ represents the surface dissolution velocity under a unit driving force. After Pharr and Ashby.[5]

4.5.1 Solution–Precipitation

This mechanism of creep deformation was first suggested by Coble as an explanation for creep deformation of monolithic materials.[73] Under an applied stress, σ, chemical potential differences develop between grain interfaces: Fig. 4.10. These differences in chemical potential act as a driving force for diffusive transport of material between grain surfaces. Coble viewed the transport process as occurring along grain boundaries. The creep rate, $\dot{\varepsilon}$, was found to be a function of the grain size, d, the grain boundary thickness, w, the diffusivity, D, and the temperature, T:[5]

$$\dot{\varepsilon} = \alpha_1 \sigma D w \Omega / (k_b T d^3) \tag{4}$$

where α_1 is a dimensionless constant, Ω is the molar volume of the diffusing species, and k_b is Boltzmann's constant. With glass at the boundary, the boundary width is now set equal to the glass layer thickness. Diffusive transport proceeds through the glass from high to low chemical potential grain surfaces: Fig. 4.10. Equation (4) can be expressed in terms of the viscosity of the glass, η, by using the Stokes–Einstein equation,[74] $D = k_b T / (6\, \eta \pi \Omega^{1/3})$:

$$\dot{\varepsilon} = \alpha_2 \sigma w \Omega^{2/3} / (\eta d^3) \tag{5}$$

From Eqn. (5) we see that the important intrinsic variables controlling creep are the boundary thickness, w, the grain size, d, and the viscosity of the boundary fluid, η. If these parameters change in the course of the deformation process, a corresponding change in the creep rate will occur, leading to transient creep behavior.

Because of its effect on matrix viscosity, devitrification of the intergranular matrix has a pronounced effect on the creep behavior of the composite. In vitreous-bonded alumina, for example, a twofold order of magnitude decrease in creep rate has been demonstrated by devitrification of the grain boundary glass.[40] In silicon nitride, devitrification of the grain boundary phase is so effective that it is one of the principal methods of improving the creep resistance of silicon nitride. When devitrification occurs gradually with time, transient creep is observed.[33] In vitreous systems, improved creep resistance can also occur by the migration of the glass modifier ions from the body to the surface of the test specimen.[75] When this happens, a more refractory glass is left behind on the grain boundaries, and the rate of deformation decreases.

Solution–precipitation theory cannot be used to justify creep asymmetry or high tensile stress exponents for ceramic matrix composites. The theory suggests that creep is symmetric in stress and that the stress exponent is equal to 1. Justification of creep asymmetry by solution–precipitation would require other parameters in Eqn. (4) to depend on the sign of the applied stress. A nonlinear dependence on stress would be required. Diffusion and devitrification may play a role in this regard: however, the data needed to support this possibility have yet to be obtained.

4.5.2 Matrix Flow

When a substantial amount of matrix is present between grains, creep can also occur by the extrusion (or percolation) of the matrix from those intergranular regions that are subjected to high pressures to those that are at lower pressures: see Fig. 4.10. This model of creep was first suggested by Drucker[76] and by Lange[77] to explain the deformation of two-phase materials. The model was extended by Dryden et al.[78] and Debschütz et al.[79] to consider transient creep behavior. Chadwick et al.[80] recently applied the theory to nonNewtonian viscous fluids. On the basis of these viscous flow models, the microstructural parameters controlling creep are grain size, d, bonding phase thickness, w, and the effective viscosity, η, of the bonding phase. The functional dependence of the strain rate, $\dot{\varepsilon}$, on these parameters is given by:

$$\dot{\varepsilon} = \alpha_3 \sigma (w/d)^3/\eta = \alpha_3 \sigma (1 - \phi)^3/\eta \qquad (6)$$

where σ is the applied stress. Although the same microstructural variables control creep as in solution–precipitation, they are grouped differently for the matrix flow mechanism. In the matrix flow model, the creep rate is proportional to the volume fraction of the matrix cubed, since $w/d \approx (1 - \phi)$. In solution–precipitation, the creep rate is proportional to the volume fraction of the matrix, and inversely proportional to the grain size squared. These differences in behavior should be amenable to experimental verification. Many of the same comments made above with regard to solution–precipitation, also apply to viscous flow. Processes, such as devitrification, will effect the viscosity, and hence the creep rate. Thus, transient creep would be expected as a consequence of devitrification.

The flow model readily explains creep asymmetry and transient creep of two-phase materials.[78–80] In this model, resistance to creep depends on the distance separating grain surfaces. As the separation distance decreases, resistance to creep increases, resulting in transient creep behavior. Creep asymmetry has its origins in the fact that approximately two-thirds of the intergranular fluid are associated with surfaces that have their normals perpendicular to the applied stress. This means there is a basic asymmetry in the creep process in tension and compression. Since the narrowest channels between grain faces provide the highest resistance to fluid flow, those surfaces that approach one another during deformation determine the creep rate. In tension, there are twice as many surfaces approaching one another as there are in compression. Therefore, the tensile surfaces change their separation distance half as fast for a given strain rate as the compressive surfaces. As a consequence, a higher stress is needed to achieve the same creep rate in compression as in tension. The mathematical details of this process are given by Dryden *et al.*[78]

Another feature of this model is the eventual contact that occurs between grains as deformation proceeds. This behavior contributes to the transient creep sometimes observed in ceramic matrix composites. As the grain surfaces make contact, creep must cease unless another mechanism of creep is achieved.† In Fig. 4.10, all of the grains are expected to make contact at the same time. However, because of the irregularity and randomness of grains in real structures, contact will occur gradually as a function of strain, some grains making contact at low strains, others at large strains. At the same time, other mechanisms of creep gradually contribute to the deformation process. Hence, resistance to creep will also build up gradually as the strain increases and a smooth creep curve will result. In a recent study of the flexural creep of silicon nitride, Chadwick *et al.*[81] provide experimental support for this model. Also, experimental evidence for direct contact between grains is obtained in the form of stress whorls between silicon nitride grains. These have been reported on specimens loaded in compression,[82] flexure[83] and in tension.[33]

4.5.3 *Dilation of Particulate Composites*

Osborne Reynolds was the first to show that tightly packed granular solids expand their volume when deformed.[84] This phenomenon is called dilatancy. It is well understood and is discussed in some detail in the literature on soil mechanics.[85–87] In vitreous-bonded structural materials such as silicon nitride, dilatancy has been suggested as a contributing factor in the formation of cavities,[88] and may be an important factor in the cavitation of ceramic matrix composites.[64] Dilatancy has also been suggested as an important factor in controlling the creep and creep relaxation of glass-ceramics.[89]

---

†In structural grades of silicon nitride, intergranular glass layers are ≈ 1 nm, whereas the grain sizes are $\approx 1\ \mu$m, leading to the conclusion that this creep mechanism is effective for strains <0.1%. As silicon nitride exhibits creep asymmetry at deformation strains greater than 0.1%, other sources of creep asymmetry are also active in the deformation of this material.

When rigid particles are densely packed, they must move into a looser configuration for deformation to occur. The centers of the particles move apart during deformation causing an initial increase in the volume of the structure with uniaxial deformation. Volume changes during deformation can be substantial. Taylor[87] has shown that a sand with a void content of 38% underwent a volume increase of ≈3% during the first 5% axial strain in compression. Loosely packed sands, by contrast, show little change in volume upon deformation; sand with a void content of 45%, or greater, is not dilatant.[87]

Asymmetric creep behavior of particulate composites can also be rationalized on the basis of dilatant behavior. In compression, the applied stress opposes the dilation, thereby enhancing contact stresses between grains and increasing the stress required for one grain to slide over another. Conversely, tensile stresses tend to pull the particles apart, reducing forces between grains. Using soil mechanics concepts, these ideas have recently been put on a more quantitative basis to rationalize the creep asymmetry in siliconized silicon carbide.[64] For a given creep rate, the theory accounted for an increase of at least a factor of two in applied stress in compression over that required for creep in tension. Dilatant behavior also gives rise to cavitation during deformation. As the particles move apart, the interstitial fluid, i.e., the matrix of a ceramic matrix composite, must flow into the newly created space between particles in order to maintain continuity within the composite.[64] If the fluid does not flow fast enough, negative pressures build up within the solid and cavitation occurs.

4.5.4 Cavity Formation

Cavitation in particulate composites often initiates from between closely spaced grains, where constraints for fluid flow are maximized:[77] Fig. 4.11. If only a small amount of bonding matrix is present, e.g., <10 vol.%, the cavity often grows from one grain to the next until failure occurs: Fig. 4.12. This type of crack growth is observed in vitreous-bonded alumina[40] and SiC_w-reinforced Si_3N_4.[53] The effect of small amounts of glass on crack growth is illustrated in a study by Dalgleish et al.[90] They showed that glass inclusions in otherwise pure alumina act as nucleation sites for cracks. Cracks grew until they reached the end of the glass-impregnated zone, at which point they arrested: Fig. 4.13.

The behavior of cavities during deformation also depends on the refractoriness of the bonding matrix.† In a recent study,[33,34] stable cavities were observed to form at the grain boundaries of a grade of silicon nitride containing 4 wt.% yttria, even though there was very little glass at these boundaries: Fig. 4.14a. The cavities observed were reminiscent of Hull–

---

†Refractoriness of the interfacial glass can be defined by the softening temperature of the glass. High softening temperatures lead to higher temperatures for deformation, and hence, a higher chemical activity of the silicon nitride during deformation.

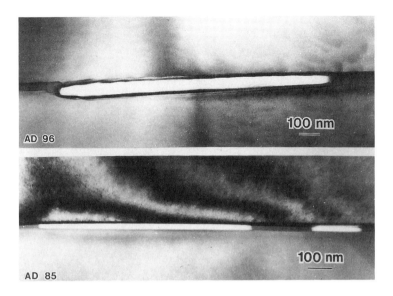

Fig. 4.11 Transmission electron micrographs showing cavity formation between narrow grain boundaries. AD96 contains about 8 vol.% glass whereas AD85 contains about 20 vol.% glass. Despite this difference in glass content, cavities nucleate between narrow boundaries where constraints are greatest. Once the cavity grows beyond the initial grain facet, it grows as a crack in the AD96, but tends to arrest in the AD85. After Wiederhorn *et al.*[95]

Rimmer cavities frequently discussed in the metals literature.[91–93] These cavities nucleated at two-grain interfaces, and then grew by diffusion of silicon nitride from the cavity, through the interfacial glass to the two-grain junction. This classical Hull–Rimmer mechanism seems to be active for the more refractory interfacial glass compositions, i.e., commercial grades of Si_3N_4 containing only Y_2O_3 as a sintering aid (NT154 and PY6). Depending on temperature, the lenticular cavities seem to grow to a maximum size and then arrest. Fewer, but larger, cavities are observed at higher temperatures. The role of these cavities in the failure process is not, at present, fully understood.

When the bonding phase in the silicon nitride is less refractory, i.e., glasses containing yttria and alumina, for example,[94] cavities spread rapidly along the boundary, quickly growing into cracks, in much the same way as is observed in alumina:[95] Fig. 4.14b. These crack-like cavities probably act as failure origins in grades of Si_3N_4 containing less-refractory bonding phase.

Penetration of both types of cavity into the silicon nitride grains, as shown in Fig. 4.14, suggests that cavity growth occurs by the diffusion of silicon nitride from the cavity surface to the grain boundary. The transition between the two types of cavity can be rationalized by the Chuang theory of cavity growth,[96] which relates the mode of cavity growth to the relative diffusion rate

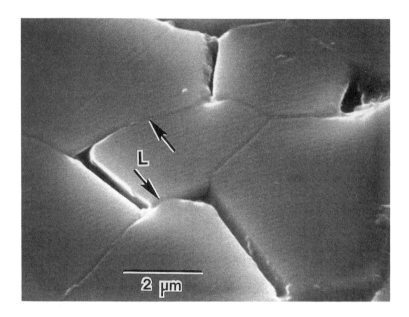

Fig. 4.12 Crack growth along grain boundaries in AD96. Because of the relatively narrow separation between grains, cavities grow into cracks once they start to propagate. Note that the crack is not completely open, but is joined across its faces by unbroken ligaments (L) that must also break for propagation to proceed. From Jakus *et al.*[105]

along the cavity surface. If diffusion is more rapid along the surface than along the grain boundary, the cavity maintains its equilibrium shape. If the converse is true, the cavity becomes crack-like.

When large amounts of boundary phase are present in a particulate composite, cavities arrest when they reach full facet size: Fig. 4.15. This behavior has been observed for a siliconized silicon carbide containing ≈33 vol.% silicon[64] and vitreous-bonded alumina containing ≈20 vol.% glass.[95] When cavities arrest, failure occurs by the coalescence of many small cavities, rather than by the propagation of one of these into a large crack.

When a large dominant crack is present initially, cavitation often occurs in front of the large crack, resulting in crack growth by direct extension of the main crack, or by linkage of the main crack within the damage region surrounding the main crack:[36,97] Fig. 4.16. Both processes were illustrated recently in vitreous-bonded alumina by Chan and Page[97–99] who showed that a critical value of cavity or microcrack density was required in front of a dominant crack to initiate crack growth. Cavitation occurred only above the crack growth threshold; below the threshold, cavities closed up as crack growth ceased. Crack growth and cavitation were closely connected in this material.

Factors governing cavity linkage have been discussed by Evans and Rana[6] and more recently by Wiederhorn *et al.*[36] The most important factor

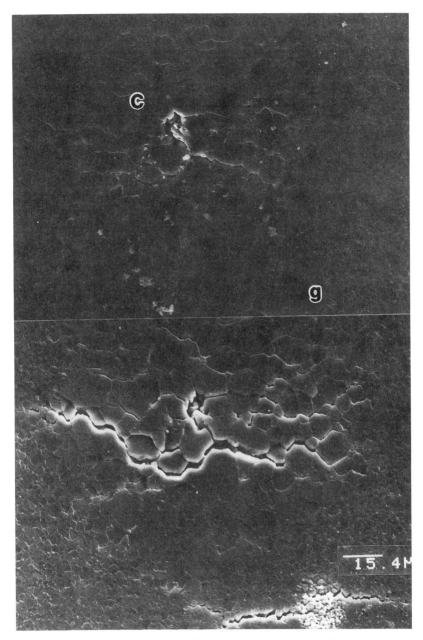

Fig. 4.13 Cavity nucleation and growth in aluminum oxide. In this figure, cavity nucleation is illustrated from either an area of coarse grains, c, in the structure, or a droplet of glass, g, that contaminated the surface. The crack that nucleated in the coarse-grain structure blunts out when it propagates into the fine-grain aluminum oxide. The cavity that nucleated from the region of glass remains sharp as long as there is glass at the crack tip; it is the more dangerous of the two cracks. From Dalgleish *et al*.[90]

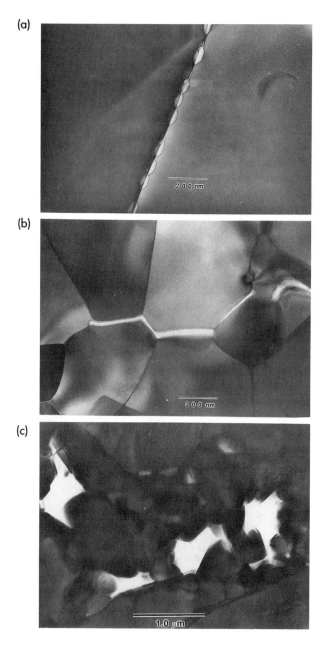

Fig. 4.14 Cavity formation in silicon nitride. (a) Lenticular cavities form at grain boundaries for grades of silicon nitride bonded by highly refractory glasses. (b) Crack-like cavities form at boundaries for grades of silicon nitride that are less refractory. (c) In addition to these interfacial cavities, more irregular, μm-size cavities are also observed at multigrain junctions in silicon nitride. From Hockey and Wiederhorn,[94] Luecke et al.[101] and Wiederhorn et al.[100]

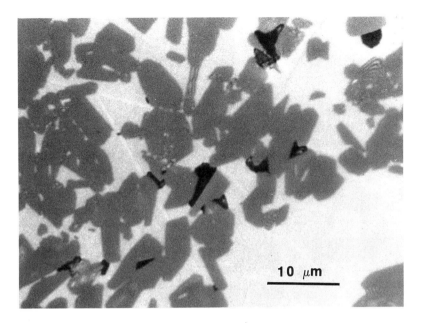

Fig. 4.15 Cavity formation in siliconized silicon carbide, KX01. Cavities are always located at Si/SiC interfaces, most often between two closely spaced SiC grains. As the cavities grew from the narrow space between the grains, they arrested on encountering a large pool of silicon. Applied tensile stress in vertical direction in the figure. From Hockey and Wiederhorn.[64]

controlling the coalescence process is the rate of damage accumulation as a function of creep strain. At a critical value of cavity density, a crack forms, which then propagates to failure. In a recent analysis of cavity coalescence in siliconized silicon carbide,[36] creep rupture lifetime was predicted to follow a Monkman–Grant equation. The slope and intercept of the Monkman–Grant curve was similar to that measured experimentally.

4.5.5 Cavitation and Creep

Cavitation enhances the creep rate by increasing the volume of the solid. Material originally present within the cavity finds its way to either the surface of the grains or spaces between the grains.[91,92] In either case, the volume of the solid increases, and the creep strain increases by the displaced volume. Several recent studies of the role of cavitation during creep indicate a linear relation between the cavitation volume and the creep strain. Figure 4.17, obtained by Carroll and Tressler,[58] illustrates this relationship for siliconized silicon carbide tested in tension at 1100°C. Data suggest a strain threshold of at least 0.1% before cavitation occurs in this material. The relative cavitation volume accounts for ≈80% of the total strain. If all of the cavitation volume goes

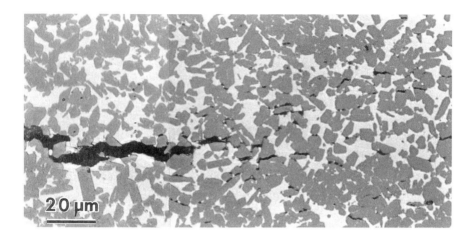

Fig. 4.16 Cavity distribution around the tip of an indentation crack in KX01. Cavities form a cloud around the crack tip, which later link up as the crack propagates. Taken from Wiederhorn *et al.*[36]

towards increasing the rate of creep, then as much as 80% of the creep rate is associated with the formation and growth of cavities. Other studies of creep cavitation also indicate a high contribution to the creep process.[98-101]

4.5.5.1 Cavity Nucleation

The linear relationship between cavitation volume and creep strain suggests a similar relationship between the cavitation process and the creep process. If cavitation is important, then both the nucleation rate and the cavity growth rate will affect the rate of creep. Unfortunately, none of the theories for cavitation can be used in a simple way to explain the creep behavior of composite materials. Stresses needed for cavity nucleation at grain boundaries or grain interfaces are generally a factor of 50 times higher than the applied stresses.[91,92,102] This being the case, cavitation requires one of three conditions to be satisfied: (1) the interface must be weak enough to cavitate at low stresses; (2) an additional stress such as a gas pressure is needed to cavitate the interface; or (3) the local stress at the interface is much higher than the applied stress through some stress magnification process such as constrained grain boundary sliding. This last explanation has been used by Page *et al.*[3,103] to explain cavitation in a number of vitreous-bonded ceramics, and is currently the most favored explanation for cavitation at interfaces. However, as grain boundary sliding occurs preferentially on boundaries oriented at large angles to the applied load, the explanation does not rationalize cavity nucleation on two grain boundaries that lie normal to the applied load. Nucleation on boundaries normal to the applied load is observed in silicon nitride,[33] aluminum nitride,[104] SiC whisker-reinforced silicon nitride[53] and vitreous-bonded alumina.[105]

Another problem with cavity nucleation as an explanation of creep

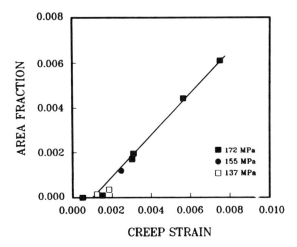

Fig. 4.17 Area density of cavities as a function of creep strain at 1100°C. The area density of cavities is proportional to the volume fraction of cavities. This figure indicates a linear relation between cavitation volume and strain. From Carroll and Tressler.[58] Additional analysis on this material suggests that most of the strain can be accounted for by cavity formation.[100]

behavior is the very strong dependence of the nucleation rate on stress. Classical nucleation theory yields the result that the cavitation rate, $\dot{N}$, is an exponential function of the applied stress:[58,106–108]

$$\dot{N} \, \alpha \, \exp[-16\pi\gamma^3/(3\sigma^2 kT)][(2 + \cos\Theta)(1 - \cos\Theta)^2/4] \tag{7}$$

where Θ is the dihedral angle between the matrix and reinforcement particle. If the creep rate is assumed to be proportional to the cavitation rate, then the stress exponent, n, of the creep rate, $\partial \ln\dot{\varepsilon}/\partial \ln\sigma = n$, will also be highly stress dependent. As noted by Riedel,[91] the dependence of the nucleation rate on the stress is sufficiently high as to give the impression of a critical stress for cavity nucleation. This conclusion is, however, inconsistent with values of the stress exponent, 3–10, found for ceramic matrix particulate composites. If nucleation is important in creep, then the rate of stress build-up for cavitation has to be constrained by the deformation rate of the material as a whole. The rate of cavity nucleation will then also be constrained by the deformation rate of the material.

4.5.5.2 Cavity Growth
Most theories of cavity growth are based on the Hull–Rimmer model[91–93] in which cavities grow by the transport of vacancies from boundaries, or interfaces to cavities that are subjected to a tensile stress. The driving force for cavity growth is the local stress in the vicinity of the cavity. During this process, grains are assumed to be rigid so that elastic stresses are not present at the

cavity perimeter. The effect of such a process is the transport of material from the cavity to the adjacent boundary where it plates-out, contributing to creep of the solid. The process is a variation on the Coble creep mechanism,[73] and, therefore, depends on the grain boundary diffusivity, i.e., the diffusivity of the matrix phase between grains. Transient creep occurs if the diffusivity decreases as a consequence of devitrification of the matrix. For this type of process, stress exponents for cavity growth range from 1 (for spherical cavities) to 3 (for rigid crack-like cavities). These exponents are too small to rationalize the values observed for the tensile creep of composite materials.

Higher values of the creep exponent are obtained by considering elastic crack-like cavities. This problem was treated by Chuang.[96] The growth of these cavities also involves diffusion of vacancies. However, stresses around the cavity differ substantially from those around a rigid crack-like cavity. Near the tip of the elastic crack-like cavity, stresses rise to high values, as they do for cracks. These stresses are relaxed by vacancy diffusion away from the crack tip. The vacancies are replaced by a wedge of solid material that relaxes the crack tip stress. Under an applied load, the growth rate, v, of these crack-like cavities can have a stress exponent as high as 12 ($v \propto \sigma^{12}$).[96] The cavity width, w, is an inverse function of the applied stress, so that narrower cavities propagate as the applied stress increases ($w \propto \sigma^{-4}$).[109] Since the rate of cavitation growth is the product of the crack width and the crack velocity, a stress exponent for creep as high as 8 might be expected. This value is large enough to include stress exponents of ≈ 6 measured on SiC_w-reinforced silicon nitride, in which crack-like cavities grew. The model does not, however, rationalize the high values of the stress exponent obtained on materials that have more equi-axed cavities. Although a model of elastic cavities might provide some justification for high stress exponents, such a model has yet to be developed.

As with cavity nucleation, the problem with most theories of cavity or microcrack growth is that the growth rate is determined by the local stress near the cavity and not the stress applied to the boundary of the specimen or component. As noted originally by Dyson,[110] these theories can apply only if the material surrounding the cavity creeps fast enough for the remotely applied stress to have its full effect on the cavities. If deformation surrounding the cavities is slow, then cavity growth is determined by the deformation process away from the cavity. In several particulate ceramic matrix composites there is evidence that constrained cavity growth (and nucleation) is the rule rather than the exception. In siliconized silicon carbide, for example, Carroll and Tressler[58,59] showed that as the cavitation volume increased during creep, the number of cavities also increased, but their size remained roughly constant. Their results suggest that after nucleation, cavities grew rapidly to a terminal size and then arrested. The lenticular interfacial cavities in silicon nitride show a similar behavior: after nucleation they grow to a terminal size. These cavities never seem to link up to form the larger cracks that can lead to failure.

A considerable amount of theoretical and experimental work on metals has led to the quantification of constrained cavity growth for a variety of failure mechanisms.[111] In many of these mechanisms, a reduction in cross-sectional area, resulting from cavity formation and growth, increases the rate of creep and leads to tertiary behavior. These theories are not applicable to particulate ceramic matrix composites, which usually fail during primary or secondary creep. Theories relating the creep behavior of particulate composites to the rate of cavity formation have yet to be developed.

4.5.5.3 Crack Growth

In the above discussion it is assumed that the volume displaced from the cavity by diffusion is the sole contribution to the creep rate. However, an additional elastic displacement associated with cavities also contributes to the creep rate. This form of cavitation creep is important in fibrous composites, in which crack-like cavities propagate along the fiber interface. Termed elastic creep, this type of creep was first analyzed by Venkateswaran and Hasselman[112] and later by Suresh and Brickenbrough.[113] As with other forms of cavitation, both the nucleation rate and the growth rate are important. From Venkateswaran and Hasselman, the creep rate, $\dot{\varepsilon}$, for a body containing penny-shaped cavities distributed throughout the volume is

$$\dot{\varepsilon} = 16(1 - \nu^2)\sigma r^2(3N\nu + r\dot{N}/(3E_0) \tag{8}$$

where ν is Poisson's ratio, r is the crack size, N is the density of cracks per unit volume, E_0 is Young's modulus of the cavity-free solid, and v is the crack velocity. If only pre-existing cavities are assumed to be present, then only crack growth will contribute to the creep process. Assuming cracks propagate according to a power law equation ($v \propto \sigma^n$), then the creep rate will also be a power function of the applied stress. As stress exponents for crack growth in ceramic composites range from ≈ 2 to ≈ 13, elastic creep offers a possible explanation for the high stress exponent observed in many ceramic matrix composites. The same restrictions concerning local and remote stresses, discussed above for cavity growth, also apply to crack growth. If the deformation rate is not fast enough, crack growth will be constrained by the deformation process. The stress exponent for creep will then be determined by the deformation process, not the fracture process.

4.6 Practical Aspects of Creep

Despite the large number of creep and creep rupture mechanisms,[91,92] the lifetime of structural materials at elevated temperatures is often a simple function of the creep rate. This relationship was first noted by Monkman and Grant[114] who presented an empirical relationship between rupture life and

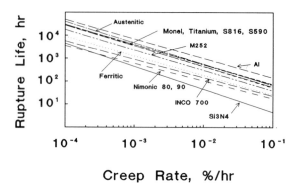

Creep Rate, %/hr

Fig. 4.18 Summary of data used to establish the validity of the Monkman–Grant equation[114] for high temperature metallic alloys. The range of rupture lifetimes corresponds to a total strain to failure of ≈ 3 to $\approx 30\%$. The line for Si_3N_4 was obtained from Fig. 4.20.

minimum creep rate. These authors summarized creep rupture data for 10 different alloy systems and showed that for each system the rupture time, t_r, and the minimum creep rate, $\dot{\varepsilon}_{min}$, were related by the following formula:

$$\log t_r + m \log \dot{\varepsilon}_{min} = c \qquad (9)$$

where m and c are empirical constants. For $\dot{\varepsilon}_{min}$ expressed in % strain h^{-1} and t_r in h, values of m ranged between 0.77 and 0.93; and values of c ranged between 0.48 and 1. For each alloy, the error bands for the rupture time covered approximately one order of magnitude in time. Furthermore, when plotted on a common graph, curves for all of the alloys tended to cluster in a band, about one order of magnitude in width along the time axis: Fig. 4.18. The thickness of the band is determined by the strain at failure. At the bottom of the band, alloys were strained $\approx 1.8\%$ before failure; whereas at the top of the band, alloys were strained $\approx 15\%$ before failure, both taken at a strain rate of 0.1% h^{-1}.[114]

Two important conclusions are reached from the Monkman–Grant equation. First, the most important engineering requirement to improve the lifetime of materials that fail by creep rupture is to improve creep behavior. Over the years, this has been the goal of much of the research on high temperature alloys. Second, theoretical treatments of creep rupture must be consistent with the Monkman–Grant equation, as has been shown by Ashby and Dyson.[111]

In ceramic composites, tensile creep rupture data are gradually being collected to show that these materials also obey the Monkman–Grant plot. Figure 4.19, for example, shows logarithmic plots of creep rate versus failure time for two grades of siliconized silicon carbide.[35,36,60] Data for each grade plot on a single curve, regardless of applied stresses and temperatures. In

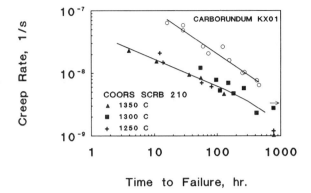

Fig. 4.19   Creep rupture behavior of siliconized silicon carbide. Both sets of data were taken over a range of temperatures and stresses. Regardless of the test conditions, data from each material tend to cluster about a single curve. Data from Refs. 36 and 60.

silicon nitride, two types of behavior are observed, depending on grade of material. For some materials, the data of the Monkman–Grant plot on a single curve, independent of temperature and applied stress;[31,53] in others, the Monkman–Grant curve is temperature dependent:[101] Fig. 4.20. Temperature-dependent Monkman–Grant curves seem to occur for the more refractory grades of silicon nitride.

In Fig. 4.21, creep rupture data from a number of different grades of silicon nitride are plotted in a Monkman–Grant format.[30,31,34,115,116] For purposes of comparison with metallic alloys, the temperature dependence of the Monkman–Grant curves has been ignored. As with the metallic alloys, the curves for all of the grades of material tend to plot within a relatively narrow band. These results imply that lifetime can be improved merely by improving creep rate: the lower the creep rate, the longer the lifetime.

In Fig. 4.18, Monkman–Grant curves for metallic alloys are compared with an average curve for silicon nitride.† The silicon nitride curve is seen to lie along the lower edge of the metal alloy curves. For a constant creep rate, this indicates less strain-at-failure for silicon nitride than for the metals. Indeed, most grades of silicon nitride fail after ≈1–2% strain, whereas metallic alloys can sustain ten times as much strain before failure. This extra strain is not, however, necessarily useful. For engineering purposes, small amounts of strain, 1–2%, often distort structural parts beyond limits of acceptability. As a rule, 1% strain is considered the outer limit of tolerance for structural applications. This being the case, silicon nitride has an adequate degree of strain tolerance for creep at elevated temperatures.

---

†This "average" curve was obtained by visually placing a straight line on the figure, so that it passed through a region that contained the bulk of the Si_3N_4 data. The data for the metallic alloys are taken from Ref. 1.

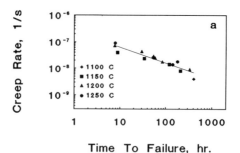

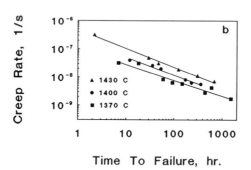

Fig. 4.20 Monkman–Grant curves for two commercial grades of silicon nitride. Some grades give curves that are temperature-independent: (a) AY6, SiC_w-reinforced; others give a series of curves depending on temperature: (b) NT154. The temperature independent curves have creep rate exponents, m, for the Monkman–Grant equation, $t_f = c\dot{\varepsilon}_{min}^{-m}$, that are approximately 1, whereas the creep rate exponent for the temperature-dependent curves are greater than 1: e.g., ≈ 1.7 for NT154.

The importance of creep resistance to high temperature reliability can be illustrated by comparing NT154, one of the more promising grades of this material, with NC132, a grade of silicon nitride that was "state of the art" ten years ago. MgO was used as the sintering aid in NC132, whereas Y_2O_3 is used for the NT154. Glasses containing Y_2O_3 are much more refractory than those containing MgO and, therefore, are more viscous and hence more creep-resistant at high temperatures. Also, Y_2O_3-containing glasses readily devitrify at high temperature, which also increases the creep resistance of the silicon nitride composite. For both materials, the minimum creep rate can be expressed as a function of the applied stress and temperature by the following equation:

$$\dot{\varepsilon}_{min} = A\sigma^n \exp(-\Delta H/RT) \tag{10}$$

Both materials also follow a Monkman–Grant curve:

$$t_f = c\dot{\varepsilon}_{min}^{-m} \tag{11}$$

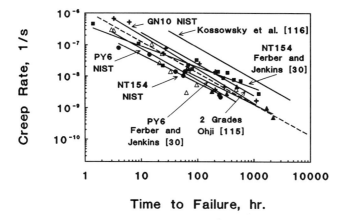

Time to Failure, hr.

Fig. 4.21 Monkman–Grant curves for experimental and commercial grades of silicon nitride. The results from seven different studies are plotted on the same graph. With the exception of the data by Kossowsky *et al.*,[116] data tend to plot within a relatively narrow band. Within a factor of three in lifetime prediction, all of these data can be represented by the dashed line. It is this line that is plotted in Fig. 4.17 with data from high temperature alloys.

Substituting Eqn. (10) into (11), the lifetime is obtained for both materials as a function of stress and temperature:[33,34,36]

$$t_r = cA^{-m}\sigma^{-mn}\exp(m\Delta H/RT) \tag{12}$$

Using Eqn. (12), the lifetime of NC132 is compared with NT154 at 1300°C: Fig. 4.22. The solid line represents actual strength data obtained on NC132 at 1300°C. The dashed lines represent extrapolations of data. As can be seen, five orders of magnitude in lifetime separate the two materials at 100 MPa applied stress. This improvement in lifetime is primarily a consequence of the development of silicon nitride with a more refractory glass at grain boundaries. Current work on this type of material is being directed towards further improvement of the refractoriness of the glass phase. Work is also being directed towards increasing the quantity and size of elongated grains within the material. By so doing, both the creep resistance and the toughness of silicon nitride will be improved.

The data in Fig. 4.22 illustrate the importance of creep to lifetime at high temperatures. Even though the NC132 had a longer lifetime at a given creep rate, the creep resistance of the NT154 was so superior that lifetime as a function of temperature and stress was favored by the NT154. This same benefit accrues to ceramic composites when creep determines lifetime, as can be shown by comparing the rupture life of a structural alloy with silicon nitride. In Fig. 4.23 the rupture strength of NT154 is compared with Mar-M246. For stresses up to 400 MPa, superior creep behavior gives the NT154 a temperature advantage of ≈350°C over the metallic alloy.

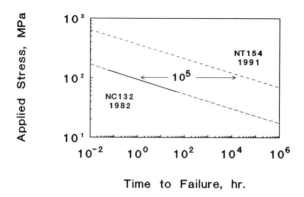

Fig. 4.22 A comparison of the measured lifetime of NC132 (solid line) with projected lifetimes of NT154 (dashed line). Test temperature, 1300°C. The improvement of lifetime of ≈5 orders of magnitude is mainly a consequence of the improved creep resistance of the NT154.

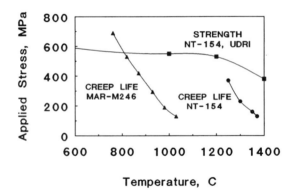

Fig. 4.23 A comparison of 100 h creep-life strength of NT154 with Mar-M246. As can be seen, NT154 exhibits a 350°C advantage over the Mar-M246 for stresses up to 500 MPa. The strength data were obtained from Hecht et al.[117]

4.7 Summary

In this paper, the importance of particle and whisker reinforcement to creep and creep rupture behavior of ceramics is discussed. Particle and whisker additions generally increase both the fracture toughness and creep resistance of structural ceramics. These additions also act as nucleation sites for cavities. Cavities form preferentially in tensile specimens. This results in a creep asymmetry, in which composites creep faster in tension than in compression. As a consequence of cavitation, the stress exponent for creep in tension: 6–10, is greater than when measured in compression: 1–3. Reasons for the higher stress exponent in tension are not yet understood.

Studies on the rheology of two-phase particulate systems suggest the existence of a deformation threshold that depends on the concentration of particles in the composite. Below this threshold (≈ 68 vol.% for spherical particles), deformation occurs primarily by the flow of the composite matrix. Particles increase the effective viscosity of the matrix by absorbing energy and by forming clusters.

Above the threshold, deformation occurs as a consequence of direct particle interaction. Several mechanisms of interaction have been suggested: solution–precipitation; flow of fluid between particles; and cavity formation at the particle matrix interface. These theories of creep suggest several rules to improve creep behavior: (1) increase the viscosity of the matrix phase in multiphase materials; (2) decrease the volume fraction of the intergranular phase; (3) increase the grain size; (4) use fiber or whisker reinforcement when possible. As the creep rupture life is inversely proportional to creep rate, lifetime can be improved by improving creep resistance.

Acknowledgment

Support of the Ceramic Technology for Advanced Heat Engines Program, U.S. Department of Energy, under Interagency Agreement No. DE-AI05-85OR21569, is gratefully acknowledged.

References

1. R. W. Hertzberg, *Deformation and Fracture Mechanics of Engineering Materials*, John Wiley and Sons, New York, NY, 1976.
2. A. G. Evans and B. J. Dalgleish, "Some Aspects of the High Temperature Performance of Ceramics and Ceramic Composites," in *Creep and Fracture of Engineering Materials and Structures*, eds. B. Wilshire and R. W. Evans, The Institute of Metals, London, U.K., 1987, pp. 929–955.
3. K. S. Chan and R. A. Page, "Creep Damage Development in Structural Ceramics," *J. Am. Ceram. Soc.*, **76**[4], 903–926 (1993).
4. A. B. Metzner, "Rheology of Suspensions in Polymeric Liquids," *J. Rheology*, **29**[6], 739–775 (1985).
5. G. M. Pharr and M. F. Ashby, "On Creep Enhanced by a Liquid Phase," *Acta Metall.*, **31**, 129–138 (1983).
6. A. G. Evans and A. Rana, "High Temperature Failure Mechanisms in Ceramics," *Acta Metall.*, **28**, 129–141 (1980).
7. P. F. Becher and G. C. Wei, "Toughening Behavior in SiC Whisker-Reinforced Alumina," *J. Am. Ceram. Soc.*, **67**[12], C267–C269 (1984).
8. G. C. Wei and P. F. Becher, "Development of SiC-Whisker-Reinforced Ceramics," *Am. Ceram. Soc. Bull.*, **64**[2], 298–304 (1985).
9. T. N. Tiegs and P. F. Becher, "Whisker Reinforced Ceramic Composites," in *Tailoring Multiphase and Composite Ceramics*, Materials Science Research Series, Plenum Press, New York, NY, 1986, pp. 639–647.

10. P. D. Shalek, J. J. Petrovic, G. F. Hurley, and F. D.Gac, "Hot-Pressed SiC Whisker/Si_3N_4 Matrix Composites," *Am. Ceram. Soc. Bull.*, **65**[2], 351–356 (1986).
11. N. Claussen, K. L. Weisskopf, and M. Rühle, "Tetragonal Zirconia Polycrystals Reinforced with SiC Whiskers," *J. Am. Ceram. Soc.*, **69**[3], 288–292 (1986).
12. T. Tiegs, "SiC Whisker-Reinforced Sialon Composites: Effect of Sintering Aid Content," *Ceram. Eng. Sci. Proc.*, **10**[9–10], 1101–1107 (1989).
13. H. W. Lee and M. D. Sacks, "Pressureless Sintering of Al_2O_3/SiC Whisker Composites," *Ceram. Eng. Sci. Proc.*, **10**[7–8], 720–729 (1989).
14. R. J. Brook (ed.), *Concise Encyclopedia of Advanced Ceramic Materials*, Pergamon Press, Oxford, U.K., 1991.
15. R. J. Schaefer and M. Linzer (eds.), *Hot Isostatic Pressing: Theory and Application*, ASM International, Materials Park, OH, 1991.
16. A. J. Whitehead, T. F. Page, and I. Higgins, "Novel Siliconized Mixed-Phase Ceramics," *Ceram. Eng. Sci. Proc.*, **10**[9–10], 1108–1120 (1989).
17. Michael C. Kasprzyk, "Large Silicon Carbide Radiant Tube Production Process," in *Silicon Carbide '87*, ed. D. Cawley, The American Ceramic Society, Columbus, OH, 1989, pp. 387–394.
18. *Industrial Heating*, Nov. 1990.
19. *Industrial Heating*, June 1990.
20. G. Ziegler, J. Heinrich, and G. Wötting, "Relationships between Processing, Microstructure and Properties of Dense and Reaction-Bonded Silicon Nitride," *J. Mater. Sci.*, **22**, 3041–3086 (1987).
21. F. F. Lange, "Silicon Nitride Polyphase Systems: Fabrication, Microstructure and Properties," *International Metals Reviews*, **1**, 1–20 (1980).
22. D. R. Clarke and G. Thomas, "Boundary Phases in a Hot-Pressed MgO Fluxed Silicon Nitride," *J. Am. Ceram. Soc.*, **60**[11–12], 491–495 (1977).
23. L. K. V. Lou, T. E. Mitchell, and A. H. Heuer, "Impurity Phases in Hot-Pressed Si_3N_4, *J. Am. Ceram. Soc.*, **61**[9–10], 392–396 (1978).
24. D. R. Clarke, "Detection of Thin Intergranular Films by Electron Microscopy," *Ultramicroscopy*, **4**, 33–44 (1979).
25. A. J. Pyzik, D. F. Carroll, C. J. Hwang, and A. R. Prunier, "Self-Reinforced Silicon Nitride—A New Microengineered Ceramic," in *4th International Symposium on Ceramic Materials and Components for Engines*, eds. R. Carlsson, T. Johansson and L. Kahlman, Elsevier Science Publishers, Barking, U.K., 1992, pp. 585–593.
26. C. W. Li and J. Yamanis, "Super-Tough Silicon Nitride with R-Curve Behavior," *Ceram. Eng. Sci. Proc.*, **10**[7–8], 632–645 (1989).
27. K. Suzuki and M. Sasaki, "Pressureless Sintering of Silicon Carbide," in *Fundamental Structural Ceramics*, eds. S. Sōmiya and R. C. Bradt, Terra Scientific Publishing Co. Tokyo, Japan, 1987, pp. 75–87.
28. T.-J. Chuang, D. F. Carroll, and S. M. Wiederhorn "Creep Rupture of a Metal–Ceramic Particulate Composite," *Seventh International Conference on Fracture*, in *Advances in Fracture Research*, Vol. 4, eds., K. Salama, K. Ravi-Chandler, D. M. R. Taplin, and P. Rama Rao, Pergamon Press, New York, NY, 1989, pp. 2965–2976.
29. M. K. Ferber, M. G. Jenkins, and V. J. Tennery, "Comparison of Tension, Compression, and Flexure Creep for Alumina and Silicon Nitride Ceramics," *Ceram. Eng. Sci. Proc.*, **11**[7], 1028–1045 (1990).
30. M. K. Ferber and M. G. Jenkins, "Evaluation of the Elevated-Temperature Mechanical Reliability of a HIP-ed silicon Nitride," *J. Am. Ceram. Soc.*, **75**[9], 2453–2462 (1992).

31. S. M. Wiederhorn, R. Krause, and D. C. Cranmer, "Tensile Creep Testing of Structural Ceramics," in *Proceedings of the Annual Automotive Technology Development Contractors' Coordination Meeting 1991*, (P-256, Dearborn, MI, October 28–31, 1991), Society of Automotive Engineers, Warrendale, PA, 1992, pp. 273–280.
32. R. M. Arons and J. K. Tien, "Creep and Strain Recovery in Hot-Pressed Silicon Nitride," *J. Mater. Sci.*, **15**, 2046–2058 (1980).
33. S. M. Wiederhorn, B. J. Hockey, D. C. Cranmer, and R. Yeckley, "Transient Creep Behavior of Hot Isostatically Pressed Silicon Nitride," *J. Mater. Sci.*, **28**, 445–453 (1993).
34. D. C. Cranmer, B. J. Hockey, S. M. Wiederhorn, and R. Yeckley, "Creep and Creep-Rupture of HIP-ed Si_3N_4," *Ceram. Eng. Sci. Proc.*, **12**[9–10] 1862–1872 (1991).
35. S. M. Wiederhorn, D. E. Roberts, T.-J. Chuang, and L. Chuck, "Damage-Enhanced Creep in Siliconized Silicon Carbide: Phenomenology," *J. Am. Ceram. Soc.*, **71**, 602–608 (1988).
36. S. M. Wiederhorn, B. J. Hockey, and T.-J. Chuang, "Creep and Creep Rupture of Structural Ceramics," in *Toughening Mechanisms in Quasi-Brittle Materials*, ed. S. P. Shah, Kluwer Academic Publishers, The Netherlands, 1991, pp. 555–576.
37. R. J. Jacodine, "Surface Energy of Germanium and Silicon," *J. Electrochem. Soc.*, **110**, 524–527 (1963).
38. K. Kromp, T. Haug, R. F. Pabst, and V. Gerold, "C* for Ceramic Materials? Creep Crack Growth at Extremely Low Loading Rates at High Temperatures using Two-Phase Ceramic Materials," in *Creep and Fracture of Engineering Materials and Structures*, eds. B. Wilshire and R. W. Evans, The Institute of Metals, London, U.K., 1987, pp. 1021–1032.
39. H. J. Frost and M. F. Ashby, *Deformation-Mechanism Maps: The Plasticity and Creep of Metals and Ceramics*, Pergamon Press, New York, NY, 1982.
40. S. M. Wiederhorn, B. J. Hockey, R. F. Krause, Jr., and K. Jakus, "Creep and Fracture of a Vitreous-Bonded Aluminum Oxide," *J. Mater. Sci.*, **21**, 810–824 (1986).
41. G. Urbain, F. Cambier, M. Deletter, and M. R. Anseau, "Viscosity of Silicate Melts," *Trans. J. Brit. Ceram. Soc.*, **80**, 139–141 (1981).
42. K. Sadananda, H. Jones, J. Feng, J. J. Petrovic, and A. K. Vasudevan, "Creep of Monolithic and SiC Whisker-Reinforced $MoSi_2$," *Ceram. Eng. Sci. Proc.*, **12**[9–10], 1671–1678 (1991).
43. S. Bose, "Creep Deformation of Molybdenum Disilicide," (Paper presented at the First High Temperature Structural Silicides Workshop, Nov. 4–6, 1991, The National Institute of Standards and Technology, Gaithersburg, MD), *Mat. Sci. Engr.* **A155**, 217–225 (1992).
44. S. M. Wiederhorn, R. J. Gettings, D. E. Roberts, C. Ostertag, and J. J. Petrovic, "Tensile Creep of Silicide Composites," *Mat. Sci. Engr.*, **A155**, 209–215 (1992).
45. A. H. Chokshi and J. R. Porter, "Creep Deformation of an Alumina Matrix Composite Reinforced with Silicon Carbide Whiskers," *J. Am. Ceram. Soc.*, **68**[6], C144–C145 (1985).
46. H.-T. Lin and P. F. Becher, "Creep Behavior of a SiC Whisker Reinforced Alumina," *J. Am. Ceram. Soc.*, **73**[5], 1378–1381 (1990).
47. P. Lipetzky, S. R. Nutt, and P. F. Becher, "Creep Behavior of an Al_2O_3-SiC Composite," in *High Temperature/High Performance Composites*, Materials Research Society Symposium Proceedings, Vol. 120, eds. F. D. Lemkey, S. G. Fishman, A. G. Evans, and J. R. Strife, Materials Research Society, Pittsburgh, PA, 1988, pp. 271–277.

48. J. R. Porter, "Observations of Non-Steady State Creep in SiC Whisker Reinforced Alumina," in *Whisker- and Fiber-Toughened Ceramics*, eds. R. A. Bradley, D. E. Clark, D. C. Larsen, and J. O. Stiegler, ASM International, Metals Park, OH, 1988, pp. 147–152.

49. K. Xia and T. G. Langdon, "The Mechanical Properties at High Temperatures of SiC Whisker-Reinforced Alumina," in *High Temperature/High Performance Composites*, Materials Research Society Symposium Proceedings, Vol. 120, eds. F. D. Lemkey, S. G. Fishman, A. G. Evans, and J. R. Strife, Materials Research Society, Pittsburgh, PA, 1988, pp. 265–271.

50. D. A. Koester, R. D. Nixon, S. Chevacharoenkul, and R. R. Davis, "High Temperature Creep of SiC Whisker-Reinforced Ceramics," in *Proceedings of the International Conference on Whisker- and Fiber-Toughened Ceramics*, eds. R. A. Bradley, D. E. Clark, D. C. Larsen, and J. O. Stiegler, ASM International, Metals Park, OH, 1988, pp. 139–145.

51. M. Backhous-Ricoult, J. Castaing, and J. L. Routbort, "Creep of SiC-Whisker Reinforced Si_3N_4," *Revue Phys. Appl.*, **23**, 239–249 (1988).

52. S. T. Buljan, J. G. Baldoni, M. L. Huckabee, J. T. Neil, and G. Zilberstein, "SiC Whisker-Reinforced Si_3N_4," in *Proceedings of the Twenty-Fifth Automotive Technology Development Contractors' Coordination Meeting* (P-209, Dearborn, MI, October 26–29, 1987), Society of Automotive Engineers, Warrendale, PA, 1988, pp. 137–144.

53. B. J. Hockey, S. M. Wiederhorn, W. Liu, J. G. Baldoni, and S.-T. Buljan, "Tensile Creep of Whisker-Reinforced Silicon Nitride," *J. Mater. Sci.*, **26**, 3931–3939 (1991).

54. R. D. Nixon, D. A. Koester, S. Chevacharoenkul, and R. F. Davis, "Steady State Creep of Hot Pressed SiC Whisker Reinforced Silicon Nitride," *Comp. Sci. Tech.*, **37**, 313–328 (1990).

55. S. T. Buljan, M. L. Huckabee, J. G. Baldoni, and J. T. Neil, Ceramic Matrix Composites, GTE Laboratories, Inc., Waltham, MA, Semiannual Report prepared for Oak Ridge National Laboratories, TN, October 19, 1989.

56. R. Morrell and K. H. G. Ashbee, "High Temperature Creep of Lithium Zinc Silicate Glass-Ceramics, Part 1, General Behavior and Creep Mechanisms," *J. Mater. Sci.*, **8**, 1253–1270 (1973).

57. J.-G. Wang and R. Raj, "Mechanism of Superplastic Flow in a Fine-Grained Ceramic Containing Some Liquid Phase," *J. Am. Ceram. Soc.*, **67**, 381–444 (1984).

58. D. F. Carroll and R. E. Tressler, "Accumulation of Creep Damage in a Siliconized Silicon Carbide," *J. Am. Ceram. Soc.*, **71**[6], 472–477 (1988).

59. D. F. Carroll and R. E. Tressler, "Effect of Creep Damage on the Tensile Creep Behavior of a Siliconized Silicon Carbide," *J. Am. Ceram. Soc.*, **72**[1], 49–53 (1989).

60. S. M. Wiederhorn, W. Liu, D. F. Carroll, and T.-J. Chuang, "Creep Rupture of Two Phase Ceramics," Paper presented at the 91st Annual Meeting and Exposition of the American Ceramic Society, Columbus, OH, April 23–27, 1989, Paper 7-JIII-89.

61. S. M. Wiederhorn and B. J. Hockey, "High Temperature Degradation of Structural Composites," *Ceramics International*, **17**, 243–252 (1991).

62. C.-F. Chen, S. M. Wiederhorn, and T.-J. Chuang, "Cavitation Damage during Flexural Creep of Sialon-YAG Ceramics," *J. Am. Ceram. Soc.*, **74**[7], 1658–1662 (1991).

63. F. F. Lange, B. I. Davis, and D. R. Clarke, "Compressive Creep of Si_3N_4/MgO Alloys, Part I, Effect of Composition," *J. Mater. Sci.*, **15**, 601–610 (1980).

64. B. J. Hockey and S. M. Wiederhorn, "Effect of Microstructure on the Creep of Siliconized Silicon Carbide," *J. Am. Ceram. Soc.*, **75**[7], 1822–1830 (1992).
65. A. Einstein, "Investigations on the Theory of Brownian Movement," *Annalen der Physik*, **4**, 289–306 (1906); corrections, ibid., **34**, 591–592 (1911); edited with notes, ed. R. Furth, Dover Publications, New York, NY, 1956.
66. D. G. Thomas, "Transport Characteristics of Suspension: VIII. A Note on the Viscosity of Newtonian Suspensions of Uniform Spherical Particles," *J. Colloid Sci.*, **20**, 267–277 (1965).
67. T. Kitano, T. Kataoka, and T. Shirota, "An Empirical Equation of the Relative Viscosity of Polymer Melts Filled with Various Inorganic Fillers," *Rheol. Acta*, **20**, 207–209 (1981).
68. C. K. Yoon and I.-W. Chen, "Superplastic Flow of Two-Phase Ceramics Containing Rigid Inclusions-Zirconia/Mullite Composites," *J. Am. Ceram. Soc.*, **73**[6], 1555–1565 (1990).
69. P. M. Adler, A. Nadim, and H. Brenner, "Rheological Models of Suspensions," *Adv. Chem. Eng.*, **15**, 1–72 (1990).
70. R. Blanc and E. Guyon, "Transport Properties of an Assembly of Spheres," *Annals of the Israel Physical Society*, **5**, 229–250 (1983).
71. J. L. Bouillot, C. Camoin, M. Belzons, R. Blanc, and E. Guyon, "Experiments on 2-D Suspensions," *Adv. Coll. Interf. Sci.*, **17**, 299–305 (1982).
72. R. Blanc, J. L. Bouillot, C. Camoin, and M. Belzons, "Cluster Statistics in a Bidimensional Suspension: Comparison with Percolation," *Rheol. Acta*, **22**, 505–511 (1983).
73. R. L. Coble, "A Model for Boundary Diffusion Controlled Creep in Ceramic Materials," *J. Appl. Phys.*, **34**, 1679–1682 (1963).
74. W. J. Moore, *Physical Chemistry*, 3rd edn., Prentice-Hall, Englewood Cliffs, NJ, 1962.
75. M. Gürtler and G. Grathwohl, "Tensile Creep Testing of Sintered Silicon Nitride," in *Proceedings of the Fourth International Conference on Creep and Fracture of Engineering Materials and Structures*, Institute of Metals, London, U.K., 1990, pp. 399–408.
76. D. C. Drucker, "Engineering and Continuum Aspects of High-Strength Materials," in *High Strength Materials*, ed. V. F. Zackay, Wiley, New York, NY, 1965, pp. 795–833.
77. F. F. Lange, "Non-Elastic Deformation of Polycrystals with a Liquid Boundary Phase," in *Deformation of Ceramic Materials*, eds., R. C. Bradt and R. E. Tressler, Plenum Press, New York, NY, 1972, pp. 361–381.
78. J. R. Dryden, D. Kucerovsky, D. S. Wilkinson, and D. F. Watt, "Creep Deformation due to a Viscous Grain Boundary Phase," *Acta Metall.*, **37**[7], 2007–2015 (1989).
79. K.-D. Debschütz, R. Danzer, and G. Petzow, "Finite Element Modeling of Ceramic Materials with a Viscous Grain Boundary Phase," in *Ceramics Today—Tomorrow's Ceramics*, ed., P. Vincenzini, Elsevier Science Publishers B.V., Amsterdam, The Netherlands, 1991, pp. 727–736.
80. M. M. Chadwick, D. S. Wilkinson, and J. R. Dryden, "Creep Due to a Non-Newtonian Grain Boundary Phase, *J. Am. Ceram. Soc.*, **75**[9], 2327–2334 (1992).
81. M. M. Chadwick, R. S. Jupp, and D. S. Wilkinson, "Creep Behavior of a Sintered Silicon Nitride," *J. Am. Ceram. Soc.*, **76**[2], 385–396 (1993).
82. F. F. Lange, B. I. Davis, and D. R. Clarke, "Compressive Creep of

Si₃N₄/MgO Alloys, Part 2, Source of Viscoelastic Effect," *J. Mater. Sci.*, **15**, 611–615 (1980).

83. G. D. Quinn and W. R. Braue, "Fracture Mechanism Maps for Advanced Structural Ceramics, Part 2, Sintered Silicon Nitride," *J. Mater. Sci.*, **25**, 4377–4392 (1990).

84. O. Reynolds, "On the Dilatancy of Media Composed of Rigid Particles in Contact. With Experimental Illustrations," *Phil. Mag. Series 5*, **20**[127], 469–481 (1885).

85. T. W. Lambe and R. V. Whitman, *Soil Mechanics*, John Wiley and Sons, New York, NY, 1969.

86. K. Terzaghi, *Theoretical Soil Mechanics*, John Wiley and Sons, New York, NY, 1943.

87. D. W. Taylor, *Fundamentals of Soil Mechanics*, John Wiley and Sons, New York, NY, 1948.

88. R. L. Tsai and R. Raj, "Creep Fracture in Ceramics Containing Small Amounts of a Liquid Phase," *Acta Metall.*, **30**, 1043–1058 (1982).

89. K. James and K. H. G. Ashbee, "Plasticity of Hot Glass-Ceramics," in *Progress in Materials Science*, Vol. 21, eds. B. Chalmers, J. W. Christian, and T. B. Massalski, Pergamon Press, Oxford, U.K., 1976, pp. 1–59.

90. B. J. Dalgleish, S. M. Johnson, and A. G. Evans, "High-Temperature Failure of Polycrystalline Alumina: I, Crack Propagation," *J. Am. Ceram. Soc.*, **67**[11], 741–750 (1984).

91. H. Riedel, *Fracture at High Temperatures*, Springer-Verlag, Berlin, Germany, 1987.

92. H. E. Evans, *Mechanisms of Creep Fracture*, Elsevier Applied Science Publishers, London, U.K., 1984.

93. D. Hull and D. E. Rimmer, "The Growth of Grain-Boundary Voids under Stress," *Phil. Mag.*, **4**, 673–687 (1959).

94. B. J. Hockey and S. M. Wiederhorn, "Tensile Creep of Y-Si₃N₄: II. Cavitation," in *Proceedings of the 49th Annual Meeting of the Electron Microscopy Society of America*, August 4–9, 1991.

95. S. M. Wiederhorn, B. J. Hockey, and R. F. Krause, Jr., "Influence of Microstructure on Creep Rupture," in *Ceramic Microstructures '86: Role of Interfaces*, eds. J. A. Pask and A. G. Evans, Plenum Press, New York, NY, 1987, pp. 795–806.

96. T.-J. Chuang, "A Diffusive Crack-Growth Model for Creep Fracture," *J. Am. Ceram. Soc.*, **65**[2], 93–103 (1982).

97. R. A. Page, K. S. Chan, D. L. Davidson, and J. Lankford, "Micromechanics of Creep-Crack Growth in a Glass-Ceramic," *J. Am. Ceram. Soc.*, **73**[10], 2977–2986 (1990).

98. K. S. Chan and R. A. Page, "Creep-Crack Growth by Damage Accumulation in a Glass-Ceramic," *J. Am. Ceram. Soc.*, **74**[7], 1605–1613 (1991).

99. K. S. Chan and R. A. Page, "Origin of the Creep-Crack Growth Threshold in a Glass-Ceramic," *J. Am. Ceram. Soc.*, **75**[3], 603–612 (1992).

100. S. M. Wiederhorn, B. A. Fields, and B. J. Hockey, "Fracture of Silicon Nitride and Silicon Carbide at Elevated Temperatures," *Mat. Sci. Eng.*, **A176**, 51–60 (1994).

101. W. Luecke, S. M. Wiederhorn, B. J. Hockey, and G. G. Long, "Cavity Evolution During Tensile Creep of Si₃N₄," in *Silicon Nitride Ceramics: Scientific and Technological Advances*, Mat. Res. Soc. Symp. Proceedings, Vol. 287, eds. I. W. Chen, P. F. Becher, M. Mitomo, G. Petzow, and T. S. Yen, Materials Research Society, Pittsburgh, PA, 1993, pp. 467–472.

102. M. D. Thouless and A. G. Evans, "Nucleation of Cavities During Creep of

Liquid-Phase-Sintered Materials," *J. Am. Ceram. Soc.*, **67**[11], 721–727 (1989).

103. R. A. Page and K. S. Chan, "Stochastic Aspects of Creep Cavitation in Ceramics," *Metall. Trans. A.*, **18A**[11], 1843–1854 (1987).
104. Z. C. Jou and A. V. Virkar, "High-Temperature Creep and Cavitation of Polycrystalline Aluminum Nitride," *J. Am. Ceram. Soc.*, **73**[7], 1928–1935 (1990).
105. K. Jakus, S. M. Wiederhorn, and B. J. Hockey, "Nucleation and Growth of Cracks in Vitreous-Bonded Aluminum Oxide at Elevated Temperatures," *J. Am. Ceram. Soc.*, **69**[10], 725–731 (1986).
106. J. C. Fisher, "The Fracture of Liquids," *J. Appl. Phys.*, **19**, 1062 (1948).
107. R. Raj, "Nucleation of Cavities at Second Phase Particles in Grain Boundaries," *Acta Metall.*, **26**, 995–1006 (1978).
108. R. Raj and M. F. Ashby, "Intergranular Fracture at Elevated Temperature," *Acta Metall.*, **23**, 653–666 (1975).
109. T.-J. Chuang, private communication.
110. B. F. Dyson, "Constraints on Diffusional Cavity Growth Rates," *Met. Sci.*, **10**[10], 349–353 (1976).
111. M. F. Ashby and B. F. Dyson, "Creep Damage Mechanics and Micromechanisms," in *Advances in Fracture Research* (Proceedings of the 6th Intl. Conf. on Fracture, New Delhi, India, Dec. 4–10, 1984), eds. S. R. Valluri, D. M. R. Taplin, P. Ramarao, J. F. Knott, and R. Dubey, Plenum Press, New York, NY, 1984, pp. 3–30.
112. A. Venkateswaran and D. P. H. Hasselman. "Elastic Creep of Stressed Solids due to Time-Dependent Changes in Elastic Properties," *J. Mater. Sci.*, **16**, 1627–1632 (1981).
113. S. Suresh and J. R. Breckenbrough, "A Theory for Creep by Interfacial Flaw Growth in Ceramics and Ceramic Composites," *Acta Metall. Mater.*, **38**[1], 55–68 (1990).
114. F. C. Monkman and N. J. Grant, "An Empirical Relationship between Rupture Life and Minimum Creep Rate in Creep-Rupture Tests," *Proc. ASTM*, **56**, 593–620 (1956).
115. Tatsuki Ohji and Yukihiko Yamauchi, "Tensile Creep and Creep Rupture Behaviors of Silicon Nitride," *J. Am. Ceram. Soc.*, **76**[12], 3105–3112 (1993).
116. R. Kossowsky, D. G. Miller, and E. S. Diaz, "Tensile and Creep Strengths of Hot Pressed Si_3N_4," *J. Mater. Sci.*, **10**, 983–997 (1975).
117. N. L. Hecht, S. M. Goodrich, L. Chuck, and D. E. McCullum, "Effects of the Environment on the Mechanical Behavior of Ceramics," in *Proceedings of the Annual Automotive Technology Development Contractors' Coordination Meeting*, (P-243, Dearborn, MI, October 22–25, 1990), Society of Automotive Engineers, Warrendale, PA, 1991, pp. 199–212.

CHAPTER 5

Elevated Temperature Creep Behavior of Continuous Fiber-Reinforced Ceramics

J. W. Holmes and Xin Wu

5.1 Introduction

Over the past decade, a considerable amount of research has been devoted to the development of ceramics reinforced with continuous ceramic fibers.[1-9] Compared with other ceramic matrix composites reinforced with particulates or whiskers, fiber-reinforced ceramics offer significantly higher fracture toughness, together with the ability to fail in a non-catastrophic manner (commonly referred to as "graceful" failure). For applications involving multiaxial loading, composites can be fabricated with 2-D cross-ply, 2-D woven, or 3-D fiber architecture. A primary driving force for the development of fiber-reinforced ceramics is the need for low-density, high temperature materials to replace the Ni-base superalloys used in power generation equipment. The density of typical ceramic matrix composites, which ranges from approximately 2 to 3 g cm^{-3}, is roughly ¼ to ⅓ that of Ni-base superalloys. Moreover, when used in advanced gas turbines, current generation Ni-base superalloys are subjected to temperatures that can exceed 90% of their incipient melting point, which is typically between 1250 and 1400°C.[10] Further improvements in the thermodynamic efficiency of engines and heat exchangers will require the use of materials with significantly higher melting points, such as ceramic matrix composites. Within the past few years, fiber-reinforced ceramics have been successfully used in aerospace applications, including rocket nozzles and exhaust flaps for advanced gas turbines.[11] Future applications for these materials include gas-turbine combustors and airfoils, heat exchangers, structural members and heat shields for aircraft and space-based structures, and containment vessels for spent fuel from nuclear power generation.[12-14] Elevated temperature creep and fatigue life are important design considerations for most of these applications.

This chapter discusses, from theoretical and practical perspectives, the current status of elevated temperature research in the area of creep deformation of fiber-reinforced ceramics. The first part of the chapter addresses the theoretical analysis of creep behavior; a simple 1-D model is used to provide insight into how the creep behavior of the constituents influences transient creep behavior and microstructural damage accumulation. In the second part of the chapter, a brief overview of the experimental techniques used to study creep behavior is provided, along with results from selected experimental investigations that provide insight into creep behavior and microstructural damage modes. The third part of the chapter deals with recent research into the cyclic creep behavior of fiber-reinforced ceramics. Lastly, the results from the analytical and experimental investigations of creep behavior are pulled together to provide practical guidelines for the microstructural design of creep-resistant composites. This latter discussion also highlights an important dilemma that exists in the microstructural design of fiber-reinforced composites. Namely, the microstructural parameters that are required for high monotonic toughness, such as low interfacial shear strength, matrix microcracking, and crack bridging by fibers, typically have a negative impact on creep and fatigue resistance.

5.2 Analytical Modeling of Creep Behavior

In this section a simple 1-D model is used to investigate the creep deformation of fiber-reinforced ceramics. The model provides insight into the transient stress redistribution that occurs between the fibers and matrix during creep loading. As discussed in detail below, this stress redistribution, which is a consequence of the coupling together of two phases having different creep behavior, has an important influence on creep damage mode and creep life. An understanding of why this stress redistribution occurs is a necessary first step for the microstructural design of composites with improved creep resistance.

5.2.1 Modeling of Creep Behavior and Transient Redistribution in Stress between Constituents

Background At elevated temperatures the rapid application of a sustained creep load to a fiber-reinforced ceramic typically produces an instantaneous elastic strain, followed by time-dependent creep deformation. Because the elastic constants, creep rates and stress–relaxation behavior of the fibers and matrix typically differ, a time-dependent redistribution in stress between the fibers and matrix will occur during creep. Even in the absence of an applied load, stress redistribution can occur if differences in the thermal expansion coefficients of the fibers and matrix generate residual stresses when a component is heated. For temperatures sufficient to cause the creep deformation of either constituent, this mismatch in creep resistance causes a progres-

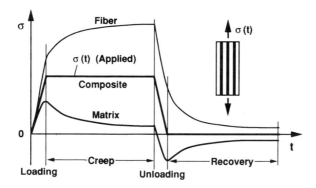

Fig. 5.1 Idealized representation of the transient change in fiber and matrix stress that occurs during the isothermal tensile creep and creep recovery of a fiber-reinforced ceramic (the loading and unloading transients have been exaggerated for clarity). It is assumed that the fibers have a much higher creep resistance than the matrix. The matrix stress reaches a maximum at the end of the initial loading transient. After full application of the creep load, the matrix stress relaxes and the fiber stress increases. Upon specimen unloading, elastic contraction of the composite occurs, followed by a time-dependent decrease in fiber stress and increase in matrix stress. Overall, creep tends to increase the difference in stress between constituents and recovery tends to minimize the difference in stress. After Wu and Holmes.[15]

sive increase in the stress within the constituent with the higher creep resistance, and a decrease in stress in the less creep-resistant constituent. Figure 5.1 illustrates the transient change in fiber and matrix stress that would occur during the initial loading, sustained creep, and unloading of a composite. It is assumed in Fig. 5.1 that the fibers have a higher creep resistance than the matrix, which is typically the case. As shown in this figure, the rapid application of a creep load causes initial elastic deformation of the fibers and matrix. At the end of the loading transient the stress level of each constituent is determined by the applied load and by the elastic moduli and volume fractions of the fibers and matrix. *It is important to note that the stress in the matrix reaches its maximum level at the end of the initial loading, and decays thereafter.* If the matrix stress is sufficiently high, matrix fracture can occur during initial loading. The matrix stress can be reduced by decreasing the initial loading rate, which allows the matrix stress to relax by creep during initial loading (see later discussion in Section 5.2.5). After application of the creep stress, load is shed from the matrix to the fibers; the matrix stress relaxes, and the fiber stress increases. Upon unloading, elastic contraction of the composite places the matrix in compression, with the fibers in residual tension. For a given composite and temperature, the residual stress state that develops after unloading is influenced by the creep temperature, stress, and the prior creep strain. As discussed in Section 5.5 on creep strain recovery, this residual stress state provides a driving force for strain recovery that is absent in monolithic ceramics. With time, the stresses in the fibers and matrix decrease as intrinsic recovery and stress relaxation occur. *Overall, it can be said that creep loading tends to increase the difference in axial stresses between the constituents which have different creep rates, while strain recovery reduces the difference in stresses.*

The degree and rate of stress redistribution that occurs during creep loading depends upon many parameters, including the initial residual stress state of the composite, mismatch in the elastic constants and creep rates of the fibers and matrix, and the degree of load transfer along the fiber–matrix interface. The initial residual stress state is determined by the processing history and the mismatch in the elastic constants and thermal expansion coefficients of the constituents. For a given temperature, the elastic constants of the fibers and matrix will also influence the stress distribution in the fibers and matrix during the application and removal of a creep load, with the mismatch in creep rates governing the transient redistribution in stresses.

The above discussion addresses only the redistribution of axial stress in the fibers and matrix that occurs during uniaxial creep loading. It should be appreciated that time-dependent changes in the radial stress at the fiber–matrix interface also occur, as well as changes in the hydrostatic stress that develops in the matrix (these stresses develop as a consequence of constraint to deformation from the fibers). This is of concern because the radial stress directly influences the debonding characteristics of the interface and the degree of load transfer between constituents. If debonding occurs, creep deformation will cause a time-dependent change in the frictional shear stress along the fiber–matrix interface. Many physical properties are influenced by the frictional shear stress along the fiber–matrix interface. Interfacial debonding or a change in the frictional shear stress will directly affect the strength, toughness, and damping behavior of fiber-reinforced ceramics. Thermophysical properties that depend upon the degree of contact between the fibers and matrix (e.g., transverse thermal conductivity) will also be influenced by a time-dependent change in radial stress. Hydrostatic stresses, which develop in the matrix because of local constraint between the fibers and matrix, will also influence creep deformation mechanisms such as matrix cavitation.

Analytical and Numerical Modeling   Ideally, one would like to model the creep behavior of a composite from knowledge of the creep behavior of the fibers, matrix and interface. These models can provide guidelines for the microstructural design of composites with improved creep resistance and can help in understanding how microstructural damage accumulates during sustained and cyclic creep loading. However, because of Poisson's ratio mismatch between the fibers, matrix, and near-interface region, and inhomogeneities in fiber packing, even the unidirectional loading of 1-D composites will cause the development of a multiaxial stress–strain state within the composite. In addition, creep deformation of the fibers and matrix is time-dependent and non-linear. This makes exact analytical solutions exceedingly difficult, if not impossible, to obtain. Numerical techniques, such as finite element methods are usually required to fully analyze the time-dependent stress and strain distributions that develop during creep loading. For example, Park and Holmes[16] used 2-D and 3-D finite element modeling to predict the changes in axial and radial stresses that occurred during the sustained and cyclic creep loading of 1-D SCS-6 SiC$_f$/HPSN composites. It was found that constraint from adjacent fibers caused non-symmetric stresses to develop normal to the fibers; this localization of the normal stress can play an important role in interfacial debonding and the load transfer that occurs between the fibers and matrix. For isothermal cyclic-creep loading, the finite element analysis also showed that the transient stress state that develops during specimen unloading provides a strong mechanical driving force for the time-dependent recovery of creep strain (see discussion in Section 5.5).

Because of the difficulty encountered in obtaining exact analytical solutions, and the time and cost of conducting a time-dependent finite element analysis, simplified 1-D concentric cylinder models have been extensively used to study the influence of constituent creep behavior on composite creep behavior. The earliest analysis of the creep behavior of continuous fiber-reinforced composites was performed by De Silva,[17,18] who investigated the transient redistribution in stress that occurs between the fibers and matrix during the creep of unidirectional metal matrix composites. In his model, the fibers and matrix were treated as two parallel cells containing elastic and creep elements in series, and were subjected to uniaxial tension. It was shown that a progressive transfer in load from the constituent with the lower creep resistance to the constituent of higher creep resistance occurs. Subsequent work by others[19-27] essentially followed similar approaches, but with different combinations of material properties (e.g., creeping matrix and elastic or creeping fibers), different interface properties, or different mathematical treatments. Particularly noteworthy is the detailed work of McLean and co-workers[23-26] who modeled the creep behavior of uniaxial metal matrix composites with short or long elastic fibers embedded in a matrix that exhibited power-law creep. Their analysis treated the interface as a third phase, either slipping (incoherent interface) or non-slipping (coherent interface); the fibers were assumed to behave in an elastic manner, which is typically the case for ceramic fibers in a metallic matrix. The work of McLean and co-workers highlighted the important role that interphase rheology can play in the creep response of composites.

The 1-D "concentric cylinder" models described above have been extended to fiber-reinforced ceramics by Kervadec and Chermant,[28,29] Adami,[30] and Wu and Holmes;[31] these analyses are similar in basic concept to the previous modeling efforts for metal matrix composites, but they incorporate the time-dependent nature of both fiber and matrix creep and, in some cases, interface creep. Further extension of the 1-D model to multiaxial stress states was made by Meyer *et al.*,[32-34] Wang *et al.*,[35] and Wang and Chou.[36] In the work by Meyer *et al.*, 1-D fiber-composites under "off-axis" loading (with the loading direction at an angle to fiber axis) were analyzed with the assumptions that the fibers and matrix behave as non-linear Maxwell solids in pure shear and that the interface obeys a rate-dependent Coulombic frictional law without a cohesive strength. Their model, used in conjunction with a finite element model, was used to predict the "off-axis" compressive creep behavior of Nicalon/CAS composites as a function of the angle between the loading direction and fibers. In the work by Wang and Chou, unidirectionally aligned short-fiber ceramic composites subjected to off-axis tension were modeled using a multiaxial creep constitutive relation and an advanced shear-lag method. The model was used to predict the creep behavior of a short-fiber SiC/Al$_2$O$_3$ composite. This model considered the fiber–matrix interface sliding effect, volume fraction, and fiber aspect ratio. An interface factor was introduced to account for the effect of interface behavior on creep deformation.

5.2.2 A 1-D Model for the Creep of Unidirectional Fiber-Reinforced Ceramics

To gain a better understanding of the creep behavior of fiber-reinforced ceramics, a simple 1-D analytical approach will be used to examine the effects of constituent behavior on composite creep deformation and changes in internal stress. Since the derivation of the model provides valuable insight into the parameters that influence composite creep behavior, the derivation of the 1-D "concentric cylinder" model will be outlined first.

Consider a continuous fiber-reinforced ceramic as a multiphase system where the individual phases are parallel to one another and to the uniaxial loading direction. The fibers (or fiber bundles), matrix, and interface zone are treated as individual phases. In general, each phase undergoes elastic-plastic (creep) deformation. In the present analysis, the creep rate of each phase, $\dot{\varepsilon}_i$, is assumed to obey a general creep law of the following form

$$\dot{\varepsilon}_i = A_i(T,t)\sigma_i^{n_i} \quad i = 1, 2, \ldots, N \quad (N \text{ is the number of phases}) \quad (1)$$

where σ_i is the *in situ* stress experienced by a given phase i within the composite and n_i is the stress exponent for phase i. If required, the factor $A_i(T,t)$ in Eqn. (1) can be separated into three terms, corresponding to primary, secondary (steady-state), and tertiary creep.[31]

Each phase in a composite will typically have different elastic and creep properties. However, assuming strong interfacial bonding as a limiting case, compatibility requires that the total strain, and the total strain rate, of each constituent be equal. The total strain rate of each constituent, $\dot{\varepsilon}_{i,tot}$, is given by the sum of the elastic strain rate, $\dot{\varepsilon}_{i,el}$, and the creep strain rate, $\dot{\varepsilon}_i$. To satisfy compatibility, this sum must equal the total creep rate of the composite $\dot{\varepsilon}_{c,tot}$:

$$\dot{\varepsilon}_{i,tot} = \dot{\varepsilon}_{i,el} + \dot{\varepsilon}_i = \dot{\sigma}_i/E_i + \dot{\varepsilon}_i = \dot{\varepsilon}_{c,tot} \quad (2a)$$

or

$$\dot{\sigma}_i = E_i(\dot{\varepsilon}_{c,tot} - \dot{\varepsilon}_i) \tag{2b}$$

From the rule of mixtures, the stress in each constituent can be expressed in terms of the volume fraction of the constituent and the applied stress σ_c:

$$\sum_{j=1}^{N} v_j\sigma_j = \sigma_c \tag{3a}$$

or, in terms of stress rates:

$$\sum_{j=1}^{N} v_j\dot{\sigma}_j = \dot{\sigma}_c \tag{3b}$$

By substituting Eqn. (2b) into Eqn. (3b) to eliminate $\dot{\sigma}_i$, one obtains the total strain rate of the composite as a sum of terms that represent the elastic and creep components of creep strain:

$$\dot{\varepsilon}_{c,tot} = \frac{1}{E_c}\left(\dot{\sigma}_c + \sum_{j=1}^{N} v_j E_j \dot{\varepsilon}_j\right) = \dot{\varepsilon}_{c,el} + \langle\dot{\varepsilon}_j\rangle_j \quad (j = 1, 2, \ldots, N) \tag{4}$$

where $\dot{\varepsilon}_{c,el}$ ($=\dot{\sigma}_c/E_c$) is the elastic component of the composite strain rate ($E_c = \sum_{j=1}^{N} v_j E_j$ is the axial modulus of the composite), and $\langle\dot{\varepsilon}_j\rangle_j$ is the creep component of the composite strain rate, $\langle\dot{\varepsilon}_j\rangle_j = \sum_{j=1}^{N}\{v_j(E_j/E_c)\dot{\varepsilon}_j\}$, representing an intermediate value of all constituent creep rates $\dot{\varepsilon}_j$ ($=A_j\sigma_j^{nj}$), with a weighted factor $v_j E_j/E_c$. This expression allows representation of the creep behavior of the composite in terms of the individual creep behavior of the constituents. *For a multiphase structure with continuous phases aligned parallel to the loading direction and subjected to elastic/creep deformation, the overall creep rate of the composite is equal to the weighted mean value of the constituent creep rates, $\dot{\varepsilon}_i$, with a weighted factor $v_i E_i/E_c$.*[31]

To find $\dot{\varepsilon}_c$ from $\dot{\varepsilon}_i$, one needs to determine the local stress, σ_i, (or stress distribution) from the applied stress. This can be achieved by solving the kinetic equation of constituent stress. By substituting Eqn. (4) into Eqn. (2b) to eliminate $\dot{\varepsilon}_{c,tot}$, one obtains a system of first-order differential equations:

$$\dot{\sigma}_i = E_i\{\dot{\varepsilon}_{c,el} + \langle\dot{\varepsilon}_j\rangle_j - \dot{\varepsilon}_i\} \quad (i = 1, 2, \ldots, N) \tag{5}$$

Since $\dot{\varepsilon}_i$ is a power-law function of σ_i (given by Eqn. (1)), and $\dot{\varepsilon}_{c,el}$ is known from the loading history $\dot{\sigma}_c(t)$, Eqn. (5) has a form of $\dot{\sigma}_i = f_i(\sigma_1, \sigma_2, \ldots \sigma_{N,t})$, which can be readily solved by an iterative method, providing that the initial condition is known. The iterative computation involves the following:

(1) With the initial conditions ($\sigma_c = \sigma_i = 0$, $\varepsilon_c = \varepsilon_{i,tot} = 0$, and $\dot{\varepsilon}_c = \dot{\varepsilon}_i = 0$) given, the right-hand side of Eqn. (5) is known. This allows the initial rate of stress change in each phase, $\dot{\sigma}_i$, to be determined.

(2) After a time increment dt, the local stresses in the fibers and matrix can be updated to obtain $(\sigma_i, \varepsilon_i, \dot{\varepsilon}_i)_t$ at any time $t + dt$. Using this information, the composite stress, strain, and strain rate $(\sigma_c, \varepsilon_c, \dot{\varepsilon}_c)_t$ can be obtained from the constituent parameters $(\sigma_i, \varepsilon_i, \dot{\varepsilon}_i)_t$ by using Eqns. (2)–(4). By iterative computation, the creep behavior of the composite and constituents can be predicted for any loading history, including cyclic creep.

If a composite is considered as a two-phase system (with fibers and matrix only), Eqn. (5) reduces to:

$$\dot{\sigma}_f = \frac{E_f}{E_c} \{\dot{\sigma}_c + v_m E_m (\dot{\varepsilon}_m - \dot{\varepsilon}_f)\} \tag{6a}$$

$$\dot{\sigma}_m = \frac{E_m}{E_c} \{\dot{\sigma}_c + v_f E_f (\dot{\varepsilon}_f - \dot{\varepsilon}_m)\} \tag{6b}$$

where the subscripts "f" and "m" denote the fiber and matrix, respectively.

Driving Force for Stress Redistribution   Examination of Eqn. 5 (or 6) indicates that the driving force for stress redistribution during transient creep originates from the mismatch in the elastic and creep properties of the constituents. During the rapid application of a creep load, stress is distributed to each constituent based on their elastic properties (Young's modulus). During subsequent sustained (static) loading, the stress tends to redistribute between the constituents based on their creep properties. Following the application of a creep load, a progressive stress redistribution occurs and will continue until a steady state is reached, at which time $\dot{\varepsilon}_i = \langle \dot{\varepsilon}_j \rangle$ and $\dot{\sigma}_i = 0$ for all the constituents. This requires that at steady-state, the stress is distributed in such a way that the intrinsic creep rates of all constituents are equal $(\dot{\varepsilon}_i = \dot{\varepsilon}_{c,tot})$, and that the elastic components of strain rate equal zero $(\dot{\varepsilon}_{i,el} = 0)$. This stress redistribution process, which is a characteristic feature of the transient creep behavior of composites, may occur over a long period of time (i.e., some hundred hours), or may be only a short period (i.e., several minutes); the precise kinetics of redistribution will depend on the combination of elastic properties (E_i), creep properties (A_i, n_j), and volume fractions (v_i) of the constituents.

The 1-D model presented above can be used to study the effect of loading history and microstructural changes on composite creep behavior. For example, using the elastic strain rate term in Eqn. (5), $(E_i/E_c)\dot{\sigma}_c$, various loading patterns can be simulated without changing the computation procedure. Using this model, the influence of constituent creep behavior, elastic properties and volume fractions can be studied in an attempt to optimize the microstructure of a composite for given temperature and loading histories. The number of constituents in the model is arbitrary, as is the behavior of each constituent. If we consider the fibers, matrix, and interfacial zone as a three-component composite system, Eqn. (5) reduces to Adami's[30] model for the uniaxial creep of a fiber-reinforced ceramic, or McLean's[25] model for metal matrix composites with long fibers (in his derivation, McLean assumed that the fibers behaved in an elastic fashion). If only the fibers and matrix are considered, the model reduces to DeSilva's[17,18] two-component model.

5.2.3 Application of the 1-D Model: Transient Creep and Stress Redistribution

Equation (5) is used in this section to provide additional insight into the transient creep behavior of fiber-reinforced ceramics and the stress redistribution that occurs between the fibers and matrix. For simplicity, consider a two-phase composite system with fibers aligned in the loading direction. Assume that the composite is loaded at an infinitely high rate to a constant creep stress σ_c. Using data for SiC fibers (SCS-6) and hot-pressed Si_3N_4 (HPSN), the influence of applied creep stress on the creep strain and creep rate of a hypothetical 0° SCS-6 SiC_f/HPSN composite with 40 vol.% fibers was determined using the 1-D model, and compared with the results from a 2-D finite element analysis (see Fig. 5.2).[37] In both analyses, microstructural damage was ignored (the influence of microstructural damage on creep behavior is discussed later). It was also assumed that the creep stress was instanteously applied and that the constituents exhibited only steady-state creep. The transient change in fiber and matrix stress that would occur during the creep experiment was also calculated (see Fig. 5.2c).

The similar results predicted by the 1-D (ROM) and 2-D (FEM) models suggest that transverse stresses (which are not accounted for in a 1-D analysis) have only a minor effect on the uniaxial tensile creep behavior of 0° composites. This is generally the case if microstructural damage is absent (e.g., if the applied creep stress is relatively low). In this case, the effect of Poisson's ratio mismatch is less important, because plastic deformation obeys volume conservation, giving a transverse strain equal to one-half of the axial strain for all the phases; only elastic strain components are responsible for the transverse stresses that arise from a mismatch in Poisson's ratio. Therefore, the minor difference between the 1-D and 2-D analysis (by FEM) which considers the Poisson contraction effect is understandable. Although transverse stresses do not significantly influence the creep behavior of 0° composites, it will be shown later that there is considerable experimental evidence that the transverse stresses that develop during the uniaxial loading of 0°/90° composites can have a significant influence on creep behavior and microstructural damage.

To illustrate a key point concerning the creep behavior of fiber-reinforced ceramics, the primary creep behavior of the constituents was purposely omitted in the above analysis. As shown in Fig. 5.2, even though it was assumed that the constituents undergo only steady-state creep, a protracted transient creep

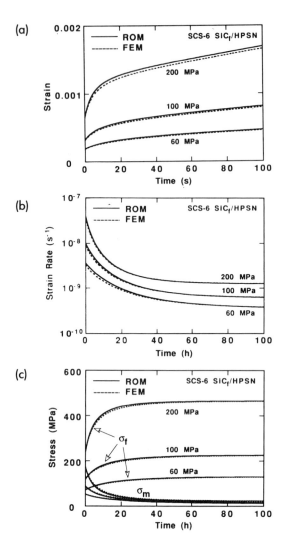

Fig. 5.2 Comparison of creep behavior and time-dependent change in fiber and matrix stress predicted using a 1-D concentric cylinder model (ROM model) (solid lines) and a 2-D finite element analysis (dashed lines). In both approaches it was assumed that a unidirectional creep specimen was instantaneously loaded parallel to the fibers to a constant creep stress. The analyses, which assumed a creep temperature of 1200°C, were conducted assuming 40 vol.% SCS-6 SiC fibers in a hot-pressed Si_3N_4 matrix. The constituents were assumed to undergo steady-state creep only, with perfect interfacial bonding. For the FEM analysis, Poisson's ratio was 0.17 for the fibers and 0.27 for the matrix. (a) Total composite strain (axial), (b) composite creep rate, and (c) transient redistribution in axial stress in the fibers and matrix (the initial loading transient has been ignored). Although the fibers and matrix were assumed to exhibit only steady-state creep behavior, the transient redistribution in stress gives rise to the transient creep response shown in parts (a) and (b). After Wu *et al.*[37]

regime is predicted. This is a general feature of the creep behavior of fiber-reinforced ceramics and is caused by the time-dependent redistribution in stress that occurs between the fibers and matrix (Fig. 5.2c).

In the absence of matrix or fiber fracture, the constituents in a composite must creep at the same *total* strain rate. The *creep* strain rate of the constituents will typically differ during the transient creep period; the mismatch in creep rates between the two constituents is compensated by the *elastic* strain rate of the fibers and matrix. This concept is schematically illustrated in Figs. 3a and b, which show the total strain rate and the elastic/creep components for the individual phases (fibers and matrix) as a function of time (Fig. 5.3a) and the local stress (Fig. 5.3b). After initial application of the creep load, the local stress on the fibers and matrix is mainly determined by the Young's moduli of the two phases (as an extreme, when the loading rate is instantaneously applied, the stress in the fibers and matrix is given by $\sigma_c E_f/E_c$ and $\sigma_c E_m/E_c$, respectively). The intrinsic creep rate of the constituents generally differs, resulting in a continuous stress redistribution; compatibility is maintained by changes in the elastic strain rate of the constituents. With reference to Fig. 5.3b, the creep rates of the constituents follow the intrinsic creep rate of the particular constituent (shown as straight lines), while the total creep rates of the constituents deviate from the assumed steady-state creep curves of the constituents in order to maintain the same total creep rate. The difference between the intrinsic and total creep rates of each constituent gives the elastic strain rate of the constituent (shown shadowed in Fig. 5.3b). After a long period of time, the elastic component of strain in each constituent will approach zero (in the absence of microstructural changes, the composite will approach a steady-state creep rate at this stage).

As can be deduced from the above discussion, there exists a transient period, during which a composite has an initial creep rate determined mainly by the elastic properties of the constituents (if the loading rate is rapid), followed by a gradual change in creep rate, and finally by a steady-state creep rate (assuming that the constituents exhibit steady-state creep behavior and that microstructural changes do not significantly influence the creep behavior). Therefore, the transient creep behavior can be characterized by the initial creep rate, $\dot{\varepsilon}_{c,0}$, final creep rate, $\dot{\varepsilon}_{c,ss}$, and the transition kinetics, which will be further discussed below.

5.2.3.1 Initial and Final Creep Rates, and Transient Process

Equation (4) can be used to compare the change in creep behavior at zero time and for an infinitely long time. Consider that a composite is initially loaded at an infinitely rapid rate to a constant creep stress σ_c. From Eqn. (4), and with $\dot{\varepsilon}_i = \{A_i(E_i/E_c)\sigma_c\}$ at $t = 0$, the stress and strain-rate relation for the composite (σ_c, $\dot{\varepsilon}_c$) becomes

$$\dot{\varepsilon}_{c,o} = \sum_{j=1}^{N} v_j \frac{E_j}{E_c}\left(\frac{E_j}{E_c}\sigma_c\right)n_j \quad \text{(for } t = 0^+) \tag{7}$$

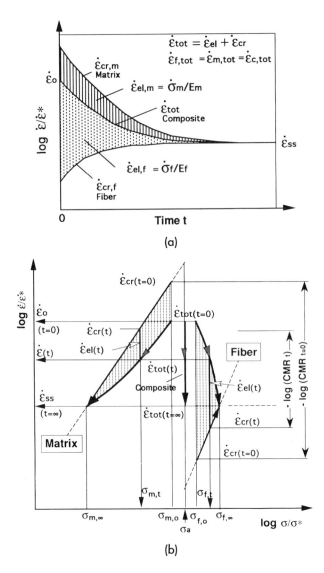

Fig. 5.3   Schematic showing the changes in strain rate and elastic/creep strains of the individual constituents that occur during creep of a composite. (a) Strain rate versus time, (b) strain rate versus *in situ* stress acting in the fibers and matrix. In both plots, the shadowed portions show the elastic strain components, which compensate the creep rate mismatch of the individual phases, such that the total creep rates of the constituents remain equal. The creep mismatch ratio (CMR) is discussed in Section 5.2.4. After Wu and Holmes.[31]

After a long period of time, the composite will approach a steady-state condition, with no further change occurring in the fiber and matrix stress ($\dot{\sigma}_i = 0$). From Eqn. (2b), and with $\sigma_{i,ss} = (\dot{\varepsilon}_{i,ss}/A_i)^{1/n_i}$ the creep rate of the composite approaches a steady-state value $\dot{\varepsilon}_{c,ss}$ ($=\dot{\varepsilon}_{i,ss}$), which is determined by the condition

$$\sum_{j=1}^{N} v_j (\dot{\varepsilon}_{c,ss}/A_j)^{1/n_j} = \sigma_c \quad \text{(for } t = \infty) \tag{8}$$

Between the two limits $\dot{\varepsilon}_{c,0}$ and $\dot{\varepsilon}_{c,ss}$, the stress–strain state of the composite can be calculated using the iterative method, as described earlier.

To more clearly understand the transient changes in constituent stress and strain rate that occurs between the limits of Eqns. (7) and (8), the transient creep and changes in fiber and matrix stress are plotted in Fig. 5.4 as a function of $\log(\sigma)$ and $\log(\dot{\varepsilon})$. For convenience in presenting the interaction of the constituents, a normalized creep equation is used for each constituent:

$$\dot{\varepsilon}_i/\dot{\varepsilon}^* = (\sigma_i/\sigma^*)^n \quad (i = 1, 2 \text{ for fiber and matrix}) \tag{9}$$

where the reference stress, σ^*, and strain rate, $\dot{\varepsilon}^*$, are chosen as the point of intersection of the creep curves of the two constituents (this can be done only for a two-phase system). In Fig. 5.4, constituents are assumed to undergo steady-state creep only. If both primary creep and steady-state creep are included, or if steady-state creep is not observed, the values of the reference stress and strain rate will change with time. The reference stresses correspond to the equilibrium point at which fiber and matrix have the same creep rate and stress (see Fig. 5.4). At this equilibrium point $\dot{\varepsilon}^* = A_1^{n_2/(n_2-n_1)} A_2^{n_1/(n_1-n_2)}$ and $\sigma^* = A_1^{1/(n_2-n_1)} A_2^{1/(n_1-n_2)}$; here A_i is given by a general power-law equation for the stress dependence of creep rate ($\dot{\varepsilon}_i = A_i \sigma_i^{n_i}$) for each phase.

In Fig. 5.4a and b, the initial creep rate of each phase (Eqn. (7)) is represented by the intersection of the monolithic creep curve for that phase and the elastic stress and strain (vertical line). After initial loading, the total strain rate (elastic + creep) of each phase, which remains equal to the total strain rate of the composite (for compatibility), decreases. The only exception arises if $\dot{\varepsilon}_{1,0} = \dot{\varepsilon}_{2,0}$ ($=\dot{\varepsilon}_{c,0}$), so that $\dot{\varepsilon}_{c,0} = \dot{\varepsilon}_{c,ss}$ (see Fig. 5.4c). In this instance, the initial condition matches the steady-state condition—the composite strain rate remains unchanged. The applied stress for this condition is given by

$$\sigma_c = E_c \left(\frac{A_1 E_1^{n_1}}{A_2 E_2^{n_2}} \right)^{1/(n_2-n_1)} \tag{10}$$

As shown in Fig. 5.4, for stresses far away from σ^* (either $\sigma_c \gg \sigma^*$ or $\sigma_c \ll \sigma^*$), the initial creep behavior (stress dependence) of the composite is determined primarily by the constituent having the *higher* creep rate. On the other hand, the final creep behavior of the composite is governed by the constituent with the *lower* creep rate. For applied creep stresses close to σ^*, a gradual change in creep stress exponent n occurs from n_1 to n_2 (or vice versa).

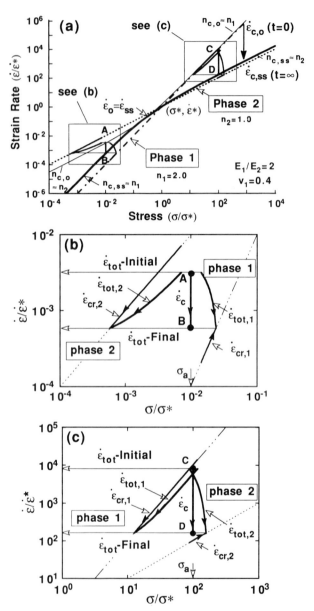

Fig. 5.4 (a, b and c) Initial and final (steady-state) strain rate of a hypothetical composite as a function of normalized stress. The dashed lines represent the creep rate of the constituents. (b) and (c), which detail the two framed regions in Fig. 5.4(a), show the transient paths of the stress and strain rate for the composite and its constituents for the two corresponding stress regimes: $\sigma > \sigma^*$ and $\sigma < \sigma^*$. The dashed lines in (b) and (c) show the creep rate of the constituents (excluding the elastic components), which follow the monolithic creep behavior of each phase; the total strain rate of the composite and the constituents must remain equal.

Although a reference stress and strain rate was used in the above discussion, practical composites may work in a regime far away from this equilibrium point; this does not affect the results of the above discussion. As discussed in detail elsewhere,[31] plots such as Fig. 5.4 are useful for estimating the constituent creep behavior from experimental studies of the initial and final creep behavior of a composite.

5.2.3.2 Parametric Studies: Influence of Constituent Moduli and Creep-Stress Exponents on Composite Creep Behavior

The stress redistribution process is influenced by the elastic and creep properties of the constituents (including the interface), the fiber/matrix architecture, volume fraction of fibers, temperature, and loading history. It is instructive to investigate through parametric studies the influence of material and microstructural changes on the redistribution of stress between the fibers and matrix. As an example, consider how changing the elastic modulus of the fibers and matrix influences the transient creep behavior of a 1-D composite. Figure 5.5a shows the effect of changing the ratio of elastic moduli E_1/E_2 ($=E_f/E_m$) between 0.2 and 5 on the stress redistribution and the initial and final creep rate of the composite (this is a practical example, since by eliminating the glassy phase, stoichiometric SiC fibers which are currently under development will have a significantly higher elastic modulus than current generation SiC fibers, such as Nicalon and SCS-6). The modulus ratio of the two phases, which determines the initial stress distribution in the fibers and matrix, influences the initial creep rate of the composite, but has no effect on the long-term creep rate (assuming matrix cracking does not occur).

The effect of changing the creep stress exponent of the constituents is shown in Fig. 5.5b for a fixed modulus ratio of $E_1/E_2 = 2.0$. For the region $\sigma > \sigma^*$, the creep stress exponent of the composite (which determines the stress dependence of composite creep rate) changes from an initial value of n_2 to a final value n_1. In the region $\sigma > \sigma^*$, the n-value of the composite changes from an initial value of n_1 to a final value of n_2. In both cases, the initial stress exponent for the composite is determined by the phase with the higher strain rate, and the final stress exponent of the composite is determined by the phase with the lower strain rate. A graceful transfer between n_1 and n_2 occurs in the intermediate region where $\sigma \approx \sigma^*$. Similar parametric studies can be performed to examine the influence that other parameter changes have on composite behavior.

5.2.4 Creep Mismatch Ratio

It is useful to define a parameter that describes the direction and magnitude of the driving force for stress redistribution. Rather than use the difference in creep rates, it is more convenient to define this parameter as the ratio between the constituent creep rates. For this purpose, a *time-dependent*

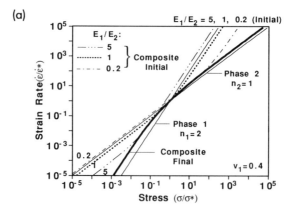

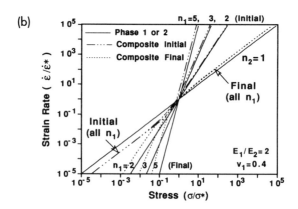

Fig. 5.5 Effect of changing the elastic modulus ratio and constituent creep stress exponents on the total strain rate of a 1-D composite subjected to tensile creep loading.[31] In both (a) and (b), the dashed lines represent the composite behavior, and the thin solid lines the constituent behavior. In the calculations, it was assumed that the creep load was applied instantaneously.

creep mismatch ratio (CMR) can be defined as the ratio of the fiber and matrix creep rates:

$$\text{CMR}(t, T, \sigma) = \frac{\dot{\varepsilon}_f(\sigma_f)}{\dot{\varepsilon}_m(\sigma_m)} \qquad (11)$$

The creep rates in Eqn. (11) refer to the *in situ* creep rates experienced by the fibers and matrix within a composite. Since the creep rates of fibers and matrix are a function of time during the stress redistribution process, the *in situ* creep mismatch ratio is time dependent. As defined here, the creep mismatch ratio has values that can be greater than, equal to, or less than, unity. If the fibers have a higher creep resistance than the matrix ($CMR_t < 1$), there is a transient increase in fiber stress and a parallel relaxation of matrix stress. If the matrix has the higher creep resistance ($CMR > 1$), the matrix stress progressively increases during tensile creep. If $CMR = 1$, there is no driving force for stress redistribution; the further away from unity that the CMR is, the higher the driving force.

Initial and Final Creep Mismatch Ratio   At low temperatures, or during rapid loading, the stress in the fibers and matrix can be estimated from a simple rule-of-mixtures approach; this gives the *elastic* stress distribution between the fibers and matrix. During creep, the stress distribution is time dependent and is influenced by both the initial elastic stress distribution and the creep behavior of the constituents. Immediately after applying an instantaneous creep load (i.e., at $t = 0^+$), the $CMR_{t=0+}$ can be found by substituting $\dot{\varepsilon}_{f,0} = A_f\{(E_f/E_c)\sigma_c\}^{n_f}$ and $\dot{\varepsilon}_{m,0} = A_m\{(E_m/E_c)\sigma_c\}^{n_m}$ into Eqn. (11):

$$CMR_{t=0^+} = \frac{A_f E_f^{n_f}}{A_m E_m^{n_m}}\left(\frac{\sigma_c}{E_c}\right)^{n_f-n_m} \tag{12}$$

For the other extreme of an exceedingly long time, the creep rate mismatch ratio approaches unity:

$$CMR_{t=\infty} = 1 \tag{13}$$

For times between these two extremes, the creep mismatch ratio is time dependent and can be determined using Eqns. (4) and (5) and an iterative process. Since, in general, $CMR \neq 1$, time-dependent load redistribution between the fibers and matrix is a general phenomenon that occurs in all fiber-reinforced ceramics.

The *in situ* creep mismatch ratio CMR_t can provide a quantitative estimate of the driving force for load transfer between constituents. However, it is a rather complicated function of the stress redistribution processes. In order to indicate the basic characteristics of stress redistribution, it is convenient to directly compare the *intrinsic* (unconstrained) creep rates using the initial elastic stress experienced by the constituents. From Eqn. (12),

$$CMR = CMR(\sigma_{applied}) = \dot{\varepsilon}_f(\sigma_{f,initial})/\dot{\varepsilon}_m(\sigma_{m,initial}) \tag{14}$$

Although time dependence is not included, the constituent with the higher creep rate will eventually experience a stress lower than the applied stress, while the constituent with the lower creep rate will experience an increase in stress. As a simple example, consider that the fibers and matrix have similar volume fractions and moduli: the following relationship between creep mismatch ratio and long-term stresses in the fibers and matrix will hold:

$$\text{if CMR} < 1, \text{ then } \sigma_f > \sigma_c > \sigma_m$$

$$\text{if CMR} = 1, \text{ then } \sigma_f = \sigma_c = \sigma_m$$

$$\text{if CMR} > 1, \text{ then } \sigma_f < \sigma_c < \sigma_m \tag{15}$$

Stress and Temperature Dependence of Stress Redistribution in Composites   Both temperature and stress influence the degree and kinetics of stress redistribution. For most creep rate equations (e.g., Eqn. (1)), the temperature dependence of creep behavior is included in a pre-exponential factor, which is commonly expressed in the form of an Arrhenius relationship between the activation energy for creep, Q, and the absolute temperature, T (e.g., $A = A_0 \exp(-Q/RT)$). The activation energy for the creep deformation of a material is determined by fixing the creep stress and conducting experiments at various temperatures. In a similar fashion, the stress exponent for creep is determined by fixing the temperature and conducting creep experiments at various stresses. In general, the fibers and matrix will have different activation energies and stress sensitivity to creep deformation. Because of these differences, it is possible that the direction of load transfer could change depending upon the regime of temperature and applied creep stress (i.e., the CMR could be >1, <1, or unity, for different combinations of temperature and stress).

Suppose that one conducts a series of experiments to determine the stress and temperature dependence of creep behavior for the fibers and matrix; these experiments would provide curves such as those shown schematically in Fig. 5.6a and b. Conducting these experiments over a range of temperatures and stresses would provide a family of curves that could be combined to provide a relationship between strain rate, stress, and temperature. Such a temperature and stress dependence of constituent intrinsic creep rates, together with the intrinsic creep mismatch ratio, is schematically illustrated in Fig. 5.6c. In this plot, the creep equations for the two constituents at a given temperature and stress are represented by planes in ($1/T$, $\log\sigma$, $\log\dot{\varepsilon}$) space, with different slopes, described by Q_f, Q_m and n_f, n_m. The intersection of the two planes represents the condition where CMR $= 1$, which separates temperature and stress into two regimes: CMR < 1 and CMR > 1.

The stress and temperature dependence of the composite creep rate is governed by the values of the activation energies and stress exponents of the constituents. The initial stress and temperature dependence of composite creep rate is governed by the values of n and Q for the constituent which has the higher creep rate; the final stress and temperature dependence is governed by the values of n and Q for the constituent with the lowest creep rate. This is illustrated in Fig. 5.6d, which compares the stress and temperature dependence of the constituent creep rate with the initial and final creep behavior of the composite.

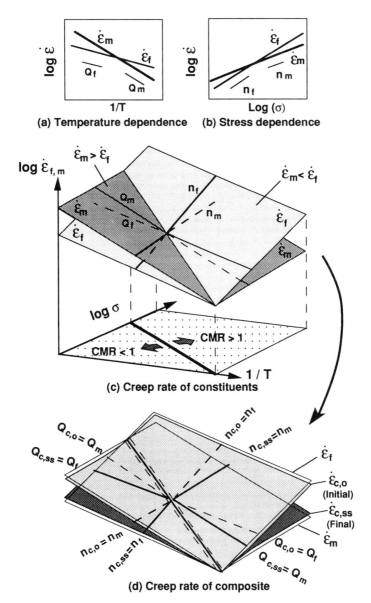

Fig. 5.6 Relationship between the creep rate of a composite and the stress and temperature dependence of the creep parameters of the constituents.[31] (a) Temperature dependence of constituent creep rate. (b) Stress dependence of constituent creep rate. (c) Intrinsic creep rate of constituents as a function of temperature and stress illustrating the temperature and stress dependence of the creep mismatch ratio. In general, load transfer occurs from the constituent with the higher creep rate to the more creep-resistant constituent. (d) Composite creep rate with reference to the intrinsic creep rate of the constituents. The planes labeled $\dot{\varepsilon}_f$ and $\dot{\varepsilon}_m$ represent the intrinsic creep rates of the fibers and matrix, respectively.

5.2.5 Influence of Creep Rate Mismatch Ratio on Microstructural Damage Mode

The nature of the stress redistribution between constituents has a direct influence on microstructural damage accumulation. Stress redistribution causes the stress acting in one of the constituents to progressively increase during sustained creep loading. This increase in stress may cause fracture of the constituent that is forced to support the majority of the applied creep load. If the axial creep rate of the fibers is lower than that of the matrix (CMR = $\dot{\varepsilon}_f/\dot{\varepsilon}_m < 1$), and if the threshold stress for fiber creep is exceeded, the axial stress in the fibers will continually increase during tensile creep. This increase in fiber stress can lead to periodic fiber fracture within the composite (i.e., fracture occurs at multiple locations along the length of each fiber—see Fig. 5.7). This damage mode has been observed experimentally in SCS-6 SiC$_f$/HPSN,[38] Nicalon SiC$_f$/MLAS,[39,40] and Nicalon SiC$_f$/CAS-II composites.[15,41] If the creep rate of the fibers exceeds that of the matrix (CMR $\dot{\varepsilon}_f/\dot{\varepsilon}_m > 1$), the axial stress in the matrix will progressively increase as the matrix sheds load to the fibers. In this case, a stress level sufficient to initiate matrix cracking may be reached, leading to the development of periodic matrix cracks perpendicular to the applied load, as illustrated in Fig. 5.7. As discussed in detail in Section 5.4, this damage mode has been observed in SCS-6 SiC$_f$/RBSN composites that were crept in tension at 1300°C.[42] This creep damage mode is only possible if the stress on the bridging fibers is low (i.e., for a low applied creep stress or high fiber volume fraction); otherwise, matrix cracking and fiber rupture may occur concurrently. In general, periodic matrix fracture is a very undesirable damage mode, since the bridging fibers must support the entire creep load; the life of the composite is determined by the creep rupture strength of the fibers. Moreover, the fibers and the fiber–matrix interface will be directly exposed to the surrounding environment, and can be rapidly degraded if oxygen or other corrosive species are present. This raises an important point for microstructural design; namely, to reduce the likelihood of matrix cracking it may be advantageous to use a matrix with a lower creep resistance than the fibers. This would allow the matrix stress to relax during creep loading, shifting the majority of the load to the fibers.

Influence of Initial Loading Rate on Creep Life   Because of the time-dependent nature of stress redistribution between the fibers and matrix, the rate at which a creep load is applied can have an important effect on creep life and the governing mechanism of creep deformation in brittle matrix composites. This is dramatically illustrated in Fig. 5.8a and b, which show the influence of initial loading rate on the tensile creep life of 0° SCS-6 SiC$_f$/HPSN at 1200°C. Rapidly loading to a creep stress of 250 MPa in 2.5 s caused specimen failure in less than 1.1 h. Applying the same creep stress at a much lower rate (1000 s) increased the creep life by two orders of magnitude, to roughly 120–170 h. The short creep life observed after rapid loading is a

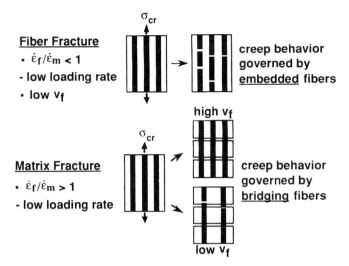

Fig. 5.7 Macroscopic damage modes that occur during the tensile and flexural creep of fiber-reinforced ceramics. It is assumed that matrix or fiber damage is avoided during initial application of the creep load (see discussion of loading rate effects in the next section). Periodic fiber fracture can occur if the creep rate of the matrix exceeds that of the fibers. Periodic matrix fracture is common when the matrix has a higher creep resistance than the fibers. In this figure, it is assumed that initial microstructural damage is avoided during application of the creep load.

consequence of matrix microcracking that occurred during, or shortly after, application of the creep stress. As illustrated in Fig. 5.9, under rapid loading, where time-dependent redistribution in stress between the fibers and matrix is minimal, the tensile stress in the matrix reaches a maximum at the end of the initial loading transient. If this stress exceeds that required for matrix fracture, microcracking and fiber bridging will occur; *creep life is then governed by failure of the highly stressed fibers that bridge the initial microcracks.* In contrast, by slowly applying the creep load, the matrix stress has sufficient time to relax during the loading transient. Depending upon the loading rate, the tensile stress in the matrix may never reach a level sufficient for matrix fracture; in this case, both the fibers and matrix contribute to overall creep resistance. *As this example illustrates, fiber-reinforced ceramics can exhibit a strong loading-path dependence of creep behavior and creep rupture time.*

The unloading rate, and the rate at which a specimen is cooled after unloading, are of equal importance if measurements of retained strength or toughness are to be made. An appreciation for this can be obtained by examination of the transient changes in fiber and matrix stress that occur upon specimen unloading (see Fig. 5.1). Upon unloading after prior creep deformation, a transient change in the residual stress state in the fibers and matrix will occur. The degree to which the stresses change will depend upon the length of

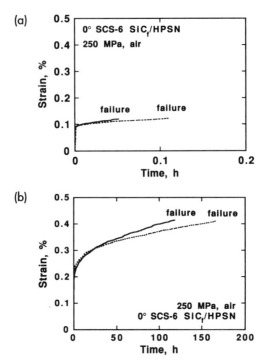

Fig. 5.8 Influence of initial loading rate on the 1200°C tensile creep life of 0° SCS-6 SiC$_f$/HPSN composites crept in air at a nominal stress of 250 MPa. The creep curves include the elastic strain on loading. For each loading rate, duplicate tests were performed to provide a rough indication of the scatter in the results. (a) Creep load applied in 1000 s, and (b) creep load applied in 2.5 s. After Holmes *et al.*[38]

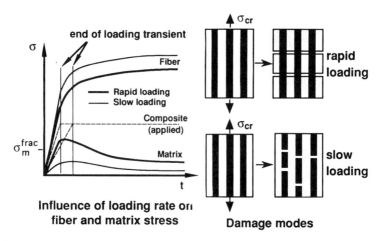

Fig. 5.9 Influence of initial loading rate on fiber and matrix stress and microstructural damage mode in materials where CMR < 1. Under rapid loading, the matrix stress may achieve a level sufficient to initiate matrix fracture.

time that a specimen is maintained at a temperature sufficient for viscous recovery to occur. It is important to note that the relaxation of internal stresses will influence the apparent microcracking threshold, proportional limit stress, and residual strength of a composite.

5.2.6 Closing Comments

The 1-D model introduced above to describe transient creep behavior neglected microstructural damage accumulation. To date, there has been only limited modeling of the creep behavior of fiber-reinforced ceramics with microstructural damage. Commonly observed microstructural damage processes include microcracking and cavitation of the matrix, slow crack growth, and time-dependent phase changes (e.g., in MLAS matrix composites).[39] Compared to metals, brittle materials are far more sensitive to the development of microstructural damage. Moreover, in fiber-reinforced ceramics, additional damage processes such as interface debonding and fiber fracture/pull-out will also contribute to the strain accumulation that occurs during creep loading.

To study the effects of microstructural damage accumulation on composite creep behavior, several damage processes may need to be modeled: (1) the development of matrix cracks and associated fiber bridging, (2) the progressive development of fiber fractures and isolated matrix cavities in an otherwise undamaged matrix, and (3) interfacial debonding which will be influenced by radial creep at the interface. Only a few models have addressed the effect of these damage modes on creep behavior. Among them, Chuang[43] considered the creep rupture of SiC whisker-reinforced Si_3N_4 by the coalescence of adjacent microcracks along fiber–matrix interfaces. Nair and Jakus[44–46] used a fracture mechanics approach to study the process of thermally activated crack-wake toughening, which was applied to crack propagation in a continuous SiC fiber-reinforced RBSN matrix composite. In their model, the fibers were allowed to deform elastically within a matrix that was either perfectly rigid or permitted to creep; the interface was allowed to slide in a Newtonian viscous fashion. The process involved crack-wake opening and crack tip extension. These models focused on the toughening effect of fibers and did not directly relate damage mode to the overall creep behavior of the composite. For the second type of damage process (periodic fiber fracture), models of short-fiber composites, such as those proposed by McLean *et al.*[26] and by Wang *et al.*[36] could be adapted. These models need to be extended however, to include the statistical nature of fiber rupture.

Although analytical models are very effective tools for understanding how changes in the elastic and creep behavior of the constituents of a composite influence overall creep behavior, one must not blindly assume that accurate predictions of composite creep behavior can be obtained based upon creep experiments conducted on the individual constituents. For instance, even if a monolithic ceramic and a composite were processed under identical conditions, the fracture and creep behavior of the monolithic ceramic may be quite different from the behavior that would occur under the constrained deformation experienced within a composite. In other words, the *in situ* creep behavior and fracture characteristics of the matrix or fibers may differ from predictions that are based upon data obtained from measurements performed on the individual constituents. The degree to which the fracture or creep behavior of the coupled constituents is changed will depend upon elastic constant mismatch, interface conditions, and packing parameters such as the volume fraction of fibers and fiber architecture (e.g., 0° or 0°/90°); processing conditions will also have a significant influence. Simple 1-D models do not account for the transverse constraint imposed by the 90° fibers; in this instance, numerical methods such as finite element analysis musts be employed. To fully appreciate the fracture behavior of the constituents within a composite, it may be necessary to conduct creep experiments using monolithic ceramics that are subjected to varying degrees of constraint (e.g., subjecting specimens to multiaxial loading or using deeply grooved specimens of various thickness to alter the degree of constraint). Chemical interaction can also lead to time-dependent changes in the creep behavior and fracture characteristics of the matrix and fibers, and in the degree of interfacial bonding; modeling these changes would be difficult (accurately measuring these changes would be exceedingly difficult).

5.3 Experimental Techniques for Creep Testing

As with monolithic ceramics, the cost and complexity of conducting creep experiments with fiber-reinforced ceramics requires careful design of the test specimen and equipment. This is particularly true of tensile creep testing, where the brittle nature of the matrix requires that the bending strains imposed on the test specimen by misalignment of the loading fixtures be kept to a minimum. Because of the transient nature of stress redistribution between the fibers and matrix, precise control over the rate of application and removal of a creep load is required. Morever, because of the low creep rates that are encountered, one must ensure that there are no temperature changes in the vicinity of the experimental setup; otherwise, fluctuations in the extensometer and load cell readings can obscure the true creep rate.

Figure 5.10 illustrates a typical experimental arrangement used for the tensile creep (or fatigue) testing of ceramic matrix composites. A servo-hydraulic load frame is used to enable accurate control over the loading and unloading rates. Moreover, the rigid nature of the load frame along with self-aligning test fixtures, allows one to readily obtain low bending strains. The temperature within the test chamber, as well as the temperature of the cooling water for the extensometer, is maintained at a constant temperature by using an isothermal water bath which maintains the temperature at $20 \pm 0.1°C$ (using city water as a cooling source is typically not a feasible alternative; in early experiments, the authors measured fluctuations of 5–10°C over a 24 h period!).

To prevent fluctuations in the load cell and extensometer readings, caused by changes in ambient temperature, it is advantageous to enclose these transducers completely within the test chamber.

An important consideration is the type of gripping arrangement that is used for specimen loading. Generally, edge-loaded or face-loaded grips are used for the creep testing of fiber-reinforced ceramics. A typical edge-loaded grip used for the creep testing of monolithic ceramics and ceramic matrix composites is shown in Fig. 5.11. As discussed elsewhere,[47] edge-loaded grips transfer the applied load to the specimen entirely along the tapered edges; there are no moving parts in the grips. For most fiber-reinforced ceramics, the simpler edge-loaded grips have proven very effective for tensile creep testing. Since these grips do not rely upon friction to transmit the applied load, specimens of large or small cross-sectional area can be tested. In contrast, face-loaded grips, which rely upon friction to transmit the creep load, are typically useful only for specimens with small cross-sectional areas (note that the transverse load required to prevent slippage of face-loaded specimens increases as the cross-sectional area increases—this increases the likelihood of specimen failure near the grips). If machined from Ni-base alloys or a ceramic (e.g., SiC), the grips can be used without cooling (the hydraulic grips typically used for face loading require water cooling in the vicinity of the grip faces, which can establish large temperature gradients in a specimen). This latter point is important, since the temperature at the ends of a specimen can influence the damage that occurs in the gauge section (e.g., by a gradient in thermal stress along the specimen axis). One last comment is in order. Namely, irrespective of the gripping arrangement that is utilized, it is important that the temperature distribution of a grip be symmetric. Non-uniform grip temperatures can develop by easily overlooked factors such as furnace insulation that is incorrectly placed, or by a hydraulic line that enters one side of a grip. This is an important consideration, since if one portion of a grip is slightly hotter than adjacent regions, non-uniform thermal expansion of the grip can directly introduce bending strains in a test specimen. Since specimen alignment is usually determined by using strain-gauged specimens at room temperature, the bending strains that are present when a specimen is heated to an elevated temperature are typically unknown.

The brittle nature of ceramics, and high testing temperatures, make the selection of an extensometer a key element in the experimental design process. There are many types of extensometers available for elevated temperature mechanical testing, including mechanical, optical and laser-based systems. No system is perfect. Optical and laser extensometers, which are non-contacting, eliminate the introduction of bending strains into a specimen, which can be of concern when mechanical contacting-type extensometers are used. However, these extensometers can be susceptible to fluctuations caused by changes in the density of air between the laser window and the test specimen (see Wiederhorn and Hockey[48]). Because of signal processing limitations, most commercially available laser extensometers have a low frequency response and are generally

Fig. 5.10(a)   Experimental arrangement used for the tensile creep testing of fiber-reinforced ceramics showing isothermal test chamber mounted on a servohydraulic load frame. The load cell and extensometer are located inside the test chamber to minimize fluctuations in the transducer readings. Ambient temperature is controlled by use of an isothermal water bath that circulates water through the chamber walls (the portion of the hydraulic ram is also cooled). (b) Schematic of induction-heated furnace, tensile creep specimen, extensometer location and edge-loaded grips (note that any type of furnace that ensures a uniform and stable temperature over long periods of time is suitable). After Holmes.[47]

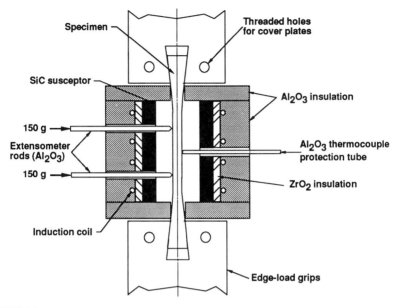

Fig. 5.10(b)

limited to loading frequencies of at most a few hertz, which may be of concern for cyclic creep or fatigue experiments (these frequency limitations can be overcome by use of optical extensometers, which can be used at loading frequencies in the kHz range). The authors have had considerable success with inexpensive mechanical extensometers. However, these extensometers are very susceptible to changes in the temperature and flow rate of the water used to cool the extensometer; for stable readings, these extensometers *must* be used with isothermal water baths and flow-rate controllers. The authors utilize commercially available water baths that maintain the temperature of the extensometer cooling water to within 0.1°C of the setpoint.[38] Also, with mechanical extensometers, one must ensure that the ceramic loading rods do not slip during specimen loading and unloading, or when matrix or fiber fracture causes a mechanical disturbance of the specimen. Since the contact force of the extensometer rods must be low to avoid introduction of bending strains, it may be necessary to use dimples on the specimen surface to minimize the likelihood of extensometer slippage. For composites such as SiC_f/CAS, $SiC_f/HPSN$, and $SiC_f/RBSN$, the use of shallow dimples (≈ 50–100 μm deep), machined with diamond-tipped tools, has proven very effective. For woven composites, non-contact extensometers are preferred; otherwise, for mechanical extensometers, the tips of the extensometer rods can be attached to the specimen with a ceramic adhesive (assuming that the adhesive does not react with the composite).

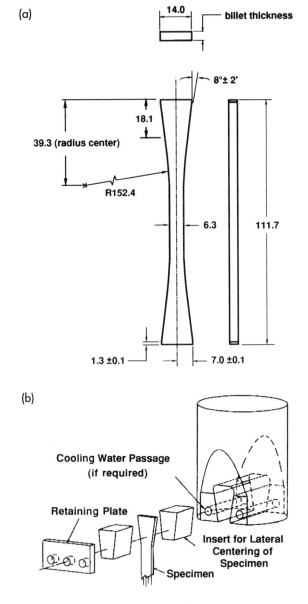

Fig. 5.11 Edge-loaded specimen geometry and grips used for tensile creep testing of fiber-reinforced ceramics.[47] (a) Single-reduction edge-loaded tensile specimen with 8° tapered ends. (b) Typical grip arrangement used with edge-loaded specimens. Edge-loaded specimens, which have many advantages over face-loaded specimen geometries, have been successfully used to test a variety of 1-D and 2-D fiber-reinforced ceramics. Because hydraulic components are not used, the grips can be used without direct cooling (if machined from Ni-base alloys or ceramics). If cooling is required, water passages can be located in the grips near the specimen ends.

The time-dependent change in stress distribution between the fibers and matrix that occurs during specimen loading and unloading, makes it crucial that the precise loading history of a creep specimen be specified. This is of particular concern, since the rate at which a creep load is initially applied to a specimen can have a profound influence on subsequent creep behavior and damage mode (see discussion in Section 5.2.5). Likewise, the stress redistribution that occurs upon specimen unloading and cooldown after a creep experiment will influence the measured strength and toughness of a composite. One must also be careful if periodic inspections of a specimen are made during a creep experiment; stress redistribution during specimen unloading and furnace cooldown may alter the damage mechanism upon subsequent reloading.

To summarize, flexural and tensile creep testing of fiber-reinforced ceramics is now routinely performed. However, several precautions are required to ensure the validity of the test data. When testing any brittle matrix composite, it is important to minimize bending strains introduced by load-train misalignment or by poorly designed grips. Because of the exceedingly low creep rates encountered with ceramic composites, fluctuations in the ambient temperature near the test setup must be minimized. This is most easily accomplished by using a test chamber connected to an isothermal water bath. The transient nature of stress redistribution, discussed earlier, and its effect on initial microstructural damage, requires that one carefully documents the loading and thermal histories used in creep testing.

5.4 Experimental Studies of Creep Behavior

Table 5.1 provides a summary of the composite systems that have been studied to date, along with the test conditions and the references that can be consulted for additional information. When discussing creep behavior, it is instructive to broadly classify the ceramic composites into groups with different creep mismatch ratios. Since the creep behavior of the individual constituents is not always known, this separation is somewhat subjective; however, as will be shown in this section, such a separation is consistent with the observed creep damage modes. Also, as discussed earlier, it should be kept in mind that the creep behavior of ceramic matrix composites is strongly dependent on the initial damage state of the composite, which is influenced by processing and prior loading history, and by the rate at which a creep load is initially applied; this information is not available for all of the studies.

Due to space limitations, a complete review of published literature is not possible. Thus, the results from selected studies that provide insight into creep behavior and microstructural damage accumulation are highlighted. For convenience, this section is divided according to the type of reinforcing fiber and matrix: (1) Nicalon fiber/glass ceramic matrix composites, (2) SCS-6 fiber/Si_3N_4 matrix composites, (3) Al_2O_3 fiber composites, and (4) carbon fiber

Table 5.1 Composite systems, expected creep mismatch ratios and test conditions for the composites that have been studied to date. The creep behavior of the composites marked with an asterisk is discussed in greater detail in this section

Composite (manufacturer)	Loading	Temperature, °C (environment)	Stress, MPa	References
(CMR <1)				
*Nicalon SiC_f/CAS (Corning)	Tensile	1100–1300 (argon, <10 ppm O_2)	60–250	Wu and Holmes[15,41]
Nicalon SiC_f/1723 (Textron)	Tensile	600, 700, 800		Khobaib and Zawada[49]
*Nicalon SiC_f/MLAS (Aérospatiale, d'Aquitaine, France)	Flexural (3-pt)	900–1200 (vacuum)	25–400	Kervadec and Chermant[39,40]
*Nicalon SiC_f/CAS (Corning Glass Works)	Flexural (4-pt), Comp.	1200 (argon), 1200 (argon)	50–150, 20–75	Weber et al.[50]
Nicalon/CAS-III (Corning Glass Works)	Comp.	1300, 1310 (argon)	35	Meyer et al.[33]
*3-D T-300 C_f/SiC (DuPont)	Tensile	1400 (argon, 100 ppm O_2)	30–100	Holmes and Morris[51]
*SCS-6 SiC_f/HPSN (dry powder lay-up) (Textron)	Tensile	1200 (air), 1350 (air)	99–135	Holmes[52,53]
SCS-6 SiC_f/HPSN^a (tape cast) (Textron)	Tensile	1200 (air), 1315 (air)	60–250, 30–150	Holmes et al.[38,54]
(CMR >1)				
*SCS-6 SiC_f/RBSN (NASA-Lewis)	Tensile	1300 (N_2, <5 ppm O_2)	90–150	Hilmas et al.[42]
(CMR ~1)				
*2-D-Al₂O₃f/CVD-SiC (SEP, Bordeaux, France)	Tensile	950–1100 (vacuum $<10^{-6}$ mbar)	90–200	Adami[30]
(CMR—not easily identified)				
**Nicalon SiC_f/SiC^b (SEP, Bordeaux, France)	Flexural	900–1200 (vacuum)	50–300	Abbé et al.[55] Abbé and Chermant[56]

a This system is representative of the hot-pressed composites that are currently produced by Textron Specialty Materials.

b Insufficient information is available in the literature concerning the creep behavior of CVI-SiC (note that there is typically from 8 to 20% porosity in the matrix of typical CVI composites).

composites. A comparison of the 1200°C creep rate of several of the composites is given in Section 5.4.6.

5.4.1 Nicalon Fiber/Glass-Ceramic Composites

5.4.1.1 Flexural Creep of Nicalon SiC$_f$/CAS Composites

The flexural creep behavior of $[0]_{16}$-Nicalon/CAS (40 vol.%) was studied by Weber *et al.*[50] at a temperature of 1200°C. The specimens had a cross section of 3 mm × 4 mm and were loaded under four-point bending with a major to minor span ratio of approximately 0.5 (21 mm/39 mm). The creep experiments, conducted in argon at flexural stresses of 50–150 MPa, were typically terminated within 50 h (prior to failure) to allow characterization of creep damage.

Figure 5.12 shows the stress dependence of flexural creep rate as a function of accumulated creep strain. A transient, decelerating creep rate was observed for all stress levels examined (note that when plotting creep rate versus creep strain, a constant creep rate would indicate the existence of steady-state creep). Similar transient creep behavior has been observed during the tensile creep of 0° and 0°/90° Nicalon/CAS-II composites (discussed below).[15,41] At 1200°C, the matrix contributes very little to the overall creep resistance of the composite (at 1200°C, and for the flexural loading discussed

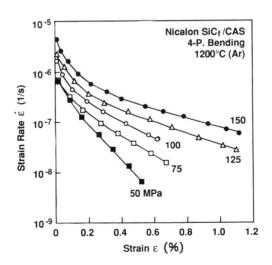

Fig. 5.12 Flexural creep rate versus accumulated creep strain and creep stress for [0]16-Nicalon/CAS crept at 1200°C in argon. For all stress levels, the creep rate decreases continuously with time. The decelerating transient with accumulated strain is attributed to grain growth in the Nicalon fibers, which increases their resistance to diffusional creep. From Weber *et al.*[50]

above, Weber *et al.*[50] estimated that approximately 95% of the creep load was carried by the fibers).

During the course of the experiments, abnormal grain growth was observed within the fibers, forming a shell that extended inward from the fiber–matrix interface; this abnormal growth is caused by the outward diffusion of excess carbon from the fibers. Within this outer shell, the average grain size had increased from the several nanometer range found in virgin fibers and in the interior of the crept fibers, to approximately 10–15 nm. Similar abnormal grain growth was found by Mah *et al.*[57] in experiments conducted to determine the influence of heat treatment on the microstructure of Nicalon fibers. For a diffusional creep mechanism, an increase in the average grain size would decrease the creep rate of the fibers, consistent with the transient decrease in creep rate found for the composite. Thus, in addition to influencing transient creep behavior, grain growth in the Nicalon fibers will also modify the rate and extent of stress redistribution that occurs between the fibers and matrix.

5.4.1.2 Transverse Compressive Creep of Nicalon SiC/CAS Composites

It is difficult to process calcium aluminosilicate (CAS) in monolithic form with the same structure that would exist in a composite.[58] To avoid these difficulties, and to provide details concerning the *in situ* creep behavior of the CAS matrix and damage mechanisms, Weber *et al.*[50] conducted transverse compressive creep experiments with $[0]_{16}$-Nicalon/CAS composites. All experiments were conducted in argon.

The stress dependence of transverse compressive creep rate is plotted in Fig. 5.13 as a function of accumulated creep strain. Note that the "rigid" fibers will tend to constrain the creep flow of the softer and less creep-resistant matrix. The initial decelerating transient in strain rate can be attributed to the development of hydrostatic stresses between the fibers, which reduce the deviatoric component of strain. At a flexural stress of 20 MPa, the creep rate approached a constant value of $6 \times 10^{-8} \, \text{s}^{-1}$ at a strain of 0.4%. In contrast, for flexural stresses of 50 and 75 MPa, the creep rate increased after reaching a minimum at strains of approximately 2% and 1.5%, respectively. The increase in creep rate, which was most apparent at 75 MPa, is caused by the development of additional damage modes in the composite; primarily the coalescence of voids that form in the vicinity of the fiber–matrix interface, which leads to interface separation. These voids coalesce in the regions of high shear between fibers that are at an angle of 45° from the compressive loading axis.

5.4.1.3 Tensile Creep of 0° and 0°/90° Nicalon SiC$_f$/CAS-II

The tensile creep behavior of 0° and 0°/90° Nicalon SiC/CAS-II composites has been studied in detail by Wu and Holmes.[15,41] The composites, which had 40 vol.% fibers, were fabricated by hot pressing (Corning Glass Works, Corning, NY). Creep experiments were conducted in a high purity argon

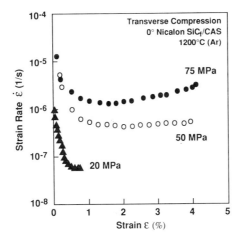

Fig. 5.13 Transverse compressive creep rate versus accumulated creep strain for 0°
Nicalon/CAS crept at 1200°C in argon. During compressive creep, matrix creep controls
the overall creep rate of the composite. After Weber *et al.*[50]

atmosphere at 1100°C, 1200°C, and 1300°C. At temperatures of 1200°C and
higher, the matrix contributes little resistance to creep deformation. Thus, this
study provided information regarding creep damage mechanisms in composites
where the creep mismatch ratio is considerably less than unity.

Influence of Stress, Temperature and Total Plies on Creep Behavior. The
1200°C creep behavior of unidirectional specimens with 16 and 32 plies is
shown in Fig. 5.14a for creep stresses of 60 MPa to 250 MPa and times up to
100 h (these stresses ranged from approximately 13% to 52% of the 1200°C
monotonic strength). There was no significant difference in the creep behavior
of the 16- and 32-ply specimens. For stresses up to 200 MPa all specimens
survived the 100 h creep test; for these stresses the composite exhibited a
decelerating creep rate—steady-state creep was not observed. Increasing the
creep stress to 250 MPa caused rapid failure within approximately 70 minutes.
Figure 5.14b shows the temperature dependence of creep behavior at a fixed
stress of 100 MPa. It is interesting to note that the accumulated creep strain at
1200°C and 1300°C is significantly higher than the failure strain typically quoted
in the literature for Nicalon fibers[59] (as discussed below, this is attributed to
fragmentation of the fibers within the composite, which allows a considerable
amount of strain without composite failure).

Differences in the Creep Behavior of 0° and 0°/90° Composites A
comparison of the creep behavior of the 0° and 0°/90° composites is shown in
Figure 5.14c. Several unexpected trends were found that provide valuable
insight into the creep behavior of fiber-reinforced composites. For example, in

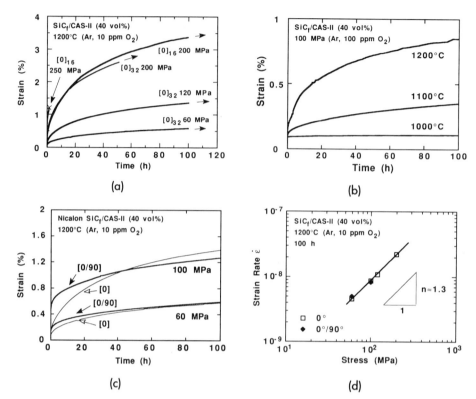

Fig. 5.14 Creep behavior of Nicalon SiC$_f$/CAS-II composites crept in a high purity argon atmosphere. (a) Influence of total plies and creep stress on the strain accumulation (elastic + creep) of 0° composites. (b) Temperature dependence of strain accumulation (elastic + creep). (c) Comparison of the 1200°C creep rate of 0° and 0°/90° SiC$_f$/CAS-II composites at stresses of 60 MPa and 100 MPa. Although the 0°/90° composite has effectively one-half the fibers in the loading direction compared to the 0° composite, the 100 h creep rate and strain accumulation was equal to, or lower than, that of the 0° composite. (d) Stress dependence of 100 h creep rate for the 0° and 0°/90° composites at 1200°C. Data taken from Wu *et al*.[15,41]

creep experiments conducted at 60 MPa and 100 MPa, the overall strain accumulation of the 0° and 0°/90° composites was similar, even though the 0°/90° composite had effectively only one-half (20 vol.%) of the fibers in the axial direction (this result is even more dramatic when it is realized that the initial elastic strain on loading is much higher in the lower modulus 0°/90° composites). The 100 h creep rate of the two composites is shown in Fig. 5.14d for a temperature of 1200°C. As with the accumulated strain, the 100 h creep rates are very similar. For both the 0° and 0°/90° composites, the creep stress exponent was approximately 1.3 at 100 h. This stress exponent is similar to that found for the creep of Nicalon fibers,[62] indicating that the composite creep rate appears to be primarily controlled by creep of the Nicalon fibers.

As noted above, for the 0°/90° specimens only 20 vol.% of the fibers are in the axial loading direction; thus, compared to the 0° composite, one would expect a much higher creep rate and strain accumulation for the same applied stress level. The similar strain accumulation, which was verified by duplicate testing, is attributed to the constraint to matrix creep provided by the rigid (non-creeping) fibers in the 90° orientation (this effect has been verified by finite element analysis modeling of the creep behavior of 0°/90° composites).[31] In effect, the transverse fibers increase the axial creep resistance of the matrix by decreasing creep flow in the matrix. These results show that transverse fibers can contribute significantly to overall creep resistance. Moreover, these results clearly show that the effect of transverse fibers cannot be neglected in analytical models of creep deformation. As discussed in Section 5.5.2 on cyclic creep behavior, the 0°/90° composites also exhibited significantly more strain recovery than the 0° composites, adding a further complication to analytical modeling of creep deformation.

Microstructural Damage and Discussion   Microstructural studies of damage accumulation were conducted on the specimens that were crept at 1200°C. After 100 h of creep at 60 MPa there was no evidence of fiber or matrix fracture. However, cavities had formed in the matrix; the density of cavities was much higher within fiber-rich regions. This cavity formation is consistent with observations made by Weber *et al.*[50] of cavity formation in 0° Nicalon SiC$_f$/CAS composites that had been crept in flexure at 1200°C. Increasing the creep stress to 120 MPa caused limited matrix microcracking and fiber fracture; this random microstructural damage can be attributed to inhomogeneities in the fiber distribution, since the microcracking generally occurred in regions of the specimen that were matrix rich. During creep at 200 MPa, the matrix cavities began to link up, forming networks within the fiber-rich regions of the specimens. Far more extensive matrix cracking, along with periodic fiber rupture, was also observed (see Fig. 5.15a and b). The matrix cracks, which were located primarily in matrix-rich regions of the composite, most likely formed during initial application of the creep load. Because of rapid relaxation of the matrix stress after the initial loading transient, the cracks do not extend further, but can open up parallel to the applied load as creep of the bridging fibers occurs.

The periodic fiber fracture that occurs during tensile creep at 200 MPa is a direct consequence of load transfer from the matrix to the more creep-resistant fibers, which leads to a time-dependent increase in fiber stress. This creep damage occurs when the creep rate of the matrix significantly exceeds that of the fibers. Although periodic fiber fragmentation occurred during creep at 200 MPa, the composite exhibited a continually decelerating creep rate. Two factors are thought to contribute to the decelerating creep rate: (1) grain growth in the fibers, and (2) realignment of off-axis fibers with the creep loading direction. Grain growth in the fibers will result in a time-dependent increase in the creep resistance, and a corresponding decrease in creep rate. Not all of the

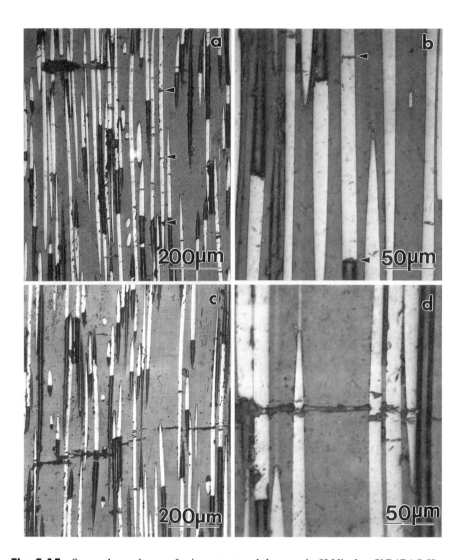

Fig. 5.15   Stress dependence of microstructural damage in 0° Nicalon SiC$_f$/CAS-II composites crept at 1200°C in argon. (a, b) Periodic fiber fracture observed after 100 h of creep at 200 MPa (the specimen exhibited a continually decreasing creep rate). The arrows mark the locations of periodic fiber fracture along one of the fibers. (c, d) Matrix fracture and rupture of bridging fibers observed after 70 min of creep at 250 MPa (the micrographs were taken approximately 5 mm from the failure location). After Wu and Holmes.[15]

fibers in the composites that were tested were aligned perfectly with respect to the tensile loading direction and will, therefore, initially not contribute as much to creep resistance compared to perfectly aligned fibers. At sufficiently high temperatures, creep deformation of the matrix allows these off-axis fibers to "straighten" parallel to the applied load. As additional fibers begin to fully share the creep load, the overall creep resistance of the composite will increase, leading to a decrease in the composite creep rate.

At 250 MPa, where the creep life was short, significantly more matrix microcracking would occur during application of the creep load. Rupture of the highly stressed fibers that bridge these matrix cracks appears to be the likely mechanism responsible for the significant reduction in creep life. Micrographs showing the matrix fracture and rupture of the bridging fibers are given in Fig. 5.15c and d. Because the creep life was short, periodic fiber fracture remote from these initial cracks was not observed. In summary, the stress and time dependence of creep damage can be summarized into three regimes: (1) low stress/long duration creep, which leads to cavity formation, (2) moderate stress/long duration creep, characterized by both cavity formation and periodic fiber fracture (without matrix fracture), and (3) high stress/short duration creep, characterized by the rupture of fibers that bridge matrix cracks formed during initial loading.

5.4.1.4 Flexural Creep of 0° Nicalon SiC$_f$/MLAS

Kervadec and Chermant[39,40] studied the flexural creep of unidirectional Nicalon SiC$_f$/MLAS composites. Specimens were fabricated by the hot pressing of fiber preforms that were infiltrated with a slurry of 0.5 MgO–0.5 LiO$_2$–1.0 Al$_2$O$_3$–4.0 SiO$_2$. The residual porosity in the matrix was below 1%. Creep experiments were performed in vacuum under three-point bending over the temperature range of 900–1273°C at stresses from 25 to 400 MPa.

Stress and Temperature Dependence of Creep Behavior The stress and temperature dependence of flexural creep is given in Fig. 5.16a and b. Kervadec and Chermant indicate that for low temperatures and stress levels the stress exponent for creep is relatively constant between approximately 0.3 and 0.6. It is not clear why the *n* value is so low, but this may be an artifact of the nonuniform stress distribution developed during flexural creep.

Microstructural Damage and Discussion For the creep loading histories examined, three microstructural damage modes were found. At temperatures of 900°C and 1000°C, and for low creep stresses, debonding along the fiber–matrix interface and matrix cracking parallel to the fibers was the primary damage mode. This cracking is caused by the shear stresses developed during flexural loading. At 1100°C, rupture of the fibers located on the tensile surface of the flexural specimens and delamination occurred. At high temperature (1200°C), creep deformation was controlled by creep of the Nicalon fibers; at this temperature the creep mismatch ratio is expected to be considerably smaller than unity; the matrix will flow by creep to accommodate the large increase in fiber strain.

Analysis of the Arrhenius plot indicated that the thermal activation energy was constant at low temperatures (900–1000°C) with a value of approximately $80 \, kJ \, mol^{-1}$. Because of the low thermal activation energy for creep, Kervadec and Chermant proposed that creep deformation for low temperatures was controlled by microcracking in the matrix. For higher temperatures (1100–1200°C), the thermal activation energy increased with both temperature and creep stress to a maximum value of $400 \, kJ \, mol^{-1}$, which is of the same order of magnitude as that found by Bunsell et al.[60] for the creep of SiC_f fibers. Thus, for high temperatures and applied stresses, the creep rate of the composite is governed primarily by creep of the SiC fibers.

5.4.2 Nicalon SiC_f/SiC Composite: Flexural Creep of 0°/90° SiC_f/SiC

Abbé et al.[55,56] investigated the flexural creep behavior of 0°/90° Nicalon SiC_f/SiC composites at temperatures of 1100–1400°C and stress levels of 50–300 MPa. The composites were manufactured by chemical vapor infiltration (CVI) of SiC into woven Nicalon-fiber preforms. The infiltrated composites had a residual matrix porosity of 10–15%. Creep experiments were conducted in vacuum under three-point bending. Test durations ranged from 50 to 200 h, which, according to the authors, was sufficient to establish a steady-state creep rate for the range of stresses examined (approximately 30–100 MPa at 1200°C).

Stress and Temperature Dependence of Creep Rate The stress and temperature dependence of flexural creep rate is shown in Fig. 5.17. For a temperature of 1100°C and a stress of 100 MPa, the creep rate of the composite was approximately $2 \times 10^{-9} \, s^{-1}$; at 1400°C, and for the same stress, the creep rate increased to $4 \times 10^{-8} \, s^{-1}$. These creep rates are higher than the creep rate of siliconized SiC[61] for similar temperatures and stress levels. For a stress of 150 MPa and temperature of 1500°C, the creep rate of monolithic α-SiC[62] is still roughly an order of magnitude lower than the creep rate of the composite. Since the composite was crept in vacuum, the higher creep rate of the composite was most likely a consequence of the high porosity content of the CVI-SiC matrix which will increase the likelihood of matrix fracture. In essence, the creep rate of the composite is controlled by the creep rate of the fibers that bridge matrix cracks. Note also that there is a high stress concentration in the matrix and fibers in the vicinity of the locations where the fiber bundles cross over one another. Because of the stress concentration near the crossover points in woven-fiber composites, the actual stress and creep rate of the fiber bundles that bridge matrix cracks may be quite high (at these locations the stress state experienced by the bridging fibers will be a combination of bending and tension).

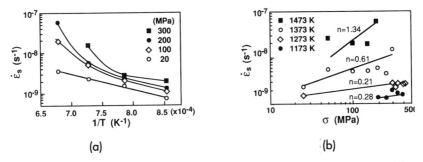

(a)   (b)

Fig. 5.16 Flexural creep behavior of 0° Nicalon SiC$_f$/MLAS composites tested in vacuum at temperatures between 900°C and 1200°C.[39,40] (a) Quasi steady-state creep rate versus temperature. (b) Stress and temperature dependence of quasi steady-state creep rate showing the change in creep stress exponents with temperature.

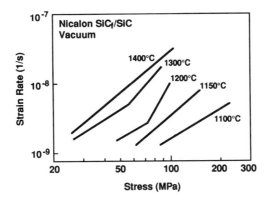

Fig. 5.17 Flexural rate of 0°/90° Nicalon SiC$_f$/SiC at 1100°C and 1400°C. The experiments were conducted in vacuum. After Abbé *et al.*[55,56]

Microstructural Damage and Discussion   According to Abbé and Chermant, microcrack growth in the matrix and the creep of bridging fibers are the dominant creep mechanisms operating in this composite system. The bilinear nature of the creep curves at temperatures of 1200°C and 1300°C appears to be caused by crack growth in the matrix and creep of the bridging fibers as the creep stress is increased. At temperatures of 1200°C and higher, grain growth in the Nicalon fibers occurs. At 1400°C, the stress exponent for creep decreases and the bilinear nature of the creep curves observed for temperatures of 1200°C and 1300°C is absent. Abbé and Chermant attribute this change in behavior to the blunting of matrix cracks by surface diffusion; this blunting decreases the likelihood of further crack extension in the matrix.

Although the creep rates of current-generation SiC$_f$/SiC composites are higher than those found for monolithic SiC, it should be appreciated that monolithic silicon carbide would not be suitable for use in structural applications requiring a material with high toughness. For engineering applications, the relatively low creep strength of Nicalon/SiC composites will limit their use to low stress applications where a low density material with high monotonic toughness and thermal shock resistance are the important design parameters. These applications include thermal shields for aerospace structures, and gas-turbine components such as exhaust flaps; heat exchangers, which operate under low internal pressure, are another potential application. Although improvements in creep resistance are necessary, SiC fiber/SiC matrix composites are not as susceptible as other ceramic matrix composites to frictional heating during cyclic loading, making them good candidates for structural applications where high frequency fatigue is encountered (frictional heating is discussed in Chapter 6). Since little is known about the creep behavior of the CVI-SiC matrix used in this composite, it is not clear if reductions in the porosity of the matrix would significantly improve creep resistance as a further reduction in the matrix creep resistance could establish an unfavorable creep mismatch ratio (CMR > 1) which would ultimately lead to matrix fracture by load transfer from the fibers. On the other hand, a decrease in matrix porosity may decrease the likelihood that matrix fracture will occur during initial application of the creep load, and have the added advantage of increasing the tensile strength and improving the oxidation resistance by closing interconnected matrix porosity. Additional research will be required to provide answers to these questions.

5.4.3 SCS-6 SiC_f/Si_3N_4 Composites

5.4.3.1 Tensile Creep of 0° SCS-6 SiC_f/HPSN

Holmes[52,53] has investigated the tensile creep behavior of 0° SCS-6 SiC_f/HPSN composites in air at temperatures of 1200–1350°C. The composites, which contained 28–30 vol.% fibers, were fabricated by hot pressing at 1700°C (Textron Specialty Materials, Lowell, MA, USA). Two versions of this hot-pressed composite were studied, an early version of the composite processed by dry powder lay up and a more mature version processed by tape casting; the tape-cast composite had a far more uniform fiber distribution and more uniform matrix composition. For brevity, the majority of this section will be devoted to the creep behavior of the tape-cast composites, although highlights of several interesting results from a study of the creep behavior of the dry powder lay up composite at a temperature of 1350°C will also be discussed.

Stress and Temperature Dependence of Creep   The tensile creep behavior of the tape-cast composite at 1200°C is shown in Fig. 5.18a for creep stresses of 0 to 200 MPa. For stresses of 75 MPa and higher, transient (primary creep) persisted for approximately 50–75 h, followed by a region of nearly

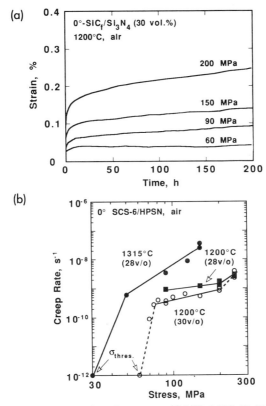

Fig. 5.18 Tensile creep behavior of tape-cast 0° SCS-6 SiC_f/Si_3N_4 composites. (a) Typical creep behavior at applied stresses from 0 to 200 MPa (the initial elastic strain on loading has been included in the curves). (b) 100 h creep rate versus applied stress for composites with 28 and 30 vol.% fibers. At 1200°C, the creep rate dropped below detectable limits at a creep stress of 60 MPa. (Data at 1315°C are from unpublished work by Holmes.)

constant creep rate. The stress dependence of 100 h creep rate is plotted in Fig. 5.18b. Several interesting results were obtained from the study:

(1) A threshold stress of 60 MPa at 1200°C and 30 MPa at 1315°C was observed (at this stress the creep rate of the composite dropped below detectable levels (the order of 10^{-11} to $10^{-12}\,s^{-1}$).

(2) The stress dependence of creep rate was low, with a stress exponent of approximately 1.0 at 1200°C, but increased to approximately 3 at 1315°C.

(3) At stresses above 250 MPa, the creep rate increases sharply (this occurs even if the specimen is slowly loaded to avoid initial matrix fracture during application of the creep stress). This increase is attributed to a combination of fiber and matrix fracture that occurs during creep.

Microstructural Damage and Discussion   As with the creep of Nicalon SiC$_f$/CAS composites discussed earlier, periodic fiber rupture (without accompanying matrix fracture) was identified as the primary microstructural damage mode for the tensile creep of the HPSN composite. For stresses of 150 MPa and 200 MPa, periodic fiber rupture (without accompanying matrix fracture) was identified as the primary creep damage mode. This damage mode is expected in composites with creep mismatch ratios ($\dot{\varepsilon}_f/\dot{\varepsilon}_m$) less than unity. As discussed in Section 5.2.5, for a creep stress of 250 MPa, the creep life was strongly influenced by the rate at which the creep stress was applied. Rapidly applying the creep stress in 2.5 s resulted in failure in approximately 0.1 h; applying the creep stress in 1000 s increased the creep life dramatically (to between 117 and 167 h). The influence of loading rate on creep life can be understood in terms of the transient stress redistribution that occurs during loading. Rapid loading does not allow sufficient time for relaxation of the matrix stress, resulting in the formation of matrix cracks during initial loading.

If the threshold stress for creep deformation of a composite is controlled by the onset of creep deformation in the fibers, it could be raised by increasing the fiber fraction. By designing a component such that the design stress is below the threshold for composite creep, the life of a component would not be creep limited. Also, by designing below a threshold stress, the need for potentially risky extrapolation of short-term creep data could be avoided (of course, this design approach ignores the possibility of other damage modes, such as impact or environmental damage, which must be accounted for irrespective of the creep properties of a composite).

As noted above, in the absence of matrix fracture, the creep rate of hot-pressed SCS-6 SiC$_f$/Si$_3$N$_4$ composites is fiber dominated. Thus, for the same matrix composition and processing parameters, the most efficient approach for increasing the creep resistance of SiC$_f$/Si$_3$N$_4$ composites would be to increase the volume fraction of fibers. Since increasing the fiber fraction will also increase the proportional limit stress, this approach has the added advantage of reducing environmental sensitivity caused by matrix cracking. Alternatively, the creep resistance of the Si$_3$N$_4$ could be improved by reducing the amount of sintering oxides used to achieve consolidation, or by using a different processing route such as reaction bonding. However, it would be a mistake to increase the creep resistance of the HPSN matrix above that of the reinforcing fibers; otherwise, unless parallel improvements in the fiber creep resistance are made, the unfavorable creep mismatch ratio will result in the development of matrix cracks during tensile or flexural creep.

5.4.3.2 Tensile Creep of SCS-6 SiC$_f$/RBSN

Hilmas *et al.*[42] investigated the tensile creep behavior of 0° SCS-6 SiC$_f$/RBSN (reaction-bonded silicon nitride) composites at 1300°C in a high purity nitrogen environment. The composites, fabricated at 1200°C, contained 24 vol.% SCS-6 SiC fibers. Because matrix oxides are not required for processing, the RBSN matrix has an intrinsically lower creep rate than that of

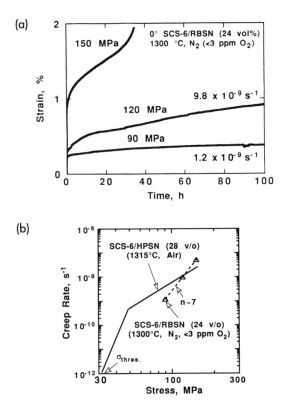

Fig. 5.19 Tensile creep behavior of 0° SCS-6 SiC$_f$/RBSN at 1300°C.[42] (a) Total strain (elastic plus creep). (b) 100 h creep rate versus applied stress for RBSN and HPSN composites (for a stress of 150 MPa, the creep rate of the RBSN composite was taken at 20 h).

current-generation SiC fibers, such as SCS-6 SiC or Nicalon SiC fibers. The results of this study provided insight into the creep damage mechanisms that occur in composites with creep mismatch ratios (CMR) greater than unity.

The tensile creep experiments were performed at applied stresses of 90, 120, and 150 MPa. Experiments were conducted for a maximum of 100 h (to determine the evolution of creep damage with time; additional experiments at 120 MPa were conducted for 1 h and 50 h). Tensile creep curves for stresses of 90, 120, and 150 MPa are shown in Fig. 5.19a. At 90 MPa and 120 MPa, the composite exhibited a continually decreasing creep rate up to the maximum creep time of 100 h. At 150 MPa, the creep rate increased sharply above 20 h, with specimen failure occurring in under 40 h. The stress dependence of 100 h creep rate is shown in Fig. 5.19b. For comparison with the RBSN composite, the 100 h creep rate of tape-cast 0° SCS-6 SiC$_f$/HPSN composites with 28 vol.%

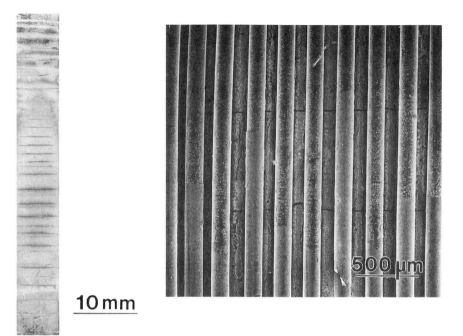

10 mm

Fig. 5.20 Typical matrix damage observed in unidirectional SCS-6 SiC$_f$/RBSN composites after 100 h of creep at 1300°C in a high purity nitrogen atmosphere. A constant crack density was achieved within the first 50 h of creep.

fibers is also included (the creep experiments with the HPSN composite were performed at the slightly higher temperature of 1315°C). At a stress of 90 MPa, the creep rate of the RBSN composite was lower than the HPSN composite. However, at higher stresses, the creep rate of the RBSN matrix composite was equivalent to, or *higher* than that of the hot-pressed composite (it should also be noted that the HPSN composites survived 100 h at 150 MPa,[54] versus less than 40 h for the RBSN composite). As discussed below, the lower creep life of the RBSN composite is attributed to matrix cracking which significantly increases the stress on the portions of fibers that bridge the cracks.

Microstructural Damage and Discussion Creep for 100 h at 90 MPa produced no observable fiber or matrix damage. At 120 MPa, periodic matrix cracking was observed throughout the specimen gauge-section (see Fig. 5.20); from experiments conducted for 1, 50 and 100 h, it was determined that matrix cracking started within the first 1 h of creep, with a saturated crack spacing achieved within 50 h (the average crack spacing at 50 h was approximately 1.3 mm). The crack spacing at 100 h was similar to that found at 50 h; the primary change in damage state was a significant opening of the matrix cracks.

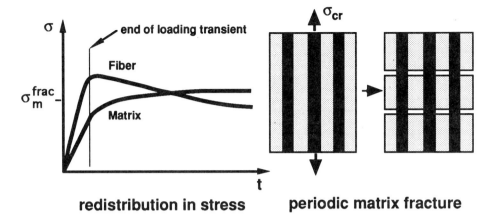

redistribution in stress   **periodic matrix fracture**

Fig. 5.21 Schematic illustrating the transient increase in stress that occurs in composites such as SCS-6 SiC/RBSN with creep mismatch ratios (CMR) greater than unity. Periodic matrix cracks can form if the matrix stress achieves a critical stress for fracture. The stress redistribution curves shown above hold prior to the development of matrix cracks.

The additional crack opening is caused by creep of the bridging fibers (fiber rupture was not observed during creep at 120 MPa). Creep at 150 MPa also resulted in periodic matrix fracture (at this stress level failure occurred in less than 40 h). As expected for the higher creep stress, the matrix cracks had a finer spacing—roughly 0.5 mm, compared to the average of 1.2–1.3 mm found after creep at 120 MPa.

It is interesting to note that creep rate of the HPSN composite, which has a less creep-resistant matrix, is lower than that of the RBSN composite for stresses above approximately 120 MPa. This can be explained in terms of the fundamental difference in creep damage mode between the two composites. During tensile creep loading, the high creep resistance of the RBSN matrix causes a progressive increase in matrix stress as load is shed from the fibers, as qualitatively illustrated in Fig. 5.21. The increase in matrix stresses causes the development of periodic matrix cracking, with load transferred across the cracks by the bridging fibers. This is an undesirable damage mode for long-term creep applications, since the creep life of the composite would be controlled by the rupture strength of the highly stressed bridging fibers. In contrast, because of its lower creep resistance relative to that of the fibers (CMR < 1), matrix fracture is avoided in the HPSN composite. *When designing for creep resistance, one should strive for a creep mismatch ratio less than unity.*

The successful use of RBSN matrix composites will require fibers with higher creep resistance. With current generation fibers, a closer match between the creep rates of the fibers and matrix could be achieved by decreasing the creep resistance of the matrix through the addition of oxides such as MgO. It

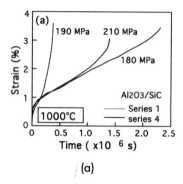
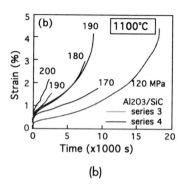

(a)   (b)

Fig. 5.22 Typical creep curves obtained during the tensile creep of 2-D $Al_2O_{3(f)}$/ CVD-SiC. (a) 1000°C, and (b) 1100°C. (Note: the series number represents different billets of material.) After Adami.[30]

may also be beneficial to increase the volume fraction of fibers in an effort to lower the stress on the bridging fibers. In addition, the inherently high porosity levels found in RBSN composites, although attractive from a density view-point, will require the use of oxidation protection schemes.

5.4.4 2-D $Al_2O_{3(f)}$/SiC Composite

Adami[30] studied the tensile creep behavior of an $Al_2O_{3(f)}$/CVD-SiC composite with a 2-D cross-ply fiber architecture (manufactured by Société Européenne de Propulsion (SEP)). This is the only study on an oxide fiber–SiC matrix system that has been conducted to date. The study is also noteworthy because of its completeness, including modeling of creep damage mechanisms. The creep experiments were performed under high vacuum ($<10^{-6}$ mbar); creep stresses were in the range of 90–200 MPa, which is above the first matrix cracking stress. From data given by Adami[30] for the unconstrained creep rates of the fibers and matrix it appears that the CMR ≈ 1.

Stress and Temperature Dependence of Creep Behavior Typical creep curves for the 2-D $Al_2O_{3(f)}$/CVD-SiC composite are shown in Figs. 5.22a and b. For all temperatures and stress levels examined, primary, secondary and tertiary creep regimes were observed. Creep rupture strains were typically over 1%; rupture times ranged from 0.6 to 640 h. An important finding was that the steady-state creep rate exhibited two distinct regimes depending upon the creep stress. At low stresses, a high stress exponent for creep (n) of approximately 9.5 was obtained, which Adami attributed to matrix creep and stress redistribution occurring under an unsaturated matrix cracking state. At high stresses, an n-value of 4.5 was found (see Fig. 5.23), which was similar to the stress exponent for the creep of monolithic Al_2O_3 fibers, suggesting that the creep rate of the composite was controlled by creep of the bridging fibers.

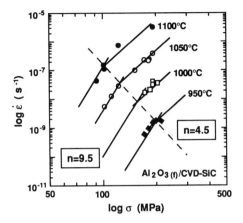

Fig. 5.23 Stress and temperature dependence of tensile creep rate for $Al_2O_{3(f)}$/CVD-SiC composites.[30] Note the two stress regimes with different values of the stress exponent n.

Microstructural Damage and Discussion Adami found that for all test conditions, matrix microcracking occurred during initial application of the creep stress. At low creep stresses, matrix cracking and accompanying stress redistribution occur until a constant microstructural damage state is obtained, which gives rise to a regime of apparently constant creep rate. Adami attributes the tertiary creep regime to the statistical rupture of fibers. For high creep stresses, matrix cracking is more extensive, which allows complete bridging of the matrix cracks; the creep rate and creep life are controlled by creep of the Al_2O_3 fibers.

At room temperature, the Al_2O_3 fibers would be under residual tension, with the SiC matrix under compression. The matrix fracture observed during initial loading may be a consequence of the loading rate used in the experiments, as well as the large stress concentrations inherent in 2-D woven composites in the vicinity of the crossover points of the fiber bundles. To achieve full potential from this composite system, it may be advantageous to fabricate the composite using hot pressing of 0° and 90° plies. This would minimize the likelihood of matrix fracture during loading; moreover, as discussed earlier for hot-pressed Nicalon SiC_f/CAS composites, the 90° fibers can significantly enhance creep resistance if matrix fracture can be avoided.

5.4.5 3-D C_f/SiC Composite

Holmes and Morris[51] examined the tensile creep behavior of 3-D T-300 C_f/SiC composites at 1400°C. To determine the sensitivity of creep life to oxygen level in the test environment, experiments were conducted in air and in an argon atmosphere that contained 1–10 ppm O_2. The composite was

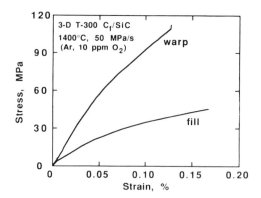

Fig. 5.24 Monotonic tensile behavior of 3-D C_f/SiC composites at 1400°C.[51] Specimens were removed from infiltrated panels such that the tensile loading axis was aligned with either the warp or fill direction.

manufactured by chemical vapor infiltration of SiC into a 3-D (angle-interlock) weave of T-300 carbon fibers. In the warp direction 3 K fiber tows were used; the fill direction had 1 K tows. The total fiber volume fraction was 45%. Specimens were cut from the billets such that their tensile axis was parallel to the warp direction. The test specimens had an average density of $2.05 \pm 0.1 \, \mathrm{g \, cm^{-3}}$. Figure 5.24 gives the 1400°C monotonic tensile behavior of the composite in the warp and fill directions. As with other C_f/SiC composites, the monotonic loading curve for this initially microcracked composite shows non-linear behavior from the onset of loading.

Representative tensile creep curves for specimens loaded parallel to the warp direction are shown in Fig. 5.25a for stresses of 45, 60, and 90 MPa and for an oxygen level of 10 ppm. For stresses of 60 MPa and lower, a quasi-steady-state creep regime was observed, followed by abrupt specimen failure (i.e., an absence of tertiary creep). At 90 MPa, only transient creep was observed, with failure in less than 2 h. The stress dependence of quasi-steady-state creep rate for specimens removed from the warp directions is shown in Fig. 5.25b for an oxygen level of 10 ppm. The creep rate ranged from approximately $1.5 \times 10^{-6} \, \mathrm{s^{-1}}$ at 30 MPa to $7.5 \times 10^{-6} \, \mathrm{s^{-1}}$ at 60 MPa. Assuming a power law dependence of creep rate on applied stress, the stress exponent for creep in the 10 ppm O_2 atmosphere was approximately 2.3. Insufficient data were available to permit determination of the influence of test environment on stress exponent.

Figure 5.26 shows the influence of oxygen level in the test atmosphere on the creep rupture time. As would be expected for an uncoated composite with carbon fibers, the rupture time increased significantly as the oxygen content was lowered. At a stress level of 50 MPa, the rupture time increased from roughly 10 h at 10 ppm O_2 to over 100 h at 1 ppm O_2. For the same stress level, the rupture time in air was less than 0.1 h. For all test environments, final

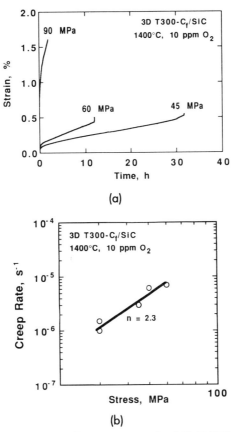

Fig. 5.25 (a) Representative tensile creep curves for 3-D C_f/SiC at 1400°C.[51] The curves shown were obtained in an argon atmosphere with an oxygen level of 10 ppm. Note the abrupt failure and lack of tertiary creep. (b) Stress dependence of tensile creep rate for 3-D C_f/SiC at 1400°C in 10 ppm O_2.

failure of the composite occurred by rupture of the fiber bundles. For experiments conducted in air, significant oxidation of the fibers was observed. The significantly improved creep life of the composite at 1 ppm provides an indication of the potential for this composite system if suitable surface coatings or internal oxidation protection schemes can be developed.

These results also show that it is very important to measure the oxygen level when conducting creep experiments in "inert" atmospheres. The level of oxygen in a test chamber rarely corresponds to the oxygen level present in, for example, a tank of argon or nitrogen gas since oxygen is readily introduced through leaks in regulators, gas lines, and chamber seals; outgassing from furnace insulation can also change oxygen levels. This comment is particularly relevant in the case of C_f/SiC composites where an increase in oxygen level from 1 to 10 ppm caused an order of magnitude reduction in creep life at

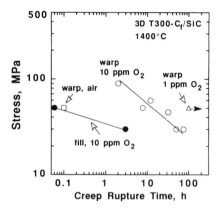

Fig. 5.26   Influence of oxygen level on the tensile rupture time of 3-D C$_f$/SiC crept at 1400°C.[51]

1400°C. As a consequence of the high porosity level of CVI composites and processing-related matrix cracking, internal oxidation protection schemes or surface coatings will be required for these composites in high temperature oxidizing environments.

5.4.6 Summary Remarks: Creep Behavior under Sustained Loading

Creep Rate and Strain Accumulation   As discussed earlier, the initial transient creep rate of fiber-reinforced ceramics is governed to a large extent by load transfer between the fibers and matrix and microstructural evolution in the fibers and matrix. In the absence of microstructural damage such as matrix fracture, the long-term creep rate and strain accumulation of a composite are controlled by the creep rate of the constituent with the higher creep resistance. Both the elastic constants of the matrix and fibers and the creep rates of the constituents, contribute to overall strain accumulation. Total strain accumulation within a given time is perhaps a more important design parameter than strain rate, since matrix fracture and associated interfacial debonding can lead to significant increases in accumulated strain. If the creep rate of the fibers is low, the overall creep rate may still be within acceptable design limits, even though the accumulated strain may be quite high. This is exemplified by the tensile creep of Nicalon SiC$_f$/CAS composites. Although the overall creep rate was relatively low (of the order of $10^{-8}\,\mathrm{s}^{-1}$ at 1200°C and 120 MPa), the total strain accumulation in 100 h was high—of the order of 2–3%. This large strain accumulation is primarily a consequence of the low creep resistance of the CAS matrix at 1200°C, which causes a large initial strain transient. On the other hand, SCS-6 SiC$_f$/HPSN composites exhibit both low creep rates and extremely low levels of total strain accumulation (Fig. 5.27).

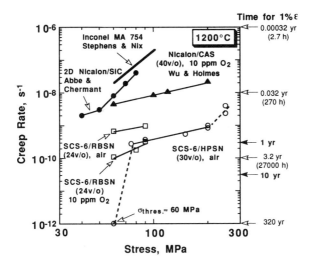

Fig. 5.27   Comparison of the 100 h tensile creep rate of various ceramic matrix composites at 1200°C with the Ni-base superalloy MA 754.[63] Data for 2-D Nicalon SiC$_f$/SiC was taken from Abbé *et al.*[55,56]

It is of interest to compare the tensile creep rate of current-generation fiber-reinforced ceramics with the creep rate of a current-generation Ni-base superalloy. Figure 5.27 compares the 100 h tensile creep rate of MA 754 (an oxide dispersion strengthened superalloy)[63] with several ceramic matrix composites (2-D Nicalon SiC$_f$/SiC, 1-D Nicalon SiC$_f$/CAS, SCS-6 SiC$_f$/RBSN, and SCS-6 SiC$_f$/HPSN). The creep rate of the composites is lower than that of MA 754; in fact the creep rate of SCS-6 SiC$_f$/HPSN is as much as three orders of magnitude lower. Although the creep rates are lower than those for MA 754, are they acceptable for practical engineering applications? To answer this it is perhaps better to think in terms of total accumulated creep strain. The acceptable level of creep strain will primarily be dictated by the end use of the composite. One of the most widely quoted requirements, which is a holdover from the days of steam-turbine design, is that the total creep strain of a component should not exceed 1% over the life of the component. Although this is probably not an appropriate reference strain for ceramic matrix composites, it is nonetheless interesting to compare the time required to reach a creep strain of 1% for the various composites. Taking a reference stress of 70 MPa, and assuming for discussion purposes that the 100 h creep rate can be maintained indefinitely (this would depend upon the microstructural stability of the composite), the time for 1% creep strain would range from roughly 300 *hours* for 2-D SiC$_f$/SiC, to over 300 *years* for HPSN. Although applications such as rocket nozzles require only short-term creep resistance, it is likely that most practical engineering designs will require creep lives from 100 h to several thousand hours.

When analyzing the creep behavior of fiber-reinforced ceramics, it is important to keep in mind that most of the experiments that have been performed to date have been of rather short duration. Thus, what may appear to be steady-state creep during a 100–200 h test may, in fact, represent transient behavior during a longer-term experiment. Moreover, what may appear to be suitable creep life in an inert atmosphere may turn out to be exceedingly short when the same experiment is conducted in an oxidizing atmosphere. It is also important to keep in mind that most of the data have been generated from static load creep experiments. Many components may experience accidental overloads when in service. Unfortunately, information concerning how the creep rate of a composite changes after being subject to tensile or flexural overload is virtually non-existent at this time (an indication of how matrix cracking influences creep life was discussed in Section 5.4.3). Moreover, most composites have been designed with monotonic toughness as a primary design goal; however, as discussed later, microstructures that provide optimal composite toughness, which requires matrix cracking and interfacial debonding, are at odds with the microstructures required for creep resistance.

Even with these cautions in mind, the creep studies that have been conducted to date clearly indicate that fiber-reinforced ceramics hold considerable promise for use in elevated temperature structural applications, where, in certain instances, the use temperature may be above the incipient melting point of Ni-base superalloys (typically 1300–1375°C). As discussed in Section 5.6, the key to the successful use of these materials is to design the microstructure for the competing requirements of high monotonic toughness and creep resistance.

5.5 Cyclic Creep and Creep Strain Recovery

Many of the potential applications for fiber-reinforced ceramics will involve cyclic loading at elevated temperatures. Interestingly, recent experimental results suggest that, when compared to sustained loading, elevated temperature cyclic loading can decrease the overall strain accumulation in fiber-reinforced ceramics; this effect is typically enhanced when one of the constituents has a high glassy-phase content. The decrease in strain is a consequence of viscous strain recovery that occurs during the unloading portion of a fatigue cycle. Although the occurrence of strain recovery in monolithic ceramics has been known for some time, this phenomenon has only recently been documented for fiber-reinforced ceramics.[15,16,38,41] This section presents an overview of the mechanisms responsible for strain recovery and the influence of loading history on creep strain recovery in fiber-reinforced ceramics.

Mechanisms for Strain Recovery in Ceramics and Composites In monolithic ceramics, where dislocation mobility is very low at temperatures of practical interest, creep typically occurs by grain boundary sliding and diffusional mechanisms. This grain boundary motion is accompanied by the development of internal stresses. If a glassy grain boundary phase is present, deformation occurring near grain boundaries can lead to the development of elastic and capillary stresses along the grain boundaries that resist further deformation during creep loading. These internal stresses provide a driving force for the *intrinsic* strain recovery during cyclic creep loading of monolithic ceramics, including ceramic fibers.

Because the creep rates and elastic constants of the fibers and matrix are generally different, a redistribution in stress between the fibers and matrix occurs during the hold time at maximum stress. This redistribution in stress, which influences the residual stress state upon unloading of a component, can have a profound influence on the recovery behavior of a composite. Thus, in addition to the intrinsic recovery process mentioned above for monolithic ceramics, the residual stresses that develop in fiber-reinforced ceramics can provide an additional driving force for strain recovery. In this section, the influence of applied stress, cyclic loading history, and hold time at maximum stress on the cyclic creep behavior of Nicalon/CAS-II and SCS-6 SiC_f/Si_3N_4 composites are discussed. The latter part of this section discusses the practical implications of creep strain recovery.

5.5.1 Nomenclature and Definitions used to Describe Creep Strain Recovery

Two recovery ratios are used to quantify the amount of strain recovered during a particular loading–unloading cycle: (1) total-strain recovery ratio (which includes the instantaneous elastic strain on loading and unloading), and (2) creep-strain recovery ratio (which considers only inelastic strains on loading and unloading). With reference to Fig. 5.28, the total-strain recovery ratio, R_t, is defined by the elastic and creep strains ($\varepsilon_{el,R} + \varepsilon_{cr,R}$) that are recovered in a given cycle, divided by the total accumulated strain (ε_t) that is present immediately prior to unloading (note that this includes the elastic strain):

$$R_t = (\varepsilon_{el,R} + \varepsilon_{cr,R})/\varepsilon_t \qquad (16)$$

In a similar fashion, the creep-strain recovery ratio, R_{cr}, is defined as the time-dependent strain recovered during the unloading segment of a particular cycle ($\varepsilon_{cr,R}$) divided by the pure creep strain (ε_{cr}) that accumulated during the loading portion of the cycle under consideration (the instantaneous elastic strain on loading and unloading is not included):

$$R_{cr} = \varepsilon_{cr,R}/\varepsilon_{cr} \qquad (17)$$

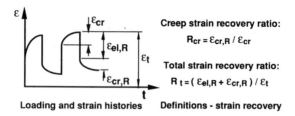

Loading and strain histories   Definitions - strain recovery

Fig. 5.28 Definition of variables used to describe the strain recovery behavior of fiber-reinforced ceramics. The creep strain recovery ratio R_{cr} considers only time-dependent creep strains during loading and unloading. The total strain recovery ratio also includes the elastic components of strain during loading and unloading. These definitions are general and can be used to compare the recovery behavior of other ceramics and metals.

Note that both R_t and R_{cr} are functions of time. During the unloading portion of a cycle, $\varepsilon_{cr,R}$ is zero or increases with time; whereas, ε_{cr} and ε_t are constant during a particular unloading period. Inspection of Eqns. (16) and (17) shows that both R_t and R_{cr} increase with time during the unloading segment of each cycle. However, since ε_t typically increases during subsequent creep loading (unless a creep threshold exists), R_t decreases as the number of loading–unloading cycles increases.

5.5.2 Creep Recovery in Nicalon/CAS-II Composites

The isothermal tensile creep recovery behavior of 0° and 0°/90° Nicalon SiC$_f$/CAS-II composites has been investigated by Wu and Holmes.[15,41] The composites were reinforced with 40 vol.% fibers. Cyclic creep experiments, performed in argon, were conducted at temperatures of 1000, 1100, and 1200°C.

As shown in Fig. 5.29a and b, the Nicalon SiC$_f$/CAS composite exhibits viscous strain recovery upon unloading after tensile creep. From Fig. 5.29a, the creep strain recovery ratio R_{cr} generally increases as the temperature decreases. At 1000°C, essentially all of the prior creep strain was recovered upon 100 h unloading. The increase in R_{cr} at low temperatures is caused, in part, by the significant reduction in microstructural damage as the creep temperature is decreased. For a given temperature, the amount of recovery is influenced by the creep stress and fiber lay up. From Fig. 5.29b, for a given ply lay up, R_{cr} decreases as the prior creep stress is increased. This reduction is, in part, caused by the much larger primary strain accumulation before unloading and also by the development of additional microstructural damage caused by the higher creep stress.[41] Comparing the 0° and 0°/90° specimens, one observes that there is significantly more strain recovery for the 0°/90° composites. For a

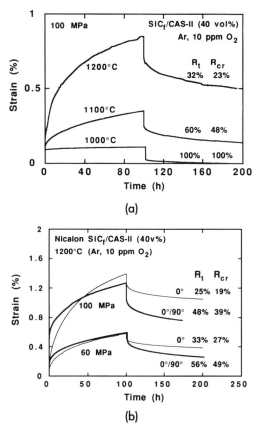

Fig. 5.29 Influence of stress and temperature on the isothermal strain recovery of $[0]_{16}$ and $[0/90]_{4S}$ Nicalon SiC_f/CAS composites. (a) Recovery of $0°$ and $0°/90°$ specimens after tensile creep at 60 MPa and 100 MPa for 100 h. (b) Influence of temperature on the strain recovery of $0°$ specimens crept at 100 MPa for 100 h.

loading history involving 100 h of creep at 60 MPa, followed by a 100 h hold at 2 MPa, the creep strain recovery ratio, R_{cr}, at 100 h was approximately 27% for the $0°$ composite and 49% for the $0°/90°$ composite. As discussed in an earlier section, the $90°$ fibers offer considerable constraint to axial creep deformation (note that this constraint is developed as the composite attempts to contract in the lateral direction as creep in the axial direction progresses). Upon specimen unloading, the residual stresses that develop in the transverse ($90°$) direction provide a driving force for axial contraction of the composite (intuitively, the $90°$ fibers will act as elastically compressed springs when the composite is unloaded).

Figure 5.30 shows the recovery behavior of a $0°/90°$ specimen subjected to two cycles involving 40 minutes of creep at 60 MPa, followed by 40 minutes of recovery at 2 MPa. Compared to the 100 h creep/100 h recovery experiments

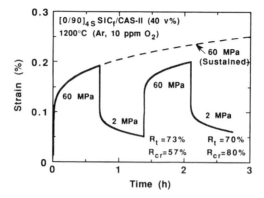

Fig. 5.30 Isothermal recovery behavior of $[0/90]_{4S}$ Nicalon SiC$_f$/CAS during short-duration cyclic creep loading at 1200°C between stress limits of 60 MPa and 2 MPa (40 min creep/40 min recovery). After Wu and Holmes.[15]

described above, the strain recovery ratios were much higher for the shorter duration creep cycles. For the first cycle, $R_{cr} = 57\%$ and $R_t = 73\%$; for the second cycle, these ratios were 80% and 70%, respectively (note that R_t decreases because of the increased amount of accumulated strain; the amount of recovered strain was similar for both cycles). A similar increase in R_{cr} for short duration cyclic loading was also observed during cyclic creep experiments conducted with hot-pressed SCS-6 SiC$_f$/Si$_3$N$_4$ composites (discussed in the next section). Inspection of Fig. 5.30 shows that the increase in R_{cr} on the second cycle was caused by a significant reduction of creep strain and in the duration of transient creep; this reduction is attributed to a change in the residual stress state of the composite after the first loading–unloading cycle (note also, more fibers will tend to equally share the creep load as matrix creep proceeds, i.e., initially, not all fibers will carry the same creep stress, since many of the 0° fibers may not be perfectly aligned with the axial loading direction).

Using an approach similar to that described in Section 5.2, the strain recovery process can be modeled analytically using a simple spring/viscous-dashpot model.[31] As will be discussed in Section 5.5.4, strain recovery that occurs in fiber-reinforced composites, and its dependence on accumulated creep strain, have important implications for cyclic creep behavior and creep life.

5.5.3 Cyclic Creep/Creep Recovery of SiC$_f$/Si$_3$N$_4$ Composites

The isothermal creep-recovery behavior of unidirectional SCS-6 SiC$_f$/Si$_3$N$_4$ composites has been investigated in air at 1200°C (the tensile creep behavior of these composites was described above). These investigations examined the influence of cycle duration on creep rate, strain accumulation, and strain recovery. Specimens were subjected to tensile creep at 200 MPa for times that ranged from 300 s to 200 h, followed by recovery at 2 MPa with hold times from 0 s to 50 h.

Rapid unloading and reloading, without a finite hold time, caused no appreciable change in the accumulated strain or creep rate. Significant strain recovery was observed for all loading histories that incorporated a finite recovery hold time. As shown in Fig. 5.31a, during multiple 50 h creep/50 h recovery cycles, the creep recovery ratio, R_{cr}, increased from 38% on the first cycle, to 82% for subsequent cycles. This recovery ratio is much higher than that typically found with monolithic Si_3N_4. For example, in single-cycle recovery experiments conducted with NC-132 Si_3N_4 at 1204°C, Arons and Tien[64] observed less than 10% total strain recovery. Haig *et al.*[65] reported maximum single-cycle recovery ratios of 40% for fine-grained monolithic Si_3N_4 that had undergone tensile creep at 1100°C. In addition to intrinsic recovery of the matrix, SCS-6 SiC fibers are also known to undergo significant anelastic recovery.[66]

The length of the recovery hold time influences creep behavior during subsequent loading cycles. Compared to sustained creep loading, allowing 300 s of recovery during each cycle resulted in a significant reduction in the duration of primary creep, from approximately 75 h for sustained loading to less than 10 h for the cyclic creep experiment (Fig. 5.31b). After 200 h of short-duration cyclic loading (300 s load/300 s recovery) between 200 MPa and 2 MPa, the total accumulated creep strain was also 60% lower than that found with sustained creep loading at 200 MPa. Even after extrapolating the 300 s creep/300 s recovery curves to 400 h (to compare the results for an equivalent time at a stress of 200 MPa), the creep strain would remain significantly lower than that found after 200 h of sustained loading.

5.5.4 Discussion of Cyclic Creep Behavior

Mechanisms of Creep Recovery: Influence of Residual Stresses. As noted earlier, for monolithic ceramics (including SiC fibers), grain boundary sliding during creep deformation causes the build-up of internal stresses which may be elastic or capillary in nature. These internal stresses provide a mechanism for intrinsic strain recovery after a partial or full reduction in the applied load. In addition to the intrinsic recovery behavior of the constituents, the residual stress state that exists in a fiber-reinforced composite after unloading can provide a further driving force for strain recovery. For a given composite, and ignoring microstructural damage, the residual stresses that exist upon unloading are governed by temperature, applied stress, accumulated creep strain, and the mismatch in elastic constants between the fibers and matrix. As a consequence of the history dependence of the residual stress state that develops by load transfer, the amount of strain recovery depends upon the details of the prior loading history. How do the residual stresses develop? With reference to the earlier discussion of creep mismatch ratio and load transfer

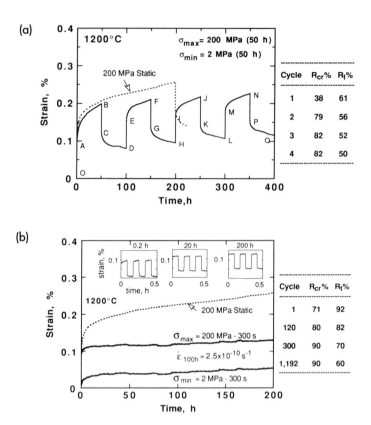

Fig. 5.31 Isothermal tensile creep/strain recovery curve for 0° SCS-6 SiC_f/Si_3N_4 at 1200°C. (a) Long-duration cyclic creep behavior (50 h at 200 MPa/50 h at 2 MPa). (b) Short-duration cyclic creep (300 s at 200 MPa/300 s at 2 MPa). For comparison, the creep behavior during sustained creep loading at 200 MPa for 200 h is also shown in each figure. Note, in particular, the significant change in primary creep behavior that occurs during short-duration cyclic loading. After Holmes et al.[38]

between constituents, suppose that the matrix creeps at a much higher rate than the fibers. For this limiting case, the axial stress in the matrix relaxes towards zero during tensile creep and the stress in the fibers increases as they must now carry more of the applied load. Upon removal of the applied load, elastic contraction of the fibers places the matrix in compression and the fibers in tension (the tensile stress in the fibers will be lower by an amount proportional to the elastic unloading strain—note that constraint by the matrix does not allow the fibers to fully recover the elastic strain). In purely mechanical terms, if the temperature of the composite is maintained after unloading, the residual stresses in the fibers and matrix will be reduced by *in situ* relaxation of the fibers and matrix. For example, if it is assumed (for discussion purposes) that the fiber stress upon unloading is below the threshold stress for fiber creep, and that the matrix is in compression, the matrix stress will relax; for stress equilibrium, there will be a parallel reduction in fiber stress by elastic unloading of the fibers (this argument assumes that the compressive stress in the matrix exceeds the threshold for matrix creep in compression). As a result, the matrix and fibers will undergo a net contraction. This recovery process will become exceedingly slow as the threshold stress for compressive matrix creep is approached. This description provides only a qualitative explanation of the mechanical component of stress recovery. In fact, the actual recovery process is far more complex, and depends upon such factors as the degree of interfacial bonding which, in turn, controls the load transfer between the fiber and matrix. Moreover, the stress state is by no means uniaxial; internal residual stresses also exist in the radial direction (normal to the fiber axis). The hydrostatic stress in the vicinity of the fibers will also influence the recovery behavior upon unloading; these stresses, in turn, are influenced by parameters such as fiber and matrix modulus, interfacial bonding, fiber packing distribution and fiber diameter.

The above argument pertains to tensile creep deformation. It is important to appreciate that the residual stress that develops upon unloading after compressive creep will be much different; elastic unloading could place the matrix in tension. This can be rationalized by considering the case where the compressive creep rate of the matrix is much higher than the compressive creep rate of the fibers. During long-duration compressive creep loading the axial stress in the matrix will relax (become less compressive), and the fibers will support an increasing percentage of the compressive load (it is assumed that constraint from the matrix is sufficient to prevent buckling of the fibers). Upon unloading, the fibers will expand elastically. If perfect interfacial bonding is assumed, this elastic unloading will place the matrix in residual tension; depending upon how the microstructure/composition of the matrix changes during creep, this tension could be sufficient to promote matrix fracture. If matrix fracture does not occur, *in situ* creep of the fibers and matrix will occur, reducing the magnitude of the compressive strain.

There are several practical consequences of creep strain recovery for the life prediction of cyclically loaded structures:

(1) The creep rate and total strain accumulation in components that experience periodic changes in load, at a constant or near constant temperature, will be significantly lower than that observed for sustained creep loading. For structural components, the accumulated creep strain is an important design consideration. Life predictions that do not include strain recovery may seriously underestimate component life. Many components are subjected to varying stresses at either a constant or fluctuating temperature. An example of the former would be a heat exchanger that operates at constant temperature but with a fluctuating internal pressure. An example of the latter is a gas-turbine airfoil or combustor in a commercial aircraft, both of which operate

under periods of near-constant stress and temperature during cruise, followed by periods of lower stress and temperature during turbine idle or shutdown. During thermomechanical fatigue loading (where the temperature and stress are varied in an in-phase or out-of-phase fashion), a certain amount of strain recovery will occur during specimen unloading. For in-phase TMF loading, where the stress and temperature are increased or decreased in parallel, recovery during unloading will occur for portions of the cycle where the temperature is above the threshold stress for creep. The amount of strain recovery, which influences the residual stress state upon unloading, and therefore, the stress–strain response during subsequent TMF cycles, will depend upon the rate at which the load and temperature are reduced. Although recovery will not occur if a component is rapidly cooled, the redistribution in internal stress that occurs during the prior high temperature deformation will directly influence the room temperature strength and fracture characteristics of the component. The effect of prior loading on the residual properties of a composite is an area of research that has received minimal attention, but is crucial for the successful use of composites in cyclically loaded structures.

(2) Isothermal heat treatment of components removed from service could potentially allow one to increase the life of components that are subjected to sustained or cyclic creep loading. Removal of a component from service, followed by stress-free exposure to high temperatures, would decrease the residual stresses that build up in the fibers or matrix during long-duration tensile creep loading. In analogy to the annealing of cold-worked metals to relieve internal stresses, it is possible that structural components manufactured from fiber-reinforced ceramics could be periodically removed from service and placed in furnaces to promote strain recovery and decrease the likelihood of creep rupture during subsequent loading.

5.6 Microstructural Design for the Competing Requirements of Monotonic Toughness and Creep Resistance

The analytical modeling and experimental observations of the damage modes that occur during creep loading have provided insight into the role of the microstructure on creep deformation, and suggest possible changes in microstructure that could improve creep life.

To date, the vast majority of the research in the area of fiber-reinforced ceramics has focused on understanding and improving monotonic toughness. While this is undeniably an important area of research, it is, nonetheless, unfortunate that so little attention has been paid to the broader picture of designing composite microstructures. *When optimizing a specific property (e.g., toughness) one must not lose sight of the fact that other desirable properties may be adversely affected.* This is particularly true for ceramic composites, which will likely see a wide range of loading histories. To be used successfully in most engineering applications, it is likely that fiber-reinforced ceramics will have to be designed to withstand accidental overloads, and at the same time be able to continue to operate under adverse conditions which may involve creep, cyclic creep, or fatigue loading. *The key to the further development of fiber-reinforced ceramics is to understand how the microstructural damage that develops under one loading history (e.g., monotonic loading) will affect the behavior of the composite under a different loading history (e.g., creep).* Unfortunately, as discussed below, microstructures that have been optimized for monotonic toughness, are generally incompatible with long-term creep resistance.

In a broad sense, macroscopic creep damage is related to the mismatch in creep properties between the constituents. The influence of creep mismatch ratio on creep damage mode was summarized earlier in Fig. 5.7. If the fibers are the more creep-resistant constituent (CMR < 1), periodic fiber rupture can occur during tensile or flexural creep loading. If the matrix has a higher creep resistance than the fibers (CMR > 1), matrix fracture can occur by load transfer from the fibers to matrix. As noted earlier, although one would like to avoid the fracture of either constituent, periodic fiber rupture is a far more desirable failure mode than matrix fracture. If matrix fracture occurs, the creep stress on the bridging fibers can be quite high and may cause fiber rupture (this is particularly true of composites with low fiber volume fractions). Moreover, matrix cracking results in direct interaction between the environment and the fiber–matrix interface, which can further reduce creep life. Also, as discussed in Chapter 6 on the fatigue behavior of fiber-reinforced ceramics, matrix cracking and interfacial debonding can lead to frictional heating if a component is simultaneously subjected to high frequency cyclic loading. One concern with the use of woven composites should also be mentioned. Namely, because of the large stress concentration near the crossover points of fiber bundles, matrix cracking at these locations can occur readily. This fracture will occur under both uniaxial and biaxial creep loading. In this sense, it may be advantageous to process 0°/90° composites as laminates with individual 0° and 90° plies. As shown by Wu *et al.*,[37] the cross-ply composites have excellent creep resistance (in some instances better than 0° composites of the same fiber volume fraction).

When designing a composite microstructure for long-term creep resist-ance, one should strive for a creep mismatch ratio close to unity, or if the fibers have a significantly higher creep resistance than the matrix, one should ensure that the fiber strength is not degraded during processing or long-term exposure. One approach to minimize fiber rupture in composites with creep mismatch ratios less than unity is to increase the fiber volume fraction, which has the effect of reducing the average fiber stress. However, when designing a composite microstructure for optimal creep resistance, it is important to strike

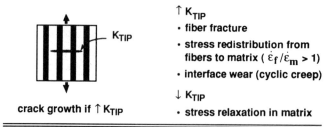

・**assume matrix cracking <u>will</u> occur (e.g., by a tensile overload)**

$\uparrow K_{TIP}$
・**fiber fracture**
・**stress redistribution from fibers to matrix ($\dot{\varepsilon}_f / \dot{\varepsilon}_m > 1$)**
・**interface wear (cyclic creep)**

$\downarrow K_{TIP}$
・**stress relaxation in matrix**

K_{TIP}

crack growth if $\uparrow K_{TIP}$

Recommendations for combined toughness and creep resistance

$\Rightarrow$ **want a creeping matrix ($\dot{\varepsilon}_f / \dot{\varepsilon}_m < 1$)**

・**bridging fibers must support the applied creep load ($\uparrow v_f$)**
・**strain recovery is a desirable attribute**

Fig. 5.32 Microstructural design for combined toughness and creep resistance. Initial microcracks formed during loading of a damage-tolerant microstructure will propagate during subsequent creep loading if the bridging force behind the crack tip decreases, (e.g., by the fracture of bridging fibers), a decrease in the normal pressure along the interface (e.g., by interface wear), or by load redistribution from the fibers to a creep-resistant matrix (CMR > 1). The driving force for crack extension will decrease if stress relaxation occurs at the crack tip, for example by transient load transfer from the matrix to fibers (CMR <1). To accommodate the competing requirements of monotonic toughness and creep resistance, it is recommended that the composite be designed such that the bridging fibers can support the entire creep load should an accidental tensile overload occur. For long-term creep life, it is also recommended that the creep mismatch ratio be greater than unity: this ensures that the matrix stress will remain low.

a balance between optimal creep resistance and the competing requirement of high toughness should an unexpected overload occur.

Toughness under monotonic loading is achieved by matrix microcracking, controlled interfacial debonding, and relative slip between the fiber and matrix. These processes dissipate mechanical energy. Unfortunately, the microstructural damage associated with these modes of toughening will negatively impact creep life. For example, as discussed earlier, there is a significant increase in stress on the fibers that bridge matrix cracks, which can lead to early fiber rupture. Interface debonding limits the degree of load transfer between the fibers and matrix, leading to an increase in fiber stress within the debonded zones; this increase in stress increases the likelihood of fiber rupture. Moreover, the decrease in load transfer will increase the amount of strain accumulation during creep.

Ideally, if one was concerned only with creep resistance, the microcracking threshold of the matrix could be increased. An increase in the degree of interfacial bonding would also improve creep resistance. However, since these changes would lower toughness, one must strike a compromise between optimal toughness and creep resistance. In the authors' opinion, one should design *first* for high monotonic toughness (which is the foremost attribute that one strives for when using fiber-reinforced composites), but at the same time ensure that the microstructure provides adequate creep resistance even in the presence of a microcracked composite. In the authors' opinion, this latter approach to microstructural design is the preferred one, since it accounts for the possibility that component may be subjected to accidental overloads during service. The guidelines for microstructural design are relatively straightforward. For example, with reference to Fig. 5.32, and assuming that a composite has been partially microcracked by a tensile overload, one wants to limit the further propagation of the microcracks during subsequent creep loading. Crack growth will occur if the bridging stress behind the crack tip decreases, leading to an increase in the stress intensity at the crack tip. For example, the bridging stress can be decreased by fracture or creep of the bridging fibers, by interface wear (if cyclic creep loading — see Chapter 6 on fatigue) and by interface debonding. Importantly, an unfavorable creep mismatch ratio, where the fibers have lower creep resistance than the matrix, will also lead to an increase in matrix stress and, hence, an increase in the driving force for crack extension. Alternatively, the crack tip stress can be reduced by stress relaxation in the matrix.

The above discussion leads to the following recommendations for microstructural design. Namely, one wants to design such that the matrix has a lower creep resistance than the fibers, allowing relaxation of the stress near matrix cracks. Assuming that matrix cracking has unavoidably occurred, one wants to ensure that the bridging fibers can support the entire creep load; this may require an increase in fiber volume fraction. This requirement also places added emphasis on the development of creep-resistant fibers. Finally, it is the authors' opinion that strain recovery is a very desirable attribute in a composite, since it can provide substantial reductions in accumulated creep strain during cyclic creep or fatigue loading (it also allows relaxation of stresses near a crack tip that is subjected to cyclic loading). Moreover, as discussed in an earlier section, strain recovery reduces the build-up of residual stresses in a component. Thus, some caution should be exercised in the development of fibers that achieve their improved creep resistance by the complete elimination of glassy grain boundary phases; a small amount of residual glassy phase may be beneficial in composites subjected to fatigue loading.

The concepts presented in Fig. 5.32 are not complete, and will undoubtedly be extended further as additional experimental and analytical research is conducted. For example, the mode of creep damage is also influenced by the fracture stress and creep rupture strain of the constituents. Prior to the occurrence of matrix or fiber fracture, the strain in the constituents is equal (unless the frictional shear stress along the interface is zero, i.e., in the absence of load transfer). Thus, if the matrix has a much higher failure strain than the fibers, the situation could arise where the matrix supports the majority of the creep load, but the fibers, with a lower failure strain, fracture first. The actual situation is, of course, far more complicated; one must also consider the flaw sensitivity of the constituents, which can be characterized in terms of a critical stress intensity for crack propagation. Moreover, the stress states of the fibers and matrix are by no means uniaxial, which influences the fracture characteristics of the constituents. It should be appreciated that time-dependent microstructural damage or chemical interaction can alter the fracture characteristics of the constituents, in particular cavitation can occur in many glass-ceramic and ceramic matrices. One must also consider the damage state that develops during initial loading of a component. This point was clearly illustrated in Fig. 5.8a and b, which showed the effect of initial loading rate on subsequent creep life of SCS-6 SiC$_f$/HPSN composites. In addition, as noted by Suresh and co-workers,[67] ceramics that contain a glassy grain boundary or interfacial phase can exhibit an intrinsic rate sensitivity; this rate dependence is a consequence of viscous deformation of the glassy phase.

References

1. K. M. Prewo and J. J. Brennan, "High Strength Silicon Carbide Fiber Reinforced Glass Matrix Composites," *J. Mater. Sci.*, **15**, 463–468 (1980).
2. J. J. Brennan and K. M. Prewo, "Silicon Carbide Reinforced Glass Ceramic Matrix Composites Exhibiting High Strength and Toughness," *J. Mater. Sci.*, **17**, 2371–2383 (1982).
3. P. J. Lamicq, G. A. Bernhart, M. M. Dauchier, and J. G. Mace, "SiC/SiC Composite Ceramics," *Am. Ceram. Soc. Bull.*, **65**[2], 336–338 (1986).
4. M. S. Newkirk, A. W. Urguhart, H. R. Zwicker, and E. Breval, "Formation of Lanxide™ Ceramic Composite Materials," *J. Mater. Res.*, **1**[1], 81–89 (1986).
5. C. A. Andersson, P. Barron-Antolin, A. S. Fareed, and G. H. Schiroky, "Properties of Fiber-Reinforced Lanxide™ Alumina Matrix Composites," in *Proceedings of the International Conference on Whisker- and Fiber-Toughened Ceramics*, ASM International, Materials Park, OH, 1988, pp. 209–215.
6. K. M. Prewo, "Fiber-Reinforced Ceramics; New Opportunities for Composite Materials," *Am. Ceram. Soc. Bull.*, **68**[2], 395–400 (1989).
7. J. J. Mecholsky, Jr., "Engineering Research Needs of Advanced Ceramics and Ceramic-Matrix Composites," *Am. Ceram. Soc. Bull.*, **68**[2], 367–375 (1989).
8. T. Yamamura, T. Ishikawa, M. Sato, M. Shibuya, H. Ohtsubo, T. Nagasawa, and K. Okamura, "Characteristics of a Ceramic Matrix Composite using a Continuous Si-Ti-C-O Fiber," *Ceram. Eng. Sci. Proc.*, **11**[9–10], 1648–1660 (1990).
9. J. R. Strife, J. J. Brennan, and K. M. Prewo, "Status of Continuous Fiber-Reinforced Ceramic Matrix Composite Processing Technology," *Ceram. Eng. Sci. Proc.*, **11**[7–8], 871–919 (1990).
10. R. A. Sprague, "Future Aerospace Materials Directions," *Advanced Materials and Processes*, **133**[1], 67–69 (1988).
11. D. R. Dryell and C. W. Freeman, "Trends in Design of Turbines for Aero Engines," in *Materials Development in Turbo-Machinery Design: 2nd Parsons International Conference*, eds. D. M. R. Taplin, J. F. Knot, and M. H. Lewis, The Institute of Metals, Parsons Press, Dublin, Ireland, 1989, pp. 90–102.

12. C. H. Henager, Jr. and R. H. Jones, "Subcritical Crack Growth in CVI-SiC Reinforced with Nicalon Fibers: Experiment and Model," *J. Am. Ceram. Soc.*, submitted.
13. S. J. Dapkunas, "Ceramic Heat Exchangers," *Am. Ceram. Soc. Bull.*, **67**, 388–391 (1988).
14. M. A. Karnitz, D. F. Graig, and S. L. Richlen, "Continuous Fiber Ceramic Composite Program," *Am. Ceram. Soc. Bull.*, **70**[3], 430–435 (1991).
15. X. Wu and J. W. Holmes, "Tensile Creep and Creep-Strain Recovery Behavior of Silicon Carbide Fiber/Calcium Aluminosilicate Matrix Ceramic Composites," *J. Am. Ceram. Soc.*, **76**[10], 2695–2700 (1993).
16. Y. H. Park and J. W. Holmes, "Finite Element Modeling of Creep Deformation in Fiber-Reinforced Ceramic Composites," *J. Mater. Sci.*, **27**, 6341–6351 (1992).
17. A. R. T. De Silva, "A Theoretical Analysis of Creep in Fiber Reinforced Composites," *J. Mech. Phys. Solids*, **16**, 169–186 (1968).
18. A. R. T. De Silva, "Creep Deformation and Creep Rupture of Fibre Reinforced Composites," in *Advances in Composite Materials*, Pergamon Press, New York, 1980, pp. 1115–1128.
19. A. Kelly and K. N. Street, "Creep of Discontinuous Fibre Composite II, Theory for the Steady-State," *Proc. Roy. Soc. London*, A.[328], 283–293 (1972).
20. H. Lilholt, "Creep of Fibrous Composite Materials," *Comp. Sci. Tech.*, **22**, 277–294 (1985).
21. J. Porter, in *Whisker- and Fiber-Toughened Ceramics* (Oak Ridge, TN, 7–9 June), eds. R. A. Bradley, D. E. Clark, D. C. Larsen, and J. O. Stiegler, ASM International, Metals Park, OH, 1988, p. 147.
22. K. Y. Donaldson, A. Venkateswaran, and D. P. H. Hasselman, "Speculation on the Creep Behavior of Silicon Carbide Whisker-Reinforced Alumina," *Ceram. Eng. Sci. Proc.*, **10**[9–10], 1191–1211 (1989).
23. D. McLean, in *"High Temperature—High Performance Composites,"* *Mat. Res. Soc. Symp. Proc.*, eds. F. D. Lemkey, A. G. Avans, S. C. Fishman, and J. R. Strife, 1988, p. 67.
24. S. Goto and M. McLean, "Modelling Interface Effects During Creep of Metal Matrix Composites," *Scripta Metall.*, **23**, 2073–2078 (1989).
25. S. Goto and M. McLean, "Role of Interfaces in Creep of Fibre-Reinforced Metal-Matrix Composites—I. Continuous Fibres," *Acta Metall. Mater.*, **39**, 153–164 (1991).
26. S. Goto and M. McLean, "Role of Interfaces in Creep of Fibre-Reinforced Metal-Matrix Composites—II. Short Fibres," *Acta Metall. Mater.*, **39**, 165–177 (1991).
27. M. Taya and R. Arsenault, in *Metal Matrix Composites*, Pergamon Press, New York, 1989, p. 123.
28. D. Kervadec and J. L. Chermant, Paper presented at the Second European Ceramic Society Conference: Augsburg, Germany, September 11–14, 1991.
29. D. Kervadec and J. L. Chermant, "Some Aspects of the Morphology and Creep Behavior of a Unidirectional SiCf-MLAS Material," in *Fracture Mechanics of Ceramics*, Vol. 10, eds. R. C. Bradt *et al.*, Plenum Press, New York, NY, 1992, pp. 459–471.
30. J. N. Adami "Comportement en Fluage Uniaxial sous Vide d'un Composite à Matrice Ceramique Bidirectional A1203-SiC, These de Docteur és Sciences Techniques, L'École Polytechnique Fédérale de Zürich, Switzerland, 1992.
31. X. Wu and J. W. Holmes, "On the Transient Creep of Continuous Fiber Ceramic Matrix Composites," to be published.
32. D. W. Meyer, M. E. Plesha, and R. F. Cooper, "A Contact Friction

Algorithm Including Nonlinear Viscoelasticity and a Singular Yield Surface Provision," *Comput. Struct.*, **42**, 913–925 (1992).

33. D. W. Meyer, R. F. Cooper, and M. E. Plesha, "High-Temperature Creep and the Interfacial Mechanical Response of a Ceramic Matrix Composite," *Acta Metall. Mater.*, **41**[11], 3157–3170 (1993).

34. D. W. Meyer, R. F. Cooper, and M. E. Plesha, "Rheological Modeling of Ceramic Composites: An Indirect Method of Interfacial Mechanical Property Measurements," *Int. J. Solids Structures*, **29**[20], 2563–2582 (1992).

35. Y. M. Wang, Y. P. Qiu, and G. J. Weng, "Transient Creep Behavior of a Metal Matrix Composite with a Dilute Concentration of Random Oriented Spheroidal Inclusions," *Composites Science and Technology*, **44**, 287–297 (1992).

36. Y. R. Wang and T. W. Chou, "Analytical Modeling of Creep Behavior of Short Fiber Reinforced Ceramic Matrix Composites," *J. Composite Materials*, **26**[9], 1269–1286 (1992).

37. X. Wu, A. C. N. Ma, and J. W. Holmes, "Modeling of Nicalon SiCf/CAS Creep Behavior Using a 1D Analytical Approach and 2D Finite Element Analysis," *Scripta Metall.*, to be published.

38. J. W. Holmes, Y. Park, and J. W. Jones, "Tensile Creep and Creep Recovery Behavior of a SiC-Fiber Si_3N_4-Matrix Composite," *J. Am. Ceram. Soc.*, **76**[5], 1281–1293 (1993).

39. D. Kervadec and J. L. Chermant, "Viscoelastic Deformation During Creep of 1-D SiCf/MLAS Composite," in *High Temperature Ceramic Matrix Composites* (Proceedings of the 6th European Conference on Composite Materials, Bourdeaux, France), September 20–24, 1993, pp. 649–657.

40. D. Kervadec and J. L. Chermant, "Some Aspects of the Morphology and Creep Behavior of a Unidirectional SiCf-MLAS Material," *Fracture Mechanics of Ceramics*, **10**, 459–471 (1992).

41. X. Wu and J. W. Holmes, "Static and Cyclic Creep Behavior of a SiC-Fiber Glass-Ceramic Matrix Composite," unpublished.

42. G. E. Hilmas, J. W. Holmes, R. T. Bhatt, and J. D. DiCarlo, "Tensile Creep Behavior and Damage Accumulation in a SiC-Fiber/RBSN-Matrix Composite," in Ceramic Transactions, Vol. 38, *Advances in Ceramic Matrix Composites*, ed. N. Bansal, American Ceramic Society, Westerville, OH, 1993, pp. 291–304.

43. T.-J. Chuang, "Estimation of Power Law Creep Parameters from Bend Test Data," *J. Mater. Sci.*, **21**, 165–175 (1986).

44. S. V. Nair, T.-J. Gwo, N. Narbut, J. G. Kohl, and G. J. Sundberg, "Mechanical Behavior of a Continuous SiC Fiber Reinforced RBSN Matrix Composite," *J. Am. Ceram. Soc.*, **74**[10], 2551–2558 (1991).

45. S. V. Nair and K. Jakus, "The Mechanics of Matrix Cracking in Fiber Reinforced Ceramic Composites Containing a Viscous Interface," *Mechanics of Materials*, **12**[3–4], 229–244 (1991).

46. S. V. Nair and T.-J. Gwo, "Role of Crack Wake Toughening on Elevated Temperature Crack Growth in a Fiber Reinforced Ceramic Composite," *J. Eng. Mater. Tech.*, to be published.

47. J. W. Holmes, "A Technique for Tensile Fatigue and Creep Testing of Fiber-Reinforced Ceramics," *J. Comp. Mater.*, **26**[6], 916–933 (1992).

48. S. M. Wiederhorn and B. J. Hockey, "High Temperature Degradation of Structural Composites," Paper presented at the Seventh World Ceramics Congress, Montecatini Terme, Italy, June 24–30, 1990.

49. M. Khobaib and L. Zawada, "Tensile and Creep Behavior of a Silicon Carbide Fiber-Reinforced Aluminosilicate Composite," *Ceram. Eng. Sci. Proc.*, **12**[7–8], 1537–1555 (1991).

50. C. H. Weber, J. P. A. Lofvander, and A. G. Evans, "The Creep Behavior of CAS/Nicalon Continuous-Fiber Composites," *Acta Metall. Mater.*, submitted.

51. J. W. Holmes and J. Morris, "Elevated Temperature Creep of a 3-D C_f/SiC Composite," Paper presented at the 15th Annual Conference on Ceramics and Advanced Composites, Cocoa Beach, FL, January, 1991, Paper No. 89-C-91F.

52. J. W. Holmes, "Influence of Stress-Ratio on the Elevated Temperature Fatigue of a SiC Fiber-Reinforced Si_3N_4 Composite," *J. Am. Ceram. Soc.*, **74**[7], 1639–1645 (1991).

53. J. W. Holmes, "Tensile Creep Behavior of a Hot-Pressed SiC Fiber-Reinforced Si_3N_4 Composite," *J. Mater. Sci.*, **26**, 1808–1814 (1991).

54. J. W. Holmes, University of Michigan, Ann Arbor, MI, 1991, unpublished work.

55. F. Abbé, J. Vicens, and J. L. Chermant, "Creep Behavior and Microstructural Characterization of a Ceramic Matrix Composite," *J. Mater. Sci. Letters*, **8**, 1026–1028 (1989).

56. F. Abbé and J. L. Chermant, "Creep Resistance of SiC-SiC Composites Under Vacuum," in *Proceedings of the Fourth International Conference of Creep and Fracture of Engineering Materials and Structures*," The Institute of Metals, London, U.K., 1990, pp. 439–448.

57. T. Mah, N. L. Hecht, D. E. McCullum, J. R. Hoenigman, H. M. Kim, A. P. Katz, and H. A. Lipsitt, "Thermal Stability of SiC Fibres (Nicalon®), *J. Mater. Sci.*, **19**, 1191–1201 (1984).

58. D. Larsen, personal communication, Corning Glass Works, Corning, New York, NY, 1991.

59. R. Bodget, J. Lamon, and R. E. Tressler, "Effects of Chemical Environments on the Creep Behavior of Si-C-O Fibers," Proceedings of the 6th European Conferance on Composite Materials, Bourdeaux, France, September 20–24, 1993, pp. 75–83.

60. A. Bunsell, G. Simon, Y. Abe, and M. Akiyama, in *Fibre-Reinforcements for Composite Materials*, Composite Materials Series 2, Elsevier, New York, 1988, pp. 427.

61. S. M. Wiederhorn, L. Chuck, E. R. Fuller, and N. J. Tighe, "Creep Rupture of Siliconized Silicon Carbide," in *Tailoring Multiphase and Composite Ceramics*, eds. R. E. Tressler, G. K. Messing, C. G. Pantano, and R. E. Newnham, Plenum Publishing Co., 1986, pp. 755–773.

62. S. M. Wiederhorn, D. E. Roberts, T.-J. Chuang, and L. Chuck, "Damage-Enhanced Creep in a Siliconized Silicon Carbide: Phenomenology," *J. Am. Ceram. Soc.*, **71**[7], 602–608 (1988).

63. J. J. Stephens and W. D. Nix, "Creep and Fracture of Inconel M754 at Elevated Temperatures," in *Superalloys 1984* (Conference Proceedings of the Metallurgical Society of AIME, Champion, PA), October 7–11, 1984, eds. M. Gell, C. S. Kortovich, R. H. Bricknell, W. B. Kent, and J. F. Radavich, Metallurgical Society of AIME, pp. 327–334.

64. R. M. Arons and J. K. Tien, "Creep and Strain Recovery in Hot-Pressed Silicon Nitride," *J. Mater. Sci.*, **15**, 2046–2059 (1980).

65. S. Haig, W. R. Cannon, P. J. Whalen, and R. G. Rateick, "Microstructural Effects on the Tensile Creep of Silicon Nitride," in *Creep: Characterization, Damage and Life Assessment*, eds. D. A. Woodford, C. H. A. Townley, and M. Ohnami, ASM International, Metals Park, OH, 1992, pp. 91–96.

66. J. A. DiCarlo, "Creep of Chemically Vapour Deposited SiC Fibres," *J. Mater. Sci.*, **21**, 217–224 (1986).

67. U. Ramamurty, A. S. Kim, and S. Suresh, "Micromechanics of Creep-Fatigue Crack Growth in a Silicide-Matrix Composite with SiC Particles," *J. Am. Ceram. Soc.*, **76**[8], 1953–1964 (1993).

CHAPTER 6

Fatigue Behavior of Continuous Fiber-Reinforced Ceramic Matrix Composites

*J. W. Holmes and B. F. Sørensen**

6.1 Introduction

The development of continuous fiber-reinforced ceramic matrix composites (CMCs) has been motivated by the prospect of obtaining damage-tolerant behavior in high temperature structural ceramics. Damage-tolerant behavior in fiber-reinforced ceramics was first documented by Prewo and Brennan[1-3] who conducted flexural testing of SiC fiber-reinforced LAS (lithium aluminosilicate) glass-matrix composites. Although many ceramic composites have been shown to possess damage-tolerant behavior under monotonic loading, most of the applications for these materials will involve cyclic loading histories,[4] particularly at elevated temperatures. Thus, a thorough understanding of damage mechanisms and microstructural stability during fatigue loading is required before ceramic composites can be used in structural applications. Moreover, as discussed in this chapter, the microstructural features and damage mechanisms that provide high monotonic toughness, such as low interfacial frictional shear stress, matrix microcracking, and long fiber pull-out lengths, often conflict with the microstructural requirements for optimal fatigue resistance.

When compared to monolithic and whisker-reinforced ceramics, the fatigue behavior of fiber-reinforced ceramics exhibits several unique features. For example, the fatigue failure of monolithic ceramics generally occurs by the

*Visiting Scientist from Materials Department, Risø National Laboratory, 4000 Roskilde, Denmark.

growth of a single crack. In contrast, in continuous fiber-reinforced ceramics, matrix cracking is far more distributed; for low fatigue stresses the bridging of these cracks by fibers prevents composite failure. As will be discussed later, matrix cracking occurs early in the fatigue damage process; fatigue life in fiber-reinforced ceramics is primarily governed by interface and fiber damage, rather than by matrix crack growth. Also, in contrast to monolithic ceramics, the fatigue life of fiber-reinforced ceramics can exhibit a very pronounced sensitivity to loading frequency. This frequency dependence of fatigue life is related to pronounced internal heating that occurs as debonded fibers slide relative to the matrix. This repeated fiber–matrix slip causes a bulk temperature increase that can exceed 100°C in composites such as unidirectional SiC fiber-reinforced CAS (calcium aluminosilicate).

The aim of this chapter is to provide a general overview of the fatigue behavior of ceramic matrix composites with continuous fiber-reinforcement. While the emphasis will be on experimental results, the data presented will be related to fatigue damage mechanisms and models that provide insight into how microstructural damage controls fatigue life. Because of space limitations, the information presented here assumes a working knowledge of fatigue nomenclature and the fatigue behavior of ceramics (for a general review of these topics the reader is referred to a comprehensive monograph by Suresh[5] on the fatigue of engineering materials).

The following topics will be discussed: monotonic tensile behavior and microstructural damage (Section 5.2), isothermal fatigue life—influence of fatigue maximum stress (Section 5.3), mechanisms of fatigue damage (Section 5.4), influence of mean stress on isothermal fatigue life (Section 5.5), influence of temperature and test environment on isothermal fatigue life (Section 5.6), influence of loading frequency on isothermal fatigue life (Section 5.7), thermal fatigue (Section 5.8), thermomechanical fatigue (Section 5.9), and summary comments (Section 5.10).

6.2 Monotonic Tensile Behavior and Microstructural Damage

The microstructural damage that forms during the initial application of a fatigue load can have a significant influence on subsequent fatigue behavior and fatigue life. If the loading rates are similar for fatigue and monotonic loading, the damage that develops during the first fatigue cycle can, in principle, be determined by conducting a monotonic tensile test to the same stress level. Thus, it is useful to briefly review the monotonic stress–strain behavior of fiber-reinforced ceramics and the variables that govern microstructural damage during monotonic loading. This review also provides an opportunity to introduce the parameters and nomenclature used in this chapter. The reader is referred to Chapter 1 (Evans *et al.*) for a detailed discussion of the mechanics of matrix cracking and the influence of interfacial properties on the monotonic behavior of fiber-reinforced ceramics.

6.2.1 Stress–Strain Behavior and Mechanisms of Damage

6.2.1.1 Unidirectional Composites

Generally, when testing materials with a nonlinear stress–strain behavior, the tests should be conducted under uniform stress fields, such that the associated damage evolution is also uniform over the gauge section where the material's response is measured. Because the stress field varies with distance from the neutral axis in bending tests, uniaxial tension or compression tests are preferred when characterizing the strength and failure behavior of fiber-reinforced composites.

Figure 6.1 shows a typical stress–strain curve for a unidirectional SiC_f/CAS composite loaded in uniaxial tension parallel to the fibers. The features of this curve are represenative of many ceramic matrix composites. In order to distinguish between the various damage states that a composite undergoes, it is convenient to divide the stress–strain curve into several sequential parts.

Within Stage I, no microstructural damage occurs, and the stress–strain response is linear elastic. With continued loading, matrix cracks begin to appear, denoting the onset of Stage II. Matrix cracking, which is statistical in nature, does not occur at a specific stress level, but rather is a progressive phenomenon that depends upon the statistical nature of matrix strength and microstructural features such as fiber volume fraction and distribution, interfacial shear stress and residual stress state. With continued loading, the crack density eventually reaches a saturated level and remains relatively constant with further increase in the applied load. The initial matrix cracks that form during monotonic loading are more or less randomly distributed throughout the gauge length, however they often initiate in matrix-rich regions of a composite.[6–9] The cracks will stop when the crack tip reaches a fiber–matrix interface which is typically engineered to have a low fracture toughness, allowing interfacial debonding to readily occur (crack deflection typically occurs parallel to the fiber axis). Such behavior is essential for damage-tolerant behavior. If fiber–matrix debonding does not take place, matrix cracks may penetrate the fibers, leading to brittle fracture with little energy dissipation. The applied stress level at which microcracking initiates is typically referred to as the *microcracking threshold stress*, σ_{mc}. For unidirectional SiC_f/CAS, and a volume fraction of fibers of 35%, the matrix cracking threshold at room temperature is approximately 120 MPa (0.1% strain), which is roughly 20% of the room temperature ultimate strength of this composite system (see Fig. 6.1). For nominally the same material, considerable scatter in σ_{mc} is commonly observed. The variation of σ_{mc} can be attributed to variability in the processing conditions, composite microstructure (e.g., differences in interface condition and porosity distribution), and to differences in specimen alignment which influence the bending strains in a test specimen. The small compliance change that accompanies initial matrix microcracking is

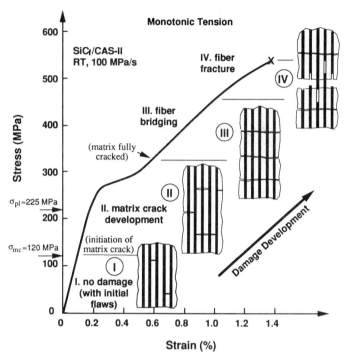

Fig. 6.1 Monotonic tensile stress–strain behavior of a unidirectional Nicalon SiC$_f$/CAS-II composite. The development of microstructural damage takes place in several diffuse stages. In Stage I, the stress–strain behavior is linear; microstructural damage is absent. In Stage II, matrix cracks initiate and grow progressively until a saturated crack spacing is reached; after a sufficient crack density is achieved a decrease in tangential modulus can be experimentally measured. Thereafter, in Stage III, the stress–strain curve is again linear, since the bridging fibers now support the applied load. Beyond Stage III, distributed fiber failure becomes pronounced, leading to failure.

usually not detectable using conventional load and strain transducers. However, it is possible to determine σ_{mc} by acoustic emission,[7,10,11] surface replication,[6–9,10–12] and internal heating measurements.[12]

Additional matrix cracks form as the stress level in a tension specimen is increased beyond σ_{mc}; the fiber-bridged matrix cracks also begin to spread across the width and thickness of a specimen. The increase in matrix crack density causes the stress–strain curve to become nonlinear; for some composites a rather abrupt decrease in tangent modulus may be observed (see Fig. 6.1). The stress level at which nonlinearity is first detectable on an engineering stress–strain curve is commonly referred to as the *proportional limit stress*, σ_{pl}. This is, unfortunately, not a clear definition (although, widely used), since the detection of nonlinearity depends strongly upon the resolution of the load and strain transducers used during tensile testing. A useful engineering definition has been proposed by Prewo,[13] who suggested that σ_{pl} should be calculated as the stress level corresponding to a 0.02% strain offset taken parallel to the initial linear portion of the tensile curve (this is in analogy to the "offset yield stress" used for metals). For the SiC_f/CAS composite shown in Fig. 6.1, the proportional limit is around 225 MPa (as with the microcracking threshold, the proportional limit of a given composite system can vary from billet to billet—these changes are primarily caused by differences in the processing conditions, fiber distribution and test technique). It is important to note that the proportional limit stress is influenced by external parameters such as temperature and loading rate, as will be elaborated upon in the next section. Using the proportionl limit stress as an indicator of the first matrix cracking stress is, in general, a poor practice for most ceramic matrix composites (for the Nicalon SiC_f/CAS composite shown in Fig. 6.1, the proportional limit stress is 225 MPa, *above* the microcracking threshold of 120 MPa). However, for certain composites systems such as SCS-6 SiC_f/HPSN (hot-pressed silicon nitride), which utilize large diameter fibers and have a relatively uniform fiber distribution, the proportional limit and the initial matrix cracking do roughly coincide.

A pattern of multiple matrix cracks having a characteristic spacing forms near the end of Stage II. The matrix crack spacing depends upon the strength characteristics of the matrix, the degree of microstructural inhomogeneities, and the interfacial sliding friction.[14-23] Once the matrix cracking saturates (the matrix is now fully cracked and the fibers are fully debonded) the stress–strain curve regains linearity (Stage III), although with a lower tangential modulus than the initial composite modulus. At this stage, the stiffness of the composite is determined mainly by the modulus of the fibers. The magnitude of the stiffness loss due to matrix cracking depends upon the fiber lay-up, fiber volume fraction, the ratio between the moduli of the matrix and fibers, and the sign and magnitude of the residual stresses. If the fibers have a very high stiffness compared to the matrix, the stiffness change due to matrix cracking will be small since, before matrix cracking, the fibers would support the majority of the applied load. The fact that the stress–strain curve is linear in Stage III indicates that the number of distributed fiber failures at this stage is very limited. This has been verified by Sørensen and Talreja[7] who used acoustic emission and surface replication to study the evolution of damage from the stress-free state until final failure of Nicalon SiC_f/CAS composites. Almost no acoustic events were recorded at Stage III, indicating that no additional matrix cracking or significant fiber failure occurred during Stage III. This result is significant, since it suggests that matrix cracking and fiber fracture are two independent phenomena under monotonic loading. Near final failure, it was observed that the number of acoustic events increases dramatically, indicating a significant increase in the number of fiber failures; in some cases a slight nonlinearity was detectable in the stress–strain curve, suggesting that distributed fiber failure (Stage IV) occurred before final localization. For the particular SiC_f/CAS composite shown in Fig. 6.1, the failure strength, σ_u, was 550 MPa and the failure strain 1.4%.

For composite microstructures designed to optimize monotonic damage tolerance, final fracture is typically accompanied by significant fiber pull-out, indicating that the fibers do not break in the final failure plane. Note that although the fiber stress is highest in the portions of the fibers that directly bridge the crack faces, one would observe fractures in the final failure plane only if the fiber strength was deterministic. Importantly, it is the strength variation inherent in ceramic fibers[24-28] that provides a finite pull-out length and damage-tolerant composite behavior. This leads to the interesting conclusion that one does not want to use a fiber with a narrow strength distribution (high Weibull modulus) if high monotonic toughness is the primary design criterion (as discussed later, optimal fatigue resistance is obtained by microstructures which have high matrix cracking thresholds and low fiber strength variability).

The above discussion pertains to unidirectional composites that are initially free of matrix cracking; examples would include Nicalon SiC_f/CAS, Nicalon SiC_f/1723 glass, Nicalon SiC_f/LAS, and SCS-6 SiC_f/HPSN. For composites such as C_f/borosilicate, where the thermal expansion coefficient of the matrix is substantially greater than that of the fiber, microcracks can develop in the matrix during fabrication. These composites do not exhibit a linear stress–strain response (Stage I), even for small applied loads.

6.2.1.2 Laminates and Woven Composites

In cross-ply laminates, the first mode of damage involves cracking of the 90° plies perpendicular to the tensile loading direction (called transverse cracks or tunnel cracks),[6,29-33] a damage mode that is well known from studies with polymer matrix composites.[34,35] Woven composites, typically processed by chemical vapor infiltration (CVI) techniques, often contain matrix cracks that initiate from existing porosity. For instance, carbon fiber-reinforced SiC (C_f/SiC), contains a dense network of processing-related matrix cracks.[36,37] For this composite system, the thermal expansion coefficient of the matrix is greater than that of the carbon fibers; as the composite cools after processing, the tensile stress generated by this mismatch can produce extensive microcracking in the relatively porous SiC matrix (typically the CVI process used to manufacture these composites results in matrix porosity levels of 8–15%). At both ambient and elevated temperatures, the presence of initial matrix cracking and associated interfacial debonding gives rise to a nonlinear stress–strain response (Stage II) from the onset of loading (see Fig. 6.2). In woven composites, cracking may also occur at the stress concentrations that exist near the crossover points of fiber bundles; this cracking allows extension of the 0° bundles in the direction of the applied load—see Shuler et al.[37] For the C_f/SiC composite shown in Fig. 6.2, an approximately linear stress–strain response (Stage III) was observed between a strain of 0.5 and 0.9%. This suggests that a saturation in the matrix cracking also occurs in woven composites. As with other composite systems, failure of cross-ply and woven composites occurs when the net stress on the remaining fibers or fiber bundles exceeds their combined load-carrying capacity.

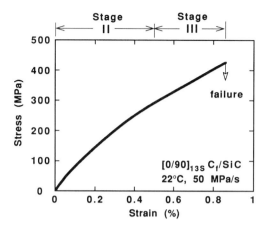

Fig. 6.2 Room temperature stress–strain behavior of a woven 0°/90° C_f/SiC composite. Because of processing-related matrix cracking and progressive fracture near the crossover points of fiber bundles, Stage II behavior (non-linear stress–strain response) is observed from the onset of loading. Above a strain of approximately 0.5% the composite exhibits Stage III (linear) behavior.

6.2.2 Loading Rate Dependence of Monotonic Stress–Strain Behavior

Even at room temperature, loading rate can have a profound effect on the proportional limit stress and ultimate strength, which generally decrease as the loading rate is reduced.[38] For composites with a glass or glass-ceramic matrix, the loading rate dependence of monotonic tensile behavior is a consequence of slow crack growth due to stress corrosion,[39] which leads to time-dependent matrix cracking (time-dependent interfacial debonding and fiber fracture may also occur). The influence of loading rate on monotonic tensile behavior is shown in Fig. 6.3 for a 0° Nicalon SiC_f/CAS composite that was loaded in tension at rates of 0.01, 1, 1.0, 10, 100 and 500 MPa/s (approximately 35 000 s to 1.0 s to failure). There is a clear decrease in proportional limit stress and ultimate strength as the loading rate is decreased below 100 MPa/s (at 100 and 500 MPa/s, time-dependent effects are negligible, with similar stress–strain behavior observed). Sørensen and Holmes[39] have shown that the matrix crack density generally increases as the loading rate decreases. These results have important implications for fatigue behavior of fiber-reinforced ceramics. For example, fatigue tests conducted at 0.1 Hz and 100 Hz would be expected to have different initial damage states after the first loading cycle (after the first cycle, the specimen subjected to 0.1 Hz would have a higher crack density).

At elevated temperatures, creep deformation and transient stress redistribution between the fibers and matrix can have a significant influence on

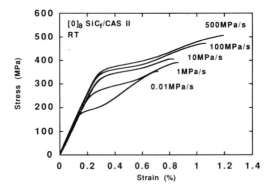

Fig. 6.3 Influence of loading rate on the monotonic stress–strain behavior of unidirectional Nicalon SiC_f/CAS-II composites tested in air at 20°C. As the loading rate decreases, the proportional limit, ultimate strength and failure strain decrease. After Sørensen and Holmes.[39]

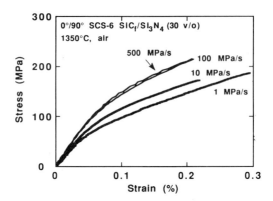

Fig. 6.4 Influence of loading rate on the monotonic stress–strain behavior of unidirectional SiC/Si_3N_4 composites tested in air at 1350°C. With decreasing loading rate, the slope of the stress–strain curve decreases, the composite strength decreases, and the failure strain increases (by creep). The stress–strain behavior is similar for loading rates of 100 MPa/s and 500 MPa/s. From Shuler and Holmes.[38]

monotonic tensile behavior. Figure 6.4 shows the 1350°C monotonic tensile behavior of a unidirectional SiC_f/HPSN composite that was loaded to failure at rates from 1 to 500 MPa/s (approximately 300 s to 0.6 s to failure). The tensile behavior was similar for loading rates of 100 and 500 MPa/s, indicating that time-dependent deformation was minimal for these rapid loading rates. At slower loading rates primary creep begins to influence tensile behavior; for example, a reduction in the loading rate from 100 to 1 MPa/s resulted in roughly a 50% decrease in the proportional limit stress; the failure strain increased from roughly 0.2% at 100 MPa/s to 0.3% at 1 MPa/s.

Based upon results obtained from monotonic tensile experiments conducted with 0° SCS-6 SiC$_f$/HPSN, 0°/90° C$_f$/SiC, and 0° Nicalon SiC$_f$/CAS-II composites, Shuler and Holmes[38] have recommended a loading rate of 20–100 MPa/s to minimize time-dependent deformation during room temperature and elevated temperature monotonic tensile or flexural testing. Equivalent times-to-failure should be used in displacement controlled tests.

Allen and Bowen[40] studied the effect of loading rate on the room temperature and elevated temperature behavior of SiC$_f$/CAS under flexural loading. The room temperature tests all exhibited a significant amount of pull-out. At 600°C and 800°C, and for slow loading rates, the failure mode was nonfibrous; under high loading rates the fracture mode was fibrous. This change in fracture mode appears to be caused by oxidation of the carbon-rich interphase layer and an increase in interface bonding (by the formation of a SiO$_2$-layer). Both of these processes are time dependent and provide likely mechanisms for the loading rate dependence of fracture behavior found at elevated temperatures.

6.2.2.1 Closing Summary

It is rarely possible to determine the microcracking threshold of a composite by examination of a monotonic tensile curve obtained using conventional load and displacement transducers. The proportional limit stress, which is influenced by loading rate and temperature, rarely coincides with the microcracking threshold. The proportional limit stress does, however, provide a useful indication of the stress level at which *significant* matrix damage begins. At elevated temperatures, it is possible that changes in matrix composition or stress redistribution between the fibers and matrix could cause a time-dependent change in the fracture resistance of the matrix (see Chapter 5), which could change both the microcracking threshold and proportional limit stress. As discussed in the next section, it is difficult to make quantitative predictions of the fatigue life of fiber-reinforced ceramics based upon monotonic tensile data.

6.3 Isothermal Fatigue Life—Influence of Maximum Fatigue Stress

6.3.1 General Fatigue Behavior of Unidirectional Ceramic Matrix Composites

Prewo and co-workers were the first to study the fatigue behavior of fiber-reinforced ceramics.[41,42] They conducted room temperature tension–tension and flexure fatigue experiments with 0° Nicalon SiC$_f$/LAS-II composites. The fatigue experiments were conducted in air at a sinusoidal loading

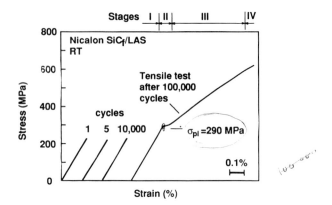

Fig. 6.5 Cyclic stress–strain behavior observed during the room temperature fatigue of unidirectional SiC$_f$/LAS-II at a maximum stress below the 10^5 fatigue limit (loading frequency = 10 Hz, $\sigma_{min}/\sigma_{max}$ = 0.1). The cyclic stress–strain curves show very limited, if any, hysteresis. Note that the strength and strain capability is retained after cyclic loading below the proportional limit stress. After Prewo.[42]

frequency of 10 Hz. Specimens survived 10^5 cycles when fatigued at maximum tensile stress levels below the 0.02% proportional limit stress of 280 MPa. When the surviving specimens were loaded statically to failure, the retained strength (500–600 MPa) and failure strain (0.85%) were similar to that found in virgin material (see Fig. 6.5). For the specimens tested in bending, the proportional limit and the fatigue limit were higher (around 500 MPa). Above this stress level, fatigue failures occurred, with fatigue life decreasing as the maximum stress was increased. Based upon these results, Prewo[42] suggested that σ_{pl} could be used to estimate the fatigue limit (σ_{fl}) of this composite system (i.e. the maximum stress level, below which fatigue failures are not expected).

A coincidence between the proportional limit, σ_{pl}, and the fatigue limit, σ_{fl}, has also been observed for other composite systems fatigued at loading frequencies of 10 Hz or lower. For example, as shown in Fig. 6.6, under tension–tension fatigue at 1000°C, 0° SCS-6 SiC$_f$/HPSN[43] composites exhibit a fatigue limit (5×10^6 cycles) at the monotonic proportional limit stress of 200 MPa.

Zawada et al.[44] showed that the proportional limit, expressed in strain (0.3%) rather than in stress, was identical for unidirectional and cross-ply laminates of SiC$_f$/1723. Moreover, the fatigue limit of the unidirectional composite, expressed in strain, corresponded well with the measured fatigue strain limit of the cross-ply laminates. This indicates that the fatigue limit of a cross-ply laminate is primarily governed by the 0° plies and that the influence of the 90° plies is minimal (this result is expected to hold only for room temperature fatigue—see Chapter 5 for a discussion of how transverse plies influence cyclic creep behavior). The 90° plies develop transverse cracks early

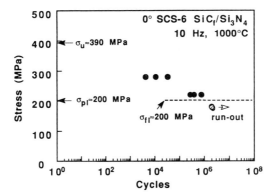

Fig. 6.6 Fatigue life diagram for the tension–tension fatigue of unidirectional SiC$_f$/Si$_3$N$_4$ at 1000°C and a stress ratio ($\sigma_{min}/\sigma_{max}$) of 0.1. Fatigue run-out (5 × 10^6 cycles) was observed when the maximum stress was below σ_{pl}. After Holmes *et al.*[43]

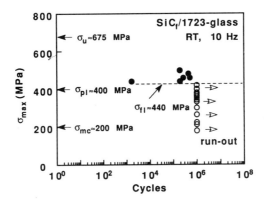

Fig. 6.7 Fatigue life diagram for the tension–tension fatigue of unidirectional SiC$_f$/1723 at room temperature (10 Hz, $\sigma_{min}/\sigma_{max}$ = 0.1). The 10^6 cycle fatigue limit of 440 MPa is higher than σ_{mc} and σ_{pl}. After Zawada *et al.*[44]

in the fatigue life and carry little of the applied load. Similar fatigue behavior has been found for SiC$_f$/LAS[30] and SiC$_f$/CAS[8,31] composites. However, the 90° plies can prevent Poisson's ratio contraction of the 0° layers, which are then loaded in biaxial tension. The effect of biaxial stress states on the fatigue life of fiber-reinforced ceramics has not yet been systematically investigated.

There are also many composites where the 10^6 cycle fatigue limit is above the proportional limit stress. Examples include Nicalon SiC$_f$/1723[44] (Fig. 6.7), Nicalon SiC$_f$/CAS[45] (Fig. 6.8), and 2-D Nicalon SiC$_f$/SiC[46] (run-out at 2.5 × 10^5 at 1 Hz) (Fig. 6.9). 2-D T-300 C$_f$/SiC, which does not possess a proportional limit, has a fatigue limit (10^6 cycles, 10 Hz) slightly above 300 MPa, which is close to 75% of the ultimate strength of this composite

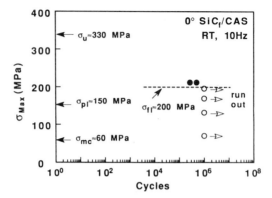

Fig. 6.8 Fatigue life diagram for the tension–tension fatigue of unidirectional SiC$_f$/CAS at room temperature (10 Hz, $\sigma_{min}/\sigma_{max} = 0.1$). The fatigue limit ($10^6$ cycles) at 200 MPa was higher than σ_{mc} and σ_{pl}. It should be noted, however, that the low proportional limit stress is most likely a consequence of the slow loading rate used by Butkus *et al.* in their monotonic tensile tests (see for example Fig. 6.3 which shows that the proportional limit and ultimate strength of this composite system are strongly influenced by the test procedure). This illustrates the difficulty in attempting to correlate fatigue behavior with time-dependent parameters such as σ_{pl} or σ_u. After Butkus *et al.*[45]

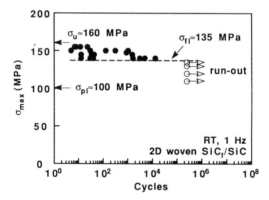

Fig. 6.9 SN curve for the room temperature fatigue of 0°/90° CVI Nicalon SiC$_f$/SiC tested at 1 Hz under a stress ratio ($\sigma_{min}/\sigma_{max}$) of zero. A distinct fatigue limit occurs at 135 MPa. Fatigue run-out was defined at 2.5×10^5 cycles, 1 Hz. After Rouby and Reynaud.[46]

system.[37] For the C$_f$/SiC composite, the tensile strength measured after fatigue run-out is higher than the tensile strength found for virgin material.[37] Also, the stress–strain curve of fatigued samples is almost linear, contrasting with the nonlinear monotonic stress–strain response of virgin C$_f$/SiC. As proposed by Shuler *et al.*,[37] for woven composites, such as C$_f$/SiC, cyclic loading may actually provide a beneficial effect by reducing stress concentrations near the crossover points of fiber bundles (the mechanism for this involves matrix cracking and wear in the vicinity of the crossover points of fiber bundles). C_f/borosilicate composites can also exhibit processing-related matrix cracking. Despite this microcracking, these composites still possess a 10^6 cycle fatigue limit of 265 MPa at a loading frequency of 10 Hz.[47] The results for the initially microcracked C_f/SiC and C_f/borosilicate composites suggest that *it is not the development of matrix cracks that controls fatigue life*. In fact, matrix cracking typically stabilizes very early in the fatigue life of fiber-reinforced ceramics.[12] Although matrix cracking is a prerequisite for fatigue failure, as discussed later, other damage modes, such as interfacial wear and fiber damage, are ultimately responsible for fatigue failure.

The above discussions lead us to the following question. Is there a relationship between the proportional limit stress and the fatigue limit? Based upon the experimental results reviewed above, the answer is no. As noted by Sørensen and Holmes[91] the fatigue limit (at 10^8 cycles or higher) does not coincide with any microstructural event that can be directly related to the monotonic tensile curve, such as the onset of matrix cracking, in particular when high loading frequencies are involved. Of course, it is equally important to point out that the process of fatigue damage can be considered to initiate once matrix cracking has occurred, since this allows interfacial slip and wear to occur; the damage that develops at these low stress levels may never reach the critical state required for failure. The fact that the proportional limit stress at times coincides with the fatigue limit is not surprising when it is realized that it corresponds to an amount of microstructural damage that is sufficient in extent that a compliance change occurs.

Another key question concerns the existence of a true fatigue limit in fiber-reinforced ceramics: namely, what would happen if fatigue tests were conducted for 10^8 cycles? Recent results by Sørensen and Holmes[91] indicate that a fatigue limit exists for SiC_f/CAS-II at 10^8 cycles, but whether it holds for an infinite number of cycles is an open question. The stress for a fatigue limit at 10^8 cycles may be considerably below the proportional limit stress. For example, fatigue run-out with Nicalon SiC_f/CAS composites occurred only when the maximum fatigue stress was 60% below the monotonic proportional limit stress.[91] It is also important to note that all of the studies discussed above (except for Ref. 91) were conducted at relatively low loading frequencies ($\leq$10 Hz). However, as elaborated upon in a later section, the fatigue limit of fiber-reinforced ceramics can decrease sharply with increasing loading frequency.[48] In fact, as discussed in Section 6.7, at loading frequencies of 25 Hz and higher, the room temperature fatigue limit of Nicalon SiC_f/CAS composites is far below the proportional limit stress (this contrasts sharply with the results shown in Fig. 6.7 where, at 10 Hz, the fatigue limit was substantially above the proportional limit stress).

Summarizing, from numerous experimental investigations of fatigue behavior, it has been established that *a fatigue limit exists* for the tension–tension fatigue loading of fiber-reinforced ceramics, at least to 10^8 cycles. In

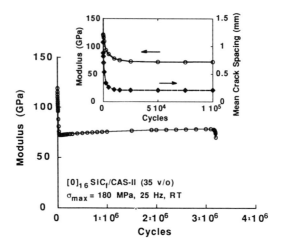

Fig. 6.10 Change in average tensile modulus during the 25 Hz fatigue of unidirectional SiC$_f$/CAS-II between fixed limits of 10 MPa and 180 MPa. The inset shows the changes in modulus and mean matrix crack spacing during the first 100 000 fatigue cycles. The modulus decays rapidly within the first 30 000 cycles, and then stabilizes; a sharp decrease in modulus is observed immediately prior to failure. The rate of stabilization of the modulus lags slightly behind stabilization of matrix cracking; this is attributed to continued interfacial debonding. After Holmes and Cho.[12]

general, the fatigue limit, σ_{fl}, is not related to the proportional limit stress, σ_{pl}. As will be further described in this chapter, in addition to the maximum applied stress level, additional external parameters such as the minimum applied stress, the test temperature, and the loading frequency, will influence fatigue life.

6.3.2 Indicators of Fatigue Damage

There are a number of indicators of fatigue damage that have attracted interest in the literature. During the life of a component subjected to fatigue, the material can exhibit changes in modulus, permanent offset strain, shape of the hysteresis loops, and temperature rise of the specimen surface. Direct evidence of matrix crack density can be obtained by surface replication, while a more detailed analysis of microstructural damage requires scanning electron microscopy (SEM).

At sufficiently high stress levels, the modulus of a composite will decrease during fatigue (usually within the first thousand cycles, dependent on the maximum applied stress) to a stable level, where it remains for a long period of time (Fig. 6.10). A small modulus recovery has been measured with increasing number of cycles.[12,45] Just prior to final failure the stiffness may decrease[12] (Fig. 6.10), but this is not always found.[31,44] The change in the modulus can be explained as follows. The initial decrease in modulus is primarily due to the formation of multiple matrix cracks and debonding, as shown by Holmes and Cho,[12] who found a good correlation between the crack density, measured by replicas, and the initial modulus decrease (Fig. 6.10). With further cycling, the crack density is saturated and remains at the same level. This may explain why the modulus stays roughly at the same level until shortly before fatigue failure. The modulus decrease just prior to failure can be attributed to a rapid increase in the number of fiber failures within the gauge section.

Holmes and Cho[12] proposed that the slight recovery in the modulus is caused by an increase in frictional shear stress, for example by the accumulation of debris along the interface caused by the wear of asperities (this debris would be effectively trapped along the fiber–matrix interface). A more pronounced modulus recovery was found by Shuler *et al.*[37] in a woven C_f/SiC matrix composite. Shuler *et al.*[37] suggested that, for woven composites, geometric stiffening is a likely mechanism for the modulus recovery. Geometric stiffening could occur by a mechanism involving matrix fracture and fiber/matrix wear near the crossover points of fiber bundles, which allows the fibers to become better aligned with respect to the tensile loading direction.

There are several practical reasons why a modulus decrease is not always observed immediately prior to final fatigue failure. Because of the tremendous amount of data storage required to store complete stress–strain loops, it is common to record only at periodic intervals; if the storage period is too long, and the modulus decrease occurs over a rather short number of cycles (see Fig. 6.10), it is easy to miss the sharp decrease that is expected near final composite failure. It is also likely that the fiber failures may be quite localized (fiber pull-out lengths are typically of the order of a millimeter or less); if the strain is measured by an extensometer over a distance of, say, 25–50 mm, the displacement from the localized strain enhancement may be quite small compared to the overall displacement from the remaining regions of the gauge section. Finally, if strain is measured by the use of strain gauges, which are typically about 5 mm in length, the localized events leading to failure may occur outside the area sampled by the strain gauge. As discussed later, damage, such as fiber fracture, localizes very rapidly over a short period at the end of the fatigue life.[91] Thus, although stiffness measurements can be used as an indicator of distributed matrix cracking and interfacial debonding, modulus measurements are not likely to be accurate predictors of remaining fatigue life.

The shape and slope of the stress–strain hysteresis curves provides important information concerning how the material behaves on a finer scale, since it is influenced by the matrix crack density, interfacial friction stress, and number of fiber fractures. The shape of the hysteresis loops generally changes during fatigue, particularly in the initial stages of fatigue where much of the initial microstructural damage takes place[12,43] (see Fig. 6.11).

It has been found[7,31,33] that a permanent offset strain, ε^*, occurs in a composite unloaded from a stress level above σ_{pl}. During fatigue, ε^* may increase as fatigue damage progresses[12,31,50] (in particular, in the initial stages of fatigue)—see Fig. 6.11. This suggests that the offset strain can provide an

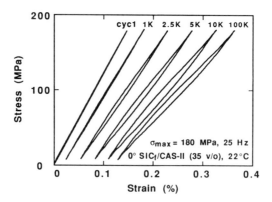

Fig. 6.11 Changes in the hysteresis behavior during the fatigue of unidirectional SiC$_f$/CAS-II. The number of cycles (in thousands) is shown above each curve. Note that the average modulus, area of the hysteresis loops, and the permanent strain offset all change during fatigue. Failure took place at 3.21×10^6 cycles. After Holmes and Cho.[12]

indication of fatigue damage accumulation. However, because it relies upon the resolution of the technique used to measure changes in strain, the strain offset will not be sensitive to small changes in the damage state of a composite, such as small increments in fiber/matrix debonding or interfacial wear damage.

The most sensitive technique found to date for determining the initiation of fatigue damage involves measurement of the temperature rise that occurs during the fatigue loading of fiber-reinforced ceramics. Holmes and Shuler[51] measured the temperature rise during the fatigue of woven C$_f$/SiC. It was observed that the surface temperature increased to a stable level, which depended on the load level and frequency (Fig. 6.12). For instance, during cycling between 10 and 250 MPa, the temperature increased by almost 10°C at a frequency of 25 Hz, while a temperature rise of 30°C was found at 85 Hz. Holmes and Shuler suggested that the energy dissipation was caused by frictional sliding along the fiber–matrix interface during loading and unloading.

6.3.2.1 Changes in Frictional Shear Stress During Cyclic Loading

Cho *et al.*[52] developed an analytical model that relates the temperature rise during fatigue to the interfacial frictional sliding stress, as elaborated later. Holmes and Cho[12] used the model to show that in SiC$_f$/CAS-II, the interfacial shear stress, τ, decreases from a value of around 15 MPa to 5 MPa within the first 25 000 fatigue cycles (Fig. 6.13). The approach used to determine τ from temperature rise data is described in greater detail in the following section.

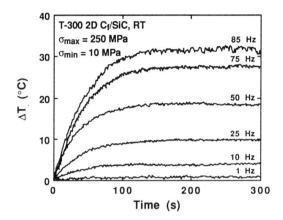

Fig. 6.12 Influence of loading frequency on the surface temperature rise measured during the tension–tension fatigue of a woven 0°/90° CVI C_f/SiC composite. The fatigue experiments were conducted at 20°C between fixed stress limits of 10 MPa and 250 MPa. After Holmes and Shuler.[51]

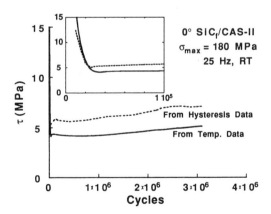

Fig. 6.13 Change in interfacial shear stress during the room temperature fatigue of SiC$_f$/CAS. The fatigue experiment was conducted at 25 Hz between stress limits of 10 MPa and 180 MPa. Note, that τ decreases within the first 25 000 cycles from a value of more than 15 MPa to approximately 5 MPa. Thereafter, it remains nearly constant (the slight increase beyond 1×10^6 cycles is attributed to debris trapped along the interface). After Holmes and Cho.[12]

6.3.3 Models for Cyclic Hysteresis Loops and Cyclic Energy Dissipation

Kotil et al.[53] modeled the effect of broken fibers and changes in the fiber–matrix interfacial sliding shear stress, τ, on the cyclic stress–strain response and formation of offset strain, ε^* (strain ratcheting), in unidirectional composites. The model was used to interpret experimental results obtained for the fatigue of unidirectional SiC_f/HPSN,[43] which exhibited a progressive increase in compliance and strain ratcheting during tension–tension fatigue at 1000°C. Assuming a uniform distribution of matrix cracks exists perpendicular to the applied stress, fatigue damage was modeled by allowing the number of fiber failures and the fiber debond length to progressively increase during fatigue. As one would expect intuitively, it was found that the magnitude of the stress–strain hysteresis was negligible when τ was either exceedingly small or large. Physically, for small τ, the fibers are free to slide within the matrix— minimal energy is expended in slip. For high levels of τ, the mechanism responsible for the lack of hysteresis is different; namely, the slip length of the fibers is small. From these results, and from later modeling by Cho et al.,[52] it can be shown that the energy dissipation during cyclic loading reaches a maximum at intermediate levels of frictional shear stress, i.e., if one were to plot the energy dissipation that occurs during cyclic loading as a function of shear stress, a bell-shaped curve would be obtained (see later discussion in Section 6.3.4 on frictional heating).

Pryce and Smith[54] developed an axisymmetric shear lag model to predict the shape of the hysteresis loops in unidirectional composites with regular matrix crack spacing and unbroken continuous fibers. Assuming a constant interfacial shear stress and partial frictional slip (i.e., the slip length, l_s, is less than half the crack spacing), the relationship between strain and stress during the first loading portion of a fatigue cycle is given by

$$\varepsilon = \frac{\sigma}{E_c} + \frac{d}{\bar{l}\tau E_f}\left[\sigma\frac{E_m(1-v_f)}{E_c v_f} - \sigma_f^{res}\right]^2 \tag{1}$$

where E_c, E_f, and E_m are the moduli of the undamaged composite, the fibers, and the matrix, respectively, d is the fiber diameter, $\bar{l}$ is the matrix crack spacing, τ is the frictional (sliding) interfacial shear stress, v_f is the volume fraction of fibers, and σ_f^{res} is the residual axial stress in the fibers. The composite modulus is given by the rule of mixtures,

$$E_c = v_f E_f + (1 - v_f) E_m \tag{2}$$

The unloading path is given by

$$\varepsilon = \frac{\sigma}{E_c} + \frac{d}{2\bar{l}\tau E_f}\left[\left(\frac{E_m(1-v_f)}{E_c v_f}\right)^2 (\sigma_{max}^2 - \sigma^2 + 2\sigma\sigma_{max}) - 4\sigma_{max}\sigma_f^{res}\right.$$
$$\left.\frac{E_m(1-v_f)}{E_c v_f} + 2\sigma_f^{res2}\right] \tag{3}$$

In Eqn. (3), σ is the applied stress and σ_{max} is the maximum stress that was applied prior to unloading. During specimen unloading, followed by reloading to the same stress level, σ_{max}, sliding will occur over a distance equal to one-half of the sliding distance encountered on the first loading cycle. The stress–strain relationship for subsequent loading cycles is given by

$$\varepsilon = \frac{\sigma}{E_c} + \frac{d}{2\bar{l}\tau E_f}\left[\left(\frac{E_m(1-v_f)}{E_c v_f}\right)^2(\sigma_{max}^2 + \sigma^2) - 4\sigma_{max}\sigma_f^{res}\right.$$

$$\left.\frac{E_m(1-v_f)}{E_c v_f} + 2\sigma_f^{res2}\right] \tag{4}$$

During further cycling, the unloading and reloading paths follow Eqns. (3) and (4). Note that the equations predict that the stress–strain traces are nonlinear (parabolic) during cyclic loading, i.e. the tangential modulus is not constant. This nonlinearity is a consequence of the change in fiber slip length that occurs during the cycle. During the start of a loading or unloading transient, the sliding direction changes, such that the sliding length is zero; the tangential modulus is given by the composite modulus, E_c. With further loading, the sliding length increases and the tangential modulus decreases. Note that the loop width (the difference between the strain during unloading and loading) depends on the applied load level, but not on the residual axial stresses. Figure 6.14 compares the measured and calculated hysteresis loops of SiC$_f$/CAS.[54] The agreement is good, indicating that the model describes the micromechanisms well and that the intrinsic material parameters were properly chosen.

The permanent offset strain, ε^*, can be found by setting σ to zero in Eqn. (4):

$$\varepsilon^* = \frac{d}{2\bar{l}\tau E_f}\left[\left(\sigma_{max}\frac{E_m(1-v_f)}{E_c v_f}\right)^2 - 4\sigma_{max}\sigma_f^{res}\frac{E_m(1-v_f)}{E_c v_f} + 2\sigma_f^{res2}\right] \tag{5}$$

Experimental investigations using SiC$_f$/CAS composites have shown that the strain offset, ε^*, increases as fatigue damage progresses.[12,31] Since a stable crack spacing is often achieved during fatigue at low stress levels (Fig. 6.10), it follows from Eqn. (5) that the increase in ε^* (Fig. 6.11) during fatigue is related, at least in part, to the decrease in τ that occurs during cyclic loading[12] (see Fig. 6.13). Note, however, that the equations above are only valid if the interfacial sliding length is less than one-half the average matrix crack spacing (i.e., for partial slip). If the applied stress during the initial loading cycle exceeds

$$\sigma_{fs} = \frac{v_f}{1-v_f}\frac{E_c}{E_m}\left(\frac{2\bar{l}}{d}\tau + \sigma_f^{res}\right) \tag{6}$$

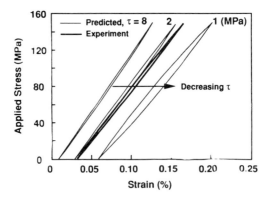

Fig. 6.14 Experimental (coarse lines) and predicted hysteresis loops (fine lines) for unidirectional SiC$_f$/CAS. The predictions, which assume partial slip along the fiber–matrix interfaces, were based upon a residual compressive stress in the fibers of 50 MPa and values of τ of 8, 2 and 1 MPa. The loop width and the permanent offset strain increase as τ decreases. After Pryce and Smith.[54]

then full slip occurs along the fiber–matrix interfaces. When full slip takes place, the stress–strain path is linear with a slope equal to $v_f E_f$. During subsequent cycles, full slip will occur during unloading and reloading if the stress range $\Delta\sigma = \sigma_{max} - \sigma_{min}$ exceeds[52]

$$\Delta\sigma_{fs}^{cyc} = 4\frac{v_f}{1 - v_f E_m}\frac{E_c}{d}\frac{\bar{l}}{\tau}\tau \tag{7}$$

In this case, the permanent offset strain is independent of the maximum stress level,[7]

$$\varepsilon^* = \frac{\bar{l}}{d}\frac{\tau}{E_f} - \frac{\sigma_f^{res}}{E_f} \tag{8}$$

Equation (8) predicts that the permanent offset strain will *decrease* as τ decreases.

As noted above, when the stress is increased during a particular fatigue cycle, a change from partial to full frictional slip can occur. Since the interfacial shear stress decreases during the initial stages of fatigue, it is possible that the stress level at which a change from partial to full frictional slip occurs will change. Moreover, for sufficiently low interfacial shear stress, it is possible that full slip could occur during most portions of a fatigue cycle. This change from partial to full frictional slip leads to important changes in the stress–strain response and energy dissipation during fatigue. For instance, for a fixed crack spacing, the tangential modulus decreases continuously during loading so long as partial slip takes place. However, the stress–strain trace will become linear when full slip begins. The effect of τ on the offset strain, ε^*, is also influenced by the interfacial slip mode. For partial slip, a decrease in τ will increase ε^*; for full slip, ε^* decreases with decreasing τ.

Sørensen *et al.*[55] have studied the cyclic stress–strain behavior of fiber-reinforced ceramics using an axisymmetric finite element model. Their analysis assumed a Coulomb friction relationship for stress transfer within the slip zones: $\tau = -\mu\sigma_{rr}$ (where σ_{rr} is the interfacial stress normal to the interface and μ is the friction coefficient). For a fixed load, it was found that τ did not change by more than 25% along the slip length. However, fiber contraction, caused by the Poisson's ratio effect, had a strong effect on the interfacial sliding stress during a fatigue cycle; at high applied stresses τ can decrease to zero. The interfacial shear stress vanishes when the Poisson's contraction of the fiber exceeds the thermal expansion mismatch. Contact between the fiber and matrix vanishes when the applied fatigue stress exceeds a value given by[7,56]

$$\sigma_{NC} = E_f \frac{v_f}{v_{f,p}}(\alpha_f - \alpha_m)\Delta T \tag{9}$$

In Eqn. (9), σ_{NC} is the stress level at which interfacial contact is lost, $v_{f,p}$ is Poisson's ratio of the fiber, α_f and α_m are the thermal expansion coefficients of the fiber and matrix, respectively, and ΔT is the temperature change from a stress-free state. When the applied stress exceeds σ_{NC}, fiber/matrix contact is lost as the fiber contracts away from the matrix; above this stress the tangential modulus is $v_f E_f$. Interestingly, Sørensen *et al.*[55] found that the tangential modulus for full sliding along the interface was lower than $v_f E_f$ during loading, and higher than $v_f E_f$ during unloading. This is one of the primary differences found when a Coulombic friction law is used rather than a constant value of τ. Use of a Coulomb friction law is, in fact, more accurate, since it accounts for the effect of Poisson's ratio mismatch on frictional shear stress. During a stress increment, the fibers elongate in the axial direction and contract in the transverse direction (Poisson's effect), lowering the interfacial mismatch and thereby lowering τ. As τ decreases during the loading portion of a fatigue cycle, the degree of load transfer to the matrix decreases. As a consequence, the stress increase in the fibers is actually higher than the imposed stress increment. This implies that the total elongation of the fiber, and thereby the composite, is also greater. Sørensen and Talreja[7] derived the following analytical expression for the tangential modulus present during full slip

$$E_{fs}^t = \frac{-v_f E_f}{-1 \pm v_f \dfrac{l}{d}\dfrac{d\tau}{d\sigma}} \tag{10}$$

where

$$\frac{d\tau}{d\sigma} = -\mu \frac{\dfrac{v_{f,p}}{v_f}\dfrac{E_m}{E_f}}{(1 - v_{f,p})\dfrac{E_m}{E_f} + \dfrac{1 + v_f}{1 - v_f} + v_{m,p} \pm \mu \dfrac{\bar{l}}{d}\left(v_{f,p}\dfrac{E_m}{E_f} + \dfrac{v_f}{1 - v_f}v_{m,p}\right)}$$

(11)

The positive sign in the denominator of Eqns. (10) and (11) is valid during loading and the negative sign is valid during unloading. Here, v_m is the Poisson's ratio of the matrix. Although the magnitude of the interfacial shear stress depends upon the test temperature (through thermal mismatch), the tangential modulus during full slip is independent of the thermal expansion mismatch and temperature. Experimental investigations[7,46] have shown that the tangential modulus at the end of the loading trace is lower than the tangential modulus at the end of the unloading trace. This suggests that the stress transfer across the interface is better described by a Coulomb friction law than by a constant interfacial shear stress (which would be independent of load level). Indeed, models with a constant value of τ may not be able to capture all of the details of the cyclic stress–strain behavior.

6.3.4 Frictional Heating During Cyclic Loading

The stress–strain response of fiber-reinforced ceramics that have undergone matrix cracking typically exhibits hysteresis, which is caused by the frictional sliding of debonded fibers. This hysteresis indicates that energy dissipation occurs. Frictional energy dissipation can lead to dramatic heating, with temperature increases in excess of 150 K measured during the fatigue of Nicalon SiCf/CAS composites.[51–53] Because the fundamental mechanisms of frictional heating do not change with temperature, it is instructive to discuss the concept of frictional heating with reference to room temperature measurements where the majority of experiments have been conducted.

From observations of the temperature rise measured during fatigue loading, Cho et al.[52] developed models that describe the frictional energy dissipation during cyclic loading of fiber-reinforced ceramics. The approach involves relating the work performed in the frictional slip of fibers to the temperature rise observed in a fatigue specimen. The frictional work is obtained by integration of the relative displacement that occurs between a fiber and matrix during a fatigue cycle. Multiplying the frictional work by the loading frequency, f, gives the rate of energy dissipation per unit volume,

$$\frac{dw_{fric}}{dt} = \frac{f}{24}\frac{d}{\bar{l}}\frac{\Delta\sigma^3}{\tau_d E_f}\left(\frac{E_m}{E_c}\frac{1 - v_f}{v_f}\right)^2$$

(12)

In Eqn. (12), τ_d represents the dynamic interfacial shear stress, which may differ from that which would be measured from fiber push-out experiments, which are typically conducted at low sliding velocities. Equation (12) holds for partial sliding along the interface. When the minimum applied stress is equal to zero, the area of the hysteresis loop can also be calculated as the integral from zero to σ_{max} of the difference between the strain paths for loading and unloading (Eqns. (3) and (4)):

$$\frac{dw_{fric}}{dt} = \frac{f}{24} \frac{d}{\bar{l}} \frac{\sigma_{max}^3}{\tau_d E_f} \left(\frac{E_m}{E_c} \frac{1 - v_f}{v_f} \right)^2 \tag{13}$$

The analysis above holds only for partial sliding along interfaces. When full sliding takes place (the applied stress exceeds $\Delta\sigma_{fs}^{cyc}$—see Eqn. (7)), the rate of energy dissipation per unit volume can be expressed as[52]

$$\frac{dw_{fric}}{dt} = 2f \frac{\tau_d}{E_f} \frac{\bar{l}}{d} \Delta\sigma - \frac{8f}{3} \frac{\bar{l}^2}{d^2} \frac{v_f}{1 - v_f} \frac{E_c}{E_m} \frac{\tau_d^2}{E_f} \tag{14}$$

Inspection of Eqns. (12) and (14) shows that the energy dissipation rate increases with increasing loading frequency and stress range, and with decreasing fiber volume fraction and fiber modulus. The mechanism by which a decrease in fiber volume fraction can lead to a higher energy dissipation can be explained as follows. For a fixed applied stress, the stress in the individual fibers increases with decreasing v_f; this leads to a higher strain in the fibers such that the fibers slide over larger distances relative to the matrix. The same argument holds for a decreasing fiber modulus.

It is worth noting the striking difference between the cases of partial and full frictional sliding. Inspection of Eqn. (12) shows that when partial sliding occurs, the energy dissipation increases in proportion to $\Delta\sigma^3$. For full slip, the energy dissipation increases only linearly with $\Delta\sigma$ (see Eqn. (14)). The effect of interfacial shear stress on energy dissipation is also dependent on interface slip mode. In the case of partial slip, the energy dissipation *increases* as τ decreases (Eqn. (12)). However, there is a limit to the increase in energy dissipation. Namely, as τ decreases, full slip can occur (note that Eqn. (12) is no longer valid when this condition is reached). From Eqn. (14), for full slip, the energy dissipation reaches a maximum and then *decreases* with decreasing τ; in the limit of zero interfacial friction, Eqn. (14) predicts the correct limit of zero energy dissipation. Finally, the ratio between the fiber diameter and the matrix crack spacing influences the energy dissipation, as follows. As long as partial sliding exists, the total slip area increases as the number of active slip zones increases, i.e. as the mean crack spacing decreases. This is not the case when full slip occurs. Instead, during full slip, the relative displacement between the fiber and matrix decreases as the crack spacing decreases, which results in a decrease in energy dissipation.

It is known that τ decreases during the initial stages of fatigue (Fig. 6.13). How does this decrease affect the energy dissipation and thereby the temperature rise during fatigue? Assume that the stress range is such that within the first few cycles, where τ is high, partial slip would take place (for a stress range lower than $\Delta\sigma_{fs}^{cyc}$). As τ decreases, due to interfacial wear, the amount of frictional heating will increase. However, in parallel, due to the decrease in τ, $\Delta\sigma_{fs}^{cyc}$ decreases (Eqn. (7)), such that the composite may enter the full-slip regime. Then, with a further decrease in τ, the energy dissipation reaches a peak, and decreases with further cycling. *Thus, with all other parameters fixed, a plot of energy dissipation versus interfacial shear stress would be bell-shaped—the energy dissipation reaches a maximum at intermediate levels of frictional shear stress, and approaches zero for exceedingly high or low levels of frictional shear stress.*

If the rate of energy dissipation due to frictional sliding can be measured experimentally, the average level of interfacial frictional shear stress can be determined using either Eqn. (12) or (14) (depending upon whether partial or full slip exists). As discussed by Cho et al.[52] there are two independent ways of determining the energy dissipation. The first method is to fit the theoretical hysteresis loops to the ones that are actually measured. For instance, an energy consistent method is to determine the energy dissipation per cycle by measuring the area of a hysteresis loop using numerical integration. Alternatively, the frictional energy dissipation can be estimated from the specimen temperature rise measured during fatigue. The energy dissipation from frictional sliding causes the specimen to heat. The total rate of heat loss per unit volume, dq/dt, is the sum of convective and radiative heat loss from the specimen surface plus the heat conducted away through the cross-section of the specimen:[52]

$$\frac{dq}{dt} = [h(T_s - T_a) + \varepsilon\beta(T_s^4 - T_a^4)]\frac{A_{surf}}{V} + \frac{2kA_{cond}}{V}\left(\frac{\Delta T}{\Delta z}\right)_{axial} \qquad (15)$$

where h is the conductive heat transfer coefficient, T_s and T_a are the temperatures of the specimen and the surrounding air, respectively, ε is the emissivity, β is the Steffen–Boltzman constant (5.67×10^{-8} W/m^2 °C^4), k is the thermal conductivity of the composite parallel to the fibers, A_{surf} and A_{cond} are the gauge-section surface area and cross-section, respectively, V is the gauge-section volume, and $\Delta T/\Delta z$ is the temperature gradient at the end of the gauge-section. Thus, by conducting an energy balance between the rate of work performed in the frictional slip of fibers and the rate of heat loss, it is possible to estimate the frictional shear stress that exists along the fiber–matrix interface.[52] The procedure used to determine the frictional shear stress present during ambient or elevated temperature fatigue loading is summarized in Fig. 6.15.

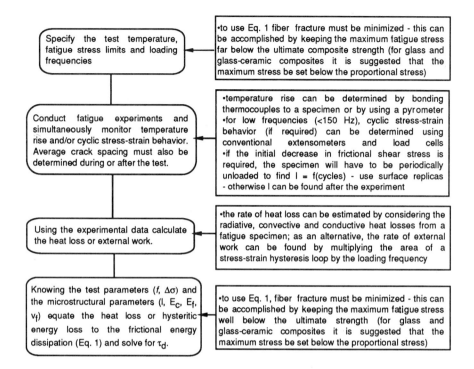

Fig. 6.15 Flow chart outlining how the dynamic frictional shear stress, τ_d, in fiber-reinforced ceramics can be calculated from cyclic loading experiments.

6.3.4.1 Changes in Interfacial Shear Stress During Fatigue

Interfacial slip and the resulting interfacial wear can lead to changes in the value of interfacial frictional shear stress. Direct evidence for cyclic interfacial wear in fiber-reinforced ceramics was obtained by Holmes and Cho[12] from fatigue experiments performed with Nicalon SiC$_f$/CAS composites. The experiments were conducted at a loading frequency of 25 Hz between fixed stresses of 10 MPa and 180 MPa. As shown in Fig. 6.13, a sharp decrease in frictional shear stress occurs during the initial stages of fatigue loading, after which an approximate plateau in frictional shear stress is achieved (for the example shown, the frictional shear stress decreased from approximately 15 MPa to 5 MPa during the first 20 000 cycles of fatigue, remained relatively constant to 1×10^6 cycles, and increased slightly thereafter. The initial decrease can be attributed to wear along the interface, which also acts to reduce the height of interface asperities.[12,57] The slight recovery in interfacial shear stress with continued cycling appears to be related to the accumulation of debris along the interface. If sufficiently high, the recovery in shear stress can cause a change from full slip to partial slip along the interface.[58]

The previous discussion and examples of frictional heating concerned room temperature behavior. How would the situation change for high temperatures? Frictional shear stress, and therefore energy dissipation, is influenced by thermal expansion mismatch between the fibers and matrix. In the case of Coulomb friction, the interfacial shear stress is related to the normal interfacial stress by $\tau = -\mu\sigma_{rr}$ (for $\sigma_{rr} < 0$). For composites where the thermal expansion coefficient of the fiber is lower than that of the matrix, σ_{rr} is negative (i.e. compression). In this case, a temperature rise would decrease the magnitude of σ_{rr}, and it follows that the frictional shear stress would also decrease. Of course, other factors can simultaneously influence the high temperature interfacial shear stress. For example, creep deformation can cause a time- and loading-history-dependent relaxation of the radial stress along the interface.

For composite systems where the thermal expansion coefficient of the matrix is greater than that of the fibers, the frictional shear stress will decrease as the bulk temperature of the composite increases. This, coupled with the temperature increase caused by frictional heating, would decrease the degree of load transfer between the fiber and matrix, leading to an increase in the net stress carried by the fibers. Because of the increased likelihood of fiber fracture, this could lead to a reduction in fatigue life caused by fiber rupture (of course, this is dependent upon the applied stress which determines whether the stress carried by the fibers exceeds the creep rupture stress of the fibers). For mechanical properties controlled by the degree of frictional shear stress, the differential thermal expansion between the fiber and matrix caused by frictional heating could lead to a frequency dependence of mechanical behavior, including fatigue life. Mechanical properties such as toughness, strength and damping will also be frequency dependent, as will thermophysical properties that depend upon the degree of frictional shear stress present in a composite (e.g., thermal expansion and thermal conductivity).

For a given fiber/matrix system and fiber lay-up, the primary microstructural parameters that control frictional heating are: (1) fiber volume fraction, (2) Weibull modulus of the fibers (which will determine the number of fiber fractures and failure location of the fibers—note that the model by Cho *et al.*[52] does not account for the presence of fractured fibers), and (3) interfacial frictional shear stress (influenced by fiber coatings and fiber/matrix roughness). It is important to note that the microstructural parameters that provide high damage tolerance during monotonic loading (e.g., low interfacial debonding strength, long fiber pull-out lengths) are precisely the parameters that exacerbate frictional heating. Thus, for applications where frictional heating is undesirable, a balance must be achieved between monotonic toughness and frictional heating.

6.3.4.2 Practical Implications of Frictional Heating

The occurrence of frictional heating has important implications for the design of structural components. One concern is that components such as gas-turbine airfoils and combustors will be subjected to creep loading combined with high frequency, low amplitude fatigue loading. The additional temperature rise caused by frictional heating could lead to accelerated microstructural damage in these components and could cause dimensional distortion in components if the damage is localized (e.g. around mounting holes where the stress and microstructural damage may be higher). Moreover, a temperature rise caused by internal heating could result in unacceptable dimensional changes. Frictional heating could also influence the mechanical and thermophysical properties of fiber-reinforced composites. The temperature increase which occurs during fatigue could alter the interfacial shear stress through differential thermal expansion between the fibers and matrix, resulting in a frequency dependence of mechanical behavior.

Summarizing, during fatigue loading the repeated frictional sliding of debonded fibers leads to the generation of heat along the numerous interfaces present in a composite. The amount of heating becomes more pronounced as the loading frequency and stress range are increased. Although the majority of the frictional heating data have been obtained during room temperature fatigue, frictional heating also occurs at elevated temperatures.[59] It is important to note that the temperature rise that occurs during elevated temperature fatigue may increase the temperature of a component to a level that could accelerate failure (for example, an increase of 50°C to 100°C in a composite designed for use at 1200°C would significantly increase the rate of creep deformation).

6.4 Mechanisms of Fatigue Damage

6.4.1 Evolution of Fatigue Damage

6.4.1.1 Basic Considerations

Talreja[60] put forth several key questions which are useful for understanding how fatigue damage initiates and accumulates in fiber-reinforced ceramics. These are: (1) What are the changes that occur in the microstructure of a material during the first fatigue cycle? (2) What can cause a crack that has been arrested at the maximum load in the first cycle to grow further in the next cycle? (3) At what rates do further changes in the microstructure occur during subsequent cycles? and (4) What is the critical damage state associated with failure? The aim of this section is to pull together the results from various investigations in an attempt to provide answers to these questions. This section also discusses how the progressive accumulation of microstructural damage during cyclic loading can influence other properties such as the monotonic strength and modulus.

(1) What are the changes that occur in the microstructure of a material in the first cycle? The earlier discussion of the microstructural damage that occurs during monotonic loading (Section 6.2.2) provides the insight needed to answer

this question. Fatigue damage will not occur if the maximum fatigue stress is below the stress level at which matrix microcracking initiates (Stage I). Thus, neglecting possible slow crack growth from pre-existing defects, cyclic loading at stress levels below the microcracking threshold (σ_{mc}) will provide an infinite fatigue life at room temperature. If the maximum fatigue stress on the first cycle is above the matrix microcracking threshold (Stage II), fiber/matrix debonding and interfacial sliding will occur on the first loading cycle (as shown in frictional heating experiments by Holmes and Cho,[12] debonding accompanies matrix cracking). In Stage III the dominant damage mechanism is interfacial sliding. The type and extent of damage that occurs on the first loading cycle sets the stage for what occurs during subsequent fatigue cycles.

(2) What can cause a crack that has been arrested at the maximum load in the first cycle to grow further in the next cycle? The answer to this question is the key to understanding fatigue damage in fiber-reinforced ceramics. Since ceramic matrix composites are made of brittle constituents that are linearly elastic, all deformation and cracking occurs instantaneously if the temperature is below that at which creep of the constituents would occur.† Assuming that matrix cracking has occurred on the first fatigue cycle, the frictional sliding and associated interfacial wear that occurs on the second and subsequent cycles changes the level of frictional shear stress and the length of the slip zones. This interfacial sliding is a nonreversible energy-dissipating mechanism which produces a different damage state during each fatigue cycle. This damage mode is path dependent and irreversible. When interfacial sliding occurs, wear damage between the debonded fibers and matrix can lead to a progressive decrease in frictional shear stress (which would increase the fiber stress within the slip zones), or may lead directly to fiber damage by abrasion of the fiber surface. To fully answer question (2), and to provide answers to questions (3) and (4), it is necessary to first examine the fiber–matrix interface at a more detailed level and to understand the role that interface damage plays in crack extension — as elaborated upon below.

As illustrated in Fig. 6.16, it is convenient to approach fatigue at three scales: (1) the macroscale, (2) the mesoscale, and (3) the microscale. At the *macroscale*, the composite is considered to be a continuum and its behavior is characterized by a constitutive law, reflecting only in an indirect way the micromechanics. The dimensions where this scale is used must be much larger than any length dimension that would affect the micromechanics. For instance, the engineering stress–strain curve is a macroscale concept, measured over a volume that is large compared to, for example, the matrix crack spacing (the concepts of stress and strain do not involve any length scale). The *mesoscale* considers a finer length scale, where the length scale includes features of the

---

†This ignores the possibility of stress corrosion crack growth (i.e. environmental driven crack growth) which can occur at room temperature in many glasses and glass-ceramics. For the sake of argument, it is assumed here that the loading frequency is sufficiently high such that time-dependent effects are absent during the first fatigue cycle.

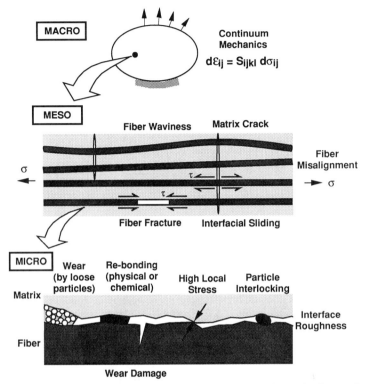

Fig. 6.16 Definition of damage at the macroscale, mesoscale, and microscale of a composite; damage is considered at three different length scales. The microscale is concerned only with the fiber–matrix interaction, The mesoscale concerns the interaction of fibers and matrix with cracks or other defects. The macroscopic properties are independent of any microstructural length dimension. When going from a finer scale to a coarser scale, the description of the finer scale is made through average quantities.

fibers and matrix. At this level, the effects of matrix cracks, bridging, sliding, and fractured fibers are considered. Effects occurring at a finer scale (such as interfacial sliding and wear) are only treated by their average values, such as an average interfacial shear stress τ or friction coefficient, μ. Observations made at the macroscale and mesoscale can provide information on the *effects* of the basic mechanisms of fatigue, but will not necessarily provide details of the *cause* of fatigue since, when going from a finer to a coarser scale, many details are omitted about changes which occur along the fiber–matrix interfaces. At the *microscale*, events along the fiber–matrix interface are considered. Several types of microdamage can occur along the fiber–matrix interface (Fig. 6.16). First, interfacial debonding and fiber slip can take place. Next, abrasive wear can cause a reduction of the interfacial roughness and, therefore, the interfacial shear and normal stresses (nonequal stresses along the fiber axis and circumference may be present because of fiber waviness at the mesoscale).

Interface wear can take place at the interphase layer, within the matrix, or directly at the fiber surface (in particular for "soft" carbon fibers in a "hard" glass or ceramic matrix). Wear will generate loose particles that may scrub around or may lock into existing roughness pockets, which can change the local stress between the fibers and matrix (e.g., the wear debris may cause mechanical locking and hinder interface slip, raising the local stresses along the interface (Fig. 6.16). It is also possible that the fracture of asperities, present on the surfaces of all fibers, could introduce additional defects into the fibers, thereby decreasing their failure strength. It should be noted that the direct wear of carbon fibers during fatigue loading has been observed by Morris *et al.*[61] during *in situ* SEM studies of fatigue damage in woven C_f/SiC composites. During forward and reverse sliding along the interface, the relative frictional slip of fibers causes energy dissipation in the form of heat. The rate of heat transfer in glass and glass-ceramic matrices is generally very low. For sufficiently high loading frequencies, the temperature rise associated with frictional slip may be sufficient to promote changes in interface chemistry or oxidation of near-interface carbon-rich layers which could alter the interfacial friction or cause rebonding, changing the fracture and fatigue characteristics of the composite.

The above discussion allows us to identify three separate types of fatigue damage: (1) matrix cracking, (2) interfacial wear, and (3) fiber failure (Fig. 6.17). We now proceed to link the micromechanical damage occurring at the fiber–matrix interface (microscale) to the evolution of fatigue damage occurring at the mesoscale. Specifically, we want to show how microscale interfacial wear controls the fatigue damage at the mesoscale, and thus the macroscopic properties of a composite.

6.4.1.2 Matrix Cracking due to Interfacial Wear

For stress levels above the microcracking threshold of the matrix, a stable crack density is typically achieved in the early stages of fatigue loading.[12] For most composites, this matrix cracking is, in part, a time-dependent phenomenon that would occur even during static loading. Although the presence of matrix cracks is required for fatigue failure, their presence does not necessarily mean that fatigue failure will occur (i.e., for low fatigue stresses and loading frequencies, microcracked composites may still exhibit fatigue run-out). Rather, matrix cracking represents only the first step in the fatigue damage process. In other words, matrix cracking and associated interfacial debonding sets the stage for the progressive damage of fibers that bridge matrix cracks. Further damage may involve interface wear which leads to an increase in fiber stress within slip zones, or it may involve mechanical damage of the bridging fibers. Final fatigue failure is related to the rate at which the fiber strength decreases. At temperatures at which creep occurs, the creep rupture of bridging fibers can limit fatigue life (note that the likelihood of fiber rupture will increase as interface wear and load transfer from the matrix to fibers

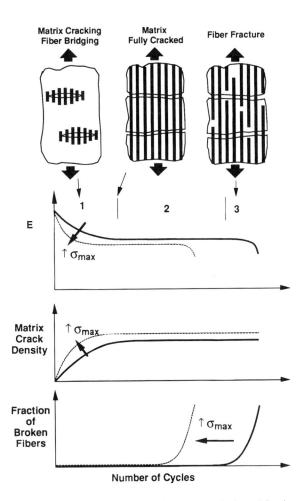

Fig. 6.17 Current understanding of fatigue damage evolution. Matrix cracking and fiber/matrix debonding occur early in the fatigue damage process. At this stage, the growth of fiber-bridged matrix cracks is governed by the bridging effect of the fibers. As the interfacial shear stress decreases by frictional wear, the bridging stresses are reduced, leading to an increase in the stress intensity at the matrix crack tip and further crack extension. The matrix cracking and debonding lead to a significant reduction in the composite modulus. When the matrix cracking and debonding have saturated, the interfacial wear continues, but no significant change in the overall stress–strain response may be observed. Finally, interfacial wear or direct mechanical damage to the fibers causes the fibers to begin failing, giving rise to a further reduction in the composite stiffness and strength. Composite failure occurs when the strength equals the applied fatigue stress.

increase the fiber stress within slip zones—see discussion in Chapter 5 dealing with the creep of continuous fiber composites).

Microdamage mechanisms may also control matrix crack growth at the mesoscale. Under cyclic loading, the decrease in interfacial sliding stress can cause the extension of matrix cracks. Since the fibers carry load across a matrix crack, they reduce the stress intensity at the tips of the matrix cracks (see Fig. 6.18).[16,56,62] A matrix crack will be arrested when the stress intensity at the crack tip, K_{tip}, falls below the critical level for crack growth, K_c[63,64] (at the fibers, the crack can be arrested by deflection along the fiber–matrix interface). As mentioned above, interfacial wear occurs between the fibers and matrix during cyclic loading, leading to a progressive reduction in interfacial frictional shear stress. This reduction in frictional shear stress decreases the crack bridging force, leading to an increase in the stress intensity at a crack tip[65–67] and the further growth of matrix cracks. However, the growth is stable since the new crack area includes some newly debonded fibers which have higher interfacial sliding stress (and, therefore, higher bridging stress) since they have not yet been exposed to cyclic wear. Further cycling will cause interfacial wear at the newly debonded fibers, reducing their bridging capacity, and leading to further growth. Thus, a decrease in the interfacial sliding stress during repeated interfacial slip provides a mechanism for the "cyclic" growth of fiber-bridged matrix cracks. If the applied stress is sufficiently high to give sliding along the entire interface, this process can continue until the fibers are fully debonded and abraded, giving a saturated crack spacing, as found experimentally.[12]

6.4.1.3 Fiber Failures due to Interfacial Wear

As discussed earlier, fatigue failure will occur when the residual strength of the composite equals the maximum cyclic stress. It is recognized that the fatigue limit is typically well above the stress level at which matrix crack initiation occurs. For example, for the fatigue of SiC$_f$/CAS at 25 Hz, Holmes *et al.*[48] observed fatigue run-out at 200 MPa, which is approximately 60 MPa above the microcracking threshold of this composite system. The fatigue limit of 200 MPa is well below the stress level that would cause significant fiber failure during a monotonic tension test (Stage IV).[7] Thus, cyclic loading must lower the strength capacity of the fibers. *More specifically, either the (local) stress in the fiber must increase or the fiber strength must decrease during fatigue loading.* Both possibilities exist. Reynaud *et al.*[68] suggested that the decrease in interfacial frictional shear stress during fatigue leads to an increase in the average axial stress in the fibers, which increases the likelihood of fiber fracture. With a decreasing value of shear stress, τ, the sliding length increases, leading to an increase in the stresses in the portion of the fibers that undergo slip, since the stress transfer to the matrix is less efficient. This leads to a higher probability of fiber fracture and a decrease in composite strength. Rouby and Reynaud[46] proposed that fatigue failure occurs when the retained strength of the composite equals the applied fatigue stress. Because of its usefulness in

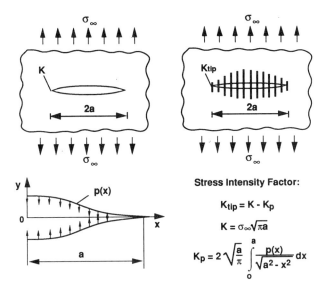

Fig. 6.18 Schematic representation showing how a reduction in the bridging force by interface wear leads to matrix crack growth. The fibers carry the applied load across the matrix cracks, reduing the crack opening displacement and the net stress intensity (K_{tip}) at the tip of matrix cracks. If the interfacial shear stress τ decreases during fatigue, then the bridging stress $p(x)$ decreases, leading to a reduction in K_p. This reduction increases K_{tip}, which can cause the further extension of a matrix crack.

understanding fatigue failure, the mathematical description of the model by Rouby and Reynaud[46] is given in the next section.

From the above discussions, we are now in a position to answer the questions posed at the beginning of this section. (1) Fatigue failure occurs only if the maximum stress level in the first cycle is such that matrix cracking, fiber/matrix debonding, and interfacial sliding take place. (2) Matrix cracks can extend as the interfacial shear stress decreases during cycling, which reduces the bridging stresses. Crack extension will continue until the matrix is fully cracked with regularly spaced cracks. (3) Continuing interfacial wear can further reduce the interfacial shear stress and cause further surface damage to the fibers. Both of these microstructural changes will reduce composite strength. (4) The critical damage state associated with fatigue failure occurs when the bridging fibers begin to fracture, reducing the residual strength of the composite to a level equal to the applied fatigue stress. Both the increase in fiber stress caused by the decreasing shear stress and the decreased inherent strength of the fibers due to direct surface damage are likely mechanisms for fiber fracture. It should be noted that the final damage process of fiber fracture can be rapid, since the fracture of even a small percentage of the bridging fibers will greatly increase the load borne by the remaining fibers (this process is expected to result in the rapid acceleration of fiber fractures within a relatively small number of cycles).

6.4.2 Model of Decreasing Strength during Repeated Loading

Rouby and Reynaud's mesoscale model[46] for the fatigue life of multiple matrix cracked composites is based on the experimental finding that the interfacial sliding shear stress, τ, decreases with the number of cycles, as observed experimentally by Holmes and Cho[12] for unidirectional SiC$_f$/CAS. The objective of the model is to find the maximum applied stress (the composite tensile strength, σ_u) that a matrix crack bridged by fibers can support. This requires taking into account the strength variability of the fibers and the accompanying fraction of broken fibers. From Rouby and Reynaud's model, the axial stress, σ_f^0, acting within intact fibers that bridge matrix cracks can be expressed as

$$\sigma_f^0 = \frac{\sigma}{v_f(1-q)} \qquad (16)$$

where σ is the applied stress, v_f is the fiber volume fraction, and q is the fraction of broken fibers. In the equation above, it has been assumed that the broken fibers do not transfer any load across matrix cracks. For the case of a large crack spacing, the axial stress acting in intact fibers decreases linearly over the sliding length, l_s, until the stress has decreased to a level that would exist in the absence of a matrix crack (Fig. 6.19). The probability of fiber failure is calculated using the same assumptions used by Thouless and Evans,[69] i.e., fiber failures that occur at a distance greater than the load transfer length are neglected. The load transfer length is given by

$$l_t = \frac{\sigma_f^0}{4\tau}d \qquad (17)$$

where τ is the interfacial shear stress and d is the fiber diameter. In the estimate for the failure probability of the fibers, the fiber stress is assumed to decrease linearly from the peak stress of σ_f^0 to zero over the load transfer length (Fig. 6.19):

$$\sigma_f(\sigma_f^0, z) = \sigma_f^0\left(1 - \frac{z}{l_t}\right) \qquad 0 \leq z \leq l_t \qquad (18)$$

The probability of fiber failure is described by Weibull statistics. The probability of failure within a gauge section of length L is given by

$$P_f[L, \sigma_f(z)] = 1 - \exp\left\{-\frac{1}{L_0}\int_0^L \left[\frac{\sigma_f(z)}{\sigma_0}\right]^m dz\right\} \qquad (19)$$

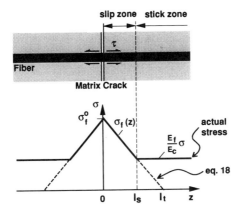

Fig. 6.19 The basis of the mesoscale model of Rouby and Reynaud. The axial stress, σ_f in the fiber decreases linearly over the slip length, from a peak value of σ_f^0 at the edge of the matrix crack, to a value of $\sigma(E_f/E_c)$ at the end of the slip-zone ($z = l_s$). It is assumed that the crack spacing $\bar{l}$ is larger than twice the load transfer length l_t. It is assumed that the fiber stress decreases linearly to zero over the load transfer length l_t. After Rouby and Reynaud.[46]

where σ_0 is the stress at which 63.2% of the fibers have failed in a gauge length of L_0 and m is the Weibull modulus, describing the variation in fiber strength. The tensile strength, σ_u, can be found by maximizing σ. It is more convenient to maximize σ with respect to σ_f^0 than with respect to q (the two approaches are equivalent since there is a one-to-one relationship between σ, σ_f^0 and q). Inserting the stress variation, $\sigma_f(\sigma_f^0, z)$ (Eqn. (18)), into Eqn. (19), integrating over z from 0 to l_t, multiplying by two (note that the stress distribution (Eqn. (18)) is only for the fiber to the right side of the matrix crack), and expressing l_t by σ_f^0 (Eqn. (17)) gives the probability of fracture

$$P_f = 1 - \exp\left[-\frac{\sigma_f^0}{\tau} \frac{d}{2L_0} \left(\frac{\sigma_f^0}{\sigma_0} \right)^m \frac{1}{m+1} \right] \tag{20}$$

In Eqn. (20), P_f corresponds to q, the fraction of broken fibers within a gauge-length of $2l_t$. Then q, expressed in terms of σ_f^0, can be inserted in Eqn. (16), such that the applied stress can be given as function of σ_f^0. By partial differentiation with respect to σ_f^0, the maximum value of σ, the tensile strength, σ_u, can be found:

$$\sigma_u = v_f \exp\left(-\frac{1}{m+1} \right) \left(\frac{2L_0 \sigma_0^m \tau}{d} \right)^{1/(m+1)} \tag{21}$$

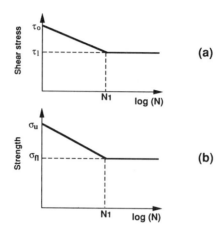

Fig. 6.20 Relationship between the decay in interfacial shear stress and the fatigue life of fiber-reinforced ceramics, as proposed by Rouby and Reynaud. (a) The interfacial shear stress τ is assumed to decrease during cyclic loading, reaching a steady-state value after N_1 cycles. (b) The composite strength decreases in parallel to the decrease in τ. An infinite fatigue life is predicted if the strength corresponding to τ_1 is above the maximum fatigue stress. After Rouby and Reynaud.[46]

When the maximum strength is reached, the fraction of broken fibers within the gauge section is given by

$$q_u = 1 - \exp\left(-\frac{1}{m+1}\right) \tag{22}$$

These equations provide insight into the strength as a function of τ. Clearly, as τ decreases, the strength of the composite will decrease. If the composite strength decreases below the maximum applied cyclic stress, then fatigue failure occurs. A first-order model for the change in τ was proposed by Rouby and Reynaud[46]

$$\tau(N) = \begin{pmatrix} \tau_0 N^{-\Omega} & N \leqslant N_1 \\ \tau_1 & N \geqslant N_1 \end{pmatrix} \tag{23}$$

where τ_0 is the initial value of the interfacial shear stress, N is the number of cycles, Ω is a constant describing the rate at which τ decreases, and τ_1 is the limiting value of τ that is reached after N_1 cycles. Inserting Eqn. (23) into Eqn. (21) gives a relationship between composite strength and number of fatigue cycles. This relationship is plotted in Fig. 6.20.

According to the Rouby–Reynaud model, fatigue failure will occur when the composite strength has decreased to the level of the maximum applied cyclic stress. The decrease in τ within the first N cycles will cause a progressive decrease in composite strength. If τ remains constant for $N > N_1$, the composite strength does not decrease further. According to the model (Eqn.

(23)), if fatigue failure occurs, it will occur in less than N_1 cycles, otherwise infinite fatigue life is predicted so long as the maximum fatigue stress does not exceed σ_{fl} (Fig. 6.20). The ratio between σ_{fl} and the monotonic tensile strength is given by[70]

$$\frac{\sigma_{fl}}{\sigma_u} = \left(\frac{\tau_1}{\tau_0}\right)^{1/(m+1)} \tag{24}$$

Note that, if the fatigue life is controlled by the decrease in the interfacial shear stress, as proposed by Rouby and Reynaud,[46] then the *rate of fatigue damage* is governed by the average *wear rate* along the fiber–matrix interface. The Rouby–Reynaud model will need to be verified using fatigue data from a variety of ceramic matrix composites. For instance, it appears that for SiC_f/CAS, τ decreases within a relatively short number of cycles (Figure 6.13) and remains constant at that level for the remaining fatigue life, yet fatigue failures are observed even after this constant level of shear is reached.[12] Since the shear stress remains constant, other mechanisms must act to promote fiber failure. For example, wear damage on the surfaces of the fibers could cause a decrease in the average fiber strength. It is apparent from Eqn. (21) that a decrease in σ_0 leads to a decrease in the composite strength. Thus, by expressing the decrease in fiber strength, σ_0, as a function of the number of fatigue cycles ($\sigma_0 = \sigma_0(N)$), Rouby and Reynaud's model could be readily extended to include the effect of fiber strength degradation on fatigue life.

The Rouby–Reynaud model neglects the load-carrying capacity of the broken fibers. This was included in Curtin's model for the ultimate tensile strength of unidirectional composites.[23] His model is based on the assumption of full sliding along the interface; an average fiber stress is considered in the analysis which neglects the change in fiber stress along the slip length (the model by Rouby and Reynaud considers this change in fiber stress). The shear stress, however, still goes into the equation for load transfer length, l_t. If a fiber fractures at a distance greater than l_t away from each side of a matrix crack at which final separation occurs (the failure locus), the portion of the fiber that bridges the crack will experience the same stress between the crack faces as that experienced by intact (unfractured) fibers. Therefore, in Curtin's model the gauge length is also set equal to $2l_t$. Then the strength is given by[23]

$$\sigma_u = v_f \frac{m+1}{m+2} \left(\frac{2}{m+2}\right)^{1/(m+1)} \left(\frac{2L_0 \sigma_0^m \tau}{d}\right)^{1/(m+1)} \tag{25}$$

At the maximum applied stress (i.e., the ultimate tensile strength, σ_u) the fraction of broken fibers within the gauge section can be expressed as

$$q_u = \frac{2}{m+2} \tag{26}$$

Interestingly, despite the significantly different assumptions used in the derivation of the models by Curtin, and by Rouby and Reynaud, their predictions of failure are similar in form. The only difference is in how final failure depends upon the Weibull modulus of the fibers. The primary reason for the similarity between the predictions is that the strength is calculated on the basis of a gauge length that depends on the load transfer length. Therefore, as τ decreases, the load transfer length increases, as given in Eqn. (17), and the composite strength decreases. Thus, the conclusions made from the Rouby–Reynaud model concerning the effects of decreasing τ and σ_0 on fatigue life will also hold for the case of full frictional slip (Curtin's model).

At the loading frequencies examined by Rouby and Reynaud (0.1 and 1 Hz), no effect of loading frequency on fatigue life was found for 2-D Nicalon SiC_f/SiC composites.[46] However, *conclusive* experimental evidence exists that the fatigue life of fiber-reinforced composites is strongly influenced by fatigue loading frequency;[48] fatigue life decreases significantly as loading frequency is increased (see later discussion). Using the insight gained from Rouby and Reynaud's model, it is interesting to speculate on the mechanisms responsible for this frequency dependence of fatigue life. If high frequency fatigue simply alters the rate and degree of interfacial wear, a model, such as that proposed by Rouby and Reynaud, can be used in its present form to predict fatigue life. However, it is likely that frequency-dependent phenomena, such as frictional heating, can alter the level of frictional shear stress through differential thermal expansion between the fiber and matrix. Future models will need to incorporate these changes.

The model proposed by Rouby and Reynaud[46] represents the first systematic approach for understanding how the microstructural damage governs the fatigue life of continuous fiber-reinforced ceramic matrix composites. This model will be used to explain various aspects of fatigue failure in the remaining portion of this chapter.

6.5 Influence of Mean Stress on Isothermal Fatigue Life

6.5.1 Experimental Findings

Experimental studies of the influence of stress ratio on elevated temperature fatigue life have been conducted by Suresh[71] for whisker-reinforced Al_2O_3 (see Chapter 7 by Suresh for a discussion of the fatigue behavior of whisker-reinforced ceramics). In this section, the influence of stress ratio on the fatigue life of continuous fiber-reinforced ceramic matrix composites is discussed.

Allen and Bowen[40] studied the fatigue behaior of unidirectional Nicalon SiC_f/CAS at room temperature by flexural testing at 10 Hz for R-ratios of 0.1 and 0.5. It was found that the specimens tested at $R = 0.1$ all failed above a certain stress level. For the specimens fatigued at $R = 0.5$ there was considerable scatter in the fatigue lifetime. While some specimens failed lifetimes similar to the specimens with $R = 0.1$, roughly half the specimens did

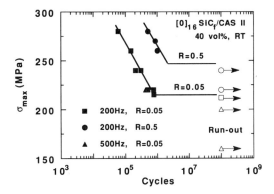

Fig. 6.21 Influence of stress ratio on the room temperature tension–tension fatigue life of unidirectional Nicalon SiC$_f$/CAS composites. After Sørensen and Holmes.[72]

not fail within 10^6 cycles. Also, the fatigue limit at $R = 0.5$ was higher than for $R = 0.1$. Building upon the results of Allen and Bowen,[40] Sørensen and Holmes[72] studied the influence of stress ratio on the tension–tension fatigue life of similar Nicalon SiC$_f$/CAS composites. The room temperature fatigue experiments were conducted at 200 Hz, using stress ratios of 0.05 and 0.5. Run-out was defined at 10^8 cycles. As shown in Fig. 6.21, the 10^8 cycle fatigue limit increased by over 40 MPa when the stress ratio was increased from 0.05 to 0.5. These results support the findings of Allen and Bowen.

Moschelle[73] conducted uniaxial fatigue tests at room temperature on $[45, 90, 0, -45]_s$ (quasi-isotropic) woven SiC$_f$/SiC composites with R-ratios of 0.1 and -1.0. The fatigue limit (10^6 cycles at 10 Hz) was found to be 105 MPa (for $R = 0.1$) and 90 MPa ($R = 1.0$), slightly above the monotonic proportional limit, $\sigma_{pl} = 50$–70 MPa. Comparing the number of cycles to failure of the failed specimens (the fatigue life), it was found that for identical maximum load, the specimens loaded with an R-ratio of -1.0 had a shorter fatigue life than the ones cycled with an R-ratio of 0.1. These results support the trend found above. As discussed in Section 6.6 dealing with elevated temperature fatigue behavior, the limited number of experiments conducted to date also show that elevated temperature fatigue life can increase at higher stress ratios.

6.5.2 Theoretical Considerations

From a theoretical point of view, fixing the maximum fatigue stress and changing the stress ratio influences the slip length of fibers. The slip length, l_s^{cyc}, can be expressed in terms of the fiber diameter, d, the stress range ($\Delta\sigma = \sigma_{max} - \sigma_{min}$), and the frictional shear stress, τ, as[52]

$$l_s^{cyc} = \frac{d}{8} \frac{(1 - v_f)}{v_f} \frac{E_m}{E_c} \frac{(\sigma - \sigma_{min})}{\tau} \tag{27}$$

It follows that, during cyclic loading, the sliding length does not depend on the *absolute* stress, but on the *difference* between the maximum and minimum applied stress, i.e., the stress range. Comparing two experiments conducted with the same maximum applied stress, but different minimum stresses (i.e. different R-ratio), the test with the highest R-ratio has the shorter cyclic slip length (assuming partial slip). In this case, a larger percentage of the interface has sticking friction all the time; no wear damage can occur there. Comparing two identical specimens, the specimen subjected to the highest R-ratio would experience the smallest decrease in τ (and σ_0) and would be expected to show the longer fatigue life. This is consistent with the experimental findings mentioned above.

If fatigue damage is controlled by interfacial wear, it is plausible that the damage is proportional to the area of the hysteresis loop (the work of the frictional stresses). The model developed by Cho *et al.*[52] indicates that the energy dissipation due to frictional sliding is proportional to $\Delta\sigma^3$ when partial slip takes place along the interface, and proportional to $\Delta\sigma$ when full slip takes place. It would be logical to assume that the interfacial wear damage also evolves with $\Delta\sigma$ rather than σ_{max}, as long as the maximum applied stress is below the stress level that causes fiber failure during the first cycle. This has not yet been studied systematically.

6.6 Influence of Temperature and Test Environment on Isothermal Fatigue Life

6.6.1 Experimental Studies of Fatigue at High Temperatures

Prewo *et al.*[30] conducted uniaxial tension–tension tests ($R = 0.1$) on cross-ply SiC_f/LAS III composites at 900°C. The experiments, conducted in both air and argon atmospheres, were performed at loading frequencies of 7–10 Hz. Specimens that were fatigued in air were found to have a significantly lower fatigue life than the room temperature fatigue life, with a fatigue limit near 86 MPa. The residual strength of specimens that survived 10^5 cycles decreased below 100 MPa. The pull-out length of fibers was slightly shorter than that found for monotonic tension tests. This suggests that oxidation damage of the carbon-rich interphase layer had taken place, forming strong bonding across the fiber–matrix interface (using transmission electron microscopy, Bischoff *et al.*[74] confirmed that the weak carbon-rich interphase layer that forms in SiC_f/LAS during processing disappears when exposed to oxygen at high temperatures, becoming replaced with strong SiO_2 bonding). In contrast, specimens that were tested at high temperature in the argon atmosphere had a much higher fatigue limit (100 MPa). The specimens that survived 10^5 cycles had a retained strength comparable to the strength measured in monotonic tension (over 200 MPa), and the fracture surface exhibited significant fiber pull-out. This suggests that if changes in the interphase layer can be avoided, either by using fiber coatings that are stable at high temperatures or by suitable protection of the composite by surface coatings, then the damage tolerant behavior and fatigue characteristics can be retained at high temperatures.

Prewo et al.[30] also conducted tensile rupture experiments using SiC_f/LAS III and found that environmental effects on material properties, such as lower strength and a decrease in fiber pull-out, were more pronounced at intermediate temperatures (900°C) than at high temperatures (1100°C). This suggests that the composite performance may, in fact, be better at high temperature (above the glass transition temperature), where the matrix deforms viscously, rather than at lower temperature, where matrix cracking allows direct access for environmental attack along the fiber–matrix interface. One approach to prevent interfacial damage is to coat the composite with a glassy layer that will deform viscously and will close matrix cracks. It is debatable, however, whether these surface coatings could be relied upon as the primary source of oxidation protection. Rather, surface coatings should be looked upon as providing secondary protection in systems that are inherently oxidation-resistant.

Allen and Bowen[40] performed flexural fatigue experiments using unidirectional Nicalon SiC_f/CAS composites at room temperature, 600°C, and 800°C. The experiments were performed in air at a loading frequency of 10 Hz. A strong temperature effect was found: the specimens tested at high temperature had a much shorter fatigue life than those fatigued at room temperature. This behavior is probably due to the oxidation damage of the interface and formation of strong interfacial bonding, such that the matrix cracks do not debond the fibers, but penetrate the fibers and cause brittle behavior.

The maximum-stress and R-ratio dependence of fatigue life for $SiC_f/$ HPSN at 1200°C are shown in Fig. 6.22. Two interesting trends are apparent: (1) stress ratio had no influence on fatigue life for maximum stresses below the scatter band for proportional limit stress (all specimens survived 5×10^6 cycles, or approximately 138 h of testing), and (2) above the scatter band, and for a given maximum stress level, fatigue life decreased as the stress ratio was lowered from 0.5 to 0.1. An increase in fatigue life as the stress ratio is increased is contrary to what would be expected if the fatigue life was controlled by creep damage. For a maximum fatigue stress of 270 MPa, the mean stress increases from 148.5 MPa at $R = 0.1$ to 202.5 MPa at $R = 0.5$, yet the fatigue life increased by over an order of magnitude for a 50 MPa increase in mean stress. This result can be explained in terms of the stress range dependence of the damage evolution. The stress range increases as the stress ratio is decreased. Interfacial wear, which may control the composite fatigue strength and lifetime, is likely influenced by the relative slip distance and, therefore, stress range. Since the composite has inherently high creep resistance, failure is controlled by crack propagation and fiber fracture. The

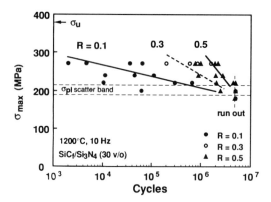

Fig. 6.22 The fatigue life diagram of SiC$_f$/HPSN for R-ratios of 0.1, 0.3, and 0.5 at 1200°C. Note that there is a clear connection between R-ratio and fatigue life: a higher R-ratio gives a longer fatigue life. However, for all loadings the fatigue limit is nearly identical, around 200 MPa, and corresponds roughly to the proportional limit. After Holmes.[49]

matrices in the early composites processed by dry powder lay-up were known to contain regions where the sintering oxides had not completely dispersed, as well as fiber-poor regions; these regions could provide paths for rapid crack propagation. It should be noted that in composite systems with lower creep resistance, or at higher temperatures, fatigue life typically decreases as the mean stress is increased. For example, unpublished work by the authors has shown that the 1200°C tensile fatigue life of $[0]_{16}$-Nicalon SiC$_f$/CAS-II composites, which have much lower creep resistance than HPSN-based composites, decreases by as much as an order of magnitude as the stress ratio is increased from 0.1 to 0.5.

A significant amount of strain ratcheting can occur during fatigue loading at stresses below the proportional limit stress. Figure 6.23a shows the strain ratcheting observed during the 1200°C tensile fatigue of hot-pressed SCS-6 SiC$_f$/HPSN at a stress level below the proportional limit stress. The ratcheting is attributed primarily to creep deformation, since the cyclic stress–strain curves showed no evidence of hysteresis, indicating an absence of significant matrix damage. Figure 6.23b shows similar stress–strain curves, but now the maximum stress is above σ_{pl}. Hysteresis and decreasing stiffness indicate progressive matrix cracking and interfacial sliding. Figure 6.24 compares the failure times observed for fatigue loading with those found for sustained creep loading (the comparison includes only those specimens fatigued at maximum fatigue stresses above the monotonic proportional limit stress, which, for this composite, corresponds to the onset of matrix cracking). To roughly simulate the initial matrix damage that would occur during intial loading of the fatigue specimens, the creep specimens were first loaded to 270 MPa, followed immediately by creep at stress levels of 150, 175, and 200 MPa (these stresses roughly correspond to the mean stress that would be present during fatigue at a

(a)

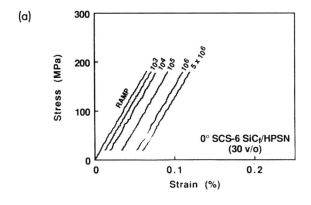

(b)

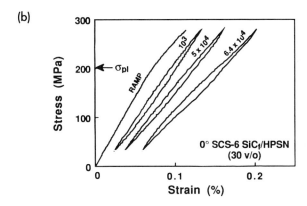

Fig. 6.23 Strain ratcheting during the tensile fatigue of 0° SCS-6 SiC$_f$/Si$_3$N$_4$ at 1200°C. (a) σ_{max} = 180 MPa, R = 0.1 (below the proportional limit stress, which is approximately 200 MPa). The strain ratcheting is primarily due to creep. Note that there is no hysteresis or change in modulus. (b) σ_{max} = 270 MPa (above σ_{pl}), R = 0.1. The stress–strain curves now display hysteresis (indicating frictional sliding). After Holmes.[49]

maximum stress of 270 MPa and stress ratios of 0.1, 0.3, and 0.5). It is clear that the failure times were shorter under cyclic loading. This decrease is most likely caused by interface wear which would increase the stress on the fibers.

Allen et al.[75] investigated the fatigue behavior of unidirectional Nicalon/ CAS composites at room temperature and 1000°C, in both air and vacuum environments. The fatigue experiments were conducted with single-edge-notched (SEN) specimens cycled at 10 Hz under a load ratio of 0.5. Crack growth was monitored by periodically obtaining surface replicas of matrix damage near the root of the machined notch. At room temperature there was clear evidence that cyclic fatigue damage occurred, with multiple matrix cracking and cyclic crack growth observed. In general, compared to sustained static loading at a stress equivalent to the maximum fatigue stress, the degree

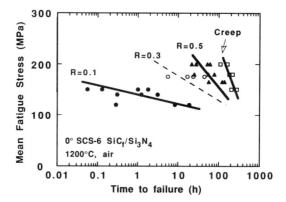

Fig. 6.24 Comparison of the failure times for fatigue loading and sustained loading at stresses approximately equal to the main fatigue stress. Although strain recovery during cyclic loading can reduce overall strain accumulation (see Chapter 5), it appears that this effect is minimal at moderate (10 Hz) loading frequencies (also note that interface wear during cycling may negate any beneficial effect of strain recovery). Cyclic loading reduces the lifetime, especially at low R-ratios. After Holmes.[49]

of matrix damage was observed to increase for cyclic loading. At 1000°C, there was little difference between the time-to-failure for statically and cyclically loaded specimens. In contrast to the room temperature behavior, the fatigue failures were governed by the growth of a dominant Mode I crack that branched ahead of the machined notch. A direct comparison between static and cyclic loading is difficult since the total time under the maximum fatigue stress would be significantly shorter than that for sustained loading. Since creep and cyclic strain recovery of the matrix will occur even at 1000°C, it is likely that creep deformation may overwhelm any fatigue damage mechanisms at 1000°C. Thus, it is possible that the lack of an apparent fatigue effect at elevated temperatures is a consequence of stress relaxation in the matrix, which reduces the driving force for additional matrix damage. Support for the role of creep comes from experiments conducted at 800°C, where creep deformation would be minimal. At this lower temperature, the life of cyclically loaded specimens is considerably shorter than for sustained loading.

6.6.2 Differences between Room Temperature and High Temperature Fatigue

There are similarities as well as important differences between fatigue behavior and damage mechanisms occurring at room temperature and at elevated temperatures. The most obvious difference is the occurrence of environmental attack by corrosive species such as oxygen or sodium salts. Oxygen penetration along matrix cracks or interconnected porosity networks can cause direct oxidation of carbon interfacial layers[73,76] or carbon fibers. The work by Allen and Bowen[40] indicates that the SiO_2 formation at the interface is of concern in glass and glass-ceramic matrix composites. Moreover, certain composites such as RBSN (reaction-bonded silicon nitride) can undergo active oxidation, which may become more pronounced at very low oxygen partial pressures where SiO, rather than a protective SiO_2 surface scale, forms.

It has also been shown that the initial processing-related matrix cracking present in C_f/SiC composites quickly stabilizes during cyclic loading and is not necessarily detrimental to fatigue life at room temperature.[37] At elevated temperatures, however, matrix cracking plays a much stronger role in composite failure, particularly in oxidizing atmospheres. For example, matrix cracking can directly expose the fibers and fiber–matrix interface to oxygen and other corrosive species such as sodium salts (detailed discussions of the oxidation and corrosion resistance of fiber-reinforced ceramics and of the degradation of mechanical properties caused by exposure to oxidizing and molten-salt environments can be found in Refs. 77–82). Corrosion by oxygen or other chemical species along a single crack, if bridged by a large number of fibers, could lead to localized attack of the fibers and interface and may serve as the initiation site for component failure. Moreover, when matrix cracking is present, the stress and local strain in the segment of a fiber that bridges a crack will be significantly higher than the stress in embedded portions. This increase in stress significantly increases the likelihood of fiber rupture—this is particularly true for high temperature loading, where creep of the highly stressed bridging fibers can occur.

Even in the absence of oxidation or creep, temperature may have an impact on the static and fatigue strength of fiber-reinforced ceramics. Referring again to the models by Curtin[23] and Rouby and Reynaud,[46] the strength is proportional to the parameter

$$\left(\frac{2L_0 \sigma_0^m \tau}{d}\right)^{1/(m+1)} \tag{28}$$

If the thermal expansion coefficient of the matrix is higher than the fibers (e.g., as in SiC_f/CAS) the matrix clamps the fiber such that the interface is in radial compression. With increasing temperature, τ will decrease, since the thermal mismatch, and thus the normal pressure across the interface, decreases. This will directly lower the fatigue limit.[70] Furthermore, σ_0 may be expected to decrease, at least at temperatures where the fibers are thermally unstable, if the fibers contain a glassy phase. If τ and σ_0 decrease with temperature, it follows that the composite strength will also decrease in static and cyclic tests. However, it is unclear if τ will decrease at the same rate in room and elevated temperature experiments.

6.7 Influence of Loading Frequency on Isothermal Fatigue Life

6.7.1 Experimental Studies

Very little information is available concerning the influence of loading frequency on the fatigue life of fiber-reinforced ceramics. The studies performed to date have examined the influence of loading frequency on the room temperature fatigue life of 2-D C_f/SiC and unidirectional Nicalon SiC_f/CAS composites. Because a much larger frequency range was studied (25–350 Hz), the discussion here will focus on results obtained by Holmes et al.[48] for Nicalon SiC_f/CAS composite. In their study, Holmes et al. subjected specimens to tension–tension fatigue at a stress ratio of $R \approx 0.05$; all experiments were conducted at room temperature in air. Figure 6.25 shows the fatigue life as a function of the loading frequency and maximum fatigue stress. Fatigue life was sharply reduced as the loading frequency was increased. For instance, at 220 MPa, the fatigue life of Nicalon SiC_f/CAS composites decreased from 5×10^6 cycles at 25 Hz to less than 10^5 cycles at 350 Hz.

Further insight into high frequency fatigue behavior was obtained from measurements of the specimen temperature rise that occurred during the fatigue experiments (this temperature rise is caused by frictional heating along the fiber–matrix interface). Figure 6.26 shows the frequency dependence of temperature rise for fixed stress limits of 220 MPa and 10 MPa. The temperature curves for specimens that exhibited run-out are bell-shaped, i.e., the temperature rises initially, reaches a maximum, and then decreases again. In the initial stages of fatigue, the energy dissipation due to frictional sliding slowly increases as the number of matrix cracks and debonded interfaces increases. As fatigue cycling continues, additional matrix cracking takes place, and the amount of interfacial sliding increases, which causes a further increase in temperature. With additional interface wear, however, the energy dissipation and temperature rise reaches a peak (within the full slip regime). For the specimens tested at higher frequencies, the temperature rise is much steeper, and the temperature continues to rise until failure. This indicates that for these specimens the damage evolution does not stabilize, but increases progressively. In fact, for high loading frequencies or applied stresses, the temperature rise appears to be self-sustaining, i.e., the high interfacial temperatures can lead to rapid interface degradation or further debonding (by differential thermal expansion), which, in turn, leads to an additional temperature increment.

It is important to note that all of the fatigue experiments in Fig. 6.25 were conducted with a maximum stress below the proportional limit on the monotonic stress–strain curve (which was around 285 MPa for this Nicalon SiC_f/CAS II composite which had a fiber volume fraction of 40%). Again, this shows that *the proportional limit cannot be used as an indicator of the fatigue limit*. Loading frequency has also been found to decrease the fatigue life of woven carbon fiber/SiC matrix composites.[37] For this composite, the fatigue

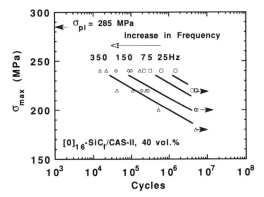

Fig. 6.25 Influence of loading frequency and maximum fatigue stress on the tension–tension fatigue life of unidirectional SiC$_f$/CAS-II. Note that the fatigue life decreases significantly as loading frequency is increased. After Holmes *et al.*[48]

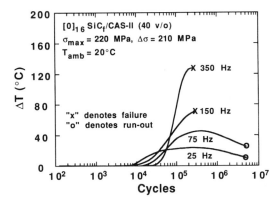

Fig. 6.26 Influence of loading frequency on the temperature rise of 0° Nicalon SiC$_f$/CAS-II composites fatigue between fixed stress limits of 220 MPa and 10 MPa. The specimens fatigued at low frequencies exhibit a bell shaped temperature rise curve; at 150 Hz and 350 Hz, the temperature continued to increase until specimen failure. After Holmes *et al.*[48]

limit (at 10^6 cycles) decreased from 325 MPa at 10 Hz to 300 MPa at 50 Hz. The modulus reduction of the samples fatigued at 50 Hz was also greater than that observed at 10 Hz.

6.7.2 Plausible Mechanisms for the Frequency Dependence of Fatigue Life

It is interesting to speculate about the mechanisms responsible for the frequency dependence of fatigue life in ceramic composites. A number of mechanisms may be responsible:

(1) Let us assume that the fatigue life is governed by an interfacial wear

mechanism. In general, the wear rate between two nonlubricated materials increases with increasing interfacial pressure and velocity. Although the carbon layer could act as an initial lubricating layer, this effect is expected to be minimal since the carbon layer would be rapidly destroyed by abrasion. A change in loading frequency would change the local velocity at the interface (note that the local velocity at the interface depends on external parameters such as the stress range and loading frequency, as well as internal parameters such as the interfacial sliding stress and the relative displacement between fiber and matrix during a cycle).

(2) For composites where the radial thermal expansion of the matrix exceeds that of the fibers (e.g., Nicalon SiC$_f$/CAS composites), the compressive radial stress at the interface decreases as the bulk temperature increases, leading to a reduction in the interfacial pressure across the interface, and reducing the interfacial shear stress. It follows from the models of composite strength[23,46] that this would reduce the composite strength, i.e., lower the fatigue limit.

(3) Differential thermal strains between the fibers and matrix, associated with local temperature increases in the vicinity of the interfaces, can provide a further driving force for the propagation of interfacial debonding. This additional debonding would increase the extent of interfacial wear, which increases the average fiber stress by reducing the degree of load transfer.

(4) The decrease in fatigue life could be due to thermochemical effects. Because of the low thermal conductivity of the CAS matrix and other glass or glass-ceramic matrices, the temperature in the vicinity of the fiber–matrix interface will be substantially higher than the bulk temperature rise. A high interfacial temperature could increase the chemical diffusion rates near the interface or cause oxidation of the carbon-rich interphase layer.

Further research must be conducted to clarify these effects.

6.8 Thermal Fatigue

6.8.1 Interaction of Thermal Fatigue and Environment Effects

Zawada and Wetherhold[83] studied thermal fatigue of a SiC$_f$/1723 glass matrix composite by subjecting specimens to controlled heating–cooling cycles between temperatures of 250°C and 700 or 800°C. The effect of static stress on thermal fatigue life was determined by subjecting some of the specimens to static stresses of 0, 28, and 138 MPa. The effect of thermal fatigue was characterized by measurement of retained flexural strength and SEM fractography. The effect of the static load was small (Fig. 6.27). The results indicated

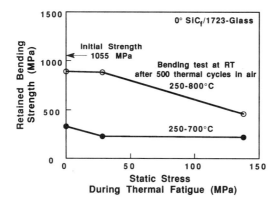

Fig. 6.27 Retained flexural strength of thermally cycled unidirectional SiC$_f$/1723 specimens as a function of temperature range and the magnitude of the static stress applied during thermal cycling. Note that the specimens cycled to 800°C exhibited a lower strength reduction than those cycled to 700°C. After Zawada and Wetherhold.[83]

that, although there was limited change of modulus and dimensional changes after thermal fatigue, the specimens subjected to thermal fatigue at 700°C showed appreciable embrittlement and degradation of flexural strength compared to those thermally fatigued to 800°C, see Fig. 6.27. Microstructural observation showed that a SiO$_2$ oxidation layer had formed at the surface of the 800°C treated specimens. This layer provided a matrix crack healing effect and reduced the sensitivity to brittle fracture. This observation confirms earlier observations of Prewo *et al.*[30] on tensile rupture of SiC$_f$/LAS-III composites. In that work it was found that the largest effect of environment occurred at 900°C (lower strength and almost no pull-out) rather than at 1100°C (higher retained strength and only limited embrittlement at the outer surfaces of the specimens).

St. Hilaire and Ertürk[84] performed thermal cycling between 500 and 1350°C at fixed loads (110, 125, and 168 MPa) on cross-ply laminates of SiC$_f$/HPSN. An aim of the experiments was to compare the thermal fatigue life of SCS-6 fibers and SCS-9 fibers. The samples were heated rapidly (within 25 seconds), the maximum temperature was held for 35 seconds, and cooling was within 60 seconds, such that one complete cycle took 2 minutes. The composites with SCS-6 fibers all failed within 10–40 cycles for an applied stress level below 110 MPa, while the composites with SCS-9 fibers all survived 1000 cycles for applied loads below 125 MPa. For specimens thermal-cycled at higher applied stresses, fatigue failures occurred. These failures were attributed to creep of the fibers. For both types of fibers, the fracture surfaces exhibited only minor pull-out, indicating that the interfacial bonding was strong or the frictional shear stress was high.

6.8.2 Models of Thermal Cycling

Cox[85] presented an analytical model describing the complicated behavior of a fiber within a matrix with different thermal and elastic properties. The geometry that was analyzed consisted of a block containing a free surface that had been cut perpendicular to the fiber direction. Load transfer was modeled by an interfacial shear stress that varied with temperature. For an interfacial shear stress that varies linearly with temperature, Cox's model revealed that the fiber sliding process is different during heating and cooling; the strain after a thermal cycle does not return to the initial strain, i.e., strain ratcheting occurs. The origin of this ratcheting can be explained as follows. Consider a situation where $\alpha_m > \alpha_f$, such that the matrix clamps the fiber with a pressure that is proportional to the temperature difference from the stress-free state (processing temperature). If Coulomb friction governs the stress transfer from the fiber to the matrix, then, as long as the temperature is below the processing temperature, the interfacial shear stress will decrease with increasing temperature. Also, remote from the free surface (i.e., the cut), a residual tension stress exists in the matrix, and a compressive stress exists in the fiber. After the free surface has been cut, the fiber end slides out of the matrix due to the relaxation of the residual stresses. This relaxation occurs over a sliding length given by an equation similar to Eqn. (27). Inside the bulk of the composite, the stress state is unaffected by the free surface, i.e., the stresses are equal to the residual stresses. During heat-up, the matrix expands more than the fibers, causing the fiber end (at the free surface) to slide into the matrix. However, during the temperature rise τ decreases, so the initial sliding length also becomes longer. During cooling, a new forward slip zone starts at the interface (from the free surface) as the matrix now contracts more than the fiber. At the remaining interface, however, no sliding takes place because now τ also increases, so sticking friction will exist. The model predicts that the ratcheting will saturate, and the hysteresis will approach an equilibrium within a number of cycles. The model was used by Cox et al.[86] to interpret data from thermal cycling tests on SiC fiber-reinforced intermetallics. The ratcheting behavior was indeed found experimentally, and values of the interfacial friction coefficient could be calculated.

Using the finite element method, Sørensen et al.[55] simulated thermal cycling of a fully cracked and debonded SiC$_f$/CAS composite. The interface shear stress was modeled by Coulomb friction. The model gave similar results to Cox's model; during cooling, interfacial sliding was found to occur along a small portion of the interface near the matrix crack. The results from these models suggest that thermal cycling, even in the absence of the thermochemical damage observed in experimental work,[83] could lead to interfacial wear damage in the composite, which could affect thermal fatigue life.

6.9 Thermomechanical Fatigue

Only a limited number of studies have addressed the combined thermal–mechanical fatigue loading of fiber-reinforced ceramics. Although precise details concerning the manner in which microstructural damage accumulates during elevated temperature fatigue is still lacking, these early studies have provided valuable insight regarding the potential for using fiber-reinforced ceramics in fatigue applications, and point the way for future investigations.

6.9.1 Thermomechanical Testing

Thermomechanical fatigue (TMF) testing involves simultaneously changing the load and temperature of a specimen (Fig. 6.28). In an in-phase TMF test the temperature and load are increased (and decreased) in parallel. Out-of-phase TMF testing involves changing the temperature and applied load in opposing directions, e.g., the temperature is increased as the load is decreased.

In the first study of the TMF behavior of fiber-reinforced ceramics, Butkus *et al.*[87] investigated the isothermal and in-phase thermomechanical fatigue behavior of unidirectional Nicalon SiC_f/CAS-II composites. In the TMF experiments, specimens were cycled between temperature limits of 500 and 1100°C. The maximum applied stress was below the proportional limit measured at 1100°C. After completion of the TMF and isothermal experiments, the specimens were immediately loaded in tension to failure at 1100°C to ascertain the residual strength and modulus of the specimens. Figure 6.29a compares the strength found after fatigue loading with the initial monotonic strength measured using virgin specimens. Considering the scatter typically encountered in the monotonic tensile testing of Nicalon/CAS-II composites, the retained strength does not appear to have been significantly changed. However, as shown in Fig. 6.29b, the tensile modulus of the composite did suffer a moderate reduction (between 15 and 20%) after isothermal and in-phase TMF testing.

Figure 6.30 shows the average mechanical strain $(\varepsilon_{max} - \varepsilon_{min})/2$ for the four loading histories that were examined. Compared to isothermal fatigue, the in-phase TMF loading showed significantly higher strain rates and overall strain accumulation, indicating a greater rate of damage accumulation.

Worthem and Ellis[88] recently conducted a more detailed investigation of the effect of in-phase and out-of-phase TMF loading on the fatigue life of unidirectional Nicalon SiC_f/CAS-II composites in air and argon atmospheres. The TMF experiments were conducted between temperature limits of 600 and 1100°C, with a 6 minute period; the various TMF loading histories with $R = 0.0$. Additional isothermal fatigue and creep experiments were performed for comparison with the TMF results; the isothermal fatigue experiments were conducted with a 5 minute load–unload period. The in-phase strain accumula-

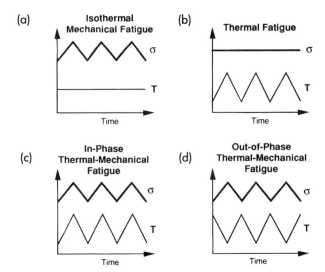

Fig. 6.28 Overview over TMF combinations. (a) Isothermal cycling, (b) thermal cycling, (c) in-phase thermomechanical cycling, and (d) out-of-phase thermomechanical cycling.

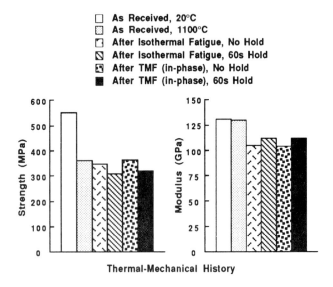

Fig. 6.29 Comparison of post-fatigue modulus and strength of Nicalon/CAS-II specimens with that found in virgin specimens. (a) The retained strength was not significantly degraded by isothermal or TMF loading. (b) The modulus showed a consistent reduction of approximately 15–20% after isothermal fatigue and TMF testing. After Butkus et al.[87]

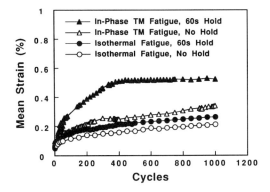

Fig. 6.30 Average mechanical strain for the four loading histories used by Butkus *et al.* in the TMF testing of $[0]_{16}$-Nicalon/CAS-II composites (40 vol.% fibers). Compared to isothermal fatigue loading, the specimen subjected to TMF loading exhibited a significantly higher strain rate and total strain accumulation. After Butkus *et al.*[87]

tion was higher than that found for out-of-phase loading. Figure 6.31 shows the stress and loading history dependence of fatigue life for the TMF experiments conducted by Worthem and Ellis,[88] and Butkus *et al.*[87] (note that the TMF experiments conducted by Butkus *et al.*[87] were not run to failure, but support the trends found by Worthem and Ellis[88]). Several important conclusions can be drawn from Fig. 6.31 regarding the TMF behavior of Nicalon SiC$_f$/CAS-II composites. (1) For both in-phase and out-of-phase loading, the fatigue life in argon was an order of magnitude higher than that observed when similar tests were performed in air. This suggests that the fatigue life in air is governed by oxidation damage. Nicalon SiC$_f$/CAS composites are particularly prone to oxidation damage because of the carbon-rich fiber–matrix interface that forms *in situ* during processing. (2) In air, the in-phase TMF life is shorter than isothermal fatigue life and creep life. (3) The shortest lives were obtained for out-of-phase TMF loading in air. (4) The threshold stress for TMF failures in air was identified as 65 MPa, roughly corresponding to the stress level below which no matrix cracks initiate during the monotonic tensile loading of Nicalon/CAS composites.[44]

6.9.2 *Simulation of Thermomechanical Cycling*

Because thermal expansion mismatch between the fibers and matrix can change the internal stress experienced by the constituents, fatigue damage under TMF loading can be more severe than that observed with isothermal fatigue or thermal fatigue. For example, suppose that the thermal expansion coefficient of the matrix exceeds that of the fibers (as in Nicalon SiC$_f$/CAS composites). In this case, after processing at high temperature, the matrix would be in axial tension (in the fiber direction), and the fiber would be in residual compression. Ignoring stress redistribution by creep, the residual

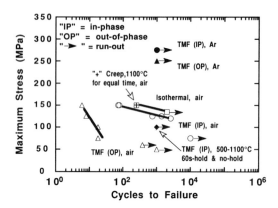

Fig. 6.31 Influence of TMF loading and temperature history on the fatigue life of Nicalon/CAS composites. "IP" denotes in-phase and "OP" denotes out-of-phase. The failure time for the creep experiment is plotted in terms of an equivalent number of isothermal fatigue cycles. Data from Worthem and Ellis[88] and Butkus et al.[87]

stresses would be highest at the lower test temperature. Any external load would cause additional stresses in the composite. It follows that the stress combination that would give the highest tensile stress in the matrix in the axial direction is the situation where the maximum applied load coincides with the lowest temperature (i.e., out-of-phase TMF). Note, however, that for composites where the thermal expansion coefficient of the matrix is less than that of the fibers (e.g., as in Nicalon SiC_f/LAS composites), the matrix will be in residual compression in the fiber direction. For such a composite, in-phase thermomechanical cycling is the load–temperature combination that gives the highest stresses in the matrix.

The trends in TMF life do not correspond to the trends in strain accumulation. For SiC_f/CAS composites, in-phase loading should give the highest strain accumulation, as found by Worthem and Ellis.[88] For out-of-phase loading, the applied load is low when the temperature is high, resulting in far less strain accumulation. The load combination giving the highest axial stress in the matrix would be the out-of-phase load case, at lowest temperature and highest applied stress. This is the point where matrix cracking could occur. Furthermore, at the lower temperature the matrix may be more brittle, i.e., it may be more sensitive to cracks. Once matrix cracks are formed (although at low temperature), oxygen can penetrate the interior of the composite and damage the interphase layer at high temperature.

Sørensen[89] simulated thermomechanical cycling of SiC_f/CAS by finite element methods. In his model, the maximum temperature was below 800°C (because of the low temperature, creep deformation was neglected). For a Coulomb friction law, it was found that out-of-phase TMF loading gave the largest hysteresis loop (indicative of frictional sliding) and ratcheting. In-phase

TMF loading exhibited even less hysteresis than isothermal cycling. The reason for this behavior is that sliding is not only induced by the strain change which accompanies a change in applied stress, but also by temperature changes. An increase in the load increases the axial strain in the fiber more than in the matrix in the regions where interfacial sliding takes place. If α_m is higher than α_f, a temperature rise will increase the strain (thermal expansion) in the matrix more than the strain in the fiber. For in-phase cycling, these two strain changes tend to balance each other, whereas during out-of-phase cycling the strain changes are additive, causing enhanced hysteresis and strain ratcheting. The trends from the model concerning ratcheting do not coincide with the experimental findings, indicating that creep was the dominant deformation mechanism in the experiments. Summarizing, while the strain accumulation during TMF is primarily caused by creep, the fatigue life is controlled by matrix cracking, leading to high temperature damage of the interphase layer. As mentioned above, the severity of in-phase or out-of-phase TMF loading will, in general, depend upon the elastic and inelastic behavior of the constituents, as well as the temperature dependence of mismatch in elastic–inelastic behavior; as such, it is too early to speculate on the impact of TMF loading on the life of other composite systems. In fact, the response to in-phase and out-of-phase TMF loading may be different for composites with different coefficients of thermal expansion, such as SiC_f/LAS and SiC_f/CAS composites. For engineers who are accustomed to designing with more forgiving metallic materials, it is important to keep in mind that small differences in thermoelastic properties of the constituents in ceramic composites could cause undesirable matrix cracking or fracture during the TMF loading of brittle matrix composites. As shown by Worthem and Ellis[88] (Fig. 6.31), for inert atmosphere operation (or presumably with suitable fiber or composite coatings), both in-phase and out-of-phase TMF life will be dramatically improved to a level that would be above the design stress of many structural components (e.g., in argon, specimens survived 1000 TMF cycles during in-phase TMF at a peak stress of 275 MPa and during out-of-phase TMF at a peak stress of 250 MPa).

It should be appreciated that gas-turbine components will be subjected to both in-phase and out-of-phase TMF loading histories in oxidizing environments. Even if protective coatings are used, the poor TMF behavior of ceramic matrix composites is of concern, since it is possible that surface coatings could be degraded by impact, cracking, or abrasion. One way to avoid TMF failure would be to ensure that the design stress is kept below the threshold stress for TMF or creep failure. For this reason, it is important that additional TMF studies are undertaken to further identify failure envelopes for various combinations of temperature and loading history. These studies must be extended to much longer times to ensure that the threshold stress is not influenced by total accumulated fatigue cycles or time-dependent microstructural changes. When designing future TMF studies it is important to take into consideration the temperature and loading transients experienced by a component. For example, in current-generation gas turbines, first stage airfoils can be subjected to temperature transients from 500°C (idle) to 1080°C and higher (peak take-off) in as little as 4–8 s. Cooling transients that are equally severe can also be experienced. Because of the low thermal conductivity of most fiber-reinforced ceramics, these rapid temperature transients will cause nonuniform temperatures and thermal stresses in components, which could significantly increase damage, leading to further reductions in TMF life.

6.10 Summary Comments

6.10.1 Current Understanding of Fatigue Damage Mechanisms and Fatigue Life

6.10.1.1 Mechanisms of Fatigue Failure

A better understanding of the mechanisms of fatigue damage in fiber-reinforced ceramics is emerging. As illustrated in Fig. 6.17, and assuming an initially crack-free matrix, the first mode of damage involves the initiation of matrix cracks; typically these cracks initiate in matrix-rich regions of the microstructure. Crack initiation is followed by fiber/matrix debonding and frictional sliding, forming fiber-bridged matrix cracks. The onset of matrix cracking and interfacial debonding initiates a process of interfacial wear and internal (frictional) heating. During repeated forward and reverse slip, the interfacial shear stress decreases by wear, reducing the crack shielding effect of the bridging fibers at the tip of matrix cracks. This decrease in crack tip shielding can provide a driving force for the further growth of matrix cracks, and provide a mechanism for the apparent "cyclic" growth of cracks. Depending upon the maximum fatigue stress, matrix cracking can continue until a saturated crack density is reached (at which point the fibers are fully debonded). In addition to microstructural features such as fiber volume fraction and fiber lay-up, the final crack density will depend upon the applied fatigue stress and the stress range, which influences the rate of interface wear. Even after matrix cracking has reached a stable level, the composite modulus can continue to decrease slightly, since additional interfacial wear may still occur—which decreases load transfer between the fibers and matrix. Beyond this damage state, several mechanisms are thought to operate and cause fatigue failure. Interfacial slip continues, reducing the interfacial sliding shear stress; the interface debris produced by the cyclic slip process could also cause abrasive wear damage to the fibers (in particular for "soft" fibers such as carbon). A decrease in τ or fiber strength will decrease the retained composite strength. Fatigue failure occurs when the retained strength decreases to the maximum fatigue stress. Although the fatigue life of a composite may be many millions of cycles, the decrease in composite strength that precipitates failure appears to occur in a relatively small number of cycles immediately prior to the observed failure (this has been verified by both modulus measurements and by measurement of residual strength of fatigue samples).[90] The strength decay likely occurs by the fracture of a few fibers, which, in turn, increases the net stress on the remaining fibers that have also been weakened by abrasive wear. At elevated temperatures, the overall damage process of matrix cracking and interface wear is the same. However, time-dependent creep rupture of the bridging fibers can act in parallel with other modes of fiber damage to cause the strength reduction that leads to fatigue failure.

At temperatures sufficient to cause the time-dependent deformation of the fibers or matrix, transient stress redistribution between the constituents occurs, as discussed in Chapter 5 on creep. If no interfacial oxidation damage takes place, and the amount of creep is negligible, it is likely that the fundamental mechanisms of fatigue damage will remain unchanged (temperature can, however, influence the rate of wear along the interface). However, if oxidation takes place, fatigue life will most likely be reduced, since the ability of the fibers to debond and bridge the initial matrix crack decreases. The fibers then break at the matrix crack tip (if no debonding takes place, the matrix cracks cut through the fibers) or in the matrix crack wake due to overstraining during matrix crack opening. This occurs in connection with matrix cracking, i.e., within a low number of cycles. The consequence of this is that matrix cracking becomes catastrophic, causing premature composite failure. The work to date on elevated temperature fatigue clearly shows that the most important problem is to develop interphase materials that are thermally stable at high temperatures. This is an essential issue in the design of ceramic matrix composites, and a challenge for interfacial chemists.

6.10.1.2 Fatigue Limit

It is worth recalling that the majority of fatigue experiments conducted to date have used a fatigue limit defined at 10^6 or fewer cycles. This has caused some confusion in the literature and has led to speculation that the monotonic proportional limit can be used to predict the fatigue limit. As described in this chapter, this assumption is generally incorrect; the correlation of the fatigue limit and the proportional limit occurs only for a few composite systems, and, even for those, the correlation only holds for low loading frequencies.[48] There is no fundamental reason why the mechanism that is responsible for fatigue damage should depend on the proportional limit. As shown by Sørensen and Holmes,[91] a true fatigue limit, where microstructural changes no longer occur, probably does not exist for fiber-reinforced ceramics fatigued at stresses above the microcracking threshold of the matrix (this is primarily a consequence of the cyclic wear along interfaces which, in principle, can continue indefinitely).

6.10.1.3 Influence of Loading History on Fatigue Life

The fatigue life of fiber-reinforced ceramics depends upon a number of external parameters, in particular the *R*-ratio, loading frequency, temperature, and environment. There is now conclusive experimental evidence that room temperature fatigue life increases as the stress ratio is increased.[49,72] This increase can be explained in terms of the interfacial wear damage, which depends upon the relative displacement between the fiber and matrix. As the load range decreases (the R-ratio increases), the amount of frictional sliding decreases and, therefore, the wear damage decreases. Fatigue life is also strongly influenced by loading frequency.[48] At high loading frequencies, frictional heating can cause an increase in the interfacial wear rate and damage. In parallel, differential thermal expansion between the fibers and matrix, associated with the temperature rise, can also directly decrease the normal pressure along the interface, increasing the stress on debonded failures and the likelihood of fiber fracture. It is clear that the degradation in fatigue life at high loading frequencies presents a serious problem for the use of ceramic matrix composites in structural applications. It must be appreciated that for practical design stresses, microcracking which is associated with frictional heating is inevitable in most fiber-reinforced ceramics (as mentioned in Section 6.2.1, at room temperature, microcracking in Nicalon SiC_f/CAS composites occurs at a stress of 60–120 MPa).

6.10.2 *Microstructural Design for Fatigue Loading*

Matrix microcracking and interfacial debonding absorb energy and, if controlled, are very desirable from a toughness and damage tolerance standpoint. However, because of the high stresses on bridging fibers, the presence of matrix cracks can negatively impact the fatigue life of fiber-reinforced ceramics. This highlights a key dilemma that exists in the micro-structural design of fiber-reinforced composites. Namely, the microstructural parameters that provide high monotonic toughness, such as low interfacial shear strength, matrix microcracking, and crack bridging by fibers, are generally undesirable if optimal fatigue resistance is desired.

As a pessimistic but conservative approach to design against fatigue, the maximum fatigue stress could be kept below the matrix cracking stress. Since the microcracking threshold can be significantly below the proportional limit stress or composite strength, this design approach would prevent the use of most current-generation CMCs. Approaches for increasing the microcracking threshold must be developed. These may involve particulate reinforcement of the matrix (hybrid composites) or changes in processing conditions. To increase the microcracking threshold it would be desirable to choose a fiber/matrix combination where the fibers have a larger thermal expansion coefficient than the matrix, such that the matrix is in residual compression in the axial direction. Such an approach is equivalent to the prestressing of concrete. Thermal mismatch between the fibers and matrix may have an important influence on fatigue life, because it can lower or raise the matrix crack initiation stress. At elevated temperatures the problem is far more complex, since the matrix stress changes by stress redistribution between the creeping fibers and matrix. Since the fibers usually have a higher modulus than the matrix, the microcracking stress of the matrix can also be raised by increasing the fiber volume fraction.[92] However, because matrix crack initia-

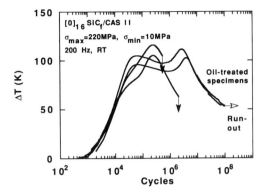

Fig. 6.32 Influence of interface condition on the frictional heating and fatigue life of unidirectional Nicalon SiC$_f$/CAS-II composites. To achieve well-defined matrix cracks, specimens were first statically loaded in air for 24 h at 220 MPa. Two of the specimens were subsequently immersed in low viscosity petroleum-based oil for 24 h at 90°C; the matrix cracks allowed oil access to the fiber–matrix interface. As the heating curves show, failure of the untreated (dry) specimens occurred at around 10^6 cycles; the oil-treated specimens survived 10^8 fatigue cycles. The second peak in the temperature rise curves of the oil-treated specimens is thought to occur when the effectiveness of the lubricating layer diminishes (near the second peak an increase in interfacial shear stress and composite modulus was measured). From Sørensen and Holmes.[58]

tion generally takes place in matrix rich regions,[6–9] it is important to ensure a uniform fiber distribution. Experimental results indicate, for instance, that large diameter SCS-6 fibers are usually much more uniformly distributed than smaller Nicalon fibers, and it is indeed found that the stress range where matrix cracking takes place is narrower for SCS-6 SiC$_f$/HPSN compared to Nicalon SiC$_f$/CAS-II composites.

For most engineering applications, the design stress will probably be above the microcracking threshold of the matrix. Thus, it is important to ensure that the stress on the bridging fibers is low enough to avoid fiber rupture, in particular at elevated temperatures where creep rupture can occur. To ensure low stresses on the bridging fibers, the fiber volume fraction should be as high as possible. This also lowers the frictional sliding distance, which can reduce frictional heating.

The development of thermally stable interphase materials, and further understanding of the underlying fatigue mechanisms, may lead to acceptance of matrix cracks and a finite fatigue life. Then, a different microstructural optimization route may emerge. If the problem of thermodynamically unstable interfaces can be successfully solved, then a significant amount of matrix cracking may be allowable. Clearly, this calls for a better understanding of fundamental fatigue damage mechanisms. The models of composite strength by Curtin[23] and Rouby and Reynaud[46] indicate that a reduction in composite strength during repeated loading (fatigue) can be caused by a reduction in τ or a decrease in σ_0. Obviously, if changes in these two parameters can be prevented, the fatigue limit can be raised. If fatigue life is controlled by interface wear, as proposed by Rouby and Reynaud,[46] then microstructural design boils down to controlling the rate of interfacial wear (to preserve τ) and controlling fiber damage caused by abrasion or by creep damage. The importance of controlling interface wear on fatigue life has been demonstrated in a series of simple experiments performed on microcracked Nicalon SiC$_f$/CAS composites that had been immersed in oil in an attempt to lubricate the interfaces—thereby decreasing the rate of interface wear.[58] As shown in Fig. 6.32, composites with lubricated interfaces exhibited a dramatic increase in fatigue life compared to untreated specimens. These results place emphasis on controlling the wear behavior of the interphase layers. The normal stress across the interface is also of importance, since it controls the magnitude of the interfacial shear stress and perhaps the interfacial wear rate. This can be controlled by suitable choice of thermal mismatch. Concerning the decrease in σ_0, it is advisable to use a fiber/coating/matrix combination where the wear resistance of the fiber is higher than that of the matrix and the interphase. Using higher strength fibers would increase the fatigue limit; strength alone, however, is only part of the picture—for elevated temperature applications, it is more important that the fibers can withstand the high creep stresses they will encounter if matrix cracking occurs. The morphology of the fibers is also important. For instance, it can be expected that fatigue life may be enhanced if fiber roughness is reduced. It should be kept in mind, however, that fibers such as the Nicalon (SiC) are not thermally stable during processing, and the post-processing fiber roughness may be considerably different from the roughness in the virgin state.

6.10.2.1 Microstructural Optimization

The current understanding of fatigue damage mechanisms suggests the following microstructural optimization.

(1) The properties of the interphase must be controlled to prevent changes in interface chemistry at high temperatures; thermodynamically stable interfaces are required.

(2) Fiber roughness should be as low as possible to minimize wear damage of the fibers and changes in τ. The interphse material should not be easily abraded by cyclic slip (this would reduce τ, and thereby the composite strength). To avoid damage to the fibers, it is preferable that interfacial slip take place within the interphase layer or at the matrix–coating interface, rather than at the fiber–coating interface. Perhaps a viscous or lubricating interphase could be used to minimize interface damage during the initial stages of fatigue.

(3) The modulus of the fibers and the fiber volume fraction should be as high as possible, raising σ_{mc} and lowering the degree of frictional heating.

(4) The fiber radius should be large, giving better control over the fiber distribution, which can increase σ_{mc}. This would also reduce the frictional heating if full slip conditions are reached (this would typically be the case for very low values of τ).

(5) For elevated temperatures, one should utilize fibers with high inherent creep resistance. This lowers the likelihood of fiber rupture by creep should matrix cracking occur. Alternatively, one could lower the fiber stress by increasing the fiber volume fraction.

References

1. K. M. Prewo and J. J. Brennan, "High Strength Silicon Carbide Fiber Reinforced Glass Matrix Composites," *J. Mater. Sci.*, **15**, 463–468 (1980).
2. J. J. Brennan and K. M. Prewo, "Silicon Carbide Reinforced Glass Ceramic Matrix Composites Exhibiting High Strength and Toughness," *J. Mater. Sci.*, **17**, 2371–2383 (1982).
3. K. M. Prewo and J. J. Brennan, "Silicon Carbide Yarn Reinforced Glass Matrix Composites," *J. Mater. Sci.*, **17**, 1201–1206 (1982).
4. J. W. Holmes, "Fatigue of Fiber Reinforced Ceramics," in *Ceramics and Ceramic Matrix Composites*, Vol. 3 in *Flight-Vehicle Materials, Structures and Dynamics—Assessment and Future Directions*, ed. S. R. Levine, ASME, New York, NY, 1992, pp. 193–238.
5. S. Suresh, *Fatigue of Materials*, Cambridge University Press, Cambridge, U.K., 1991.
6. P. G. Karandikar and T.-W. Chou, "Microcracking and Elastic Moduli Reductions in Unidirectional Nicalon-CAS Composites under Cyclic Loading," *Ceram. Eng. Sci. Proc.*, **13**[9–10], 881–888 (1992).
7. B. F. Sørensen and R. Talreja, "Analysis of Damage in a Ceramic Matrix Composite," *Int. J. Damage Mechanics*, **2**, 246–272 (1993).
8. P. G. Karandikar and T.-W. Chou, "Damage Development and Moduli Reductions in Nicalon-CAS Composites under Static Fatigue and Cyclic Fatigue," *J. Am. Ceram. Soc.*, **73**, 1720–1728 (1993).
9. P. G. Karandikar and T.-W. Chou, "Characterization and Modeling of Microcracking and Elastic Moduli Changes in Nicalon/CAS Composites," *Composite Sci. Tech.*, **46**, 253–263 (1993).
10. R. Y. Kim and N. J. Pagano, "Crack Initiation in Unidirectional Brittle-Matrix Composites," *J. Am. Ceram. Soc.*, **74**[5], 1082–1090 (1991).
11. B. Harris, F. A. Habib, and R. G. Cooke, "Matrix Cracking and the Mechanical Behavior of SiC-CAS Composites," *Proc. Roy. Soc. Lond.*, **A437**, 109–131 (1992).
12. J. W. Holmes and C. Cho, "Experimental Observations of Frictional Heating in a Fiber Reinforced Ceramic," *J. Am. Ceram. Soc.*, **75**[4], 929–938 (1992).
13. K. M. Prewo, "Tension and Flexural Strength of Silicon Carbide Fiber-Reinforced Glass-Ceramics," *J. Mater. Sci.*, **21**, 3590–3600 (1986).
14. J. Aveston, G. A. Cooper, and A. Kelly, "Single and Multiple Fracture," in *The Properties of Fiber Composites*, Conference Proceedings, IPC Science and Technology Press, Guildford, U.K., 1971, pp. 15–26.
15. J. Aveston and A. Kelly, "Theory of Multiple Fracture of Fibrous Composites," *J. Mater. Sci.*, **8**, 352–362 (1973).

16. D. B. Marshall, B. N. Cox, and A. G. Evans, "The Mechanics of Matrix Cracking in Brittle-Matrix Fiber Composites," *Acta Metall.*, **33**[11], 2013–2021 (1985).

17. D. B. Marshall and A. G. Evans, "Failure Mechanisms in Ceramic-Fiber/Ceramic-Matrix Composites," *J. Am. Ceram. Soc.*, **68**[5], 225–231 (1985).

18. B. Budiansky, J. W. Hutchinson, and A. G. Evans, "Matrix Fracture in Fiber-Reinforced Ceramics," *J. Mech. Phys. Solids*, **14**[2], 167–189 (1986).

19. M. D. Thouless, O. Sbaizero, L. S. Sigl, and A. G. Evans, "Effect on Mechanical Properties of Pullout in a SiC-Fiber-Reinforcement Lithium Aluminum Silicate Glass-Ceramic," *J. Am. Ceram. Soc.*, **72**[4], 525–532 (1989).

20. R. Y. Kim, "Experimental Observations of Progressive Damage in SiC/Glass-Ceramic Composites," *Ceram. Eng. Sci. Proc.*, **13**[7–8], 281–300 (1992).

21. X. F. Yang and K. M. Knowles, "The One-Dimensional Car Parking Problem and its Application to the Distribution of Spacing between Matrix Cracks in Unidirectional Fiber-Reinforced Brittle Materials," *J. Am. Ceram. Soc.*, **75**[1], 141–147 (1992).

22. C. Cho, J. W. Holmes, and J. R. Barber, "Distribution of Matrix Cracks in a Uniaxial Composite," *J. Am. Ceram. Soc.*, **75**[2], 316–324 (1992).

23. W. A. Curtin, "Theory of Mechanical Properties of Ceramic Matrix Composites," *J. Am. Ceram. Soc.*, **74**, 2837–2845 (1991).

24. K. Goda and H. Fukunaga, "The Evaluation of the Strength Distribution of Silicon Carbide and Alumina Fibres by a Multi-Modal Weibull Distribution," *J. Mater. Sci.*, **21**, 4475–4480 (1986).

25. D. B. Fischbach, P. M. Lemoine, and G. V. Yen, "Mechanical Properties and Structure of a New Commercial SiC-Type Fibre (Tyranno)," *J. Mater. Sci.*, **23**, 987–993 (1988).

26. M. Sutcu, "Weibull Statistics Applied to Fiber Failure in Ceramic Composites and Work of Fracture," *Acta Metall.*, **37**[2], 651–661 (1989).

27. D. B. Marshall and B. N. Cox, "Tensile Fracture of Brittle Matrix Composites—Influence of Fiber Strength," *Acta Metall.*, **35**[11], 2607–2619 (1987).

28. P. S. Steif and H. R. Schwietert, "Ultimate Strength of Ceramic-Matrix Composites," *Ceram. Eng. Sci. Proc.*, **11**[9–10], 1567–1576 (1990).

29. J. Pernot and L. P. Zawada, "Tensile Behavior of Continuous Fiber Ceramic Matrix Composites," Paper presented at the 16th Annual Conference on Ceramics and Advanced Materials, Cocoa Beach, FL, January 14–17, 1992.

30. K. M. Prewo, B. Johnson, and S. Starrett, "Silicon Carbide Fiber-Reinforced Glass-Ceramics Tensile Behavior at Elevated Temperature," *J. Mater. Sci.*, **24**, 1373–1379 (1989).

31. C. Q. Rousseau, "Monotonic and Cyclic Behavior of a Silicon Carbide/Calcium-Aluminosilicate Ceramic Composite," in *Thermal and Mechanical Behavior of Metal Matrix and Ceramic Matrix Composites*, ASTM STP 1080, eds. J. M. Kennedy, H. H. Moeller, and W. S. Johnson, American Society for Testing and Materials, Philadelphia, PA, 1990, pp. 136–151.

32. S.-W. Wang and A. Parvizi-Majidi, "Experimental Characterization of the Tensile Behavior of Nicalon Fibre-Reinforced Calcium Aluminosilicate Composites," *J. Mater. Sci.*, **27**, 5483–5496 (1992).

33. D. S. Beyerle, S. M. Spearing, F. W. Zok, and A. G. Evans, "Damage and Failure in Unidirectional Ceramic-Matrix Composites," *J. Am. Ceram. Soc.*, **75**, 2719–2725 (1992).

34. K. W. Garrett and J. E. Bailey, "Multiple Transverse Fracture in 90° Cross-Ply Laminates of a Glass Fibre-Reinforced Polyester," *J. Mater. Sci.*, **12**, 157–168 (1977).

35. A. Parvizi, K. W. Garrett, and J. E. Bailey, "Constrained Cracking in Glass Fibre-Reinforced Epoxy Cross-Ply Laminates," *J. Mater. Sci.*, **13**, 352–362 (1978).

36. G. A. Bernhart, M. M. Danchier, and P. J. Lamicq, "SiC/SiC Composite Ceramics," *Am. Ceram. Soc. Bull.*, **65**, 336–338 (1986).

37. S. F. Shuler, J. W. Holmes, X. Wu, and D. Roach, "Influence of Frequency on the Rate of Damage Accumulation in a C-Fiber SiC-Matrix Composite," *J. Am. Ceram. Soc.*, **76**, 2327–2336 (1993).

38. S. F. Shuler and J. W. Holmes, "Influence of Loading Rate on the Monotonic Tensile Behavior of Fiber Reinforced Ceramics," Research Memorandum No. 102, September 1990, Available through: Ceramic Composites Research Laboratory, Dept. of Mechanical Engineering and Applied Mechanics, 2250 GGBL, The University of Michigan, Ann Arbor, MI 48109–2125.

39. B. F. Sørensen and J. W. Holmes, "Effect of Loading Rate on the Monotonic Tensile Behavior and Matrix Cracking of a Fiber-Reinforced Ceramic," *J. Am. Ceram. Soc.*, in prep.

40. R. F. Allen and P. Bowen, "Fatigue and Fracture of a SiC/CAS Continuous Fiber Reinforced Glass Ceramic Matrix Composite at Ambient and Elevated Temperatures," *Ceram. Eng. Sci. Proc.*, **14**, 265–272 (1993).

41. E. Minford and K. M. Prewo, "Fatigue of Silicon Carbide Reinforced Lithium-Alumino-Silicate Glass-Ceramics," in *Tailoring Multiphase and Composite Ceramics*, eds. C. G. Patano and R. E. Messing, Plenum Publishing Corporation, New York, NY, 1986, pp. 561–570.

42. K. M. Prewo, "Fatigue and Stress Rupture of Silicon Carbide Fiber-Reinforced Glass-Ceramics," *J. Mater. Sci.*, **22**, 2695–2701 (1987).

43. J. W. Holmes, T. Kotil, and W. T. Foulds, "High Temperature Fatigue of SiC Fiber-Reinforced Si_3N_4 Ceramic Composites," in *Symposium on High Temperature Composites*, Technomic Publishing Co., Lancaster, PA, 1989, pp. 176–182.

44. L. P. Zawada, L. M. Butkus, and G. A. Hartman, "Room Temperature Tensile and Fatigue Properties of Silicon-Carbide Fiber-Reinforced Alumino-silicate Glass," *Ceram. Eng. Sci. Proc.*, **11**[9–10], 1592–1606 (1990).

45. L. M. Butkus, L. P. Zawada, and G. A. Hartman, "Room Temperature Tensile and Fatigue Properties of Silicon-Carbide Fiber-Reinforced Ceramic Matrix Composites," Paper presented at Aeromat '90, Long Beach, CA, May 21–24, 1990.

46. D. Rouby and P. Reynaud, "Fatigue Behavior Related to Interface Modification during Load Cycling in Ceramic-Matrix Fibre Composites," *Composites Science and Technology*, **48**, 109–118 (1993).

47. P. G. Karandikar, T.-W. Chou, R. Talreja, and O. Chen, "Static and Fatigue Characterization of Ceramic Matrix Composites," Paper presented at ICAMP '90, International Conference on Advanced Materials Mechanical Properties, '90, Utsunomiya, Japan, August 6–9, 1990.

48. J. W. Holmes, X. Wu, and B. F. Sørensen, "Frequency Dependence of Fatigue Life and Internal Heating of a Fiber-Reinforced Ceramic Matrix Composite," *J. Am. Ceram. Soc.*, **77**[12], 3238–3286 (1994).

49. J. W. Holmes, "Influence of Stress-Ratio on the Elevated Temperature Fatigue of a SiC Fiber-Reinforced Si_3N_4 Composite," *J. Am. Ceram. Soc.*, **74**[7], 1639–1645 (1991).

50. Z. G. Wang, C. Laird, Z. Hashin, W. Rosen, and C. F. Yen, "The Mechanical Behaviour of a Cross-Weave Ceramic Matrix Composite, Part II, Repeated Loading," *J. Mater. Sci.*, **26**, 5335–5341 (1991).

51. J. W. Holmes and S. F. Shuler, "Temperature Rise During Fatigue of Fiber-Reinforced Ceramics," *J. Mater. Sci. Lett.*, **9**, 1290–1291 (1990).
52. C. Cho, J. W. Holmes, and J. R. Barber, "Estimation of Interfacial Shear in Ceramic Composites from Frictional Heating Measurements," *J. Am. Ceram. Soc.*, **73**, 2802–2808 (1991).
53. T. Kotil, J. W. Holmes, and M. Comninou, "Origin of Hysteresis Observed during Fatigue of Ceramic-Matrix Composites," *J. Am. Ceram. Soc.*, **73**, 1879–1883 (1990).
54. A. W. Pryce and P. A. Smith, "Matrix Cracking in Unidirectional Ceramic Matrix Composites under Quasi-Static and Cyclic Loading," *Acta Metall. Mater.*, **41**, 1269–1281 (1993).
55. B. F. Sørensen, R. Talreja, and O. T. Sørensen, "Micromechanical Analysis of Damage Mechanisms in Ceramic Matrix Composites during Mechanical and Thermal Loading," *Composites*, **24**, 124–140 (1993).
56. L. N. McCartney, "Mechanics of Matrix Cracking in Brittle-Matrix Fibre-Reinforced Composites," *Proc. Roy. Soc. Lond.*, **A409**, 329–350 (1987).
57. P. D. Jero and R. J. Kerans, "The Contribution of Interfacial Roughness to Sliding Friction of Ceramic Fibers in a Glass Matrix," *Scripta Met. Mater.*, **24**, 2315–2318 (1990).
58. B. F. Sørensen and J. W. Holmes, "Improvement in the Fatigue Life of Fiber-Reinforced Ceramics by Use of Interface Lubrication," *Scripta Metallurgica*, accepted for publication.
59. J. W. Holmes, unpublished work, 1993.
60. R. Talreja, "Fatigue of Fibre-Reinforced Ceramics", in *Proceedings of Conference on Processing, Structural Ceramics, Microstructure and Properties*, eds. J. J. Bentzen, J. B. Bilde-Sørensen, N. Christiansen, A. Horsewell, and B. Ralph, Risø National Laboratory, Denmark, 1990, pp. 145–159.
61. W. L. Morris, B. N. Cox, D. B. Marshall, R. V. Inman, and M. R. James, "Fatigue Mechanisms in Graphite/SiC Composites at Room Temperature," *J. Am. Ceram. Soc.*, **77**, 792–800 (1994).
62. R. M. McMeeking and A. G. Evans, "Matrix Fatigue Cracking in Fiber Composites," *Mechanics of Materials*, **9**, 217–227 (1990).
63. R. F. Allen and P. Bowen, "Effects of Test Temperature and Loading Rate on the Fatigue and Fracture Behavior Resistance of a Continuous Fibre Reinforced Glass Ceramic Matrix Composite," in *Proceedings of ICCM-9*, Ninth International Conference on Composite Materials, July 12–16, 1993, Madrid, Spain.
64. R. H. Jones and C. H. Heneger, Jr., "Fatigue Crack Growth in SiC/SiC at 1100 C," in Fusion Reactor Materials Semiannual Progress Report for Period Ending September 30, 1992, Battle Pacific Laboratory.
65. R. O. Ritchie, "Mechanisms of Fatigue Crack Propagation in Metals, Ceramics and Composites: Role of Crack Tip Shielding", *Mater. Sci. Eng.*, **103**, 15–28 (1988).
66. K. S. Chan, "Effects of Interface Degradation of Fiber Bridging of Composite Fatigue Cracks," *Acta Metall. Mater.*, **41**, 761–768 (1993).
67. G. Bao and Y. Song, "Crack Bridging Models for Fiber Composites with Slip Dependent Interfaces," *Acta Metall. Mater.*, **41**, 1425–1444 (1993).
68. P. Reynaud, D. Rouby, and G. Fantozzi, "A Model Describing the Changes in Ceramic-Ceramic Fibre Composites under Cyclic Fatigue Loading," in *Proceedings of Developments in the Science and Technology of Composite Materials*, eds. A. R. Bunsell, J. F. Jamet, and A. Messiah, ECCM-5, Fifth European Conference on Composite Materials, Bordeaux, France, 1992, pp. 597–602.

69. M. D. Thouless and A. G. Evans, "Effects of Pull-out on the Mechanical Properties of Ceramic-Matrix Composites," *Acta Metall.*, **36**, 517–522 (1988).
70. P. Reynaud, G. Fantozzi, and M. Bourgeon, "Modeling of Temperature Effects on the Cyclic Fatigue Behavior of Ceramic Matrix Composites," in *High Temperature Ceramic Matrix Composites*, eds. R. Naslain, J. Lamon, and D. Doumeingts, Woodhead Publishing, U.K., 1993, pp. 659–666.
71. S. Suresh, "Mechanics and Mechanisms of Fatigue Crack Growth in Brittle Solids," *Int. J. Fract.*, **42**, 41–56 (1990).
72. B. F. Sørensen and J. W. Holmes, "Influence of Stress Ratio on the Fatigue Life of a Continuous Fiber-Reinforced Ceramic Matrix Composite," *J. Am. Ceram. Soc.*, in prep.
73. W. R. Moschelle, "Load Ratio Effects on the Fatigue Behavior of Silicon Carbide Fiber Reinforced Silicon Carbide Composite," *Ceram. Eng. Sci. Proc.*, **15**[4], 13–22 (1994).
74. E. Bischoff, M. Rühle, O. Sbaizero, and A. G. Evans, "Microstructural Studies of the Interface Zone of a SiC-Fiber-Reinforced Lithium Aluminum Silicate Glass-Ceramic," *J. Am. Ceram. Soc.*, **72**, 741–745 (1989).
75. R. F. Allen, C. J. Beevers, and P. Bowen, "Fracture and Fatigue of a Nicalon/CAS Continuous Fibre Reinforced Glass Ceramic Matrix Composite," *J. Composites*, accepted.
76. L. A. Bonney and R. F. Cooper, "Reaction Layer in SiC-Fiber-Reinforced Glass-Ceramics: A High-Resolution Scanning Transmission Electron Microscope Analysis," *J. Am. Ceram. Soc.*, **73**, 2916–2926 (1990).
77. C. H. Heneger, Jr. and R. H. Jones, "The Effects of an Aggressive Environment on the Subcritical Crack Growth of a Continuous-Fiber Ceramic Matrix Composite," *Ceram. Eng. Sci. Proc.* **13**[7–8], 411–419 (1992).
78. D. A. Woodford, D. R. Van Steele, and J. Brehm, "Effect of Test Temperature, Oxygen Attack, Thermal Transients and Protective Coatings on Tensile Strength of Silicon Carbide Matrix Composites," *Ceram. Eng. Sci. Proc.*, **13**[9–10], 752–759 (1992).
79. R. T. Bhatt, "Oxidation Effects on the Mechanical Properties of SiC-Fiber-Reinforced Reaction-Bonded Si_3N_4-Matrix Composites," *J. Am. Ceram. Soc.*, **75**[2], 406–412 (1992).
80. S.-W. Wang, R. W. Kowalik, and R. Sands, "Strength of Nicalon Fiber Reinforced Glass-Ceramic Matrix Composites after Corrosion with Na_2SO_4 Deposits," *Ceram. Eng. Sci. Proc.*, **13**[9–10], 760–765 (1992).
81. C. H. Heneger, Jr. and R. H. Jones, "Molten Salt Corrosion of Hot-Pressed Si_3/N_4/SiC Composites and Effects of Molten Salt Corrosion on Slow Crack Growth of Hot-Pressed Si_3N_4," in *Corrosion and Corrosive Degradation of Ceramics, Ceramic Transactions*, Vol. 10, eds. R. E. Tressler and M. McNallen, American Ceramic Society, Westerville, OH, 1990, pp. 197–210.
82. D. S. Fox and J. L. Smialek, "Burner Rig Hot Corrosion of Silicon Carbide and Silicon Nitride," *J. Am. Ceram. Soc.*, **73**[2], 303–311 (1990).
83. L. P. Zawada and R. C. Wetherhold, "The Effects of Thermal Fatigue on a SiC Fibre/Aluminosilicate Glass Composite," *J. Mater. Sci.*, **26**, 648–654 (1991).
84. G. M. St. Hilaire and T. Ertürk, "Thermomechanical Fatigue of Crossply SiC_f Si_3N_4 Ceramic Composite under Impinged Kerosene-Based Flame," *Ceram. Eng. Sci. Proc.*, **14**, 416–425 (1993).
85. B. N. Cox, "Interfacial Sliding near a Free Surface in a Fibrous or Layered Composite during Thermal Cycling," *Acta Metall. Mater.*, **38**, 2411–2424 (1990).
86. B. N. Cox, M. S. Dadkhah, M. R. James, D. B. Marshall, W. L. Morris, and M. Shaw, "On Determining Temperature Dependent Interfacial Shear

Properties and Bulk Residual Stresses in Fibrous Composites," *Acta Metall. Mater.*, **38**, 2425–2433 (1990).

87. L. Butkus, J. W. Holmes, and T. Nicholas, "Thermomechanical Fatigue of a SiC-Fiber Calcium Aluminosilicate Matrix Composite," *J. Am. Ceram. Soc.*, **76**, 2817–2827 (1993).

88. D. W. Worthem and J. R. Ellis, "Thermomechanical Fatigue of Nicalon/CAS under In-Phase and Out-of-Phase Loadings," *Ceram. Eng. Sci. Proc.*, **14**[7–8], 292–300 (1993).

89. B. F. Sørensen, "Simulation of Thermomechanical Cycling of a Multiple Matrix Cracked Ceramic Composite," to be published.

90. B. F. Sørensen and J. W. Holmes, "Rate of Strength Decrease of Fiber-Reinforced Ceramic Matrix Composites during Fatigue," *J. Am. Ceram. Soc.*, in prep.

91. B. F. Sørensen and J. W. Holmes, "Does a True Fatigue Limit Exist for Continuous Fiber-Reinforced Ceramic Matrix Composites?" *J. Am. Ceram. Soc.*, in prep.

92. A. S. D. Wang, X. G. Huang, and M. Barsoum, "Matrix Crack Initiation in Ceramic Matrix Composites, Part II: Models and Simulation Results," *Composites Science and Technology*, **44**, 271–282 (1992).

CHAPTER 7

High Temperature Crack Growth in Unreinforced and Whisker-Reinforced Ceramics under Cyclic Loads

S. Suresh

7.1 Introduction

In recent years, ceramic matrix composites have been the focus of increasing scientific and applied research as candidate materials for high performance structural components, owing to their superior high temperature properties compared to structural metals and alloys. A large volume of experimental and theoretical work on the room temperature mechanical properties of ceramics and ceramic composites has become available in the open literature during the past two decades (e.g., Refs. 1–4). However, significantly less fundamental understanding exists about the static and fatigue (cyclic) crack growth characteristics of ceramic materials in elevated temperature environments which are representative of potential service conditions. Since structural components made of ceramic composites are invariably subjected to cyclic loads during high temperature service conditions, the study of creep–fatigue interactions and damage tolerance under cyclic loading is essential for ultimate success in the use of these materials in high temperature applications.

This chapter provides an overview of recent advances in our understanding of the mechanics and micromechanisms of creep–fatigue crack growth in discontinuously reinforced ceramics. (Discussions of fatigue in continuously reinforced ceramics can be found in Chapter 5 of this volume.) The chapter is arranged in the following sequence. Section 7.2 begins with a description of the micromechanisms of damage and fracture at elevated temperatures in monolithic and reinforced ceramics. Particular attention is devoted to the effects of viscous interfacial glass films (which form *in situ* in the elevated temperature environment as a consequence of the addition of certain reinforcements to ceramics) on subcritical crack growth in the elevated temperature environment. Characterization of creep crack growth on the basis of fracture mechanics is addressed in Section 7.3. Here, emphasis is placed on the identification of conditions under which different crack tip parameters provide descriptions of creep crack growth under cyclic loading conditions. Sections 7.4 and 7.5 address the mechanics, mechanisms and subcritical fracture characteristics of unreinforced ceramics and ceramic composites at elevated temperatures. Wherever feasible, results of detailed transmission electron microscopy (TEM) of damage, including TEM of crack tip damage, are presented to illustrate the connection between microstructure and damage evolution. The effects of various mechanical, microstructural and environmental factors on high temperature fatigue fracture are also reviewed in Sections 7.4 and 7.5. The chapter concludes with Section 7.6 where a brief summary is provided of the current state of the field and some recommendations for future research.

The issues which are reviewed in some detail in this chapter include:

- Changes in the fatigue crack growth resistance of ceramic materials arising from discontinuous reinforcements.
- Approaches to characterizing the creep–fatigue fracture of ceramic composites.
- Differences in the macroscopic and microscopic fatigue fracture behavior of ceramic composites under static and cyclic loading at high temperatures.
- The *in situ* formation of glass films by the oxidation of Si-containing matrices and reinforcements, and the effects of such *in situ*-formed glass phase on high temperature damage tolerance *vis-à-vis* the effects of preexisting glass films (remaining in the material from the processing stage) on creep fracture resistance.
- Interfacial cavitation, microcracking, diffusion and crack branching due to the viscous flow of the glass phase during creep–fatigue.
- Competition between matrix plasticity and grain boundary/interfacial separation due to the flow of the glass phase.

Since the significant majority of the published literature on high temperature crack growth under static and cyclic loads is predicated upon experiments conducted on alumina and alumina matrix composites, the examples cited in the present review have centered around oxide ceramics and their composites. However, the implications of the results to other classes of ceramics, intermetallics, and brittle matrix composites are also described, wherever feasible, along with any available information in an attempt to illustrate the generality of the concepts developed here.

In view of the variety of terms used in the ceramic literature to refer to failure under fluctuating stresses, it is appropriate at the outset to define the terminology employed in this chapter. In the ceramic literature, progressive damage and crack growth under sustained loads is commonly referred to as "static fatigue," while damage and failure under cyclic loading conditions is denoted as "cyclic fatigue." In keeping with the universal terminology well established in the engineering literature as well as in metal and polymer science, and in an attempt to avoid confusion, the term "fatigue" is employed in this chapter only in the context of cyclic loading. Fracture under a sustained load at high temperatures is referred to here as "static crack growth" or "creep crack growth" and that under a fluctuating load is referred to as "creep–fatigue crack growth."

7.2 Brief Review of Damage Mechanisms

7.2.1 Diffusional and Dislocation-Based Processes

Fine-grained materials, when subjected to high temperatures and low applied stresses, deform by mutual accommodation of grains assisted by grain boundary sliding and transport of matter (diffusion). Under conditions where lattice diffusion dominates, the diffusional creep rate is reasonably well characterized by the Nabarro–Herring creep process. (For a review of this and other classical creep mechanisms, see Refs. 5 and 6.) Here the strain rate is expressed as

$$\dot{\varepsilon} = \frac{\lambda \sigma \Omega D}{k T d_g^2} \tag{1}$$

where D is the diffusion coefficient, d_g the grain size, Ω the atomic volume, k the Boltzmann constant, and λ a factor which is affected by the shape of the grains. If diffusion occurs along grain boundaries, rather than transgranularly, the so-called Coble creep mechanism would be expected, which also estimates a linear variation of $\dot{\varepsilon}$ with σ. If the grains deform primarily as a result of diffusion, the ensuing creep response by grain boundary sliding is only a slight modification of that predicted by the Nabarro–Herring or Coble creep processes. The more complex situation involving the sliding of grain boundaries as a consequence of power law creep of grains has been modeled by Raj and Ashby[7] (and others[8,9]) who considered the sliding in a planar array of hexagonal tiles with nonplanar boundaries.

Two types of basic creep mechanisms have been identified in models for dislocation creep. (1) In glide-controlled creep, the obstacles to dislocation motion are on the scale of the dislocation core; the obstacles are overcome by stress-assisted thermal agitation. In this case, the creep rate can be written as (see Refs. 1 and 10 for a review)

$$\dot{\varepsilon} = \dot{\varepsilon}_0 \frac{\sigma^2}{\mu^2} \exp\left(-\frac{Q(\sigma)}{RT}\right) \qquad (2)$$

where $\dot{\varepsilon}_0$ is the reference strain rate, μ the shear modulus, $Q(\sigma)$ the activation enthalpy, R the universal gas constant, and T the absolute temperature. (2) In recovery-controlled creep, the obstacles are too large to be overcome by thermal agitation, but are surpassed by diffusion-aided recovery. It is known that the stress dependence for any climb-controlled recovery creep is a power law of the form[11–14]

$$\dot{\varepsilon} = C\sigma^n \qquad (3)$$

where C is a material parameter, and n is a quasi-steady-state creep exponent. For steady-state, secondary creep which is governed by climb-controlled dislocation motion, n is of the order of 3 or 4.[1,11] In pipe diffusion, rather than lattice diffusion, n is found to be in the range 5–6.[10] A detailed discussion of these mechanisms of creep can be found in Refs. 1, 2, and 10 as well as in Chapter 4 by Wiederhorn and Fuller, in this volume.

7.2.2 Viscous Flow of Amorphous Phase

The synthesis of structural ceramics often involves the use of additives which, in addition to the second phases formed by impurities, become a viscous glass phase at the sintering temperature and improve densification of the ceramic. Upon cooling, the second phase exists as an amorphous or partly crystalline film along grain facets and triple junctions. During subsequent high temperature deformation of the ceramic, the glass phase can influence the creep deformation (e.g., Refs. 15–21) and creep–fatigue crack growth[22–24] (under both static and cyclic loads) in many ways:

- The glass phase provides preferential sites for the nucleation, growth and coalescence of cavities during high temperature deformation. During static and cyclic fracture at high temperatures, this cavitation process aids in the development of a diffuse microcrack zone as well as crack tip branching.
- The glass phase may serve as a path for rapid diffusion of matrix atoms; in this way, crystalline matrix atoms can dissolve in the viscous glass phase in local grain boundary or interface regions which are in compression, and then redeposit in regions which are locally in tension, after being transported rapidly through the glass film.
- As shown later in this chapter, the presence of glass phase in structural ceramics appears to enhance the propensity for subcritical creep crack

growth under static and cyclic loads, although the threshold stress intensity factor for the onset of such stable fracture can be reduced in some materials.

- In microstructures capable of undergoing dislocation plasticity, the presence of the amorphous films along grain boundaries and interfaces promotes excessive interfacial cavitation which ostensibly suppresses the dislocation activity within the grains.
- During cyclic loading at elevated temperature, the viscous flow of glassy films ahead of the crack tip makes the fatigue crack growth rate susceptible to loading rate (i.e., cyclic frequency) and waveforms, including hold-times (e.g., Refs. 22–25).

While preexisting glass phase left from the processing stage leads to such viscous-flow-assisted effects at high temperatures in most monolithic oxide and nitride ceramics, the formation of glass phase *in situ* during (post-processing) high temperature mechanical loading is also known to result in similar glass phase effects in brittle materials, such as those containing Si and SiC. In the latter case, the oxidation of SiC typically above 1250°C in oxygen-containing atmospheres creates copious glass pockets along grain junctions and interfaces. The resultant creep fracture and creep–fatigue mechanisms in these systems (even if they don't contain appreciable concentrations of a glassy phase *a priori*) are, as shown in detail later in this chapter, similar to those observed in liquid-phase sintered monolithic ceramics.

7.3 Choice of Parameter for Fracture Characterization

One of the key issues in the study of creep–fatigue fracture is the choice of an appropriate fracture parameter which uniquely characterizes the crack growth response for a wide range of specimen geometries, crack geometries, environmental conditions and mechanical loading parameters. Inherent in the selection of such a characterizing parameter is the need for the realization of its region of dominance.

Let us begin by considering an elastic, nonlinear viscous material whose total strain rate under uniaxial tension loading conditions is characterized by the following equation:

$$\dot{\varepsilon} = \frac{\dot{\sigma}}{E} + \dot{\varepsilon}_0 \left(\frac{\sigma}{\sigma_0} \right)^n \tag{4}$$

where E is Young's modulus, $\dot{\sigma}$ the stress rate, σ_0 and ε_0 the reference stress and strain, respectively, and n the power-law creep exponent. For creep deformation at the tip of a crack or a notch with a large stress concentrating effect, the nonlinear creep strains dominate over the elastic strains. Consequently, Eqn. (4) may be approximated by the pure power law,

$$\frac{\dot{\varepsilon}}{\dot{\varepsilon}_0} = \alpha \left(\frac{\sigma}{\sigma_0}\right)^n \tag{5}$$

where α is a constant related to the material parameters in Eqn. (4).

The choice of an appropriate parameter for characterizing high temperature fatigue crack growth is generally dictated by the relative magnitudes of the fatigue cycle time, t_c (i.e., the duration of the fatigue cycle, $t_c = 1/\nu_c$ where ν_c is the cyclic loading frequency), and the transition time, t_T, from small-scale creep to extensive creep at the crack tip.[4,26] The analysis of Riedel and Rice[27] provides an estimate for the transition time:

$$t_T = \frac{(1 - \nu^2) K^2}{(n + 1) E C^*} \tag{6}$$

where ν is Poisson's ratio, K is the maximum stress intensity factor of the fatigue cycle, n is the power-law creep exponent, E is Young's modulus, and C^* is the rate form of Rice's J-integral[28] under steady-state conditions which is path independent everywhere. Kumar et al.[29] and Shih and Needleman[30] have reported expressions for computing C^* for various specimen geometries and loading configurations with expressions of the form

$$C^* = \dot{\varepsilon}_0 \sigma_0 (W - a) \left(\frac{P}{P_0}\right)^{n+1} F(n, geometry) \tag{7}$$

where W is the width of the specimen, a is the crack length, P is the applied load, P_0 is the reference load, and F is a function of n and the specimen geometry.

Analogous to the arguments of small-scale yielding in linear elastic fracture mechanics, small-scale creep conditions (i.e., the situation where the size of the creep zone ahead of the fatigue crack tip is small compared to the characteristic dimensions of the test specimen, including the size of the crack and that of the uncracked liagment) can be assumed to exist when

$$t_c \ll t_T \tag{8}$$

Under these conditions, typical of low to moderate temperatures and high ν_c, fatigue crack growth is essentially load and cycle dependent, and it is reasonably well characterized by the stress intensity factor range, ΔK.

On the other hand, very high temperatures and low cyclic loading frequencies (high cycle times) promote near-tip creep–fatigue conditions where

$$t_c \gg t_T \tag{9}$$

In this case, crack growth is essentially a time-dependent process and there is no unique method available for characterizing fatigue fracture under extensive creep conditions. Various investigators (e.g., Refs. 31–33) have proposed C^*, $C(t)$ (the time-dependent amplitude of the Hutchinson, Rice, Rosengren or HRR-type fields ahead of the creep crack), and C_t (which is the instantaneous power release rate per unit crack advance per unit thickness of the specimen) as possible characterizing parameters for fracture under extensive creep conditions. A detailed discussion of the available methods for high temperature fatigue crack growth analyses is given by Suresh.[4]

For the four-point bend geometry used in the present experiments and under plane strain conditions, the C^* integral is given by:[30]

$$C^* = \sigma_0 \dot{\varepsilon}_0 (W - a) h_1\left(\frac{a}{b}, n\right)\left(\frac{M}{M_0}\right)^{n+1} \tag{10}$$

where $(W - a)$ is length of the uncracked ligament, M is the applied bending moment per unit thickness, $h_1(a/b, n)$ is a nondimensional function of the power-law creep exponent n and the ratio a/b, and M_0 is the bending moment per unit thickness. The value of M_0 is given by:[30]

$$M_0 = 0.364\sigma_0(W - a)^2 \tag{11}$$

Consider a four-point bend specimen (of dimensions $50\,\text{mm} \times 10\,\text{mm} \times 5\,\text{mm}$, with a through-thickness precrack length of 3.3 mm) of a polycrystalline aluminum oxide ($E = 390\,\text{GPa}$, $\nu = 0.23$). Taking typical material parameters to be $n = 1.3$ and $\alpha = 4 \times 10^{-18}\,\text{MPa}^{-n}\,\text{s}^{-1}$ and noting that $h_1 = 1.5$, it is seen that the transition time for a maximum stress intensity factor $K_{max} = 4\,\text{MPa}\sqrt{\text{m}}$ is $t_T = 1911\,\text{s}$. This would imply that for cyclic loading frequencies, $\nu_c > 0.005\,\text{Hz}$, small-scale creep conditions will prevail at the crack tip and ΔK will be a reasonably good characterizing parameter for high temperature fatigue crack growth. Similarly, it can be shown that for $K_{max} = 10\,\text{MPa}\sqrt{\text{m}}$, the transition time t_T is 824 s and 8985 s for Si_3N_4 and $Al_2O_3/33$ vol.% SiC_W composite, respectively; the corresponding values of ν_c above which ΔK is expected to hold are approximately 0.012 and 0.001 Hz, respectively.

For the data on high temperature crack growth presented in this chapter, the validity of the use of ΔK for characterizing creep–fatigue fracture was examined using the above analyses. Furthermore, this choice was also checked by monitoring the crack growth rate da/dN at different crack lengths (i.e., different specimen geometry), but at the same ΔK. Such checks revealed essentially the same values of da/dN, thereby supporting the choice of ΔK for describing fatigue crack growth for the set of experiments described here.

In the static crack growth experiments reported in this chapter, the time period for the application of the sustained load (before the interruption of the test for crack observations) was also typically smaller than the transition time, t_T. This permits the stress intensity factor K to be used as a characterizing parameter for static crack growth.

7.4 High Temperature Fatigue Crack Growth

In this section, the results of some recent work[22–24,34–36] on elevated temperature fatigue crack growth in oxide and nitride ceramics, with and without SiC whisker reinforcements, are presented. These results were gathered on four-point bend specimens containing through-thickness fatigue precracks. (Typical specimen dimensions were $50.8\,mm \times 10\,mm \times 5\text{–}9\,mm$, with the cyclic fatigue precrack introduced at room temperature spanning about one-third of the specimen width.) In all cases, the mechanics and micromechanisms of crack growth under cyclic loads are compared and contrasted with those seen under static loads, with a view to identifying possible effects of cyclic fatigue. Furthermore, in all cases, results of transmission electron microscopy of crack tip damage are presented in an attempt to elucidate the local failure processes. For reasons of brevity, discussions of experimental methods are not included here. The interested reader is encouraged to consult Refs. 22 and 34–36 for full descriptions of experimental procedures including precracking, heating and crack growth monitoring techniques, TEM specimen preparation, and basic physical and mechanical properties, as well as relevant processing details of all the materials discussed in this chapter. The choice of cyclic loading frequency and waveform (sinusoidal in all cases) was such that linear elastic fracture mechanics could be invoked with reasonable accuracy to characterize high temperature fatigue crack growth, as discussed in the preceding section.

7.4.1 Unreinforced Ceramics

The rate of tensile crack growth per fatigue cycle, da/dN, is plotted in Fig. 7.1 as a function of the stress intensity factor range, ΔK, for a 90% pure aluminum oxide.† The ceramic was subjected to cyclic loads at frequencies of $0.13\,Hz$ and $2\,Hz$ in $1050°C$ air at a load ratio $R = 0.15$. (The load ratio is defined as the ratio of the minimum stress to the maximum stress of the fatigue cycle.) At $0.13\,Hz$, the fatigue crack growth commences (at a threshold growth rate of at least $10^{-10}\,m/cycle$) at the threshold stress intensity factor range, $\Delta K \approx 1.5\,MPa\sqrt{m}$. Despite the usual scatter observed in the experimental data, the crack growth rate appears to follow a Paris-type expression

$$\frac{da}{dN} = C(\Delta K)^m \tag{12}$$

where C is a material constant, and m is the Paris exponent with a value of approximately 10 for $\nu_c = 0.13\,Hz$. (By comparison, m for ductile metals and

†Commercially available as grade AD 90 from Coors Ceramics, Golden, CO. Average grain size $= 4\,\mu m$ and range of grain sizes $= 2\text{–}10\,\mu m$. This material contains silica, magnesia and trace amounts of iron oxide, sodium oxide and potassium oxide as impurities, primarily along grain facets.

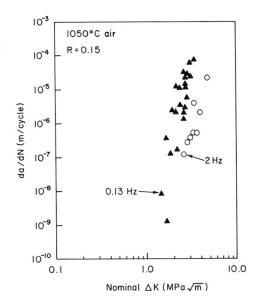

Fig. 7.1 Fatigue crack growth rate, da/dN, plotted as a function of stress intensity factor range, ΔK in AD 90 alumina in 1050°C air at cyclic frequencies of 0.13 Hz and 2 Hz at $R = 0.15$. After Ref. 34.

alloys at room temperature falls in the range 2–4.) Increasing the frequency to 2 Hz results in a slight, but noticeable, reduction in crack propagation rates.

A comparison of the crack velocities measured under static and cyclic loads is illustrated in Fig. 7.2. For this purpose, the crack velocity under cyclic loads, $da/dt = da/dN \times \nu_c$, plotted against the maximum stress intensity factor of the fatigue cycle, $K_{max} = \Delta K/(1 - R)$, from the results shown in Fig. 7.1. The static crack velocity da/dt is also plotted against the stress intensity factor K_I corresponding to the applied load. In the intermediate range of crack growth, the static crack velocity generally follows the power-law relationship

$$\frac{da}{dt} = B(K_I)^p \tag{13}$$

where B is a material constant, and p is an exponent with a range 5–8. At the same magnitude of (maximum) stress intensity factor, the crack growth rates under static loads are substantially faster than those measured under cyclic loads. This effect of load fluctuations reducing the crack velocities is particularly accentuated at lower stress intensity levels and higher values of ν_c.†

†This trend is at variance with the crack growth rates at room temperature in many nominally single-phase ceramics and transformation-toughened ceramics where fluctuations in the imposed compressive and tensile loads are seen to accelerate the rate of fracture.[36–43] While the behavior at room temperature appears to be influenced especially significantly by crack bridging and other wake effects, the trends seen at elevated temperatures are also markedly affected by creep deformation ahead of the crack tip. See later sections of this chapter for further details.

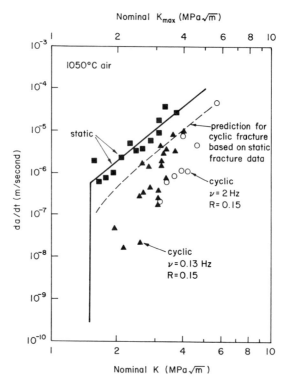

Fig. 7.2 Cyclic crack growth velocity, $da/dt = (da/dN) \times v_c$, plotted as a function of maximum stress intensity factor, K_{max}, and static crack growth velocity, da/dt, plotted as a function of applied stress intensity factor, K, in AD 90 alumina in 1050°C air. After Ref. 34. Experimental conditions are the same as those in Fig. 7.1. Also shown are da/dt values predicted for cyclic loads on the basis of static fracture data using Eqn. (14).

Early work (e.g., Refs. 44 and 45) on silicon nitride ceramics for a limited range of high temperature cyclic loading conditions led to the hypothesis that the mechanisms of cyclic and static fracture at elevated temperature are identical, and that the cyclic crack growth rates can be predicted on the basis of static fracture data. One of the techniques commonly used to derive cyclic crack growth rates solely on the basis of static load fracture data involves integration of the relationship in Eqn. (13) over the duration of the fatigue cycle such that

$$\left(\frac{da}{dt}\right)_{cyclic} = v_c \int_0^{1/v_c} \left(\frac{da}{dt}\right)_{static} dt = v_c \int_0^{1/v_c} B(K_I)^p \, dt \qquad (14)$$

Figure 7.2 also includes the fatigue crack growth rates predicted on the basis of sustained load fracture data (dashed line), which are clearly in excess of those actually observed, especially at low ΔK and high v_c.

Recent experimental results[36] on hot-pressed silicon nitride (containing about 4 vol.% Y_2O_3) covering a wide range of applied cyclic frequencies (sinusoidal waveform) reveal a decrease in fatigue crack growth rates with increasing cyclic frequencies, similar to the trends shown in Figs. 7.1 and 7.2 for Al_2O_3. Such effects, as well as the differences between static and cyclic crack growth rates, indicate that cyclic crack growth rates cannot be rationalized solely on the basis of static fracture data.

It is interesting to note that the beneficial effects of cyclic loading, as compared to sustained loads, on high temperature damage and fracture resistance are also seen during stress–life experiments (S–N curves) conducted on Al_2O_3 at 1200°C under tension–tension cycles[46] and on Si_3N_4 under fully reversed tension–compression cycles.[47]† Further discussion of such results is taken up in Section 7.5.4.

7.4.2 Reinforced Ceramics

The resistance of the ceramic material to subcritical creep crack growth under both static and cyclic loading conditions is significantly enhanced by the incorporation of SiC whiskers, which also elevate the temperature at which mechanical loads can be imposed on the ceramic. The example of Al_2O_3/ 33 vol.% SiC whisker composite‡ for which considerable creep fracture information has been obtained (e.g., Refs. 21, 49–51) over the temperature range 1300–1500°C is considered here. Figure 7.3a shows an example of a crack profile in the ceramic composite subjected to static loads (range of $K_I \approx 5\,MPa\sqrt{m}$) in 1400°C air. This figure indicates the formation of micro-cracks around the crack, as well as periodic deflection/branching of the crack. A similar macroscopic crack profile is seen in Fig. 7.3b for the fatigue specimen which was subjected to $\Delta K \approx 3.5$–$5\,MPa\sqrt{m}$ at $R = 0.15$ and $\nu_c = 0.1\,Hz$ in 1400°C air.

Han and Suresh[22] studied the elevated temperature crack growth characteristics of the alumina/SiC composite subjected to several load ratios and test frequencies. Figure 7.4, taken from their work, shows the fatigue crack growth rates, da/dN, plotted against ΔK for $R = 0.15$, 0.40 and 0.75 in 1400°C air at a test frequency of 0.1 Hz. Increasing the load ratio causes an increase in crack growth rates and a decrease in the threshold stress intensity factor range for the composite. Increasing the frequency to 2 Hz at $R = 0.15$ decreases the crack growth rates, as seen for AD 90 alumina at 1050°C in Fig. 7.1.

---

†There have also been reports that in some Si_3N_4 ceramics, predictions of cyclic fatigue lifetime on the basis of static fatigue data severely underestimate experimentally observed cycles to failure.[48] This has been attributed to an intrinsic difference in the very mechanism of fracture between static and cyclic loading experiments.

‡Commercially available as Grade WG-300 from Greenleaf Corporation, Saegertown, PA. This material contains SiC whiskers which are 0.2–0.7 μm in diameter and up to 25 μm in length. The average grain size of the alumina matrix is about 1.3 μm.

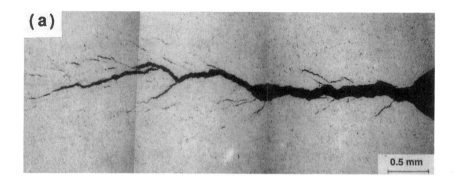

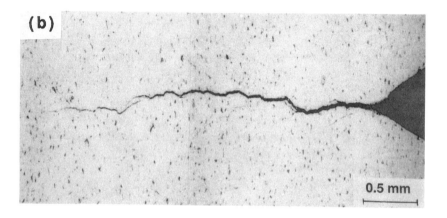

Fig. 7.3 Examples of crack profiles in Al_2O_3/33 vol.% SiC whisker composite subjected to static and cyclic tensile loads (in a four-point bend configuration) in 1400°C air. From Ref. 22. Crack growth direction is from right to left. (a) Static crack growth at $K \approx 4$–5 MPa$\sqrt{m}$. (b) Cyclic crack growth at $R = 0.15$ and $\nu_c = 0.1$ Hz in the ΔK range 3.5–5 MPa$\sqrt{m}$.

Figure 7.5 shows a comparison of the static and cyclic crack velocities for the alumina/SiC composite, similar to that illustrated in Fig. 7.2 for alumina. As seen for monolithic alumina, the ceramic composite undergoes slower rates of fracture when (1) fluctuations are introduced in the tensile loads and (2) the cyclic frequency is raised. The differences between static and cyclic crack growth rates are maximum at the lower K_{max} levels. As in the case of unreinforced alumina, crack velocities estimated for cyclic loads on the basis of sustained load crack growth data (assuming identical failure mechanisms) are higher than those measured experimentally.

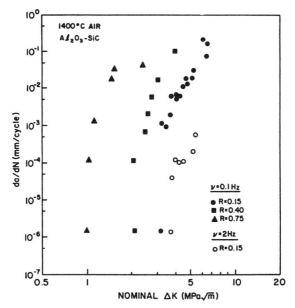

Fig. 7.4 Fatigue crack growth rate, *da/dN*, plotted as a function of stress intensity factor range, ΔK in Al$_2$O$_3$/33 vol.% SiC whisker composite in 1400°C air at cyclic frequencies of 0.1 Hz and 2 Hz at $R = 0.15$, 0.40 and 0.75. After Ref. 22.

7.5 Micromechanisms of Damage and Failure at Crack Tip

7.5.1 Role of Preexisting Glass Phase

During high temperature fracture, ceramic materials containing viscous glass films along grain boundaries exhibit predominantly intergranular separation under both static and cyclic loading. Figure 7.6a shows the static fracture characteristics of AD 90 alumina in 1050°C air. Intergranular fracture and bridging of the grain facets by glassy ligaments can be seen in this figure. Figure 7.6b shows a grain facet which is covered with a molten glass film in the AD 90 alumina subjected to cyclic loads ($R = 0.15$ and $\nu_c = 0.13$ Hz) at 1050°C. Transmission electron microscopy analyses of damage near the main crack tip revealed distributed microcracks populating grain facets. Figure 7.6c shows an example of intergranular microcracking about a fatigue crack in the alumina.

Microcracks are also known to occur in the elevated temperature creep of alumina with little or no preexisting glass phase. For example, the development of microcracks during creep fracture of two hot-pressed aluminas, which were free of grain boundary glass films, was studied by Wilkinson *et al.*[52] who employed tensile and four-point bend specimens. They found that the concentrations and morphology of the cavities and microcracks were strongly

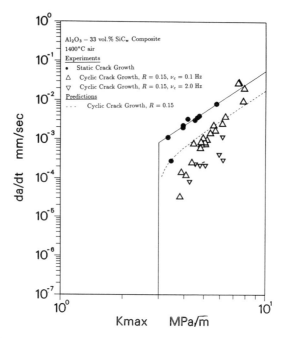

Fig. 7.5   Data for $R = 0.15$ from Fig. 7.4 replotted in terms of da/dt versus K_{max} and compared with experimentally measured static crack velocity data for $Al_2O_3/33$ vol.% SiC whisker composite in 1400°C air. Also indicated are cyclic crack velocities computed from static fracture data using Eqn. (14).

dependent on the microstructure, test temperature, specimen geometry and applied stress magnitude. While "shear bands" were observed on the tensile side of the bend specimens (which had earlier been reported by Dalgleish et al.,[53] also in a study of the creep rupture of alumina), they were not found in the direct tension specimens.

The significant role of preexisting glass phase in influencing subcritical creep crack growth in monolithic ceramics is also evident from experiments where the fatigue characteristics of different purities of alumina were investigated.[34] When a fine-grained alumina of high purity (such as the 99.9% pure AD 999 alumina from Coors Ceramics, Golden, CO) with very little glass phase content is subjected to creep crack growth, it was found that in the four-point bend geometry, subcritical fracture is essentially suppressed under both static and cyclic loads at 1050°C air. (Recall that the same specimen geometry and loading configuration at the same temperature produces substantial levels of stable fracture in AD 90 alumina: Figs. 7.1 and 7.2.) Unstable fracture occurred in the AD 999 alumina as the maximum stress intensity factor of the fatigue cycle approached the fracture toughness of the material at the test temperature, and no subcritical crack growth could be detected. This result serves to illustrate that stable crack growth at high temperature is very strongly

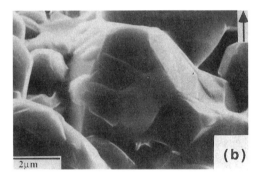

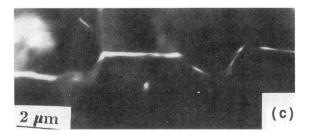

Fig. 7.6 (a) Static fatigue characteristics of AD 90 alumina at 1050°C. (b) A grain facet on the fatigue fracture surface covered with molten glass film. ($R = 0.15$ and $\nu_c = 0.13$ Hz.) Arrow indicates crack growth direction. (c) Grain boundary microcracking around the fatigue crack tip in the alumina. From Ref. 34.

mechanistically dependent on the presence of amorphous films along grain boundaries. The viscous flow of these glassy films during the creep fracture experiments imparts apparent plasticity which, in turn, stabilizes the fracture process. Furthermore, as noted in Section 7.1, the glass phase is likely to serve as a preferential path for diffusion of matrix atoms.

7.5.2 Role of In Situ-Formed Glass Phase

In some ceramics and ceramic composites, glassy films can also form *in situ* during the high temperature fatigue experiment as a consequence of the oxidation of the matrix and/or the reinforcement. Examples of such ceramics include those which contain SiC reinforcements. The high temperature fatigue crack growth characteristics of SiC-reinforced ceramics also qualitatively resemble those of unreinforced alloys. The enhanced level of glass phase production at the elevated temperature, however, provides significantly greater levels of stable crack growth in the SiC-reinforced composite (over a broader range of temperatures and stress intensity levels) than in the unreinforced ceramics where only preexisting glassy films influence creep–fatigue.

When exposed to oxygen-containing environments at temperatures typically higher than 1250°C, SiC undergoes an oxidation reaction whose primary product is silica glass. A number of independent studies have shown that the oxidation of SiC whiskers initially proceeds according to the reaction:[54,55]

$$2SiC + 3O_2 \rightarrow 2SiO_2 + 2CO \tag{15}$$

Thermal oxidation studies of unnotched and uncracked specimens of Al_2O_3/SiC composites also reveal the formation of carbon-rich and silicon-rich reaction layers via solid-state oxidation of the SiC whisker according to the reaction:[56]

$$SiC + O_2 \rightarrow SiO_2 + C \tag{16}$$

which later leads to

$$3Al_2O_3 + 2SiC + 3O_2 \rightarrow 3Al_2O_3 \cdot 2SiO_2 + 2CO \tag{17}$$

The oxidation reaction represented in Eqn. (16) is known to have the lowest standard-state Gibbs free energy of any of the reactions for SiC for temperatures of up to 1500°C.[57] During the early stages of this thermodynamically favored reaction, graphitic carbon layers are produced near the interface of SiC whiskers with the ceramic matrix. In the advanced stages of the reaction, CO, aluminosilicate glasses, and mullite are produced, depending on the test temperature, exposure time, alumina matrix composition, and the partial pressure oxygen in the environment. Creep experiments conducted in dry nitrogen also show significantly lower creep strain rates than in air.[21]

In an unnotched or uncracked specimen of the Al_2O_3/SiC composite, the oxidation of SiC leading to the formation of SiO_2, which is rate-limited by the diffusion of oxygen in the material, is confined to the near-surface regions of the specimen.[58] However, in the high temperature fatigue crack growth experiments conducted with specimens containing through-thickness cracks, the transport of the oxidizing species to the highly stressed crack tip region promotes glass phase formation through the thickness of the specimen (e.g., Ref. 22). Furthermore, microstructural changes in the creep fracture specimens are confined to a well-defined region in the immediate vicinity of the stress concentration provided by the crack tip.

The viscous flow of the silica glass causes extensive interfacial cavitation in the composite.† Figure 7.7a shows a transmission electron micrograph of the as-received microstructure of the Al_2O_3/33 vol.% SiC_w composite. Note the mechanically bonded interfaces and the absence of any interfacial cavities or cracks in this microstructure; the as-received material is of high purity and it contained very little glass phase at grain boundaries and whisker–matrix interfaces. Figure 7.7b–e shows transmission electron micrographs of the region ahead of a fatigue crack in the ceramic composite ($T = 1400°C$), which show the spread of the glass phase and the resulting void formation and debonding at the whisker–matrix interface. The formation of cavities within the damage region at the fatigue crack tip is similar to that seen in the near-surface regions of a (unnotched and uncracked) creep specimen (see Chapter 8 by Nutt, on creep, in this volume).

The growth and coalescence of the interfacial cavities under the influence of static and cyclic tensile loads results in extensive microcracking ahead of the main crack tip. Figure 7.8a shows an example of the formation of a diffuse microcrack zone ahead of a main crack in the air environment at 1500°C. Figure 7.8b is an example of microcracking damage at 1400°C.

Detailed transmission electron microscopy analyses of crack tip damage reveal that the oxidation of SiC is strongly influenced by the loading rate (cyclic frequency), test temperature, and stress intensity factor. In general, higher values of K or ΔK, higher temperatures, and lower loading rates (lower ν_c) result in a greater amount of glass at the interface between alumina and SiC.[22-24]

It is of interest to note here that certain intermetallic materials and intermetallic-matrix composites, which contain glassy films along grain boundaries and interfaces, also exhibit crack growth characteristics at high temperatures which resemble the trends seen in monolithic and composite ceramics containing glass phases. For example, experimental work by Ramamurty *et al.*[59] on powder-processed $(Mo, W)Si_2$ alloys, with and without SiC particle reinforcements, indicates that both preexisting and *in situ*-formed glass films in the silicide alloys create cavitation and microcracking whose microscopic and macroscopic effects are qualitatively the same as those in the ceramics. Here, oxidation takes place according to the following reaction:

$$5(Mo, W)Si_2 + 7O_2 \rightarrow (Mo, W)_5Si_3 + 7SiO_2 \tag{18}$$

---

†The mechanics of creep and cavitation as a result of the viscous flow of an amorphous intergranular film have been the subject of many theoretical models, which are not reviewed here because of space restrictions. Full details can be found in Refs. 22–25, 54–58.

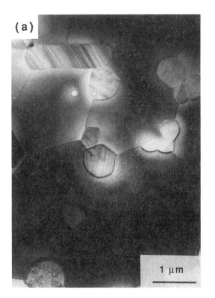

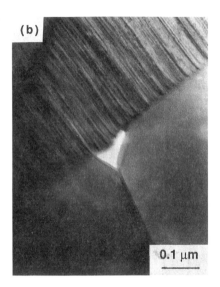

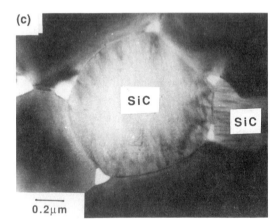

Fig. 7.7 (a) Transmission electron micrograph of the as-received, untested microstructure of Al_2O_3/33 vol.% SiC whisker composite. (b–d) Transmission electron micrographs of the fatigue crack tip region in the composite in 1400°C air ($R = 0.15$ and $v_c = 0.1$ Hz) showing the nucleation of interfacial cavities. Part (e) shows the development of a diffuse cavitation zone ahead of the fatigue crack tip. From Ref. 25.

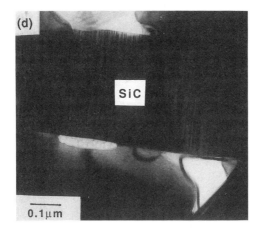

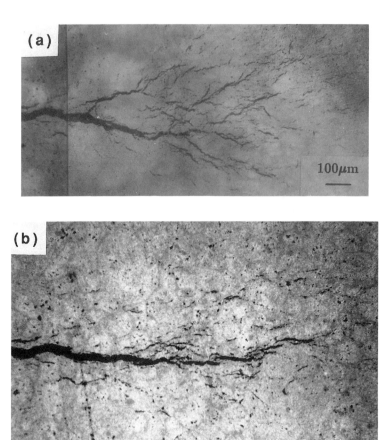

Fig. 7.8 Profiles of cracks in Al_2O_3/33 vol.% SiC whisker composite. (a) An example of the formation of a diffuse microcrack zone ahead of a main crack in the air environment at 1500°C. (b) An example of microcracking damage at 1400°C. From Refs. 22 and 25.

When the silicide alloy is reinforced with SiC, additional glass phase may form at the crack tip as a result of the oxidation of SiC, as per Eqn. (15). The cavitation process instigated by the viscous flow of interfacial glass films in the $(Mo, W)Si_2$ matrix composites can also reduce matrix dislocation activity, even at temperatures well above the ductile–brittle transition temperature. It is found that when the processing conditions of the silicide-based composites reduce the concentration of interfacial glass films, the role of matrix dislocation plasticity in influencing high temperature deformation is correspondingly increased.

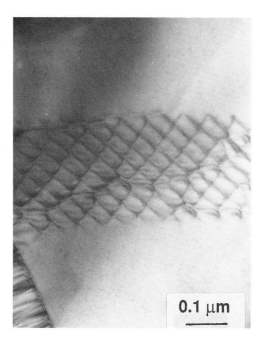

0.1 μm

Fig. 7.9 TEM micrograph taken from the fatigue crack tip ($R = 0.15$, $T = 1400°C$, and $\nu_c = 0.1$ Hz) showing the formation of dislocation networks in the alumina matrix reinforced with SiC whiskers. The dislocation network is pinned by the SiC whiskers.

7.5.3 Other Mechanisms

As mentioned earlier, cavitation and microcracking arising from amorphous glass films dominate over any possible effects of dislocation plasticity on high temperature crack growth in the ceramic materials, for the conditions of the experiments reviewed here. There is, however, sporadic evidence of hexagonal dislocation networks and subgrain formation during the high temperature fatigue fracture of ceramic materials (e.g., Ref. 22). Figure 7.9 shows an example of the formation of dislocation networks in the alumina matrix reinforced with SiC whiskers, where the dislocation network is pinned by the SiC whiskers. (It is known that dislocation networks can form in sintered ceramics. For the present material, such networks were detected only in the specimens tested at high temperatures and they could not be found in the as-received condition.)

TEM observations of thin foils taken from the *fatigue crack tip regions* have confirmed the existence of several cavitation micromechanisms which are related to diffusion and grain boundary sliding.

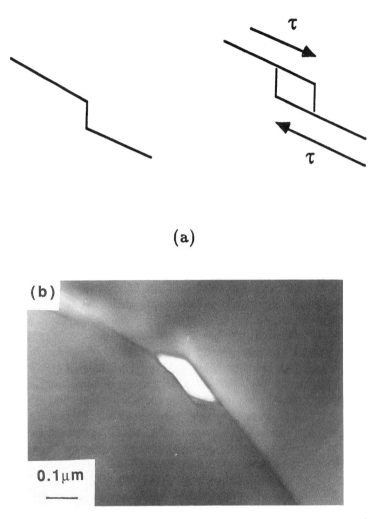

(a)

Fig. 7.10   (a) Schematic illustration of void nucleation from grain boundary sliding. (b) TEM micrograph taken from the fatigue crack tip ($R = 0.15$, $T = 1400°C$, and $\nu_c = 0.1$ Hz) showing the formation of the void in the alumina/SiC composite. From Han and Suresh, MIT.

7.5.3.1 Grain Boundary Sliding and Triple-Point Separation

This mechanism is commonly found in unreinforced ceramics and reinforcement-poor regions of ceramic composites. As illustrated in Fig. 7.10a, stress concentration is produced at triple-grain junctions when grain boundary sliding occurs; a negative pressure is created within the glass phase secreted at triple-grain junctions. Since the viscous glass phase has a smaller bulk modulus compared to the matrix grains, it cavitates quite readily as shown in Fig. 7.10b.

(Note the meniscus of the glass phase within the triple-junction cavity.) Quantitative estimation of the conditions of void nucleation due to the negative pressure of a constrained fluid has been given in Refs. 16 and 60.

7.5.3.2 Whisker Debonding and Pull-out

The debonding of the whisker–matrix interface ahead of the fatigue crack facilitates the advance of the fatigue crack around the whiskers while the faces of the fatigue crack are bridged by the intact whisker. Figure 7.11a shows an example of a SiC whisker which bridges a fatigue crack in the Al_2O_3/SiC_w composite at 1400°C. The sliding of the whisker is clearly evident in this figure from the rectangular-shaped void left at the end of the whisker. Whisker pull-out is also illustrated in the transmission electron fractograph of the fatigue fracture surface in the composite shown in Fig. 7.11b. Here, pulled-out whiskers, which have been converted to glass, can be seen protruding from the fracture surface.

7.5.3.3 Inhibition of Grain Boundary Sliding by Reinforcements

In many whisker-reinforced ceramic composites, the ratio of the grain diameter to the diameter of the whisker is typically in the range 5–10. The whiskers often populate along grain boundaries and serve as obstacles to the sliding motion of grains during high temperature creep.[61] When grain boundary motion is uninhibited by reinforcements, crack-like cavities and wedge-shaped flaws form along the grain facets (see, e.g., Fig. 7.10). In ceramics where grain boundary motion is obstructed by the presence of whiskers, only small, discontinuous flaws develop. Figure 7.12a is a schematic sketch of the whisker-inhibited grain boundary sliding and the attendant nucleation of intergranular/interfacial cavities. Figure 7.12b is a transmission electron micrograph of the fatigue crack tip region showing cavities formed by the process schematically shown in Fig. 7.12a. Increases in test temperature and applied stresses enhance the damage process shown in Fig. 7.12b.

7.5.3.4 Crack Deflection

Debonding along grain boundaries due to the viscous flow of amorphous phases also promotes periodic deflections of the crack tip in the unreinforced ceramics subjected to static and cyclic loads. An example of a periodically deflected crack profile as a result of intergranular fatigue fracture in the 90% pure alumina at 1050°C is shown in Fig. 7.14b in the next section. In reinforced ceramics, preferential failure along the whisker–matrix interface (which is the primary site for the formation of *in situ* glass films) also gives rise to crack deflection. In addition, the development of a diffuse microcrack zone, and the coalescence of microcracks ahead of the creep crack under both static and cyclic loads, caused the main crack to deflect and bifurcate, as seen in Figs. 7.3 and 7.8. As a consequence of crack deflection,[62] the effective driving force for crack growth is diminished (as compared to that of a straight crack of the same projected length subjected to the same far-field loading). While deflections in

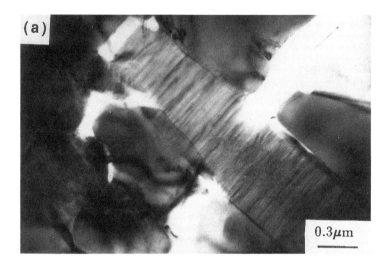

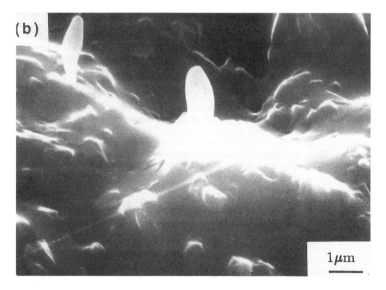

Fig. 7.11   (a) An example of a SiC whisker bridging the faces of a fatigue crack in an alumina/SiC composite ($R = 0.15$, $T = 1400°C$, and $\nu_c = 0.1$ Hz). (b) Scanning electron fractography showing pulled-out SiC whiskers which have been oxidized to form glass. From Refs. 4 and 61.

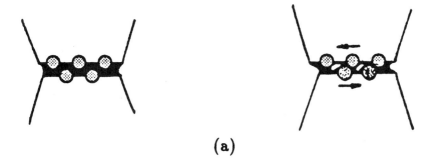

(a)

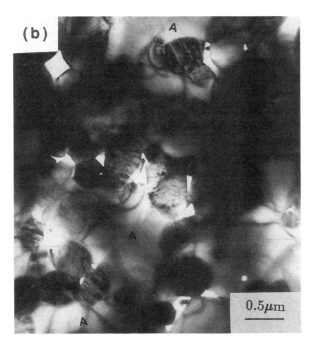

Fig. 7.12   (a) A schematic representation of the pinning of grain boundaries by whiskers. (b) An example of the process shown in (a) in the fatigue crack tip region of the alumina/SiC composite subjected to fatigue fracture at 1400°C. From Ref. 61. "A" refers to alumina grains.

elastic cracks can give rise to reductions in effective driving force for fracture, the presence of an inelastic damage zone encompassing the deflected crack tip is known to enhance the beneficial effects of crack branching (see, e.g., Suresh and Shih[63]).

7.5.4 Differences between Mechanisms of Creep Crack Growth and Creep–Fatigue Crack Growth

Whereas the primary mechanisms of failure, both ahead of the crack tip and in the wake of the crack, are apparently similar under static and cyclic fatigue, there also exist some noticeable differences between specimens subjected to static and cyclic loads in the high temperature environment in both monolithic ceramics and ceramic composites. Figure 7.13a shows an example of the development of a microscopically tortuous crack path resulting from intergranular failure in the AD 90 alumina under static loading at high temperature. Note the absence of any debris particles within the crack walls in this micrograph. Figure 7.13b shows the profile of a crack subjected to cyclic loading in the same alumina at the same temperature. In this case, repeated loading and unloading of the crack, along with recurring frictional sliding along the grain facets, leads to the formation of debris particles of the alumina between the crack faces. Furthermore, the pumping action of the crack walls under cyclic loading tends to "squeeze out" the debris particles and the glass films on the crack faces, an example of which can be seen in Figure 7.13b.

In the SiC whisker-reinforced alumina, the kinetics of microstructural transformations in SiC whiskers located within the crack tip region can also be influenced by whether static or cyclic loads are imposed on the material in the elevated temperature environment. When the applied far-field tensile stress is held constant for prolonged time periods, large fractions of the SiC whiskers located within the crack tip region are converted to amorphous glass pockets. Figure 7.14a shows an example of such "bulk conversion" of SiC whiskers into glassy regions within the damage zone of a crack in the Al_2O_3/33 vol.% SiC whisker composite subjected to a constant stress intensity factor of approximately $3.5\,MPa\sqrt{m}$. By contrast, repeated loading and unloading of the crack appears to preclude such large-scale conversion of the SiC whisker into glass pockets. Instead, cyclic loading causes the whiskers to break, as illustrated in the transmission electron micrograph of a whisker ahead of a fatigue crack in the alumina matrix composite at 1400°C. The meniscus of the molten glass film flowing within the broken whisker can be seen in Fig. 7.14b.

In addition to the intrinsic differences in the mechanisms of near-tip damage between static and cyclic fatigue, there are other more macroscopic differences which arise as a consequence of crack wake contact or rate sensitivity. Viscous deformation of the intergranular or interfacial glass films is sensitive to the loading rate. As a result, fracture along the glass-covered grains

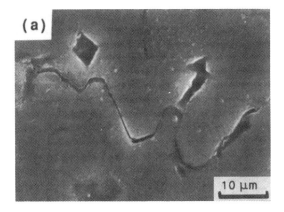

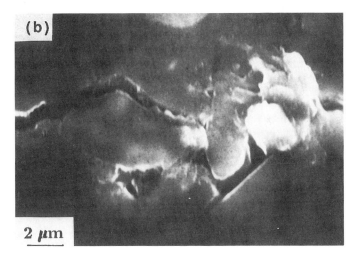

Fig. 7.13 Differences in crack wake contact and bridging mechanisms seen between static and cyclic fracture in AD 90 alumina at 1050°C. From Ref. 34. (a) Deflected crack path during static crack growth with no debris formation. (b) Deflected crack path with debris particles formed at a result of repeated rubbing between the crack faces under cyclic loading. Also seen are the debris and glass films which are squeezed out of the crack due to the pumping action of the crack walls.

or whiskers would be expected to be strongly affected by whether the loading is static or cyclic, as well as by the frequency and waveform (including hold times, if any) of the imposed far-field loads. Consistent with this expectation, Lin *et al.*[64] found that fatigue lifetimes are dependent upon the shape of the load cycle, with the hold time at the peak load of the cycle determining the overall fatigue life in stress-controlled cyclic experiments on alumina. They show that cyclic loading with a short duration of peak stress exhibited a larger

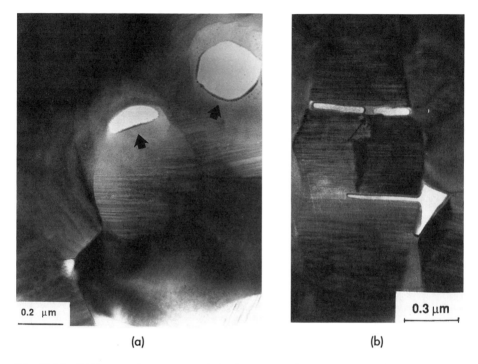

(a)   (b)

Fig. 7.14 Differences between crack tip damage mechanisms for static and cyclic fracture in alumina/SiC composite at 1400°C. From Ref. 22. (a) Crack tip TEM image showing conversion of large fractions of SiC whiskers into glass pockets (denoted by arrows) during static loading. (b) Crack tip TEM image showing breakage of SiC whiskers under cyclic loading. Note the meniscus of the glass film inside the broken whisker (indicated by the arrow).

number of cycles to failure than the one with a longer hold time at the peak stress. The beneficial effect of cyclic loading (as compared to static loading) was attributed to the mitigating effect of glass films on the grain boundaries. In agreement with the results of Fig. 7.4, the number of cycles to failure was also found to increase with increasing cyclic frequency (for a fixed cycle shape and hold time).[64]

One possible reason for the beneficial effect of cyclic loading on crack growth resistance and failure life may lie in the bridging of the crack walls by the viscous glass phase (see Fig. 7.6). When an elastic crack subjected to plane stress is bridged by a single viscous ligament, the effective stress intensity factor, K_{eff}, at the crack tip can be written as[65,66]

$$K_{eff} = K_{app}\left\{1 - \exp\left(-\frac{\pi E_m t}{8\eta_l}\right)\right\} \tag{19}$$

where K_{app} is the applied stress intensity factor, E_m is Young's modulus of the matrix, t is the time since the ligament begins to grow, and η_l is the viscosity of the ligament, which is taken to be three times the shear velocity. The rate sensitivity of deformation of the viscous ligament is expressed in terms of a characteristic time, which is defined as

$$t_c = \frac{8\eta_l}{\pi E_m} \tag{20}$$

Taking the governing equations for K_{eff} to be of the form

$$\dot{K}_{eff}(t) + \frac{K_{eff}(t)}{t_c} = \frac{K_{app}(t)}{t_c} \tag{21}$$

Lin et al.[64] examined the effect of glass phase in influencing K_{eff} for different cyclic frequencies and waveform for the same applied applied stress intensity factor range (i.e., same K_{max} and R). Taking $t_c = 1.25$ s, $E_m \approx 260$ GPa and $\eta_l \approx 1.2 \times 10^{12}$P, Fig. 7.15 shows the variation of K_{eff}/K_{max} with t/t_c for static loading ($K = K_{max}$) and cyclic loading (between $K_{min} = 0.1K_{max}$ and K_{max}) with sinusoidal waveform (2 Hz), square waveform (2 Hz) and two trapezoidal waveforms: Trapezoid I (minimum to peak rise time of $t_r = 0.01$ s, hold time at peak stress of $t_h = 0.25$ s, and the time for load drop from the peak to the minimum of $t_u = 0.01$ s), and Trapezoid II ($t_r = 0.01$ s, $t_h = 2.5$ s, and $t_u = 0.01$ s). For $t > t_c$, the mean effective stress intensity factor for the square and sine waveforms (for which the effective stress intensity factor is significantly reduced by the bridging of the crack walls by the glass ligament) is $0.55K_{max}$, which is the same as the applied mean stress intensity factor, $0.5(K_{min} + K_{max})$. On the other hand, the trapezoidal waveform II with a longer hold time at peak exhibits essentially the same K_{eff} as a function t as static loading. Trapezoid I with a shorter hold time at peak stress has a slightly smaller K_{eff}. These predictions are qualitatively consistent with the stress–life data measured at 1200°C for polycrystalline alumina with grain boundary glass films, and with the crack growth rate data presented in Fig. 7.2.

Experimental studies of compressive creep in ceramics (susceptible to extensive cavitation) have shown the possibility of resintering of cavities.[67] Cyclic tensile loading of a crack leads to the formation of a reversed damage zone at the crack tip upon unloading. Residual compressive stresses are generated within this cyclic/reversed damage zone which spans a distance equal to a large fraction (up to 25%) of the monotonic damage zone directly ahead of the fatigue crack. It is possible that some resintering of cavities occurs within the reversed damage zone at the fatigue crack tip. This resintering of cavities under cyclic loading may serve as one of the contributing factors for the apparently beneficial crack growth response observed in cyclic fatigue as compared to static fracture.

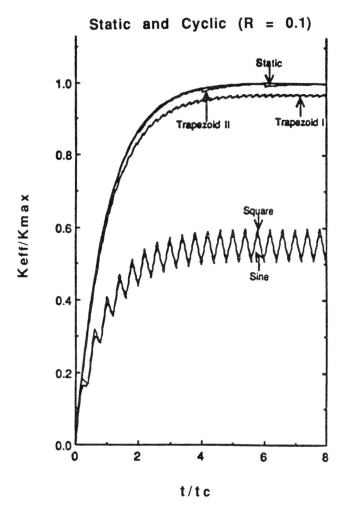

Fig. 7.15 A comparison of the normalized effective stress intensity factor, K_{eff}/K_{max}, as a function of the normalized time, t/t_c, for static loading and for different waveforms in cyclic loading. After Lin et al.[64] See text for details.

7.5.5 Unreinforced versus Reinforced Ceramics

The experimental results shown in Figs. 7.1, 7.2, 7.4, and 7.5 indicate that the addition of a reinforcing phase, such as SiC whiskers, significantly enhances the damage tolerance of ceramic materials. Similar results have also been recorded for hot-pressed Si_3N_4 reinforced with SiC whiskers[36] and $(Mo, W)Si_2$ alloys (which, similar to the ceramics, are also susceptible to interfacial cavitation by the viscous flow of glassy films) reinforced with SiC particles.[59,68] This improved crack growth resistance is seen for creep crack growth under both static and cyclic loads, although cyclic loading at higher frequencies, lower stress intensity ranges, and shorter hold times promotes a significantly lower crack growth rate and longer failure life than static loading.

Many of the toughening mechanisms discussed in detail in the literature for enhancing the quasi-static fracture toughness of ceramics through reinforcements also play an important role in improving the crack growth resistance at elevated temperatures. These mechanisms include: (1) enhanced crack bridging[1,4,25,49,69] by unbroken whiskers (Fig. 7.11), (2) crack deflection and branching[4,22,25,36] (Figs. 7.3, 7.8, and 7.13), (3) reduced grain boundary sliding as whiskers inhibit the motion of grain facets (Fig. 7.12), (4) enhanced blunting of the crack,[22,25,34,70,71] and (5) main crack–microcrack interactions and dissipation of energy at the crack tip by the development of a diffuse microcrack zone[4,25,72] (Fig. 7.8).†

In addition, discontinuous reinforcements added to ceramic matrices result in an extension of the stress intensity range over which subcritical crack growth occurs (compare, for example, Figs. 7.1 and 7.3, and Figs. 7.2 and 7.5). This effect may arise from the toughening mechanisms described in the preceding paragraph as well as from the production of glass phase *in situ* through microstructural transformations within the reinforcement phase. In their study of creep crack growth in a magnesium aluminosilicate glass at 750–775°C, Chan and Page[73] postulate the existence of a critical density of cavities and microcracks which directly determines the existence of a static crack growth threshold. Cavitation-aided fracture over an extended range of K or ΔK is aided by the increased production of glassy films. Furthermore, the increased toughness of the ceramic as a consequence of the reinforcement increases the K or ΔK at which final fracture occurs. (Note that final fracture under static loading conditions occurs when the applied stress intensity factor approaches the final (plateau) K of the resistance curve. Under cyclic loading conditions, final fracture occurs when $\Delta K \rightarrow K_{Ic}/(1 - R)$.)

7.6 Summary

In this chapter, we have sought to provide a state-of-the-art review of the mechanics and micromechanisms of high temperature crack growth in ceramics and discontinuously reinforced ceramic composites. Because of the limited amount of experimental data available in the literature which pertains primarily to oxide cermics and SiC reinforcements, the discussions of crack growth rates and fracture mechanisms have centered around alumina ceramics, with and without SiC reinforcements. However, the generality of the mechan-

---

†Han *et al.*[25] have carried out toughness measurements on SiC whisker-reinforced alumina at room temperature after introducing different amounts of controlled microcrack damage ahead of the main precrack. Their results show that some toughness gains can be achieved by controlled microcracking at the main crack tip. However, profuse microcracking can result in reductions in effective toughness by promoting fracture by the easy coalescence of microcracks.

isms presented here to significantly broader classes of ceramics and intermetallics has been pinpointed wherever appropriate.

Particular attention has been devoted in this chapter to identifying the differences between static and cyclic load failure processes in discontinuously reinforced ceramic materials at high temperatures. There is still controversy surrounding the mechanistic origins of cyclic load failure in ceramic materials at room temperature. Furthermore, early studies on ceramics hypothesized that cyclic crack growth mechanisms at elevated temperatures could be predicted on the basis of static fracture data because of the seemingly similar fracture response for the two types of loading. However, as shown in this chapter, there is emerging information which clearly shows that for both monolithic and reinforced alumina subjected to high temperature fracture in certain temperature regimes, there can be some noticeable differences between the micromechanisms of static and cyclic fracture both ahead of the crack tip and behind the crack tip.

The issue of preexisting or *in situ*-formed glass films influencing deformation in the crack tip and crack wake region was given considerable attention in this chapter. The presence of the glass film along grain boundaries and interfaces imparts to the creep fracture response marked sensitivity to strain rate, cyclic frequency, and cyclic waveform. Furthermore, the glass phase contributes to the differences between the failure mechanisms seen under static and cyclic loads. While the existence of the glass phase enhances the range of stress intensity factor over which subcritical crack growth occurs in ceramics at elevated temperatures, it can also reduce the threshold stress intensity factor for the onset of fracture by promoting early nucleation of voids. Control of the extent of glassy films in the microstructure, therefore, presents an important challenge in the processing of advanced ceramic materials with improved damage tolerance at high temperatures. It is interesting to note that in the synthesis of molydisilicide-matrix composites, which exhibit high temperature fracture characteristics similar to those of ceramics, attempts are now being made to control the concentration of glass films present in the microstructure by proper additions of carbon. Carbon reduces the *in situ* formation of SiO_2 by aiding in the following reaction:

$$2SiO_2 + 6C \rightarrow 2SiC + 4CO \qquad (22)$$

Analogous approaches to the control of glass films in ceramic materials would provide one route for the proper manipulation of crack growth resistance at high temperatures.

Considerable further work is necessary to develop a complete understanding of creep crack growth and creep–fatigue interactions in ceramic materials. First, a broad database needs to be established in a variety of ceramic microstructures and composites to develop a general theory of cyclic fracture at elevated temperatures. Second, comparisons of experimental work from different researchers require proper methodology for characterizing fracture, including specimens which are sufficiently "large" to provide unambiguous characterizations in terms of fracture mechanics. Third, microstructures require thorough characterization both in the as-received condition and after high temperature fracture to develop a proper understanding of the micromechanisms. Fourth, systematic experiments covering a wide variety of controlled microstructural variables (such as grain size, reinforcement content, and concentration of glass phase), mechanical variables (such as load ratio, cyclic frequency, and cyclic waveform), and environmental variables (such as inert and aerobic ambients, and different partial pressures of oxygen) should be conducted to enhance current understanding. Although high temperature fatigue testing remains a major challenge both in terms of difficulty and cost, recent advances in mechanical testing provide encouraging possibilities for significant advances in this area in the near future.

Acknowledgments

This work has been supported by the Basic Science Division of U.S. Department of Energy under Grant DE-FG02-84ER-45167. The author is grateful to his present and former graduate students, particularly U. Ramamurty, L. X. Han, and L. Ewart, who have contributed to the concepts reviewed in this chapter.

References

1. P. F. Becher and G. C. Wei, "Development of SiC-Whisker-Reinforced Ceramics," *Am. Ceram. Soc. Bull.*, **64**[2], 298–304 (1985).
2. K. K. Chawla, *Ceramic-Matrix Composites*, Chapman & Hall, London, 1993.
3. B. R. Lawn, *Fracture of Brittle Solids*, 2nd edn, Cambridge University Press, Cambridge, U.K., 1993.
4. S. Suresh, *Fatigue of Materials*, Cambridge University Press, Cambridge, U.K., 1991.
5. J. Freidel, *Dislocations*, Pergamon Press, Oxford, U.K., 1964.
6. J. P. Poirier, *Creep of Crystals*, Cambridge University Press, Cambridge, U.K., 1982.
7. R. Raj and M. F. Ashby, "On Grain Boundary Sliding and Diffusional Creep," *Metall. Trans. A*, **2A**, 1113–1127 (1971).
8. F. W. Crossman and M. F. Ashby, "The Non-Uniform Flow of Polycrystals by Grain Boundary Sliding Accommodated by Power Law Creep," *Acta Metall.*, **23**, 425–440 (1975).
9. I.-W. Chen and A. S. Argon, "Creep Cavitation in 304 Stainless Steel," *Acta Metall.*, **29**, 1321–1333 (1981).
10. H. J. Frost and M. F. Ashby, *Deformation Mechanisms Maps: The Plasticity and Creep of Metals and Ceramics*, Pergamon Press, New York, NY, 1982.
11. J. Weertman, "Steady-State Creep of Crystals," *J. Appl. Phys.*, **28**, 1186–1189 (1957).
12. R. L. Stocker and M. F. Ashby, "On the Empirical Constants in the Dorn Equation," *Scripta Metall.*, **7**, 115–120 (1973).
13. S. Suresh and J. R. Brockenbrough, "A Theory for Creep by Interfacial Flaw Growth in Ceramics and Ceramic Composites," *Acta Metall. Mater.*, **38**[1], 55–68 (1990).

14. J. Weertman, "Theory of Steady-State Creep Based on Dislocation Climb," *J. Appl. Phys.*, **26**, 1213–1217 (1955).
15. D. R. Clarke, "High-Temperature Deformation of a Polycrystalline Alumina Containing an Intergranular Glass Phase," *J. Mater. Sci.*, **20**, 1321–1332 (1985).
16. R. L. Tsai and R. Raj, "Creep Fracture in Ceramics Containing Small Amounts of a Liquid Phase," *Acta Metall.*, **30**, 1043–1058 (1982).
17. J. E. Marion, A. G. Evans, M. D. Drory, and D. R. Clarke, "High Temperature Failure Initiation in Liquid Phase Sintered Materials," *Acta Metall.*, **31**, 1445–1457 (1983).
18. M. M. Chadwick, D. S. Wilkinson, and J. R. Dryden, "Creep due to a Non-Newtonian Grain Boundary Phase," *J. Am. Ceram. Soc.*, **75**[9], 2327–2334 (1992).
19. K. Jakus, S. M. Wiederhorn, and B. Hockey, "Nucleation and Growth of Cracks in Vitreous-Bonded Aluminum Oxide at Elevated Temperatures," *J. Am. Ceram. Soc.*, **69**[10], 725–731 (1986).
20. A. G. Evans and A. Rana, "High Temperature Failure Mechanisms in Ceramics," *Acta Metall.*, **28**, 129–141 (1978).
21. P. Lipetzky, S. R. Nutt, D. A. Koester, and R. F. Davis, "Atmospheric Effects on Compressive Creep of SiC-Whisker-Reinforced Alumina," *J. Am. Ceram. Soc.*, **74**[6], 1240–1478 (1991).
22. L. X. Han and S. Suresh, "High Temperature Failure of an Alumina-Silicon Carbide Composite under Cyclic Loads: Mechanisms of Fatigue Crack-Tip Damage," *J. Am. Ceram. Soc.*, **72**[7], 1233–1238 (1989).
23. S. Suresh, "Mechanics and Micromechanisms of Fatigue Crack Growth in Brittle Solids," *Intl. J. Fract.*, **42**, 41–56 (1990).
24. S. Suresh, "Fatigue Crack Growth in Brittle Materials," *J. Hard Mater.*, **2**[1–2], 29–54 (1991).
25. L. X. Han, R. Warren, and S. Suresh, "An Experimental Study of Toughening and Degradation due to Microcracking in a Ceramic Composite," *Acta Metall.*, **40**[2], 259–274 (1992).
26. H. Riedel, *Fracture at High Temperatures*, Springer-Verlag, Berlin, Germany, 1987.
27. H. Riedel and J. R. Rice, "Tensile Cracks in Creeping Solids," in *Fracture Mechanics: Twelfth Conference*, ASTM Technical Publication 700, ed. P. C. Paris, American Society for Testing and Materials, Philadelphia, PA, 1980, pp. 112–130.
28. J. R. Rice, "A Path-Independent Integral and the Approximate Analysis of Strain Concentrations by Notches and Cracks," *J. Appl. Mech.*, **35**, 379–386 (1968).
29. V. Kumar, M. D. German and C. F. Shih, "An Engineering Approach for Elastic-Plastic Fracture Analysis," EPRI Technical Report NP-1931, Electric Power Research Institute, Palo Alto, CA, 1981.
30. C. F. Shih and A. Needleman, "Fully Plastic Crack Problems, Part I: Solutions by a Penalty Method," *J. Appl. Mech.*, **51**[3], 48–56 (1984).
31. J. Landes and J. Begley, "A Fracture Mechanics Approach to Creep Crack Growth," in *Mechanics of Crack Growth*, ASTM Technical Publication 590, The American Society for Testing and Materials, Philadelphia, PA, 1976, pp. 128–148.
32. K. Ohji, K. Ogura, and S. Kubo, "Stress Field and Modified J Integral near a Crack Tip under Condition of Confined Creep Deformation," *J. Soc. Mater. Sci. Jpn.*, **29**, 465–471 (1980).
33. J. L. Bassani, D. E. Hawk, and A. Saxena, "Evaluation of the C_t Parameter for Characterizing Creep Crack Growth Rate in the Transient Regime," in

Nonlinear Fracture Mechanics, ASTM Technical Publication 995, American Society for Testing and Materials, Philadelphia, PA, 1986, pp. 7–26.

34. L. Ewart and S. Suresh, "Elevated-Temperature Crack Growth in Polycrystalline Alumina under Static and Cyclic Loads," *J. Mater. Sci.*, 27, 5181–5191 (1992).
35. L. Ewart, "Ambient and Elevated Temperature Fatigue Crack Growth in Alumina," Ph.D. Thesis, Brown University, Providence, RI, 1990.
36. U. Ramamurty, T. Hansson, and S. Suresh, "High-Temperature Crack Growth in Monolithic and SiC$_w$-Reinforced Silicon Nitride Under Static and Cyclic Loads," *J. Am. Ceram. Soc.*, 77[11], 2985–2999 (1994).
37. L. Ewart and S. Suresh, "Crack Propagation in Ceramics under Cyclic Loads," *J. Mater. Sci.*, 22[4], 1173–1192 (1987).
38. M. J. Reece, F. Guiu, and M. F. R. Sammur, "Cyclic Fatigue Crack Propagation in Alumina under Direct Tension–Compression Loading," *J. Am. Ceram. Soc.*, 72[2], 348–352 (1990).
39. S. Suresh, L. X. Han, and J. J. Petrovic, "Fracture of Si$_3$N$_4$-SiC Whisker Composites under Cyclic Loads," *J. Am. Ceram. Soc.*, 71[3], C158–C161 (1988).
40. S. Horibe, "Fatigue of Silicon Nitride Ceramics under Cyclic Loading," *J. Eur. Ceram. Soc.*, 6, 89–95 (1990).
41. J.-F. Tsai, C.-S. Yu, and D. K. Shetty, "Fatigue Crack Propagation in Ceria-Partially-Stabilized Zirconia (Ce-TZP)-Alumina Composites," *J. Am. Ceram. Soc.*, 73[10], 2992–3001 (1990).
42. D. L. Davidson, J. B. Campbell, and J. Lankford, "Fatigue Crack Growth through Partially Stabilized Zirconia at Ambient and Elevated Temperatures," *Acta Metall. Mater.*, 39[6], 1319–1330 (1991).
43. R. H. Dauskardt, D. B. Marshall, and R. O. Ritchie, "Cyclic Fatigue Crack Propagation in Magnesia-Partially-Stabilized Zirconia Ceramics," *J. Am. Ceram. Soc.*, 73[4], 893–903 (1990).
44. A. G. Evans and E. R. Fuller, "Crack Propagation in Ceramic Materials under Cyclic Loading Conditions," *Metall. Trans. A*, 5A[1], 27–33 (1974).
45. A. G. Evans, L. R. Russell, and D. W. Richerson, "Slow Crack Growth in Ceramic Materials at Elevated Temperatures," *Metall. Trans. A*, 6[4], 707–716 (1975).
46. C.-K. J. Lin and D. F. Socie, "Static and Cyclic Fatigue of Alumina at High Temperatures," *J. Am. Ceram. Soc.*, 74[7], 1511–1518 (1991).
47. M. Masuda, T. Soma, M. Matsui, and I. Oda, "Fatigue of Ceramics (Part 3)—Cyclic Fatigue Behavior of Sintered Si$_3$N$_4$ at High Temperature," *J. Ceram. Soc. Jpn*, 97[6], 612–618 (1989).
48. T. Fett, G. Himsolt, and D. Munz, "Cyclic Fatigue of Hot-Pressed Si$_3$N$_4$ at High Temperatures," *Adv. Ceram. Mater.*, 1[2], 179–184 (1986).
49. P. F. Becher and T. N. Tiegs, "Temperature Dependence of Strengthening by Whisker Reinforcement: SiC Whisker Reinforced Alumina in Air," *Adv. Ceram. Mater.*, 3[2], 148–153 (1988).
50. J. R. Porter, "Dispersion Processing of Creep-Resistant Whisker-Reinforced Ceramic-Matrix Composites," *Mater. Sci. Eng.*, A107[1–2], 127–131 (1988).
51. S. R. Nutt, P. Lipetzky, and P. F. Becher, "Creep Deformation of Alumina–SiC Composites," *Mater. Sci. Eng.*, A126, 165–172 (1990).
52. D. S. Wilkinson, C. H. Caceras, and A. G. Robertson, "Damage and Fracture Mechanisms during High Temperature Creep in Hot-Pressed Alumina," *J. Am. Ceram. Soc.*, 74[5], 922–933 (1991).
53. B. J. Dalgleish, S. M. Johnson, and A. G. Evans, "High Temperature Failure of Polycrystalline Alumina: I, Crack Nucleation," *J. Am. Ceram. Soc.*, 67[11], 741–750 (1984).

54. A. H. Chokshi and J. R. Porter, "Creep Deformation of an Alumina Matrix Composite Reinforced with Silicon Carbide Whiskers," *J. Am. Ceram. Soc.*, **68**[6], C144–C145 (1985).

55. K. L. Luthra and H. D. Park, "Oxidation of Silicon Carbide-Reinforced Oxide Matrix Composites at 1375° to 1575°C," *J. Am. Ceram. Soc.*, **73**[4], 1014–1023 (1990).

56. P. Lipetzky, "Creep Deformation of Whisker-Reinforced Alumina," Ph.D Thesis, Brown University, Providence, RI, 1992.

57. L. A. Bonney and R. F. Cooper, "Reaction Layer Interfaces in SiC-Fiber-Reinforced Glass Ceramics: A High Resolution Scanning Transmission Electron Microscopy Analysis," *J. Am. Ceram. Soc.*, **73**[10], 2916–2921 (1990).

58. F. Lin, T. Marieb, A. Morrone, and S. R. Nutt, "Thermal Oxidation of Alumina-SiC Whisker Composites: Mechanisms and Kinetics," *Mater. Res. Soc. Symp. Proc.*, **120**, 323–332 (1988).

59. U. Ramamurty, A. S. Kim, S. Suresh, and J. J. Petrovic, "Micromechanisms of Creep-Fatigue Crack Growth in a Silicide-Matrix Composite with SiC Particles," *J. Am. Ceram. Soc.*, **76**[8], 1953–1964 (1993).

60. D. S. Wilkinson and V. Vitek, "The Propagation of Cracks by Cavitation: A General Theory," *Acta Metall.*, **30**, 1723–1732 (1982).

61. L. X. Han, "High-Temperature Fatigue Crack Growth in SiC-Whisker-Reinforced Alumina," Ph.D. Thesis, Brown University, Providence, RI, 1990.

62. H. Kitagawa, R. Yuuki, and T. Ohira, "Crack Morphological Aspects in Fracture Mechanics," *Eng. Fract. Mech.*, 7, 515–529 (1975).

63. S. Suresh and C. F. Shih, "Plastic Near-Tip Fields for Branched Cracks," *Intl. J. Fract.*, **30**, 237–259 (1986).

64. C.-K. J. Lin, D. F. Socie, Y. Xu, and A. Zangvil, "Static and Cyclic Fatigue of Alumina at High Temperatures: II, Failure Analysis," *J. Am. Ceram. Soc.*, **75**[3], 637–648 (1992).

65. M. D. Thouless, "Bridging and Damage Zones in Crack Growth," *J. Am. Ceram. Soc.*, **71**[6], 408–413 (1988).

66. M. D. Thouless, C. H. Hsueh, and A. G. Evans, "A Damage Model of Creep Crack Growth in Polycrystals," *Acta Metall.*, **31**, 1675–1687 (1983).

67. B. J. Hockey and S. M. Wiederhorn, "Effect of Microstructure on the Creep of Siliconized Silicon Carbide," *J. Am. Ceram. Soc.*, **75**[7], 1822–1830 (1992).

68. M. Suzuki, S. R. Nutt, and R. M. Aikin, Jr., "Creep Behavior of a SiC-Reinforced XD^{TM} MoSi$_2$ Composite," *Mater. Sci. Eng.*, **A162**, 73–82 (1993).

69. K. Jakus and S. V. Nair, "Nucleation and Growth of Cracks in SiC/Alumina Composites," *Comp. Sci. Technol.*, 37, 279–297 (1990).

70. W. Blumenthal and A. G. Evans, "High Temperature Failure of Polycrystalline Alumina: II, Creep Crack Growth and Blunting," *J. Am. Ceram. Soc.*, **67**[11], 751–759 (1984).

71. S. M. Johnson, B. J. Dalgleish, and A. G. Evans, "High Temperature Failure of Polycrystalline Alumina: III, Failure Times," *J. Am. Ceram. Soc.*, **67**[11], 759–763 (1984).

72. M. Kachanov, "On Modelling of a Microcracked Zone by a 'Weakened' Elastic Material and Statistical Aspects of Crack-Microcrack Interactions," *Intl. J. Fract.*, 37, R55–R62 (1988).

73. K. S. Chan and R. A. Page, "Origin of the Creep Crack Growth Threshold in a Glass-Ceramic," *J. Am. Ceram. Soc.*, **75**[3], 603–612 (1992).

Environmental Effects

Environmental Effects on High Temperature Mechanical Behavior of Ceramic Matrix Composites

S. R. Nutt

8.1 Oxidation Behavior

8.1.1 Reaction Mechanisms

Ceramic composites are likely to contain non-oxide phases as reinforcements, coatings, or even matrices. In fact, by far the most common reinforcing fibers, whiskers, and particles currently employed in ceramic matrix composites (CMCs) are made of silicon carbide. Like other nonoxide phases, silicon carbide is susceptible to thermal oxidation and this can have a deleterious effect on the mechanical properties of the composite. However, even for composites with similar reinforcements but different matrices, the reaction mechanism of oxidation is likely to be strongly system dependent. Two relevant systems are SiC-reinforced oxide matrix composites and SiC-reinforced nonoxide (SiC or Si_3N_4) matrix composites. In the first category, only the reinforcements are susceptible to oxidation, and critical issues concern the transport of oxygen through the matrix and the reaction products. In contrast, for materials in the second category, both the matrix and the reinforcement phase can undergo oxidation. Counter to intuition, composites in this category can often be more resistant to oxidation and less susceptible to oxidation-induced damage. Materials in this category can often self-passivate and heal cracks by the oxidation of crack flanks, as described later in the chapter. In this section, we will identify some of the mechanisms by which oxidation can occur in ceramic matrix composites and delineate some of the factors that influence the oxidation reactions.

In order for oxidation to occur in ceramic matrix composites, there must first be transport of oxygen to the nonoxide phases. Transport can occur either by gas-phase diffusion through porosity and microcracks, or by solid-state diffusion through an oxide matrix phase. Often in the case of CMCs reinforced with nonoxides such as SiC, the oxidation product can be either solid, liquid, or a multiphase mixture of the two. Diffusion of oxygen must then occur through the oxidation product(s). Furthermore, the reaction product can subsequently react with the matrix, producing an additional phase. In all such cases involving a reaction product, the reaction rate of the composite can be different from that of either of the constituents. If the reaction products are gaseous, internal pressures develop that can be sufficient to nucleate cavities and/or initiate cracking. All of these processes can potentially have deleterious effects on the mechanical properties of the composite. These aspects are now considered in detail in the following subsections.

8.1.1.1 Gas-Phase Diffusion Through Microcracks

Composites, like many ceramics, often contain microcracks and other surface flaws generated by machining or by damage sustained in processing and/or service. The effects of oxygen diffusion through cracks depend on the matrix and reinforcement materials. If the matrix is an oxidizable material such as SiC, which getters oxygen, oxygen diffusing in through the cracks will react with the matrix and possibly seal the crack. However, if the matrix is an oxide that cannot getter oxygen, oxygen will diffuse rapidly through the crack. In this situation, the oxidation rate of the fiber (or fiber coating) will depend upon the nature of the oxidation product and its interaction with the matrix oxide. If a protective oxide is not formed, as in the case of carbon oxidation, the rate of oxidation will be governed by oxygen transport through the crack.

The effects of oxygen transport to carbon through a protective film with cracks have been considered quantitatively by Luthra.[1] The results for the case of a barrier layer 0.1 mm thick, an oxygen partial pressure of 1 bar, and a total pressure of 10 bars are shown in Fig. 8.1. For large cracks, assumed here to be 1 μm wide, oxygen transport will occur via normal gas-phase diffusion for which the diffusivity is independent of the crack width. However, when the crack size is small, such as 10 nm, the crack width is the same order of magnitude as the mean free path of the gas, and oxygen diffusion occurs by Knudsen diffusion. The oxygen diffusivity then is somewhat lower and decreases with decreasing crack width. Figure 8.1 also shows acceptable oxidation rates based on an average substrate loss of 0.35 mm in 2000 and 20 h, respectively, for long-term and short-term applications. The maximum oxidation rate of carbon, based on the cross-sectional area of the crack, is extremely high even for a small crack 10 nm wide. However, the overall oxidation rate is perhaps more relevant than the rate at the crack tip. In this situation, the average oxidation rate will depend on the crack width and crack distribution. Figure 8.1 shows that even the average oxidation rate will be unacceptably high

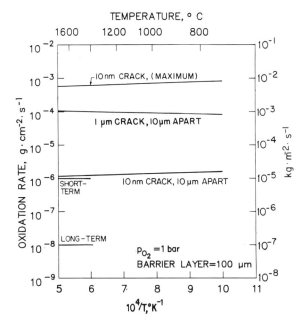

Fig. 8.1 Oxidation of a carbon substrate beneath a protective film containing cracks. The maximum rate represents the oxidation rate beneath the crack and is based on the cross-sectional area of the crack. The average rate depends on the crack size and crack distribution and is shown here for two selected examples. For the 10 nm crack, oxygen transport occurs via Knudsen diffusion, and the oxidation rate is essentially independent of the total pressure. For the 1 μm crack, oxygen transport occurs via normal gas-phase diffusion, and the oxidation rate varies inversely with the total pressure; the results shown here are for a total pressure of 10 bars.[2]

for cracks 1 μm wide and 100 μm apart. Even small cracks 10 nm wide and spaced 10 μm apart cause rapid oxidation of the fiber or coating.

This analysis demonstrates that oxygen diffusion can occur rapidly through even extremely small cracks in a matrix and will oxidize nonoxide fibers and/or coatings unless oxygen is gettered by the formation of a protective oxide within the crack.

8.1.1.2 Condensed-Phase Diffusion

Discussion of condensed-phase diffusion must first recognize the distinction between the tracer or the chemical diffusivity and the oxygen transport rate under an oxygen potential gradient. From an oxidation standpoint, it is the latter that is most relevant. Unfortunately, to calculate the transport rate, one requires a knowledge of the structure and/or concentration of defects responsible for oxygen transport, in addition to the tracer or the chemical diffusivity.

Oxidation of nonoxide phases in composites will generally result in the formation of condensed (solid or liquid) oxide(s), and continued oxidation involves oxygen diffusion through the reaction product(s). If the composite consists of an oxide matrix, oxidation of nonoxide fibers and/or fiber coatings can still occur by oxygen diffusion through the matrix. In the following discussion (taken largely from the work of Luthra[2]), the factors influencing oxygen diffusion through alumina, silica, and boron oxide will be considered.

The oxygen flux J across a thickness fx of an oxide is expressed as[1]

$$J = -\frac{C_O}{2n\Delta x} D_O^{O,*} [(P_{O_2}^g)^n - (P_{O_2}^i)^n] \tag{1}$$

where C_O is the oxygen concentration in the oxide matrix, and D_O^{O*} is the oxygen tracer diffusivity at an oxygen partial pressure (P_{O_2}) of 1 bar. The superscripts g and i refer to the gas–oxide and oxide–substrate interfaces, respectively. The value of the exponent n depends on the defect structure. For example, when the dominant defect is a doubly charged oxygen vacancy, $n = -1/6$.

If the oxygen rate is controlled by oxygen diffusion, $P_{O_2}^g \gg P_{O_2}^i$. Using this fact, the relation above can be simplified.
for $n > 0$,

$$J = -\frac{C_O}{2n\Delta x} D_O^{O,*} (P_{O_2}^g) \tag{2}$$

for $n < 0$,

$$J = +\frac{C_O}{2n\Delta x} D_O^{O,*} (P_{O_2}^i)^n \tag{3}$$

Although it is often assumed that tracer diffusivity can be used as a measure of oxygen permeability, the relationships above show that even for the same tracer diffusivity value at 1 bar, oxygen permeability may be significantly increased if n is less than unity. For example, if n is $-1/6$ and $P_{O_2}^i$ is 10^{-18} bar, the permeability as given by Eqn. (3) would be increased by a factor of 1000.

Both aluminum oxide and silica are known to have very low oxygen permeabilities, and another attractive oxide for composites applications is B_2O_3. Borate has a low melting temperature (450°C) and can be used as a crack sealant.[3] For example, boron particulate has been used for oxidation protection of C/C composites.[4-7] Oxygen diffuses through fused silica and boron oxides as a dissolved molecular species,[8,9] and n for these oxides is therefore 1. However, the oxygen diffusion mechanism in alumina is not well resolved, although the parabolic rate constants for oxidation of alumina-former alloys indicate that the value of n is $-1/6$.[10,11] This n value suggests that oxygen diffuses as a doubly charged vacancy.

Equation (2) indicates that the oxygen permeability through silica and boron oxide is independent of $P^i_{O_2}$ and thus of the substrate (or fiber). On the other hand, for alumina (in which $n = -1/6$), the oxygen permeability is proportional to $(P^i_{O_2})^{-1/6}$ and thus depends on the nature of the reinforcement. For calculations presented here, the substrate is assumed to be carbon. Two separate cases are considered for fixing the oxygen partial pressure ($P^i_{O_2}$): (1) a three-phase equilibrium between Al_4C_3, C, and Al_2O_3, and (2) equilibrium between solid carbon and CO at a partial pressure of 1 bar. The former represents the case when complete equilibrium is achieved. (In actuality, Al_4O_4C will also form, although its effect is negligible.) The oxygen partial pressures are also typical of those expected during oxidation of aluminum-containing alloys. Situation (2) represents the minimum P_{CO} and P_{O_2} needed to form gas bubbles.[2] For example, at 1600°C, the oxygen partial pressures are 6×10^{-20} and 4×10^{-16} bars, respectively, for cases (1) and (2).

Figure 8.2 compares the oxygen tracer diffusivity and the oxygen permeability through the three oxides. At temperatures below 1525°C, the oxygen tracer diffusivity value through alumina is lower than through silica. However, the oxygen *permeability* though alumina becomes lower than through silica only at temperatures below ~1000–1200°C. The difference arises because of the differences in the n values, as discussed previously.

The acceptable permeability depends on the application and on the ratio of the areas of oxide through which diffusion is occurring and the nonoxide reinforcement undergoing oxidation.[1] Consider an example in which both the areas are the same (corresponding to a fiber volume fraction of 50%), the oxide layer thickness is 100 μm, and the acceptable reinforcement loss is 0.35 mm. Thus, for a required lifetime of 2000 h, the acceptable oxygen permeability is 10^{-11} kg m^{-1} s^{-1}. Using this criterion, systems relying on protection by continuous alumina and boron oxide films would be useful up to temperatures of ~1400°C and ~600°C respectively. Systems relying on protection by passive silica films would be useful up to much higher temperatures.

8.1.1.3 Formation of Gaseous Reaction Products

Many nonoxide ceramics form gaseous reaction products when oxidized. For example, when an alumina/silicon carbide composite is exposed to an oxidizing environment, SiC will oxidize, forming carbon monoxide via the following reaction:

$$SiC(s) + \frac{3}{2} O_2(g) = \underline{SiO_2}(s, l) + CO(g) \qquad (4)$$

The underline beneath SiO_2 indicates that it can exist at subunit activities as either mullite (discussed in the next section) or an aluminosilicate liquid. While reaction (4) is written in terms of $O_2(g)$ and $CO(g)$, this does not mean that they exist as gaseous species at the reaction site or that oxygen diffuses as a

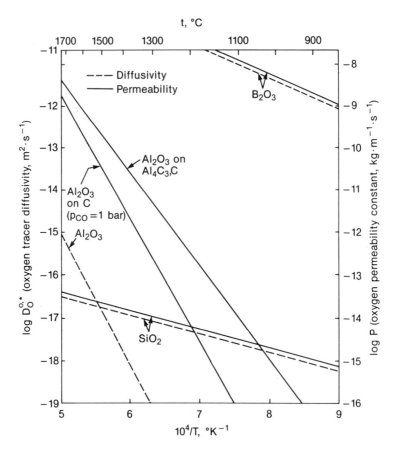

Fig. 8.2 Oxygen tracer diffusivities and oxygen permeabilities through liquid boron oxide, fused silica, and a polycrystalline alumina with a grain size of 5 μm. Tracer diffusivity values for B_2O_3 and SiO_2 were obtained from Refs. 9 and 19, respectively. Tracer diffusivity values for Al_2O_3 were obtained from data reported in Ref. 18 using a grain size of 5 μm (Ref. 13). Oxygen permeability constants were obtained using the procedure outlined in the text.[2]

molecular species. Irrespective of the form in which oxygen diffuses through the reaction product, the expression simply represents an overall reaction. The reaction occurs in a series of steps, leading ultimately to the formation of a reaction scale that consists of a complex mixture of aluminosilicate glass, alumina, and mullite.[12–14]

The initial reaction step involves the deposition of carbon and silica, as shown experimentally by Lin et al.[14] In their work, they examined the reaction front separating the reaction product from the unreacted composite using transmission electron microscopy (TEM). During the initial stages of reaction, graphitic carbon and silica glass were formed at the SiC–alumina interface, as

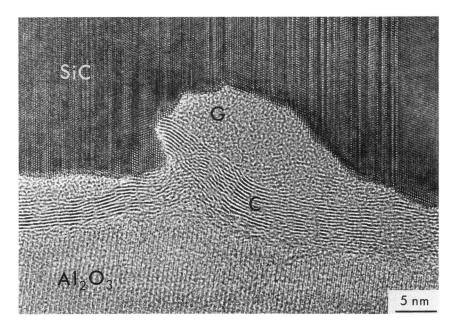

Fig. 8.3 Phase-contrast TEM image of SiC whiskers in an alumina matrix after initial stage of oxidation in which graphitic carbon (C) and silicate glass (G) are formed at the interface.

shown in Fig. 8.3. The image shows a partly oxidized SiC whisker and a nearby alumina grain situated at the reaction front. The whisker and the alumina grain are separated by thin layers of graphitic carbon, which exhibits wavy lattice fringes, and non-crystalline silica. Diffraction patterns from similar phases led to identification of graphitic carbon.

The observation of carbon at the fiber–matrix interface is significant because continued oxidation inevitably leads to the production of gaseous carbon monoxide via

$$C(s) + \tfrac{1}{2}O_2(g) = CO(g) \tag{5}$$

The escape of CO gas at the reaction site can occur via diffusion through the reaction products. However, if CO diffusion through the reaction products is slow, the CO partial pressure can increase to levels sufficiently high to cause void nucleation and cracking. The development of cracks greatly accelerates the outward diffusion of CO and also inevitably degrades the mechanical properties of the composite.

The conditions under which CO gas bubbles form and the reaction scale fractures depend on the relative permeabilities of CO and oxygen through the reaction scale, which typically consists of a three-phase mixture of mullite, alumina, and an aluminosilicate glass. If the CO permeability is greater than that of oxygen, CO will migrate out through the reaction scale before the CO partial pressure at the reaction site is sufficient to generate gas bubbles. However, if the CO permeability through the reaction scale is comparable to, or lower than, that of oxygen, gas bubbles can form.[13] In either case, the partial pressure of oxygen at the reaction site is substantially lower than that in the gas phase, and the reaction rates are expected to be limited by inward diffusion of oxygen and/or outward diffusion of CO.

The detrimental effects of gas formation can be avoided by preventing the oxygen pressure from building up at the reactant–oxide interface. One means of accomplishing this is through the use of non-oxide getters that do not form gaseous oxidation products. This and other approaches are discussed in Section 8.1.3.

8.1.1.4 Interaction of Reaction Products with the Matrix

In Section 8.1.1.2, we discussed the effects of solid-state diffusion through the protective oxide formed during oxidation. In most ceramic composites, these diffusion processes are modified by interaction of the oxidation product(s) with the other constituents of the composite, particularly the matrix. As a result, it is often the case that while the oxidation resistance of each of the individual constituents of the composite is excellent, the oxidation resistance of the composite is poor.

For example, consider the oxidation of SiC-reinforced alumina. The oxidation resistance of both constituents, SiC and alumina, is good. When oxidized, SiC forms a protective film of silica glass that is relatively impermeable to oxygen diffusion and thus prevents further oxidation. However, in the composite, the silica formed by oxidation can subsequently react with alumina in the surrounding matrix to form mullite and/or an aluminosilicate glass. Formation of either of these phases increases the oxygen permeability compared to pure silica, and degrades the oxidation resistance of the material. This phenomenon has also been observed in other SiC-reinforced oxide/matrix composites, such as alumina/SiC, mullite/SiC.[12,13,15] The resulting oxidation rates can be an order of magnitude or more higher than SiC alone.[16]

In the case of the oxidation of alumina/SiC, the final reaction product generally consists of three phases: alumina, mullite, and a non-equilibrium aluminosilicate glass, the proportions of which vary according to the composition of the composite and the oxidation conditions. However, the initial step of the reaction generally involves deposition of graphitic carbon and silica, as described previously. The silica dissolves some of the alumina to form an aluminosilicate glass, and mullite then forms by reaction of alumina with the glass. Details of these initial reactions have been observed by high-resolution TEM, as shown in Figure 8.4. The image shows the region around a whisker end that has undergone oxidation at the reaction front. A mullite nucleus has formed in the aluminosilicate glass, and it is surrounded by graphitic carbon and additional glass. The carbon subsequently oxidizes, producing CO, and when sufficient partial pressure builds up, a gas bubble nucleates. On the other

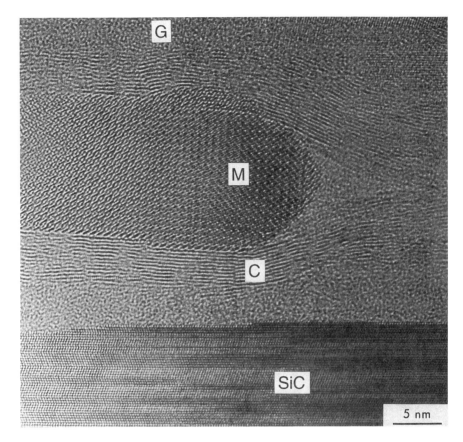

Fig. 8.4 Phase-contrast TEM image of SiC whisker in alumina matrix that has undergone oxidation. The silicate glass (G) has begun to react with alumina to form a mullite nucleus (M), and graphitic carbon (C) surrounds the crystallite.[14]

hand, the mullite nucleus grows, producing a composite scale consisting of mullite, aluminosilicate glass, and residual alumina, all of which have higher oxygen permeability than pure silica. Thus, the SiC inclusions are eventually oxidized, and the thickness of the reaction scale is much larger than for the oxidation of SiC.[13]

TEM observations of the oxidized scale have revealed mullite grains with transgranular cracks, a phenomenon that is not surprising when one considers that the oxidation of SiC produces a volume expansion of ~100%. When the reaction product contains a solid as well as a liquid product, as in the present situation, the volume expansion can be accommodated by squeezing out the liquid phase, resulting in a liquid "cap" on top of the solid reaction products. This has been observed by Luthra and Park,[13] and is apparent in the micrograph shown in Fig. 8.5.

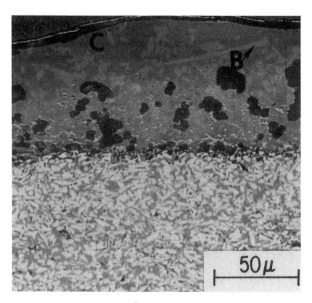

Fig. 8.5 Optical micrograph of a cross-section of the reaction product layer formed on Al_2O_3/SiC composites (50 vol.%) oxidized for 72 h at 1475°C. Phases are (B) mullite, and (C) aluminosilicate liquid.[13]

These considerations indicate that the oxidation rates of composites may be influenced by the interaction of oxidation products with other constituents of the composite. Such interactions should thus be considered when selecting the composite constituents.

8.1.2 Reaction Kinetics

8.1.2.1 Reaction Rates and Phase Equilibria

The oxidation of ceramic composites often obeys parabolic rate kinetics, implying diffusion control. Consider again the oxidation of alumina/SiC composites, one of the few ceramic composite materials for which oxidation data is available. Typical plots of (mass gain/area)^2 are shown in Fig. 8.6 for alumina reinforced with different volume fractions of SiC. The parabolic rate constants can be defined by the relation

$$m^2 = K_m t \tag{6}$$

where m is the mass gain that occurs during time t, and K_m is the mass gain rate constant. These rate constants increase with the SiC content of the composite, as shown in Fig. 8.6b.[13] This is not unexpected, because the amount of SiC per unit area of composite increases with the volume fraction of SiC. However, for a given period of oxidation at a constant temperature, the *thickness* of the reaction product can actually be *less* for composites with *higher*

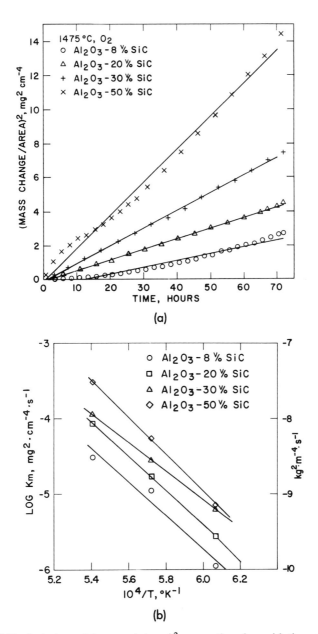

Fig. 8.6 (a) Typical plots of (mass gain/area)^2 versus time for oxidation of Al$_2$O$_3$/SiC composites. (b) Plots of parabolic rate constants (K_m) as a function of $1/T$ for oxidation of Al$_2$O$_3$/SiC composites.[13]

SiC contents, as shown by Lin et al.,[14] and this has important implications with respect to crack initiation behavior, as described in Section 8.2.1.1.

Activation energies for oxidation can be obtained from Arrhenius plots of the parabolic rate constants or scale thicknesses versus inverse temperature, and these typically yield activation energies of ~500 kJ mol^{-1}.[13,14] These activation energies are similar to activation energy values reported for oxygen tracer diffusivity through polycrystalline alumina,[17,18] and are considerably higher than the activation energy for oxygen diffusivity through silica.[19]

The oxygen process is undoubtedly complex in the case of alumina/SiC composites. As described in Section 8.1.1.3, the oxidation process involves evolution of carbon monoxide, and there is also loss of carbon that occurs simultaneously with the gain of oxygen. Furthermore, the nature of the reaction scale changes dynamically during oxidation, as first carbon and silica are formed, followed by CO, aluminosilicate glass, and mullite (at sufficiently high temperatures). Consequently, the diffusivities of various species through the reaction scale (particularly oxygen and CO) can change also, and the rate-limiting step can be elusive.

Three possible situations are considered which depend on the relative diffusion rates or permeabilities of oxygen through the matrix oxide and the reaction scale, irrespective of the form in which oxygen diffuses. In case (1), oxygen diffusion through the oxide matrix is much faster than through the oxidation product(s), and the oxidation of SiC inclusions is similar to the oxidation of SiC particles in air, only the product can be different to pure silica. All of the SiC inclusions develop a film of oxidation product that thickens with time, as shown schematically in Fig. 8.7a. Oxidation is controlled by oxygen diffusion through the film that forms at the interface. If the oxidation rate is expressed in terms of the surface area of the composite, the oxidation rate will be a function of the particle size. In case (2), oxygen diffusion through the reaction product(s) is comparable to or faster than through the matrix oxide, resulting in the situation depicted in Fig. 8.7b. All of the SiC particles would be completely oxidized in the reaction product, and the oxidation rate would be independent of SiC particle size. The boundary between the reaction product and the unreacted composite would be sharp, and oxygen diffusion would occur along the path that was most rapid. Observations have shown that this is likely to be along interfaces and grain boundaries in solid reaction products, although the presence of cracks and pores would provide accelerated pathways.[14,20]

A third situation that is possible is intermediate between the two extreme cases shown in Figs. 8.7a and b. In this case, the boundary between the reacted and unreacted composite is not sharp, and a region of semi-reacted SiC particles separates the two (Fig. 8.7c). This case can occur in two circumstances. When the oxygen diffusion rate through the oxidation product is comparable to, but slightly slower than, through the matrix, a semi-reacted zone will appear, the thickness of which depends on the SiC particle size. A similar situation will arise when the initial oxidation production reacts with the

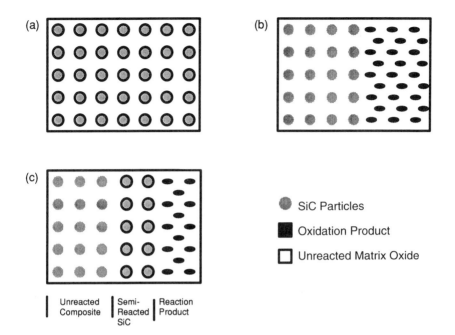

Fig. 8.7 Schematic representations of morphologies of reaction products expected during oxidation of oxide matrix composites containing SiC particles. Although only one oxidation product has been shown here, the reaction scale may contain two oxidation products (e.g., mullite and aluminosilicate glass). In some cases, the reaction product may not contain unreacted matrix, as in the case of alumina with high volume fractions of SiC. (a) Case 1 would occur when the oxygen diffusion rate through the oxide matrix is much faster than that through the oxidation product. (b) Case 2 would occur when the oxygen diffusion rate through the oxidation product is comparable to, or faster than, that through the oxide matrix. (c) Case 3 represents an intermediate situation.[13]

matrix and changes composition. In the present example, that product may be siliceous glass, through which oxygen permeability is extremely low. However, the silica may react with alumina to form an aluminosilicate liquid and mullite, through which oxygen permeabilities are much higher. The kinetics of these intermediate reactions can determine the extent to which this third situation is attained. The morphology of the reaction products observed for alumina/SiC oxidation resemble the schematic shown in Fig. 8.7b and c. Oxygen diffusion rates through the reaction products (alumina, mullite, and aluminosilicate liquid) are expected to be much higher than diffusion through silica, the customary reaction product of the oxidation of pure SiC. Thus, the rate of oxidation of silicon carbide particles in an alumina matrix is much faster than the oxidation of silicon carbide alone. Large semi-reacted zones have also been reported for mullite/SiC composites.[13]

8.1.2.2 Effects of Other Atmospheres

Although the primary driving force behind the development of ceramic composites is the potential for use in severe service environments that involve high temperatures, there are also applications that involve exposure to gases such as CO_2, SO_3, and H_2O, at more moderate temperatures. Many of these potential applications require extended exposure to gaseous environments generated by combustion of liquid fuels. For example, combustion of a typical fuel such as $CH_{1.7}$ normally produces combustion gases containing about 14% CO_2 and 12% H_2O.[16] In gas turbines, the pressure of these combustion gases can reach 2 MPa. If these gases are diluted (3%) by providing excess air, the partial pressures of CO_2 and H_2O could approach 0.1 MPa. For example, if the air/fuel weight ratio is 30 (corresponding to approximately 100% excess air) and the total pressure is 1.2 MPa, the partial pressures of CO_2 and H_2O would be 0.08 and 0.07 MPa, respectively. At these partial pressures, carbonate and hydroxide formation could be important, particularly at lower temperatures. Also, liquid fuels typically contain sulfur impurities amounting to 0.2–1 wt.% sulfur. Combustion of such fuels produces sulfur oxides which can lead to the formation of sulfates at temperatures that are low relative to the oxidation temperatures discussed in previous sections. The effects of gases produced by fuel combustion could be important for certain ceramic composites in engine applications, even at moderate temperatures.

8.1.3 Inhibition of Reaction

Oxidation reactions in ceramic matrix composites can initiate at interfaces, at nonoxide reinforcements exposed at surfaces, or in the matrix. Consequently, approaches that address the problems of oxidation are logically directed at inhibiting the reaction at the site of initiation. For example, if the interface undergoes oxidation by diffusion of oxygen from the matrix or along the interface, one might consider protective coatings for the reinforcements. Alternatively, if the composite matrix undergoes oxidation when exposed to an oxidizing atmosphere, the entire component may require a coating. Successful coatings for composites, much like ceramics, often entail a silicon-based ceramic which, when oxidized, forms a protective layer of silica that is relatively impermeable to oxygen, thus preventing oxidation of the underlying composite. The necessary characteristics of a stable oxide are that it be: (1) fully dense and impermeable to oxygen, (2) slow-growing, (3) self-healing, (4) resistant to cracking and decohesion from thermal cycling, and (5) reasonably resistant to attack from other compounds, such as water vapor, carbonaceous gases, and sulfur compounds. Both silica and alumina are slow-growing stable oxides that protect against oxidation at high temperatures. However, service environments are generally complex, involving different gases and temperature ranges, as well as stresses, which can exacerbate otherwise benign effects of the environment. Therefore, one must consider these other factors when selecting an approach to inhibit environmental effects.

Oxidation is only one of several environmental problems that face ceramic matrix composites. For example, in combustor environments, which represent some of the most severe applications, vaporization and reactions which produce volatile products are also major degradation routes for ceramic composites. These processes can occur by hydrogen reduction, by reaction with water vapor, or by simple vaporization. Furthermore, some of the most oxidation-resistant ceramics, such as SiO_2 and Al_2O_3, are susceptible to these types of vaporization reactions. Fortunately, it is possible to take advantage of some of the intrinsic properties of ceramics to intelligently allow for the types of severe service environments such as combustion engines.

The particular approach selected to protect against environmental degradation is governed by the particular reaction mechanism, the site at which the oxidation reaction initiates, and the constraints of the material system, which require compatibility between constituents of the composite. An additional factor that makes the task of inhibiting environmental degradation particularly daunting in the case of ceramic composites is cost. Implementing the means to inhibit undesirable reactions, be it through coatings or other means, often adds prohibitive cost to the production of composites, which generally are already more expensive than competitive monolithic materials.

Despite the severity of the problems associated with environmental degradation of mechanical properties, it is only recently that the ceramic composites community has begun to address them. In the following section, some of the approaches that attempt to inhibit the adverse effects of service environments are discussed for specific composite systems, and the relative effectiveness is assessed.

8.1.3.1 Oxygen Scavengers

The detrimental effects of gas formation can be avoided by preventing the oxygen pressure from building up at the reactant–oxide interface. One way in which this can be accomplished is by using nonoxide getters that scavenge dissolved oxygen but do not form gaseous reaction products. This approach has been successfully employed to inhibit oxidation of C/C composites through the addition of particulate boron. When boron oxidizes, it ultimately forms borate glass (B_2O_3), which effectively seals cracks and prevents catastrophic oxidation. Viscous flow of the borate glass is believed to be the mechanism by which cracks are sealed.[5] An additional benefit of the boron inhibitors is the scavenging of oxygen that may inadvertently penetrate the composite. This is a critically important function because scavenged oxygen would otherwise react with nonoxide constituents such as the carbon fibers, producing CO gas and generating internal pressure. The use of such oxidation inhibitors in C/C composites has been investigated extensively.[1–5] Similar approaches have not been reported for other ceramic composite systems, despite the potentially beneficial effects.

8.1.3.2 Protective Coatings

Coatings applied to composite fibers generally must satisfy a set of requirements, and it is often challenging to meet these requirements without sacrificing cost, simplicity, and other attractive features of composites without fiber coatings. One of the foremost functions of a fiber coating is to prevent diffusion of species both inward (toward the fiber) and outward (toward the matrix). In the case of environmental attack, the diffusion direction of primary concern is inward. However, the concentration of oxidizing species is often highest at the surface of the component, and fiber coatings also must be resistant to "pipeline" diffusion of oxidizing species that initiates at fiber ends exposed at the surface and proceeds along the fiber interface. This particular diffusion pathway is especially important for continuous fiber composites such as silicon carbide fiber-reinforced glass-ceramics. Finally, the coating must possess the right combination of mechanical, chemical, and physical properties over a large temperature range. This usually includes a relatively weak cohesive strength that is sufficient to allow load transfer *and* fiber pull-out, oxidative stability, and a relatively close thermal expansion match such that debonding does not occur during a thermal cycle.

As an example of a protective coating applied to a fiber, let us consider the case of Nicalon SiC fibers in a glass-ceramic matrix. The SiC fibers fortuitously form a carbon-rich surface layer when incorporated into glass-ceramic matrices at elevated temperatures. The exact mechanism by which this surface layer is formed is not well understood, although it may be related to diffusion of oxygen and silicon down the silica activity gradient that exists across the fiber–matrix interface, as proposed by Cooper and co-workers.[21,22] The formation of this weak interface zone greatly enhances the fracture toughness of the composite by allowing crack deflection to occur along the interface, while maintaining sufficient load transfer from the matrix to the fibers such that traditional composite strengthening occurs. Unfortunately, the composites are embrittled by exposure to high temperature oxidizing environments, particularly when stress is also applied. The embrittlement occurs when the interface is exposed to oxygen through matrix microcracking caused by the applied stress, or by "pipeline" oxidation of the carbon-rich interface layer that initiates from fiber ends exposed at the surface. A glassy oxide layer forms at the oxidation sites, bonding the fibers tightly to the matrix and causing brittle composite behavior and a concomitant loss of toughness and strength.

The interface can be controlled or "engineered" to achieve the desired cohesive strength and oxidative stability. One way that this can be accomplished is by doping the matrix with oxygen scavengers, which tie up oxygen and prevent volatilization of reaction products. However, this approach reduces the useful application temperature of the composite.[23] An alternate approach is to apply coatings to the fibers which achieve approximately the same function as the carbon-rich layer but exhibit improved oxidation resistance. Initial efforts along these lines focused on boron nitride (BN) coatings applied to the fibers by chemical vapor deposition (CVD). While the coatings were relatively weak and oxidation resistant, boron diffusion into the matrix prevented complete crystallization of the glass-ceramic, compromising the high-temperature properties of the composite.

Recent work has attempted to prevent boron diffusion into the matrix by applying a thin overlayer of SiC to the BN-coated fibers. The SiC overlayer effectively prevents diffusion of matrix species into the fiber (and BN) as well as diffusion of boron into the matrix. The resulting composites have superior high temperature properties and oxidative stability compared to the BN-coated fiber composites.[24] The only factor limiting the potential application temperatures for such composites appears to be the softening temperature of the matrix, which is typically in the range of 1000–1200°C, depending on matrix composition. However, the dual coating approach is not without problems. For example, the different thermal expansions of the coatings has caused the SiC overlayer to debond in certain cases, allowing liquid glass to penetrate during consolidation and reducing the effectiveness of the protective overlayer. It now appears that the relative thicknesses of the two coatings must be chosen so as to limit the thermally induced stresses and prevent debonding. In addition, controlling coating thickness and uniformity has been a continual problem, as well as reproducibility. Finally, a problem of considerable concern is cost. The coatings are expensive, and the additional cost associated with the coatings could prove to be prohibitive in all but the most demanding composite applications.

8.2 Influence on Mechanical Behavior

8.2.1 Short-Term Behavior

Ceramic matrix composites generally exhibit improved strength, modulus, and/or toughness compared to monolithic ceramic matrices. These properties collectively can be classified as "short-term" mechanical behavior because of the time frame in which the parameters are measured. In ceramic matrix composites, even short-term mechanical properties can be susceptible to environmental effects, particularly if there is some microstructural damage that occurs during the test. When present, these effects become apparent by conducting parallel tests in oxidizing and inert ambients, and, in general, higher temperatures, more aggressive ambients, and longer test times lead to more prevalent effects. In the following section, we examine some of the different kinds of environmental effects that can occur in some typical composite systems, focusing attention on the simplest properties such as strength and toughness.

8.2.1.1 Strength

Environmental effects on composite strength can be apparent in short-term tests at high temperatures. For example, Prewo *et al.* observed pronounced differences in the stress–strain response of continuous fiber/glass-

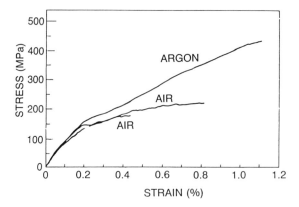

Fig. 8.8 Tensile stress–strain curves for three specimens of $[0,(0,90)_2]SiC_f/LAS\text{-}III$ glass-ceramic composite tested at 1000°C in air and in argon.[25]

ceramic composites.[25] The composites consisted of an LAS (lithium aluminosilicate) glass-ceramic matrix reinforced with ~40 vol.% SiC (Nicalon) fibers arranged in 0°/90° plies. Tensile tests conducted at 1000°C in air and in argon showed pronounced differences (Fig. 8.8). Composites tested in air showed lower strength levels that corresponded to the proportional limit stress of the 0° ply. The proportional limit stress for this composite is generally accepted to be the stress at which matrix microcracking initiates. The observed strength loss observed in air was attributed to oxygen diffusion through microcracks that developed at the proportional limit stress. The cracks exposed fiber interfaces to the test ambient, and the carbon-rich surface of the fiber was locally oxidized, causing a loss of load transfer function and local embrittlement of the interface. Thus, even in short-term tests, test atmosphere can have a pronounced effect on the high temperature properties of ceramic matrix composites.

Environmental degradation of composite interfaces can cause similar deleterious effects on other short-term properties, although this normally requires prior long-term exposure to an oxidizing high temperature environment. For example, Bhatt studied the effects of oxidation on the mechanical properties of RBSN (reaction-bonded silicon nitride) reinforced with continuous SiC fibers.[26] Exposure to oxygen at 400–1000°C for 100 h resulted in decreases in elastic modulus, matrix cracking stress, and ultimate tensile strength. The degradation of properties was attributed to decreases in interfacial shear strength caused by oxidation of a carbon-rich layer on the fiber surface. The oxidation of the interface caused the coating to debond from the fiber and degraded the fiber strength. Both of these factors were responsible for loss of mechanical properties. Interestingly, after exposure to oxidizing ambients at 1200°C and above, the composites showed better strength retention than after exposures at lower temperatures. This was attributed to the growth of an oxide (silica) scale at the higher temperatures that sealed off surface porosity and inhibited diffusion of oxygen through the vapor phase. Because of the absence of applied stress during oxidation, the oxidation product was confined to the surface and was continuous, thus passivating the material. However, as pointed out in Section 8.2.1.3, such passivated surface scales can reduce the threshold for crack initiation, much like a population of surface flaws.

The effects of high temperature exposure on the mechanical strength of whisker-reinforced alumina composites has been documented recently.[27–31] Most of the exposures have been to strongly oxidizing atmospheres such as air, in which both of the constituents, alumina and SiC, are relatively stable. In such environments, strength degradation is usually observed and is attributed to oxidation of SiC and the microstructural damage that results from the formation of intergranular siliceous glass. However, the corrosion behavior of such composites has been shown to be very sensitive to the level of oxidant in the atmosphere.[13,32] Exposure to atmospheres with low oxidant levels can cause severe degradation of strength because the oxidation of SiC becomes active as opposed to passive.[32] In this regime, the oxidation products are volatile, resulting in severe loss of material and reduction in strength. Kim and Moorhead studied the corrosion and strength of whisker-reinforced alumina after exposure to H_2-H_2O atmospheres at high temperatures.[32] They found that the fracture strength was strongly dependent on the level of oxidant (P_{H_2O} in H_2) in the ambient gas. When the P_{H_2O} in the H_2 was lower than a critical value (2% 10^{-5} MPa), active oxidation of the SiC whiskers occurred, resulting in severe reductions in the weight and strength of the samples. In the worst case, the strength was reduced to less than half that of the untreated material. As the P_{H_2O} in the H_2 was increased, the reduction in strength (and weight) became less severe, as the oxidation shifted to a passive mode in which aluminosilicate glass was formed at the surface. When the P_{H_2O} levels were greater than 5% 10^{-4} MPa, increases in strength (and weight) were observed after 10 h at 1400°C. This surprising result is shown in Fig. 8.9. The increases in strength were attributed to the healing or blunting of pre-existing cracks or flaws as a result of glass formation at the sample surface. Removal of the glassy scale by etching with HF solution erased the gains in strength, confirming this explanation of the observed strengthening.

8.2.1.2 Toughness

It is well established that improvements in the room-temperature fracture toughness of ceramics can be achieved by whisker reinforcement, and various mechanisms have been proposed to explain the phenomenon. Recently, two studies have examined the fracture behavior of SiC whisker-reinforced alumina composites at high temperatures to determine the dependence of fracture toughness on temperature, the effect of test atmosphere on fracture toughness, and mechanisms of fracture.[33,34] In both studies, single-edged-notched bars

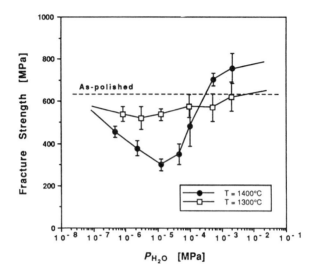

Fig. 8.9 Average strengths of Al_2O_3/SiC composites following exposure for 10 h to various H_2-H_2O atmospheres were significantly affected by P_{H_2O} during exposure at 1400°C, but much less significantly during exposure at 1300°C.[32]

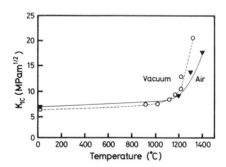

Fig. 8.10 The effect of temperature on the apparent toughness of Al_2O_3/SiC (33 vol.%) composites tested in air and in vacuum.[33]

were precracked and loaded in flexure to measure the toughness. The measurements revealed an increase in apparent fracture toughness at high temperatures (>1000°C), as shown in Fig. 8.10. Examination of samples interrupted prior to fracture revealed extensive microcracking and branching in the vicinity of the crack tip. The increase in apparent toughness observed at high temperatures was attributed to this configuration of cracks, which reduced the local stress concentration.

The experiments distinctly showed effects of test atmosphere on fracture toughness and crack propagation. Figure 8.10 shows that while fracture toughness increased at high temperatures in both air and vacuum, the increase was a slightly more pronounced in vacuum and occurred at slightly lower temperature. Furthermore, in air, stable crack growth was more extensive than in vacuum, although the increase in toughness per unit crack extension was less. Hansson *et al.* identified three distinct processes that were occurring in the crack tip region—microcracking, stable crack growth, and oxidation (glass formation).[33] Microcracking and crack growth in the composite are stress- and time-dependent processes that were thermally activated at temperatures greater than 1000°C. Stable crack growth was attributed to the proliferation of microcracks which effectively shielded any newly formed or growing crack from intensified stress. Thus, the degree of microcracking increased with increasing temperature, resulting in crack growth, deflection, and branching, as well as increased opportunity for interactions between the primary crack and the microcracked zone.[32]

These experiments showed the effect of test atmosphere on fracture processes during the measurement of fracture toughness. Metallographic examination of the samples indicated that microcracking occurred more readily in vacuum than in air, an observation that was consistent with the lower temperature at which the onset of toughness increase occurred in vacuum. However, stable crack growth was more extensive in air than in vacuum. The formation of cracks during loading inevitably exposed SiC whiskers to the oxidizing atmosphere, allowing oxygen to diffuse to whiskers at the crack tip. This led to the formation of an aluminosilicate glassy phase at the exposed whiskers which subsequently penetrated adjoining grain boundaries. In doing so, the glass filled the sites for potential microcracks and possibly even healed microcracks as they developed. On the other hand, the glass was undoubtedly weak and provided little resistance to cracks as they extended. Thus, in air, stable cracks could propagate through the microstructure with a lower degree of microcracking than in vacuum.

Of particular interest is the decrease in *nominal* fracture toughness that reportedly occurred above 1400°C in air.[33] In a strongly oxidizing atmosphere such as air, oxidation kinetics are likely to be more rapid as the temperature is increased. Consequently, SiC whiskers are likely to oxidize more rapidly at such high temperatures, and furthermore, the viscosity of the glassy reaction product is likely to be lower. The combination of these factors translates into more extensive microcracking, a larger microcracked zone, and faster stable crack growth in air than in vacuum. This, in turn, leads to an increase in interaction between the primary crack and the microcrack zone. As discussed by Han *et al.*, such interaction can lead to either increases or decreases in nominal toughness, depending on the nature of the interaction.[34] This is, in fact, what is observed during crack growth in air. Below 1400°C, thermally activated microcracks shield the crack tip by linking and deflecting the crack, then by complex interactions with the crack. Above ~1400°C in air, however, the acceleration of oxidation kinetics leads to an increase in glass phase formation in the vicinity of the crack tip, resulting in a decrease in toughness.

The magnitude of the increase in apparent toughness of the composite in vacuum is almost threefold compared to the toughness at room temperature, a remarkable phenomenon. However, the physical meaning of conventional toughness parameters is not entirely clear in this case. The nominal toughness K_{Ic} indicates a survival load for an initial precrack of specific dimension. However, once the network of microcracks begins to develop, many cracks exist and the dimension of the crack is no longer well defined. Microcrack interactions, as discussed by Han *et al.*, predominate over conventional toughening mechanisms such as deflection and branching. Consequently, linear elastic fracture mechanics or small-scale yield fracture mechanics no longer provide a valid description of the complex process interactions. These considerations need to be borne in mind when attempting to parameterize or measure fracture toughness of composites at high temperatures.

8.2.1.3 Crack Initiation Threshold

Long-term exposure of composites to oxidative environments can have deleterious effects on short-term mechanical behavior, such as resistance to crack initiation. This is particularly true in the case where the composite oxidizes to form an oxide surface scale. Although such reactions can be beneficial in limiting oxidation reactions, when the composite is subsequently cooled to room temperature, the reaction product can be a source of flaws and increase the composite's susceptibility to crack initiation. The following example illustrates this point.

In the case of alumina/SiC$_w$, oxidation leads to the formation of a multiphase surface scale consisting of primarily a mullite and aluminosilicate glass, with residual alumina. The scale is often porous because of the evolution of carbon monoxide during oxidation, and some of the pores may be closed and under pressure. Also, because of the different thermal expansivities of the composite and the surface scale, there is undoubtedly a degree of thermally induced stress, and cracks in the mullite grains are not uncommon.[14] As the exposure time to oxidizing ambients is increased, the surface scale thickens, and the susceptibility to crack initiation increases. Lin *et al.* investigated this phenomenon in cyclic compressive loading of single-edge-notched bars.[14] Cyclic loads were applied along the axis of the bars, perpendicular to the plane of the notch, and the load amplitude was implemented by 3% every 5000 cycles until the nucleation of a fatigue crack was observed. The applied compressive load necessary to initiate a crack at the base of the notch was used to calculate the applied stress intensity factor for composites that had been oxidized for different periods of time to produce a range of scale thicknesses. The results, shown in Fig. 8.11, reveal that once the scale exceeds about 60 μm in thickness, the load necessary to initiate a crack drops by almost 20%. Furthermore, once a crack initiates in the scale, continued cyclic loading causes it to propagate into the unreacted material. Thus, surface reactions forming

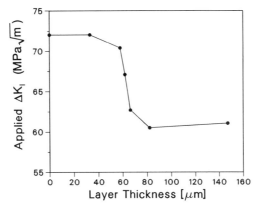

Fig. 8.11 Decrease in applied load amplitude for crack initiation versus scale thickness in Al_2O_3/SiC composites.[14]

passive layers which may be beneficial, if not essential, to the capacity of a material to withstand exposure to hostile high temperature environments, may reduce a composite's resistance to crack initiation under cyclic compressive loads. This is a potentially important issue for ceramic composites because many of the service applications envisioned for these materials involve both exposure to high temperature oxidizing environments and variable amplitude compressive loads.

8.2.2 Long-term Behavior

Because of the inherent thermal and chemical stability of ceramics and the added benefits of composite reinforcement, it is natural to consider the mechanical behavior of ceramic composites at high temperatures. In ceramic composites, an added complexity is imparted because while the constituents alone may exhibit good oxidation resistance, the composite may not, as discussed in the previous section. This is particularly true when nonoxide reinforcements such as silicon carbide are employed as reinforcing materials. For example, by itself, SiC is relatively resistant to oxidation, and in fact silicon-based ceramics are often used as protective coatings against oxidation. However, when SiC is a dispersed phase in a ceramic matrix composite, oxidation presents a problem because of the evolution of carbon monoxide gas, the formation of silica glass, and the subsequent reaction of silica with the matrix. All of these phenomena have potentially deleterious effects on the mechanical behavior of the composite. Thus, while ceramics are normally regarded as extremely inert materials that are stable at high temperatures, environmental effects can significantly influence the long-term mechanical properties of ceramic composites. In the following section, some examples of ceramic composite systems in which the creep and fatigue behavior is affected by the test atmosphere are examined.

8.2.2.1 Creep

Use of ceramic materials in high temperature structural applications is often limited by creep resistance. However, several recent studies have shown that composite reinforcement can drastically reduce the creep rates compared to the unreinforced ceramic matrix.[12,20,27–31] Most of these studies have been conducted in air, which is a strong oxidizing environment. In a few cases, investigators have attempted to isolate the effects of oxidation from creep by conducting parallel experiments in both air and inert atmospheres.[27,28] The following is a description of the salient points made in these investigations.

Al_2O_3/SiC_w Composites Let us consider the example of alumina/Sic_w, a composite that has found commercial applications in wear-critical components such as cutting tool inserts, extrusion dies, and certain valve and pump components. There have been several studies reporting the creep behavior of these composites in flexure and in compression.[12,20,27–31] In two of these studies, compressive creep experiments were conducted in air and in inert atmospheres.[27,28] Although the strain rates in both atmospheres generally decreased with increasing whisker content, tests in air caused degradation of the whiskers, resulting in increased strain rates compared to those in inert atmospheres. The oxidation of the whiskers in air led to the formation of siliceous glass, which subsequently became aluminosilicate glass and crystalline mullite.[12–14] Microstructural observations indicated that the glassy phase tended to coat interfaces and penetrate grain boundaries, thereby facilitating intergranular sliding. In cases where the applied stress was high, or at sites of local stress concentration, such as whiskers in close proximity, the glass phase facilitated damage processes such as cavitation and crack nucleation.[27,35] A closer look at the findings reveals the effects of the oxidizing atmosphere on the creep behavior.

The creep response of whisker-reinforced alumina often shows a low temperature and a high temperature regime.[27,30,31,35] Both the stress dependence of the strain rate (the stress exponent) and the activation energy are reportedly higher at higher temperatures, although the critical temperature appears to vary with whisker content and material source. Figure 8.12 shows an example of data taken from Lipetzky et al. showing the two regimes.[27] The stress exponent (for experiments conducted in inert atmosphere) increases from a value of ~1 at 1200°C to a value of ~3.5 at higher temperatures. Furthermore, for a given stress, the strain rates are substantially higher in air than in inert atmosphere, although the stress dependence is virtually independent of test ambient. The temperature dependence of the creep rate is consistent with these measurements. An Arrhenius plot of the creep rates shows what appears to be a critical temperature at ~1280°C, above which the activation energy for creep increases by a factor of 2–4, depending on the atmosphere.[27] Normally, changes in stress exponent and activation energy imply the activation of an additional deformation mechanism. However, although the nature of the mechanism can sometimes be inferred from the

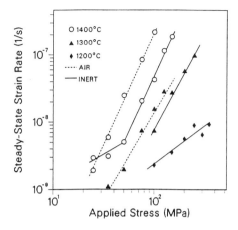

Fig. 8.12 Steady-state creep rates versus applied stress for compressive creep experiments performed in nitrogen and in air at 1200–1400°C for Al_2O_3/SiC composites. The data conform to a power-law type constitutive relation.[27]

value of the stress exponent and/or activation energy, such inferences are tenuous in the case of complex multiphase materials such as these. Fortunately, accurate assessments of the creep mechanisms can be made by microstructural observations using electron microscopy.

Microstructural observations of samples tested in inert atmosphere from the two regimes show cavitation, particularly at grain boundary interface (GBI) junctions, as well as at triple grain junctions. Cavitation is more extensive at temperatures above 1280°C, even when comparing samples with similar total strains. The evidence from this and other investigations suggests that the creep behavior of the composites in the low-stress regime is characterized by viscous flow ($\dot{\varepsilon} \propto \sigma$). Deformation in the viscous regime occurs primarily by grain boundary sliding accommodated by diffusion. Indeed, recent measurements of grain shape changes via a form factor support this assertion.[36] The increased creep resistance of the composite appears to result from the pinning of grain boundaries, which restricts the sliding of matrix grains.

Creep in the high stress and high temperature regime is characterized by damage accumulation, particularly creep cavitation. Above a critical stress–temperature regime, the process of grain boundary sliding becomes increasingly unaccommodated, as diffusion fails to "keep up," and damage ensues. The process of creep cavitation is sensitive to several parameters, including test atmosphere, stress level, total strain, temperature, whisker distribution and loading, and the amount and composition of intergranular glass phase. All of these parameters can potentially affect the extent of cavitation that occurs in the composite. For example, temperature will affect not only the diffusion processes that can accommodate intergranular sliding, but also the viscosity of the intergranular glass phase. This is significant because the glass phase is relatively weak and appears to facilitate cavitation at sites such as GBI junctions, as shown by Lipetzky *et al.*[27,35] Their observations showed that even in the pretested state, thin films of glassy phase were often present at whisker interfaces, a possible consequence of native oxide on the as-grown SiC whiskers. Redistribution of the glassy phase during creep was proposed to account for the apparent ease with which cavities nucleated GBI junctions. (It has been shown by Tsai and Raj that the presence of even a small amount of glass phase at intercrystalline interfaces can promote damage.[37])

Of particular interest in the present chapter is the effect of test atmosphere on creep and creep damage mechanisms. While there are undoubtedly several factors that can promote creep cavitation and contribute to the observed changes in stress exponent and activation energy, the fact remains that the strain rates are substantially higher in air than in inert atmospheres, as shown in Fig. 8.12. This phenomenon is a direct consequence of the topotactic oxidation reaction of SiC whiskers exposed at the surface. As described by Porter and Chokshi,[38] and subsequently by others,[21,22] at high temperatures in air, a carbon-condensed oxidation displacement reaction occurs in which graphitic carbon and silica are formed at the whisker interface via

$$SiC + O_2 \rightarrow SiO_2 + C \tag{7}$$

The silica takes the form of glass, as shown previously in Fig. 8.3, a phase-contrast TEM image of a SiC whisker during the initial stages of oxidation. The graphitic carbon phase is manifest by the wavy dark lattice fringes, while the non-crystalline silica phase appears light and is virtually featureless. As described by Bonney and Cooper, the rate-limiting process in this reaction is the diffusion of oxygen to the interface.[22] The silica subsequently dissolves some of the adjoining alumina to form an aluminosilicate glass, which can form mullite under some conditions. The carbon, meanwhile, is oxidized to produce gaseous CO, a phenomenon that has particularly deleterious consequences. The evolution of volatile gaseous species such as CO can raise internal pressure at the reaction site, thus contributing to the nucleation of cavities, particularly at sites of stress concentration such as GBI junctions.

The effect of the oxidation reaction on the creep response involves the evolution of additional glass phase and CO gas, both of which appear to be instrumental in facilitating cavity nucleation. Nearly all of the cavities observed after creep in air were lined with a thin film of glass phase, implying that the cavities nucleated from glass accumulations at GBI junctions. Such accumulations cannot be attributed solely to redistribution of pre-existing glass phase, and must be caused by oxidation resulting from diffusion of oxygen from the test ambient. A typical glass-lined cavity at a GBI junction is shown in Fig. 8.13. Glass ligaments bridge a portion of the wedge that is opening along the interface. The cavity has a distinctive triangular morphology that was characteristic of all such GBI cavities, indicating that while diffusion was not able to fully accommodate the sliding of grains, it was sufficient to permit cavity

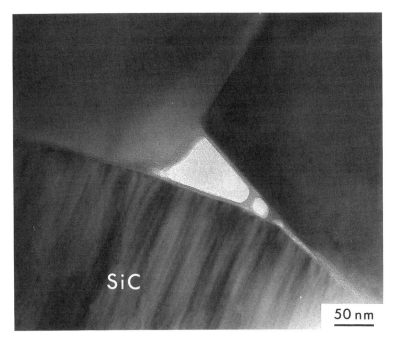

Fig. 8.13 Microstructural damage during creep of Al_2O_3/SiC composite at 1400°C in air. Cavitation occurs within glass phase accumulated at a GBI junction. Glassy ligaments bridge the separated interface.[27]

faceting. Such glass-lined cavities were ubiquitous in air-crept samples, particularly ones that had undergone strains of 1% or more. In contrast, cavitation was less extensive in samples crept in inert atmosphere, and glass films lining these cavities were virtually nonexistent. Thus, the abundance of glass phase in the air-crept samples was undoubtedly caused by oxidation and was responsible for promoting cavitation, grain boundary sliding, and ultimately the higher strain rates. The fact that the cavitation was the prevalent damage mode in compressive creep experiments is a major concern. Application of loads with tensile components will certainly make the problem of cavitation more severe, and it is unlikely that these composites will be used in anything but oxidizing environments. Tensile loads are likely to open microcracks, exposing whiskers to oxygen-rich atmospheres, and cavities are likely to grow and coalesce more quickly than in compression.

$MoSi_2$/SiC Composites An obvious approach to the problem of thermal oxidation of composites during creep is to select a matrix and reinforcement that are both oxidation-resistant phases such as oxides or silicon-based ceramics. Because oxide reinforcements are not widely available, and those that are available are prohibitively expensive, the latter approach has proven more tractable. Consider for example, composites of $MoSi_2$ reinforced with SiC_p. Molybdenum disilicide has good oxidation resistance in certain temperature ranges, as does oxidation carbide, and composites of these materials have shown promising improvements in room temperature fracture toughness.[39–42] The oxidation resistance of both phases derives from the formation of a passive surface layer of silica during the initial stages of oxidation. As the silica layer thickens, it becomes relatively impermeable to oxygen diffusion, and oxidation virtually ceases. Although mixtures of oxidation-resistant phases are not necessarily oxidation-resistant, a recent study of two $MoSi_2$-based composites showed slight improvement in oxidation resistance when compared to monolithic $MoSi_2$.[43] Thus, it is reasonable to expect some improvement in the creep resistance of $MoSi_2/SiC$ composites.

Atmospheric effects on the creep of $MoSi_2/SiC$ composites were investigated and reported by Suzuki et al.[42] Creep experiments were conducted in compression in both air and dry nitrogen using composites containing 30 vol.% SiC particles ~1–15 μm in diameter. The results showed that the test atmosphere had only a negligible effect on the creep behavior of the composite (Fig. 8.14). The creep rates were only marginally higher in air than in the inert atmosphere. The stress exponents for both cases were approximately 3, implying that the creep deformation was dominated by dislocation glide.[44] This assertion was consistent with TEM observations of crept specimens, which showed increased density of dislocations in crept samples, particularly around SiC particles.[42] Metallographic examination of the samples after creep in air showed a distinct oxide scale that was apparently effective in protecting the composite from extensive oxidation. TEM observations of the same samples revealed a slight increase in the amount of glassy phase at internal boundaries, particularly at particle–matrix interfaces. This was accompanied by a noticeable penetration of the surface oxide along near-surface grain boundaries. Although the effect on the creep behavior was negligible, the penetration of glassy oxide appeared to be enhanced by the applied stress. The passivating layer that formed on the composite during high temperature creep successfully inhibited the potentially deleterious effects of thermal oxidation.

Surprisingly, the atmospheric effects on creep tended to be more pronounced in the base alloy (without SiC reinforcements) than in the composite. The base alloy showed substantially higher creep rates in air than in nitrogen.[42] Metallographic examination of the base alloy samples crept in air showed an increase in the number of glassy silica particles present in the microstructure, as well as an oxide scale. From this, one can conclude that the addition of SiC reinforcements actually *enhances* the oxidation resistance of $MoSi_2$. Similar results have been reported for $MoSi_2$-based composites reinforced with TiB_2 and alumina particles, where the oxidation resistance of the composites is superior to the unreinforced matrix.[43] However, in the present case, the issue is complicated by the presence of higher levels of impurities in the base alloy that may contribute to the accelerated behavior. Nevertheless, the prospect of improving the oxidation resistance of the matrix

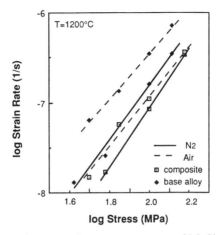

Fig. 8.14 Comparison of compressive creep response of $MoSi_2/SiC$ composites tested in air and in inert atmospheres.[42]

through judicious selection of reinforcing phases is a promising route to limiting and/or controlling atmospheric effects on the mechanical behavior of composites.

8.2.2.2 Effectiveness of Coatings

The fiber geometry has important consequences with regard to susceptibility to oxidation and ultimately the long-term mechanical behavior of the composite. When reinforcements are *discontinuous*, diffusion of oxygen through condensed phases must occur primarily through the matrix phase, and/or through the products of the reaction. However, when the reinforcements are *continuous*, fibers provide a potential pathway for rapid diffusion from the surface to the interior of the material. This is particularly true when fiber ends are exposed at the surface. Fortunately, the degree to which this occurs can be limited by component geometry. For example, in plate and sheet configurations, fibers are laid up in the plane of the component, and fiber ends are exposed at only a small portion of the surface area (the edges). This minimizes the extent of diffusion along the fiber "pipelines." Nevertheless, control of these diffusion pathways is critical to the viability of the material in high temperature applications. In addition, matrix cracking can expose portions of fibers to oxidizing atmospheres, and oxygen can then diffuse along fiber interfaces. For this reason, extended (long-term) operation above the matrix microcracking stress is likely to cause degradation of composite properties. One way to achieve some control of fiber interfaces is through oxidation-resistant coatings applied either directly to fibers, or to the composite. In the following discussion, the effectiveness of oxidation resistant fiber coatings is examined for one composite system.

Glass–ceramic/SiC$_f$ Composites. Thus far, the examples of atmospheric effects on composite behavior have been limited to composites with discontinuous reinforcements. However, some of the most promising composite properties have been achieved with *continuous* fiber reinforcement, particularly in glass-ceramic matrices.[45–50] Continuous SiC-fiber reinforcement has been demonstrated to increase the toughness and strength of silicate glasses and glass-ceramics.[46–49] Improvements in mechanical properties over monolithic silicate materials have been attributed to crack deflection and fiber pull-out that result from a weak interfacial bond between the fibers and the silicate matrix.[51] Analysis of the interface has revealed the presence of a (fortuitous) carbon-rich layer. The formation of the carbon layer is a result of an *in situ* solid-state reaction that occurs between the fiber and matrix phases during high temperature consolidation of the composites.[21,22,52] Unfortunately, the same carbon-rich layer that is so instrumental in improving the room temperature mechanical properties of the composite is susceptible to oxidation at high temperatures. When this happens, either by gas-phase diffusion through matrix microcracks, or by condensed-phase diffusion along the interface, strength and toughness are degraded, and the composite becomes brittle. Exactly how this occurs is not completely clear, although it is conjectured that the weak interface becomes strong when a glassy oxide forms at the fiber surface at sites where the fiber is exposed to oxygen by microcracks.

In recent years, several studies have reported the use of fiber coatings as a means of controlling the fiber–matrix interface properties in continuous fiber glass-ceramic composites.[24,53] While most studies have dealt with coatings that bonded too strongly to achieve tough composite behavior, some have attempted to replace the carbon-rich zone with a similar weakly bonded coating that also had improved oxidative stability. CVD coatings, such as BN[54] and SiC on BN,[55] were deposited on small-diameter fibers such as the SiC-based Nicalon fiber (Nippon Carbon Co.) and the Si-N-C based HPZ fiber (Dow Corning Corp.). Incorporating these coated fibers into glass-ceramic matrices such as lithium aluminosilicate (LAS) and barium magnesium aluminosilicate (BMAS), resulted in strong, tough, and oxidatively stable high temperature structural composites.

Previous studies on BN-coated fibers have shown that good handleability and high tensile strength can be obtained if the BN coating satisfies the following conditions: (1) it is amorphous or partly turbostatic in nature, and (2) it has a composition that is close to stoichiometric BN but contains excess carbon.[56] It has also been reported that the LAS matrices exhibit incomplete crystallization caused by boron diffusion during composite fabrication.[55] The residual boron-containing glassy regions contributed to poor high temperature strength of the composites.[55] In order to prevent boron diffusion, a thin SiC overcoat was applied to the BN coating as a diffusion barrier. This SiC layer prevented both boron diffusion into the matrix, and matrix element diffusion into the BN coating, resulting in composites with superior elevated temperature properties compared to BN-coated Nicalon fiber composites. It appears that glass-ceramic matrix composite systems with both Nicalon and HPZ fibers coated with SiC over BN have significant potential as tough, thermally stable, structural ceramic composites for applications to >1000°C (provided the matrix can be properly crystallized to barium osumilite). However, in order to develop such composites for use in high temperature structural applications, it is necessary to understand the effects of long-term exposure to oxidizing atmospheres on the microstructure and chemistry of the interfacial region, and how these factors influence the interfacial bonding and mechanical behavior of the composites.

In a recent study, the long- and short-term mechanical properties of a SiC fiber-reinforced BMAS glass-ceramic composite were investigated and corre-lated with observations of the corresponding interface microstructures.[24] The fibers were dual-coated with a layer of nominal BN and an overlayer of SiC designed to prevent boron out-diffusion into the matrix. The BN layer consisted of approximately 40% B and 40% N, and 20% C. Composites were consolidated in 0°/90° lay ups to achieve a fiber loading of ~55 vol.% Four thermomechanical histories were considered: (1) as-pressed, (2) ceramed, (3) tensile stress-rupture tested at 1100°C in air to a maximum stress of 165 MPa for a total time of 344 h, and (4) thermally aged at 1200°C in air for 500 h. Mechanical properties of the composites were evaluated by flexural and tensile testing at 20–1200°C, and indentation fiber push-out measurements were used to determine the debond energies and frictional sliding stresses of the fiber interfaces.

The flexural strength of the composite is excellent from RT (room temperature) to 1200°C, as shown in Table 8.1. Even at 1200°C, the strength and elastic modulus are quite high (79 MPa and 69 GPa, respectively.) However, at 1300°C, the BMAS matrix softens and consequently, the composite properties decline substantially. Atmospheric effects on short-term properties become apparent only after long-term exposure to air at high temperature. After 500 h at 1200°C in air, the frictional sliding stress and debond energies decrease by 20% and 55%, respectively, and there is a decline of 20% in the RT flexural strength. Also, tensile stress rupture occurs after 344 h at 1100°C at a stress of 165 MPa, a reduction of ~50% compared to the

Table 8.1 Flexural (three-pt loading) properties of BMAS matrix/SiC/BN/Nicalon fiber composites[24]

	Room temperature σ (MPa)				High temperature σ (MPa)		
As-pressed	*Ceramed*	*550°C, O$_2$, 100 h*	*1200°C, air, 500 h*		*1100°C*	*1200°C*	*1300°C*
655	675	620	510		648	565	248

tensile strength measured at the same temperature. The cause(s) of the decline in strength after long-term exposure to air can be ascertained by observations of the interface microstructure after tensile stress-rupture.

Microscopic observations of tensile stress-rupture samples revealed cracks in the 90° plies (in which the fibers were perpendicular to the tensile axis), but not in the 0° plies. The cracks generally occurred through the matrix and along the fiber–matrix interface. TEM observations indicated that cracks frequently propagated through the BN layer, as expected, given that the BN layer was designed to be a weakly bonded coating. However, a distinct change was observed at the BN/fiber interface in the 90° plies, where two nanoscale sublayers appeared, as shown in Fig. 8.15. The sublayer on the fiber side appeared light, while the one on the BN side appeared dark. Both sublayers were continuous and noncrystalline, and were about 15 nm wide. Composition profiles were acquired by stepping a focused probe across the fiber interface. The light sublayer (near the fiber) was carbon-rich, while the dark sublayer (near the BN) was silica-rich. These observations were consistent with previous studies of similar composites, as reported by Bonney and Cooper, and Naslain *et al.*[22,52] The formation of dual sublayers was attributed to a solid-state oxidation displacement reaction in which SiC is oxidized to produce glassy silica and free carbon. No distinct sublayers were observed in the 0° plies in which the fibers were aligned with the tensile stress axis. These observations suggest that the microcracking that occurred within the 90° plies provided the supply of oxygen necessary to oxidize the fiber interfaces.

In samples that were thermally aged for 500 h at 1200°C, the changes in interface microstructure were more pronounced. Figure 8.15b shows a region of the interface after thermal aging, revealing subtle but distinct sublayers separating the BN coating from the fiber. The dark sublayer is thinner than in Fig. 8.15a, implying a less advanced stage of the oxidation reaction. However, nearby regions showed more drastic effects of the reaction. Figure 8.15c shows a region in which interface voids have been created, a probable consequence of the evolution of gaseous reaction product(s). This region is near the composite surface, and the voids ostensibly result from the oxidation of carbon to produce gaseous carbon monoxide. However, unlike the near-surface regions, the interior of the composite sample generally did not show such dramatic changes in the interface, and the changes were more subtle. Plausible sources of the oxidized carbon include the BN layer, which contains ~20% carbon by design, and the SiC fiber, which develops a carbon-rich surface layer during composite fabrication. Regardless of the source of carbon, such reactions can cause degradation of interface properties such as sliding stress and debond energy which, in turn, diminish the mechanical properties of the composite.[24]

8.2.2.3 Effects of Cracks

As mentioned previously, one of the major concerns for glass-ceramics reinforced with continuous fibers is the effect of matrix cracks in oxidizing environments. An important parameter in this kind of composite is the so-called matrix cracking stress, which typically precedes fiber fracture and causes the first deflection of the elastic portion of the stress–strain curve.[51] Weakly bonded interfaces allow crack deflection and interface debonding to occur, thereby dissipating the stress concentrations and imparting toughness to the composite. However, such matrix cracks expose sites along the fiber lengths to ambient conditions, which can be oxidizing at high temperatures. Oxygen can then diffuse along interfaces, causing degradation of the properties of the interface and of the composite. However, in certain systems, such as SiC/SiC composites, matrix cracks can actually be sealed by oxidation along the crack flanks. Oxidation of SiC produces silica glass, which can undergo viscous flow at high temperatures, and thereby fill the cracks.[5] This process is generally not possible in oxide matrix composites.

Given the effects of cracking on susceptibility to oxidation in ceramic matrix composites, a critical question to address is whether long-term operation of such materials is possible above the matrix cracking stress. It is probably not realistic to expect the material to function above the matrix cracking stress for extended periods of time under oxidizing conditions. Furthermore, one must consider the effects of loading histories which involve stresses sufficient to cause matrix cracking at low temperatures followed by long-term service at high temperatures. Even at stresses below the nominal matrix cracking stress, pre-existing cracks will almost surely accelerate oxidation and lead to damage and/or premature failure of the composite. These concerns underscore the importance of matrix properties (as well as interface properties) in the design of composites for high temperature applications.

8.2.2.4 Fatigue

The resistance of ceramic composites to crack growth under cyclic loads is important for a wide range of potential structural applications. However, remarkably little is known about the fatigue crack growth characteristics in composites, particularly at high temperatures. Cyclic loading of composites at high temperatures is likely to involve not only creep–fatigue interactions, but environmental effects as well. While this is a complex and challenging problem to investigate, there is no escaping the importance of environmental effects and the relevance of cyclic loading.

One recent investigation dealt with the high temperature fatigue crack growth behavior of an Al_2O_3/SiC whisker composite.[34,57] Single-edge-notched bar specimens were precracked and subjected to cyclic tensile loads (in bending) at 1400°C. The fatigue crack velocity, da/dt, versus applied stress intensity factor, K_I, is plotted in Fig. 8.16. Stable fatigue crack growth occurred at a stress intensity level far below the fracture toughness value. Data sets are shown for two different v values, where v represents the frequency of the load cycle. For a fixed R value, where R represents the ratio of the maximum to minimum load applied, an increase in cyclic frequency (from 0.1 to 2 Hz) increases the crack propagation rates by approximately three orders of magnitude. When compared to static load creep under otherwise similar

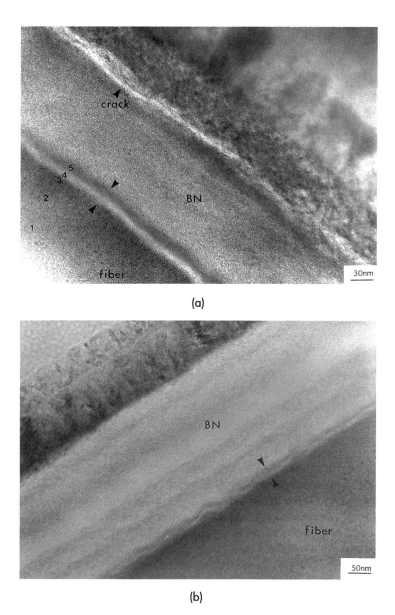

Fig. 8.15 Changes in interface microstructure in SiC fiber-reinforced BMAS glass-ceramic composites induced by exposure to high temperature oxidizing environments. (a) After tensile stress-rupture experiment at 1100°C, the 90° fibers show a distinct dual layer at the BN coating–fiber interface. (b) After thermal aging for 500 h at 1200°C, a subtle double layer appears at the same site. (c) Near the composite surface, the effects of thermal aging (and oxidation) are more pronounced.[24]

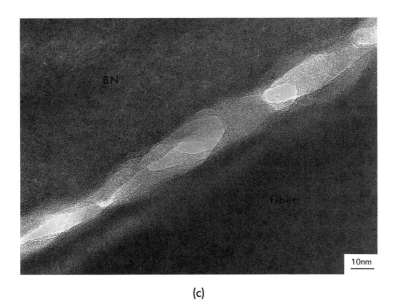

(c)

conditions, it was found that the threshold stress intensity for crack growth under cyclic loads was up to ~30% higher than the static crack growth threshold, and the static load crack growth rates were three orders of magnitude higher. Furthermore, comparison of the creep crack growth behavior of the composite with that of unreinforced alumina with a comparable grain size revealed that the composite material had a superior threshold stress intensity factor for creep crack growth compared to the monolithic material.

Examination of the crack profiles after cyclic fatigue revealed extensive branching at the crack tip and a substantial damage zone in the vicinity of the (macro)crack tip. Increases in test temperature caused an increase in the size of the zone. Close examination revealed bifurcation of cracks and ligaments extended across the crack faces. An example of a typical fatigue crack profile appears in Chapter 7 by Suresh.

The principal mechanism of damage caused by cyclic loading at 1400°C was found to be the growth of interfacial microcracks. Microscopic observations of the material near the crack tip revealed that oxidation of SiC whiskers exposed by microcracking played a major role in the nucleation and growth of the microcracks. As in the case of creep (see Section 8.2.2.1), oxidation of the SiC reinforcements results in the formation of a glass phase that tends to accumulate at sites of stress concentration such as GBI junctions. When this happens, the material is susceptible to void nucleation because of (1) the evolution of CO gas from the oxidation of SiC, generating internal pressure, (2) the minimal load-carrying capacity of the glass phase, and (3) the large hydrostatic stresses in the crack tip region. A striking example of void

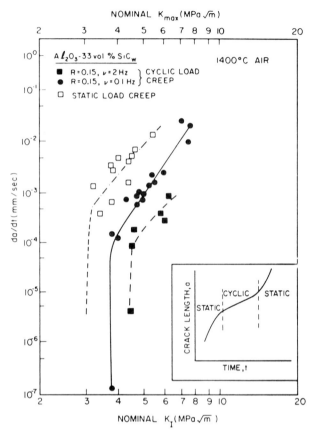

Fig. 8.16 Variation of static and cyclic fatigue crack velocity, da/dt, with the applied (maximum) stress intensity factor, K_I, for fatigue tests on Al_2O_3/SiC_w composites conducted at 1400°C. The inset shows a schematic of the change in crack velocity for a change from static → cyclic → static loading at fixed K_I.[57]

nucleation is shown in Fig. 8.17, an image of a SiC whisker in the vicinity of the fatigue crack tip. Voids have formed at both of the sites where grain boundaries intersect the whisker interface. An additional (although less prevalent) form of damage observed only in fatigue conditions was the breakage of SiC whiskers and subsequent oxidation along {111} cleavage planes.[57]

The kind of damage described above evolves during cyclic fatigue because of the close proximity of gaseous oxygen ensured by microcracking and the accelerated diffusion pathways that are available, particularly along interfaces and grain boundaries. As the glass phase is generated by oxidation, cavities inevitably nucleate and coalesce, causing cracks to branch and bifurcate. This can effectively shield the material from the nominal near-tip stress intensity and reduce the effective stress-intensity-factor range during

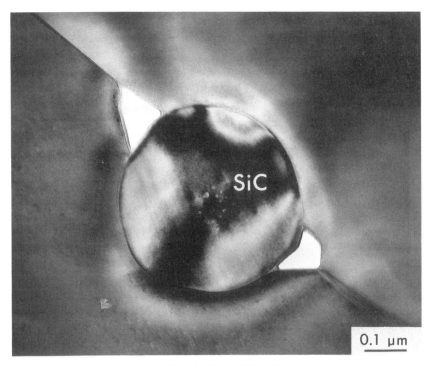

Fig. 8.17 Cavitation at GBI junctions in the vicinity of a crack tip damage zone created by cyclic fatigue loading in Al_2O_3/SiC composite tested at 1400°C in air.[57]

fatigue crack growth. Although monolithic ceramics with residual glass phase show similar cavitation mechanisms, the branching and shielding observed during high temperature fatigue of these composites appear to be more pronounced because of the large density of interfacial cavities created as a result of SiC oxidation. The combination of crack branching and microcrack shielding effects can account for a reduction of a factor of two in the near-tip driving force for fatigue crack growth from the nominal K_I. This "shielding" effect of the diffuse microcrack zone at the macrocrack tip, as well as severe crack bifurcation, leads to apparently higher fatigue thresholds and slower crack growth rates in the Al_2O_3/SiC composites relative to the monolithic matrix. Nevertheless, the irreversible nature of the crack tip deformation arising from the oxidation of SiC, viscous flow of glassy phase, interface cavitation, and microcrack growth, are indications of fatigue effects in the ceramic composite that are both chemical (oxidative) and mechanical in nature. As in the case of creep Al_2O_3/SiC, the phenomenon of fatigue (in oxidizing ambients) may best be considered a convolution of oxidation (or stress corrosion) and mechanical fatigue. Separation of these phenomena in experimental research is recommended in order to arrive at a clear understanding of the fundamental mechanisms involved.

8.3 Summary

The mechanical properties of ceramic composites are often adversely affected by exposure to hostile environments. Most of the adverse effects encountered to date stem from the high temperature oxidation of nonoxide constituents, particularly nonoxide reinforcements and coatings. The reaction products can be volatile or unstable with respect to other constituents, rendering the composite less resistant to oxidation than the independent constituents. These reactions and the associated adverse effects are compounded by the presence of stress, which can cause microcracking. Gaseous diffusion of oxygen through even very small (10 nm) cracks can lead to rapid oxidation of nonoxide constituents of composites. The gases that evolve as a result of oxidation of nonoxide constituents can facilitate cavitation and microstructural damage to the composite, thus degrading the mechanical properties. Silicon-based ceramics are attractive for high temperature applications because they oxidize to form glassy silica, which has an extremely low oxygen permeability and can, in some cases, heal cracks. However, the glassy phases generated by oxidation can also facilitate boundary sliding processes, thereby *accelerating* creep and crack growth. Fiber reinforcements, while attractive for composite strengthening, pose peculiar oxidation problems because fiber ends exposed at the composite surface provide access for oxygen to accelerated diffusion pathways along interfaces. Thus, fiber coatings designed to impart particular mechanical properties to composite interfaces must also resist oxidation and diffusion along the interface.

The problems cited above present a formidable challenge to materials scientists and engineers, as well as opportunities for important contributions. Some of the issues that need to be addressed in order to move composites into engineering applications are listed below.

(1) Separation of high temperature plastic deformation and high temperature oxidation in experiments, and understanding the synergism of these phenomena.
(2) Effects of oxidation on interface chemistry and correlation with interface mechanical properties.
(3) Development of oxide reinforcements.
(4) Oxidation at exposed fiber ends in continuous fiber composites.
(5) Interface engineering to develop low-cost, oxidation-resistant fiber coatings and/or surface modifications that impart the desired interface properties.
(6) Fundamental understanding of mechanisms and kinetics of oxidation in model composite systems.
(7) Micromechanical models that incorporate oxidation phenomena and allow quantitative prediction of effects of oxidation on interface properties.

While there are undoubtedly many more issues that must be addressed in order to develop composite materials for high temperature structural applications, the approaches that are likely to have the greatest success (and impact) are those that incorporate the perspectives of multiple disciplines, such as materials science, mechanics, surface science, and chemistry. Because of the complexity of composite materials, integrated interdisciplinary approaches are required to solve the multidimensional problems associated with deformation and damage that occur at high temperatures.

Acknowledgment

The author gratefully acknowledges the support of the Office of Naval Research under contract N00014-91-J-1480. Important contributions to the manuscript were made by Dr. K. Luthra of General Electric Company.

References

1. K. L. Luthra, "Oxidation of Carbon/Carbon Composites—A Theoretical Analysis," *Carbon*, **26**, 217–224 (1988).
2. K. L. Luthra, "Oxidation of Ceramic Composites," in *Corrosion and Corrosive Degradation of Ceramics*, eds. R. E. Tressler and M. McNallan, *Ceramic Transactions*, Vol. 10, 1990, American Ceramic Society, Westerville, OH, p. 183.
3. D. W. McKee, "Borate Treatment of Carbon Fibers and Carbon-Carbon Composites for Improved Oxidation Resistance," *Carbon*, **24**[6], 737–741 (1986).
4. J. R. Strife and J. E. Sheehan, "Ceramic Coatings for Carbon-Carbon Composites," *Am. Ceram. Soc. Bull.*, **67**[2], 369–374 (1988).
5. J. E. Sheehan, "Oxidation Protection for Carbon-Fiber Composites," *Carbon*, **27**[5], 709–715 (1989).
6. I. Jawed and D. C. Nagle, "Oxidation Protection in Carbon-Carbon Composites," *Mater. Res. Bull.*, **21**[11], 1391–1395 (1986).
7. T. D. Nixon and J. D. Cawley, "Oxidation Inhibition Mechanisms in Coated Carbon-Carbon Composites," *J. Am. Ceram. Soc.*, **75**, 703–708 (1992).
8. F. J. Norton, "Permeability of Gaseous Oxygen through Vitreous Silica," *Nature*, **4789**, 701 (1961).
9. T. Tokuda, T. Ito, and T. Yamaguchi, "Self-Diffusion in a Glassformer Melt Oxygen Transport in Boron Oxide," *Z. Naturforschung*, **26A**, 2058–2060 (1971).
10. T. A. Ramanarayanan, M. Raghaven, and R. Petkovic-Luton, "The Characteristics of Alumina Scales Formed on Fe-Based Yttria-Dispersed Alloys," *J. Electrochem. Soc.*, **131**, 923–931 (1984).
11. G. B. Abderazzik, F. Millit, G. Moulin, and A. M. Huntz, "Determination of Transport Properties of Aluminum Oxide Scale," *J. Am. Ceram. Soc.*, **68**, 307–314 (1985).
12. A. H. Chokshi and J. R. Porter, "Creep Deformation of an Alumina Matrix Composite Reinforced with Silicon Carbide Whiskers," *J. Am. Ceram. Soc.*, **68**[6], C144–C145 (1985).
13. K. L. Luthra and J. D. Park, "Oxidation of Silicon Carbide-Reinforced Oxide-Matrix Composites at 1375° to 1575°C," *J. Am. Ceram. Soc.*, **73**[4], 1014–1023 (1990).

14. F. Lin, T. Marieb, A. Morrone, and S. R. Nutt, "Thermal Oxidation of Alumina-SiC Whisker Composites: Mechanisms and Kinetics," *Mater. Res. Soc. Symp. Proc.*, **120**, 323–332 (1988).
15. N. Wang, Z. Wang, and G. C. Weatherly, "Formation of Magnesium Aluminate (Spinel) in Cast SiC Particulate-Reinforced Al (A356) Metal Matrix Composites," *Metall. Trans.*, **23A**[5], 1423–1430 (1992).
16. K. L. Luthra, "Chemical Interactions in High-Temperature Ceramic Composites," *J. Am. Ceram. Soc.*, **71**[12], 1114–1120 (1988).
17. Y. Oishi and W. D. Kingery, "Self-Diffusion of Oxygen in Single Crystal and Polycrystalline Aluminum Oxide," *J. Chem. Phys.*, **33**, 480–486 (1960).
18. J. L. Smialek and R. Gibala, "Diffusion Processes in Al_2O_3 Scales: Void Growth, Grain Growth, and Scale Growth in High Temperature Corrosion"; in *NACE-6*, ed. R. A. Rapp, National Association of Corrosion Engineers, Houston, TX, 1983, pp. 274–283.
19. K. Muehlenbachs and J. A. Schaeffer, "Oxygen Diffusion in Vitreous Silica—Utilization of Natural Isotopic Abundances," *Can. Mineral.*, **15**, 179–184 (1977).
20. S. R. Nutt and P. Lipetzky, "Creep of Whisker-Reinforced Alumina," *Mater. Sci. Engr.*, **A166**, 199–209 (1993).
21. R. F. Cooper and K. Chyung, "Structure and Chemistry of Fibre-Matrix Interfaces in Silicon Carbide Fiber-Reinforced Glass-Ceramic Composites: An Electron Microscopy Study," *J. Mater. Sci.*, **22**, 3148–3160 (1987).
22. L. A. Bonney and R. F. Cooper, "Reaction-Layer Interfaces in SiC-Fiber-Reinforced-Glass-Ceramics: A High-Resolution Scanning Transmission Electron Microscopy Analysis," *J. Am. Ceram. Soc.*,**73**[10], 2916–2921 (1990).
23. J. J. Brennan, "Interfacial Studies of SiC Fiber Reinforced Glass-Ceramic Matrix Composites," Final Report R87-917546-4 on ONR Contract N00014-82-C-0096, October, 1987.
24. E. Y. Sun, S. R. Nutt, and J. J. Brennan, "Interfacial Microstructure and Chemistry of SiC/BN Dual-Coated Nicalon Fiber Reinforced Glass-Ceramic Matrix Composites," *J. Am. Ceram. Soc.*, **77**[5], 1329–1339 (1994).
25. K. M. Prewo, F. Johnson, and S. Starrett, "Silicon Carbide Fibre-Reinforced Glass-Ceramic Composite Tensile Behaviour at Elevated Temperature," *J. Mater. Sci.*, **24**, 1373–1379 (1989).
26. R. T. Bhatt, "Oxidation Effects on the Mechanical Properties of a SiC-Fiber-Reinforced Reaction-Bonded Si_3N_4 Matrix Composite," *J. Am. Ceram. Soc.*, **75**[2], 406–412 (1992).
27. P. Lipetzky, S. R. Nutt, D. A. Koester, and R. F. Davis, "Atmospheric Effects on Compressive Creep of SiC Whisker-Reinforced Alumina," *J. Am. Ceram. Soc.*, **74**[6] 1240–1247 (1991).
28. A. R. DeArellano-Lopez, F. L. Cumbrera, A. T. Dominguez-Rodriguez, K. C. Gorretta, and J. L. Routbort, "Compressive Creep of SiC-Whisker-Reinforced Al_2O_3," *J. Am. Ceram. Soc.*, **73**[5], 1297–1300 (1990).
29. P. F. Becher, P. Angelini, W. H. Warwick, and T. N. Tiegs, "Elevated-Temperature-Delayed Failure of Alumina Reinforced with 20 vol% Silicon Carbide Whiskers," *J. Am. Ceram. Soc.*, **73**[1], 91–96 (1990).
30. H. T. Lin and P. F. Becher, "High-Temperature Creep Deformation of Alumina-SiC-Whisker Composites," *J. Am. Ceram. Soc.*, **74**[8], 1886–1893 (1991).
31. H. T. Lin and P. F. Becher, "Creep Behavior of a SiC-Whisker-Reinforced Alumina," *J. Am. Ceram. Soc.*, **73**[5], 1378–1381 (1990).
32. H. E. Kim and A. J. Moorhead, "Corrosion and Strength of SiC-Whisker-Reinforced Alumina Exposed at High Temperatures to H_2-H_2O Atmospheres," *J. Am. Ceram. Soc.*, **74**[6], 1354–1359 (1991).

33. T. Hansson, A. H. Swan, and R. Warren, "High Temperature Fracture of a SiC Whisker Reinforced Alumina in Air and Vacuum," *J. Euro. Ceram. Soc.*, **13**[5], 427–436 (1994).
34. L. X. Han, R. Warren, and S. Suresh, "An Experimental Study of Toughening and Degradation due to Microcracking in a Ceramic Composite," *Acta Metall. Mater.*, **40**[2], 259–274 (1992).
35. S. R. Nutt, P. Lipetzky, and P. F. Becher, "Creep Deformation of Alumina-SiC Composites" *Mater. Sci. Engr.*, **A126**, 165–172 (1990).
36. A. R. De Arellano-Lopez, A. Dominguez-Rodriguez, K. C. Goretta, and J. Routbort, "Plastic Deformation Mechanisms in SiC-Whisker-Reinforced Alumina," *J. Am. Ceram. Soc.*, **76**[6], 1425–1432 (1993).
37. R. L. Tsai and R. Raj, "Creep Fracture in Ceramics Containing Small Amounts of a Liquid Phase," *Acta Metall. Mater.*, **30**, 1043–1058 (1990).
38. J. R. Porter and A. Chokshi, "Creep Performance of Silicon Carbide Whisker-Reinforced Alumina"; in *Ceramic Microstructures '86: The Role of Interfaces*, eds. J. Pask and A. Evans, Plenum Press, New York, NY, 1987, p. 919.
39. F. D. Gac and J. J. Petrovic, "Feasibility of a Composite of SiC Whiskers in an $MoSi_2$ Matrix," *J. Am. Ceram. Soc.*, **68**[8], C200–C201 (1985).
40. R. M. Aikin, "Strengthening of Discontinuously Reinforced $MoSi_2$ Composites at High Temperatures," *Mater. Sci. Engr.*, **A155**, 121–133 (1992).
41. A. K. Vasudevan and J. J. Petrovic, "A Comparative Overview of Molybdenum Disilicide Composites," *Mater. Sci. Engr.*, **A155**[1–2], 1–17 (1992).
42. M. Suzuki, S. R. Nutt, and R. M. Aikin, "Creep Behaviour of an SiC-Reinforced XD $MoSi_2$ Ceramic," *Mater. Sci. Engr.*, **A162**, 73–82 (1993).
43. P. J. Meschter, "Oxidation of $MoSi_2/TiB_2$ and $MoSi_2/Al_2O_3$ Mixtures," *Scripta Metall.*, **25**[5], 1065–1069 (1991).
44. W. R. Cannon and T. G. Langdon, "Review—Creep of Ceramics, Part 1: Mechanical Characteristics," *J. Mater. Sci.*, **18**, 1–50 (1983).
45. K. M. Prewo and J. J. Brennan, "High Strength Silicon Carbide Fiber-Reinforced Glass Matrix Composites," *J. Mater. Sci.*, **15**, 463–468 (1980).
46. K. M. Prewo and J. J. Brennan, "Silicon Carbide Fiber Reinforced Glass-Ceramic Matrix Composites Exhibiting High Strength and Toughness," *J. Mater. Sci.*, **17**, 2371–2383 (1982).
47. K. M. Prewo and J. J. Brennan, "Silicon Carbide Yarn Reinforced Glass Matrix Composites," *J. Mater. Sci.*, **17**, 1201–1206 (1982).
48. T. Mah, M. G. Mendiratta, A. P. Katz, R. Ruh, and K. S. Mazdiyasni, "Room Temperature Mechanical Behavior of Fiber-Reinforced Ceramic-Matrix Composites," *J. Am. Ceram. Soc.*, **68**[1], C27–C30 (1985).
49. K. M. Prewo, "Tension and Flexural Strength of Silicon Carbide Fiber-Reinforced Glass-Ceramics," *J. Mater. Sci.*, **21**, 3590–3600 (1986).
50. K. M. Prewo, "Fatigue and Stress Rupture of Silicon Carbide Fiber-Reinforced Glass-Ceramics," *J. Mater. Sci.*, **22**, 2695–2701 (1987).
51. J. J. Brennan, "Glass and Glass-Ceramic Matrix Composites," in *Fiber Reinforced Glass-Ceramics*, ed. K. S. Mazkiyasni, Noyes Publications, Park Ridge, NJ, 1990, Chapter 8.
52. R. Naslain, O. Dugne, A. Guette, J. Sevely, C. R. Brosse, J.-P. Rocher, and J. Cotteret, "Boron Nitride Interphase in Ceramic-Matrix Composites," *J. Am. Ceram. Soc.*, **74**[10], 2482–2488 (1991).
53. O. Dugne, S. Prouhet, A. Guette, R. Naslain, *et al.*, "Interface Characterization by TEM, AES, and SIMS in Tough SiC (ex-PCS) Fibre-SiC (CVI) Matrix Composites with a BN Interphase," *J. Mater. Sci.*, **28**[13], 3409–3422 (1993).

54. R. W. Rice, "BN Coating of Ceramic Fibers for Ceramic Fiber Composites," U.S. Patent 4,642,271, February 10, 1987.
55. J. J. Brennan, "Interfacial Studies of Coated Fiber Reinforced Glass-Ceramic Matrix Composites," Ann. Rept. R90-918185-2 on AFOSR Contract F49620-88-C-0062, September 30, 1991.
56. J. J. Brennan, B. Allen, S. R. Nutt, and E. Y. Sun, "Interfacial Studies of Coated Fiber Reinforced Glass-Ceramic Matrix Composites," Ann. Rept. R92-970150-1 on AFOSR Contract F49620-92-C-0002, November 30, 1992.
57. L. X. Han and S. Suresh, "High-Temperature Failure of an Alumina-Silicon Carbide Composite under Cyclic Loads: Mechanisms of Fatigue Crack-Tip Damage," *J. Am. Ceram. Soc.*, **72**[7], 1233–1238 (1989).

Modeling

CHAPTER 9

Models for the Creep of Ceramic Matrix Composite Materials

R. M. McMeeking

9.1 Introduction

This review is intended to focus on ceramic matrix composite materials. However, the creep models which exist and which will be discussed are generic in the sense that they can apply to materials with polymer, metal or ceramic matrices. Only a case-by-case distinction between linear and nonlinear behavior separates the materials into classes of response. The temperature-dependent issue of whether the fibers creep or do not creep permits further classification. Therefore, in the review of the models, it is more attractive to use a classification scheme which accords with the nature of the material response rather than one which identifies the materials *per se*. Thus, this review could apply to polymer, metal or ceramic matrix materials equally well.

Only fiber- and whisker-reinforced materials will be considered. The fibers and whiskers will be identified as ceramics but with different characteristics from the matrix. As noted above, at certain temperatures the reinforcement phase will not be creeping and then it will be treated as elastic or rigid as appropriate to the model. At higher temperatures, the reinforcement phase will creep, and that must be allowed for in the appropriate model. On the other hand, the case of creeping fibers in an elastic matrix will not be considered, although certain of the models have a symmetry between fiber and matrix which permits such an interpretation. The models reviewed will be for materials with long fibers, broken long fibers, and short fibers or whiskers. Aligned fibers and two- and three-dimensional reinforcement by long fibers will be discussed. However, general laminate behavior will not be a subject of this review.

The material behaviors considered will include linear elasticity plus linear or nonlinear creep behavior. The nonlinear case will be restricted to power-law rheologies. In some cases the elasticity will be idealized as rigid. In ceramics, it is commonly the case that creep occurs by mass transport on the grain boundaries.[1] This usually leads to a linear rheology. In the models considered, this behavior will be represented by a continuum creep model with a fixed viscosity. That is, the viscosity is strain rate independent, although it will, in general, be temperature dependent. Thus, the mass transport *per se* will not be explicit in the models. In some situations, even though the mechanism is mass transport, the creep behavior involves a power-law response with a low exponent. Such a case is polycrystalline alumina at certain temperatures.[1] This explains the inclusion of power-law models in this review. An additional constitutive feature considered in this review is mass transport on the interface between the fiber and the matrix. This path can be a faster route for diffusion than the grain boundaries within the matrix. Therefore, it merits separate treatment as a mechanism for creep. A rudimentary model for the progressive breaking of reinforcements will be discussed. Creep void growth and other types of rupture damage in the matrix and the fiber will, however, be excluded from consideration.

Because the creep behavior of a ceramic composite often has a linear rheology, the behavior of the composite usually can be represented by an anisotropic viscoelastic constitutive law. Thus, a rather general model for such composites involves hereditary integrals with time-dependent creep or relaxation moduli[2,3] with a general anisotropy. The parameters for the law can be determined through creep and relaxation tests, but a multiplicity of experiments are required to evaluate all the functions appearing in a general anisotropic law. As a consequence, some guidance from micromechanics is essential for the generalization of the results. In this review, the focus will be on the micromechanics-based models and the hereditary integral methods will not be considered. However, the micromechanics models can, if desired, be recast in the classical viscoelastic form. It should be noted that there exists a vast literature on the linear elastic properties of reinforced materials. These elasticity models can be converted into creep models by use of standard methods of linear viscoelasticity.[2] This approach will be avoided in this review even though it can provide effective creep models for ceramic composites. Instead, the focus in this chapter will be on models which involve non-linearities or have features such as interface diffusion which are not accounted for when linear elastic models are converted to linear viscoelastic constitutive laws.

9.2 Material Models

All phases of the composite material will be assumed to be isotropic. The creep behavior of a ceramic will be represented by the law

$$\dot{\varepsilon}_{ij} = \frac{1}{2G}\dot{S}_{ij} + \frac{1}{9K}\delta_{ij}\dot{\sigma}_{kk} + \frac{3}{2}B\bar{\sigma}^{n-1}S_{ij} + \alpha\delta_{ij}\dot{T} \tag{1}$$

where $\dot{\varepsilon}$ is the strain rate, σ is the stress, $\dot{\sigma}$ is the stress rate, G is the elastic shear modulus, K is the elastic bulk modulus, δ_{ij} is the Kronecker delta, B is the creep rheology parameter, n is the creep index, α is the coefficient of thermal expansion, $\dot{T}$ is the rate of change temperature, $\underline{S}$ is the deviatoric stress, and the effective stress $\bar{\sigma}$ is defined by

$$\bar{\sigma} = \sqrt{\frac{3}{2} S_{ij} S_{ij}} \tag{2}$$

In all expressions the Einstein repeated index summation convention is used. x_1, x_2 and x_3 will be taken to be synonymous with x, y and z so that $\sigma_{11} = \sigma_{xx}$ etc. The parameter B will be temperature-dependent through an activation energy expression and can be related to microstructural parameters such as grain size, diffusion coefficients, etc., on a case-by-case basis depending on the mechanism of creep involved.[1] In addition, the index will depend on the mechanism which is active. In the linear case, $n = 1$ and B is equal to $1/3\eta$ where η is the linear shear viscosity of the material. Stresses, strains, and material parameters for the fibers will be denoted with a subscript or superscript f, and those for the matrix with a subscript or superscript m.

Various models will be used for the interface between the fiber and the matrix. For bonded interfaces, complete continuity of all components of the velocity will be invoked. The simplest model for a weak interface is that a shear drag equal to τ opposes the relative shear velocity jump across the interface. The direction of the shear drag is determined by the direction of the relative velocity. However, the magnitude of τ is independent of the velocities. This model is assumed to represent friction occurring mainly because of roughness of the surfaces or due to a superposed large normal pressure on the interface. Creep can, of course, relax the superposed normal stress over time, but on a short time scale the parameter τ can be assumed to be relatively invariant. No attempt will be made to account for Coulomb friction associated with local normal pressures on the interface.

On the other hand, a model for the viscous flow of creeping material along a fiber surface is exploited in some of the cases covered. This model is thought to represent the movement of material in steady state along a rough fiber surface and is given by

$$v_i^{Rel} = \bar{B}\bar{\sigma}^{n-1} n_j \sigma_{jk}(\delta_{ki} - n_k n_i) \tag{3}$$

where v_i^{Rel} is the relative velocity of the matrix material with respect to the fiber, $\bar{B}$ is a rheology parameter proportional to B but dependent also on roughness parameters for the fiber, and n_i is the unit outward normal to the fiber surface and the stress is that prevailing in the creeping matrix material. The law simply says that the velocity is in the direction of the shear stress on the interface but is controlled by power-law creep.

When there is mass transport by diffusion taking place in the interface between the fiber and the matrix, the relative velocity is given by[1]

$$\underline{v}^{Rel} = -\underline{n}(\underline{\nabla} \cdot \underline{j}) \tag{4}$$

where j is the mass flux of material in the plane of the interface and $\underline{\nabla}$ is the divergence operator in two dimensions also in the plane of the interface. The mass flux in the interface is measured as the mass per unit time passing across a line element of unit length in the interface. The flux is proportional to the stress gradient so that

$$\underline{j} = \mathscr{D} \, \underline{\nabla} \, \sigma_{nn} \tag{5}$$

where $\mathscr{D}$ is an effective diffusion coefficient and

$$\sigma_{nn} = \underline{n} \, \underline{\sigma} \, \underline{n} \tag{6}$$

is the normal stress at the interface. Combination of Eqns. (4) and (5) for a homogeneous interface gives

$$\underline{v}^{Rel} = -\underline{n} \mathscr{D} \, \nabla^2 \sigma_{nn} \tag{7}$$

The diffusion parameter $\mathscr{D}$ controls mass transport in a thin layer at the interface and so its relation to other parameters can be stated as[1]

$$\mathscr{D} = \frac{\sigma D_b \Omega}{kT} \tag{8}$$

where δ is the thickness of the thin layer in which diffusion is occurring, D_b is the diffusion coefficient in the material near or at the interface, Ω is the atomic volume, k is Boltzmann's constant, and T is the absolute temperature. The diffusion could occur in the matrix material, in the fiber, or in both. The relevant diffusion parameters for the matrix, the fiber, or some weighted average, would be used respectively.

It is worth noting that the "rule of mixtures" for stress, stress rate, strain, and strain rate is always an exact result in terms of the averages over the phases.[4] That is

$$\sigma_{ij} = f\sigma_{ij}^f + (1-f)\sigma_{ij}^m \tag{9}$$

$$\varepsilon_{ij} = f\varepsilon_{ij}^f + (1-f)\varepsilon_{ij}^m \tag{10}$$

and so on, where the unsuperscripted tensor variables are the averages over the composite material and the superscripted variables are the averages over the fibres (f) and the matrix (m) respectively. The volume fraction of the fibrous phase is f. The result applies irrespective of the configuration of the composite material, e.g., unidirectional or multidirectional reinforcement. However, an allowance must be made for the contribution arising from gaps which can appear, such as at the ends of fibers. The difficulty in the use of the rule of mixtures is the requirement that the average values in the fibers and in the matrix must be known somehow.

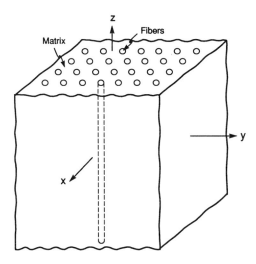

Fig. 9.1   A uniaxially reinforced fiber composite.

9.3 Materials with Long Intact Fibers

Creep laws for materials with long intact fibers are relevant to cases where the fibers are unbroken at the outset, and never fracture during life. As a model, it also applies to cases where some but not all of the fibers are broken so that some fibers remain intact during service. Obviously these situations would occur only when the manufacturing procedure can produce composites with many or all of the fibers intact.

In the problem of the creep of materials with intact unidirectional fibers, as shown in Fig. 9.1, most of the insights arise from the compatibility of the strain rates in the fibers and in the matrix. When a stress σ_{zz} is applied to the composite parallel to the fibers, the strains and strain rates of the fibers and the matrix in the z-direction must be all the same.[5] This gives rise to a creep law of the form

$$\dot{\varepsilon}_{zz} = \frac{\dot{\sigma}_{zz}}{E_L} + \dot{\varepsilon}_{zz}^c + \alpha_L \dot{T} \tag{11}$$

and

$$\dot{\varepsilon}_{xx} = \dot{\varepsilon}_{yy} = -\frac{\nu_L \dot{\sigma}_{zz}}{E_L} + \dot{\varepsilon}_{xx}^c + \alpha_T \dot{T} \tag{12}$$

where E_L is the longitudinal composite modulus, $\dot{\varepsilon}_{zz}^c$ is the longitudinal creep strain rate, α_L is the longitudinal coefficient of thermal expansion, ν_L is the Poisson's ratio for the composite relating transverse elastic strain to longitudinal stress, $\dot{\varepsilon}_{xx}^c$ is the transverse creep strain rate, and α_T is the transverse coefficient of thermal expansion. The temperature is taken to be uniform throughout the composite material. Evolution laws for the creep rates are required and these laws involve the stress levels in the matrix and fibers. Thus, in turn, evolution laws are required for the matrix and fiber stresses.

The exact laws, based on continuum analysis of the fibers and the matrix, would be very complicated. The analysis would involve equilibrium of stresses around, and in, the fibers and compatibility of matrix deformation with the fiber strains. Furthermore, end and edge effects near the free surfaces of the composite material would introduce complications. However, a simplified model can be developed for the interior of the composite material based on the notion that the fibers and the matrix interact only by having to experience the same longitudinal strain. Otherwise, the phases behave as two uniaxially stressed materials. McLean[5] introduced such a model for materials with elastic fibers and he notes that McDanels et al.[6] developed the model for the case where both the fibrous phase and the matrix phase are creeping. In both cases, the longitudinal parameters are the same, namely

$$E_L = fE_f + (1-f)E_m \tag{13}$$

$$\alpha_L = [fE_f\alpha_f + (1-f)E_m\alpha_m]/E_L \tag{14}$$

$$\dot{\varepsilon}^c_{zz} = [fE_fB_f\sigma^{nf}_f + (1-f)E_mB_m\sigma^{nm}_m]/E_L \tag{15}$$

When the fibers do not creep, B_f is simply set to zero. The longitudinal stress σ_{zz} in the fibers and the matrix are denoted σ_f and σ_m respectively. To accompany Eqns. (13)–(15), evolution laws for the fiber and the matrix stresses are required. These are

$$\dot{\sigma}_f = E_f(\dot{\varepsilon}_{zz} - B_f\sigma^{nf}_f - \alpha_f\dot{T}) \tag{16}$$

and

$$\dot{\sigma}_m = E_m(\dot{\varepsilon}_{zz} - B_m\sigma^{nm}_m - \alpha_m\dot{T}) \tag{17}$$

Indeed, combining these by the rule of mixtures, Eqn. (9) leads to Eqns. (13)–(15).

Since the fibers and the matrix do not interact transversely, the model implies that no transverse stresses develop in the matrix or the fibers. The rule of mixtures, Eqn. (10), then leads to

$$\nu_L = f\nu_f + (1-f)\nu_m \tag{18}$$

$$\alpha_T = f\alpha_f + (1-f)\alpha_m + f(1-f)(\alpha_f - \alpha_m)(\nu_fE_m - \nu_mE_f)/E_L \tag{19}$$

and

$$\dot{\varepsilon}^c_{xx} = -\tfrac{1}{2}(1-f)B_m\sigma^{nm}_m - \tfrac{1}{2}fB_f\sigma^{nf}_f$$
$$+ f(1-f)(B_m\sigma^{nm}_m - B_f\sigma^{nf}_f)(\nu_fE_m - \nu_mE_f)/E_L \tag{20}$$

The data suggest that the elastic parameters in this model are reasonably good to first order,[7] and experience with plasticity calculations[8–10] indicates that there is little plastic constraint between fibers and matrices at low volume fractions. Thus, the model should work reasonably well for any creep exponents at low volume fractions of fibers. Indeed, McLean[5] has used the isothermal version of the model successfully to explain longitudinal creep data for materials with noncreeping fibers.

Of interest, is the prediction of the uniaxial stress model when the applied stress and the temperature are held constant. The governing Eqns. (19), (16), and (17) then have the feature that as time passes, the solution always tends towards asymptotic values for stress in the fibers and the matrix. The evolution of the matrix stress occurs according to

$$\left(\frac{1}{E_m} + \frac{1-f}{fE_f}\right)\dot{\sigma}_m = B_m\sigma_m^{n_m} - B_f\left(\frac{\sigma - (1-f)\sigma_m}{f}\right)^{n_f} \tag{21}$$

and it can be shown that for any initial value of matrix stress, the matrix stress rate tends to zero. Therefore, the matrix stress tends toward the value which makes the right hand side of Eqn. (21) equal to zero. This can be solved easily for four common ceramic cases. One is when both matrix and fibers creep with a linear rheology so that both creep indices are equal to 1. In that case, the stresses tend towards the state in which

$$\sigma_m = \frac{B_f\sigma}{fB_m + (1-f)B_f} \tag{22}$$

and

$$\sigma_f = \frac{B_m\sigma}{fB_m + (1-f)B_f} \tag{23}$$

Another case is when the fibers creep linearly and the matrix creeps with an index of 2. Then the matrix tends towards a stress

$$\sigma_m = \left(\frac{B_f}{B_m}\right)\left[\sqrt{\left(\frac{B_m}{B_f}\right)\frac{\sigma}{f} + \frac{(1-f)^2}{4f^2}} - \frac{1-f}{2f}\right] \tag{24}$$

and, of course, $\sigma_m = [\sigma - (1-f)\sigma_m]/f$. The opposite case of a linear matrix and quadratic fibers is such that the fibers tend towards the stress

$$\sigma_f = \left(\frac{B_m}{B_f}\right)\left[\sqrt{\left(\frac{B_f}{B_m}\right)\frac{\sigma}{1-f} + \frac{f^2}{4(1-f)^2}} - \frac{f}{2(1-f)}\right] \tag{25}$$

and $\sigma_m = [\sigma - f\sigma_f]/(1-f)$. Finally, when the fibers do not creep, the matrix stress tends towards zero and the fiber stresses approach $\sigma/(1-f)$.

In the latter case, the transient stress can be stated as well. The isothermal result for constant σ is[5]

$$\sigma_m(t) = \left\{\frac{(n-1)fE_fE_mBt}{E_L} + \frac{1}{[\sigma_m(0)]^{n-1}}\right\}^{1-n} \tag{26}$$

when $n \neq 1$ and

$$\sigma_m(t) = \sigma_m(0) \exp(-f E_f E_m Bt/E_L) \tag{27}$$

when $n = 1$. The subscript on the creep rheology parameter for the matrix has been dropped and the unscripted B refers to the matrix henceforth. In both cases, $\sigma_f = [\sigma - (1-f)\sigma_m]/f$ and the composite strain is σ_f/E_f. The stress at time zero would be computed from the prior history with $t = 0$ being the time when both the temperature and the applied stress become constant. For example, if the temperature is held constant at creep levels until equilibrium is achieved and then the load is suddenly applied, $\sigma_m(0) = \sigma E_m/E_L$. To the extent that there are any thermal residual stresses at $t = 0$, they will contribute to $\sigma_m(0)$. However, Eqns. (26) and (27) make it clear that thermal residual stresses will be relaxed away by creep.

9.3.1 Steady Transverse Creep with Well-Bonded Elastic Fibers

The previous paragraph has made it clear that if there are elastic fibers and a constant macroscopic stress is applied, the longitudinal creep rate will eventually fall to zero. With constant transverse stresses applied as well, the process of transient creep will be much more complicated than that associated with Eqns. (27) and (28). However, it can be deduced that the longitudinal creep rate will still fall to zero eventually. Furthermore, any transverse steady creep rate must occur in a plane strain mode. During such steady creep, the fiber does not deform further because the stress in the fiber is constant. In addition, any debonding which might tend to occur would have achieved a steady level because the stresses are fixed.

For materials with a strong bond between the matrix and the fiber, models for steady transverse creep are available. The case of a linear matrix is represented exactly by the effect of rigid fibers in an incompressible linear elastic matrix and is covered in texts on elastic materials.[7,11,12] For example, the transverse shear modulus, and therefore the shear viscosity, of a material containing up to about 60% rigid fibers in a square array is approximated well by[11]

$$G_T = \frac{1 + 2f}{1 - f} G_m \tag{28}$$

It follows that in the coordinates of Fig. 9.1, steady transverse creep with well-bonded fibers obeys

$$\dot{\varepsilon}_{yy} = -\dot{\varepsilon}_{xx} = \frac{3B}{4}\left(\frac{1-f}{1+2f}\right)(\sigma_{yy} - \sigma_{xx}) \tag{29}$$

and

$$\dot{\varepsilon}_{xy} = \frac{3B}{2}\left(\frac{1-f}{1+2f}\right)\sigma_{xy} \tag{30}$$

with $\dot{\varepsilon}_{zz} = 0$. A material with fibers in a hexagonal array will creep slightly faster than this. Similarly, creep in longitudinal shear with fibers in a square array can be approximated well by

$$\dot{\varepsilon}_{xz} = \frac{3B}{2}\left(\frac{1-f}{1+f}\right)\sigma_{xz} \tag{31}$$

and

$$\dot{\varepsilon}_{yz} = \frac{3B}{2}\left(\frac{1-f}{1+f}\right)\sigma_{yz} \tag{32}$$

There are few comprehensive results for power-law matrices. Results given by Schmauder and McMeeking[10] for up to 60% by volume of fibers in a square array with a creep index of 5 can be represented approximately by

$$\dot{\varepsilon}_{xx} = -\dot{\varepsilon}_{yy} = 0.42B|\sigma_{xx} - \sigma_{yy}|^4(\sigma_{xx} - \sigma_{yy})/S^5 \tag{33}$$

where $\dot{\varepsilon}_{zz} = \dot{\varepsilon}_{xy} = \sigma_{xy} = 0$.

$$S = (1+f^2)/(1-f) \tag{34}$$

is the creep strength, defined to be the stress required for the composite at a given strain rate divided by the stress required for the matrix alone at the same strain rate. The expression in Eqn. (34) is only suitable for $n = 5$. The result in Eqn. (33) when $f = 0$ is the plane strain creep rate for the matrix alone. Results for $\sigma_{xy} \neq 0$ are not given because of the relative anisotropy of the composite with a square array of fibers. Relevant results for other power-law indices and other fiber arrangements are not available in sufficient quantity to allow representative expressions to be developed for them.

9.3.2 Three-Dimensional Continuous Reinforcement

This configuration of reinforcement can be achieved by the use of a woven fiber reinforcement or interpenetrating networks of the two phases. Another possibility is that random orientation of whiskers produces a percolating network and even if the whiskers are not bonded together, this network effectively forms a mechanically continuous phase. In the case of woven reinforcements, there may be some freedom for the woven network to reconfigure itself by the straightening of fibers in the weave or because of void space in the matrix. Such effects will be ignored and it will be assumed that the fibers are relatively straight and that there is little or no void space in the matrix. A straightforward model for these materials is that the strain rate is homogeneous throughout the composite. The response is then given by

$$\dot{\sigma}_{ij} = 2\bar{G}\dot{\varepsilon}_{ij} + (\bar{K} - \tfrac{2}{3}\bar{G})\delta_{ij}\dot{\varepsilon}_{kk} - 3\bar{K}\bar{\alpha}\dot{T}\delta_{ij} - 3fG_fB_f\bar{\sigma}_f^{n_f-1}S_{ij}^f$$
$$- 3(1-f)G_mB_m\bar{\sigma}_m^{n_m-1}S_{ij}^m \tag{35}$$

where

$$\bar{G} = fG_f + (1-f)G_m \tag{36}$$
$$\bar{K} = fK_f + (1-f)K_m \tag{37}$$

and

$$\bar{\alpha} = f\alpha_f K_f + (1-f)\alpha_m K_m \tag{38}$$

The evolution of the fiber and matrix average stresses appearing in the last two terms in Eqn. (35) is given by Eqn. (35) with $f = 1$ and $f = 0$ respectively. It is of interest that the constitutive law in Eqn. (35) is independent of the configuration of the reinforcements and the matrix. As a consequence, the law is fully isotropic and, therefore, may be unsuitable for woven reinforcements with unequal numbers of fibers in the principal directions. In addition, the fully isotropic law may not truly represent materials in which the fibers are woven in three orthogonal directions. Perhaps these deficiencies could be remedied by replacing the thermoelastic part of the law with an appropriate anisotropic model. A similar alteration to the creep part may be necessary but no micromechanical guidance is available at this stage.

   If the composite strain rate is known, the composite stress during steady-state isothermal creep can be computed from the rule of mixtures for the stress, Eqn. (9). This gives

$$S_{ij} = \frac{2}{3}\left[\frac{f}{B_f}\left(\frac{\dot{\bar{\varepsilon}}}{B_f}\right)^{(1-n_f)/n_f} + \frac{1-f}{B_m}\left(\frac{\dot{\bar{\varepsilon}}}{B_m}\right)^{(1-n_m)/n_m}\right]\dot{\varepsilon}_{ij} \tag{39}$$

where $\dot{\underline{\varepsilon}}$ must be deviatoric (i.e., $\dot{\varepsilon}_{kk} = 0$) and

$$\dot{\bar{\varepsilon}} = \sqrt{\frac{2}{3}\dot{\varepsilon}_{ij}\dot{\varepsilon}_{ij}} \tag{40}$$

A hydrostatic stress can be superposed, but it is caused only by elastic volumetric strain of the composite. The result in Eqn. (39) is, perhaps, not very useful since it is rare that a steady strain rate will be kinematically imposed. When both fiber and matrix creep, the steady solutions for a fixed stress in isothermal states are quite complex but can be computed by numerical inversion of Eqn. (39). The solution can, however, be given for the isothermal case where the fibers do not creep. (For non-fiber composites, this should be interpreted to mean that one of the network phases creeps while the other does not.) The matrix deviatoric stress is then given by

$$S_{ij}^m(t) = \frac{S_{ij}^m(0)}{\bar{\sigma}_m(0)} \{3(n-1)fG_fG_m \, Bt/\bar{G} + [\bar{\sigma}_m(0)]^{1-n}\}^{1/(1-n)} \qquad (41)$$

when $n \neq 1$; and for $n = 1$

$$S_{ij}^m(t) = S_{ij}^m(0)\exp(-3fG_fG_m Bt/\bar{G}) \qquad (42)$$

The subscripts on B and n have been dropped since only the matrix creeps. The interpretation of time and the initial conditions for Eqns. (41) and (42) are the same as for Eqns. (26) and (27). The fiber deviatoric stresses are given by

$$S_{ij}^f = [S_{ij} - (1-f)S_{ij}^m]/f \qquad (43)$$

and the composite deviatoric strain, e_{ij}, is therefore

$$e_{ij} = S_{ij}^f/2D_f \qquad (44)$$

The volumetric strains are invariant and given by

$$\varepsilon_{kk} = \sigma_{kk}/3\bar{K} \qquad (45)$$

As expected, the matrix deviatoric stresses will be relaxed away completely. Thereafter, the "fiber" phase sustains the entire deviatoric stress. As a consequence, in the asymptotic state

$$S_{ij}^f = S_{ij}/f \qquad (46)$$

and the composite strain will be given by Eqns. (44)–(46) as

$$\varepsilon_{ij} = \frac{\sigma_{ij}}{2fG_f} + \left(\frac{1}{3\bar{K}} - \frac{1}{2fG_f}\right)\frac{1}{3}\sigma_{kk}\delta_{ij} \qquad (47)$$

It follows that in uniaxial stress, with $\sigma_{zz} = \sigma$ and $\varepsilon_{zz} = \varepsilon$, the asymptotic result will be

$$\varepsilon = \left(\frac{1}{3fG_f} + \frac{1}{9\bar{K}}\right)\sigma \qquad (48)$$

This result indicates that the composite will have an asymptotic modulus slightly stiffer than fE_f because the matrix phase is capable of sustaining a hydrostatic stress.

9.3.3 Two-Dimensional Continuous Reinforcement

This configuration of reinforcement occurs when fibers are woven into a mat. It could also represent whisker-reinforced materials in which the whiskers are randomly oriented in the plane, especially if uniaxial pressing has been used to consolidate the composite material. In the case of the whisker-reinforced material, it is to be assumed that their volume fraction is so high that they touch each other. The whiskers have either been bonded together, say by diffusion, or the contact between the whiskers acts, as is likely, as a bond even if there is no interdiffusion.

In a simple model for this case, which, as in the 3-D case, ignores fiber straightening and anisotropy of the fibrous network, a plane stress version of Eqn. (35) can be developed. As such, it can only be used for plane stress states. Consider the x–y plane to be that in which the fibers are woven or the whiskers are lying. The strain rates in this plane are taken to be homogeneous throughout the composite material and σ_{zz}, σ_{xz} and σ_{yz} are taken to be zero. The resulting law is

$$\dot{\sigma}_{\alpha\beta} = 2\bar{G}\left[\dot{\varepsilon}_{\alpha\beta} + \frac{\bar{\nu}}{1-\bar{\nu}}\delta_{\alpha\beta}\dot{\varepsilon}_{\gamma\gamma} - \frac{1+\bar{\nu}}{1-\bar{\nu}}\hat{\alpha}\dot{T}\delta_{\alpha\beta}\right]$$
$$-3fG_f B_f \bar{\sigma}_f^{n_f-1}\left[S_{\alpha\beta}^f + \frac{\nu_f}{(1-\nu_f)}\delta_{\alpha\beta}S_{\gamma\gamma}^f\right]$$
$$-3(1-f)G_m B_m \bar{\sigma}_m^{n_m-1}\left[S_{\alpha\beta}^m + \frac{\nu_m}{(1-\nu_m)}\delta_{\alpha\beta}S_{\gamma\gamma}^m\right] \quad (49)$$

where the Greek subscripts range over 1 and 2 and where

$$\frac{\bar{\nu}}{1-\bar{\nu}} = \left[fG_f\frac{\nu_f}{1-\nu_f} + (1-f)G_m\frac{\nu_m}{1-\nu_m}\right]\Big/\bar{G} \quad (50)$$

and

$$\hat{\alpha} = \left[f\alpha_f G_f\frac{1+\nu_f}{1-\nu_f} + (1-f)\alpha_m G_m\frac{1+\nu_m}{1-\nu_m}\right]\Big/\left(\frac{1+\bar{\nu}}{1-\bar{\nu}}\bar{G}\right) \quad (51)$$

The fiber and matrix evolution laws for stress are identical to Eqn. (49) with $f = 0$ and $f = 1$, respectively. Being isotropic in the plane, this law suffers from the same deficiencies as the 3-D version regarding the orthotropy of the woven mat and any inequality between the warp and the woof. As before, this could be remedied with an anisotropic version of the law.

In steady-state isothermal creep, the relationship between in-plane components of stress and in-plane components of strain rate are given by

$$\sigma_{\alpha\beta} = \frac{2}{3}\left[\frac{f}{B_f}\left(\frac{\dot{\bar{\varepsilon}}}{B_f}\right)^{(1-n_f)/n_f} + \frac{1-f}{B_m}\left(\frac{\dot{\bar{\varepsilon}}}{B_m}\right)^{(1-n_m)/n_m}\right](\dot{\varepsilon}_{\alpha\beta} + \dot{\varepsilon}_{\gamma\gamma}\delta_{\alpha\beta}) \quad (52)$$

with $\sigma_{xz} = \sigma_{yz} = \sigma_{zz} = 0$, and with $\dot{\bar{\varepsilon}}$ given by Eqn. (40) but with $\dot{\varepsilon}_{xz} = \dot{\varepsilon}_{yz} = 0$. As in the 3-D case, this must be inverted numerically to establish a steady-state isothermal creep rate for a given imposed stress.

When the fibers are elastic and non-creeping, the isothermal behavior at fixed applied plane stress is given in terms of the deviatoric stress by Eqns. (41)

or (42) and Eqn. (43). The expression for the deviatoric composite strain, Eqn. (44), still applies. However, the composite strain obeys

$$\dot{\varepsilon}_{\gamma\gamma} = \frac{3}{2}(1-f)\frac{G_m}{\overline{G}}\ B\overline{\sigma}_m^{n-1}\ \frac{(1+\nu_m)(1-\overline{\nu})}{(1+\overline{\nu})(1-\nu_m)}\ S_{\gamma\gamma}^m \tag{53}$$

and

$$\dot{\varepsilon}_{zz} = -\left[f\frac{\nu_f}{1-\nu_f} + (1-f)\frac{\nu_m}{1-\nu_m}\right]\dot{\varepsilon}_{\gamma\gamma} - \frac{3}{2}(1-f)\frac{(1-2\nu_m)}{(1-\nu_m)}\ B\overline{\sigma}_m^{n-1}S_{\gamma\gamma}^m \tag{54}$$

The latter result indicates that the volumetric strains can be relaxed to some extent by matrix creep. This contrasts with the 3-D case where complete compatibility of strains precludes such relaxation. The extent to which the relaxation occurs has not yet been calculated. However, if it is assumed that the relaxation can be complete so that the matrix volumetric strain is zero, then the fiber stress tends towards $\sigma_{\alpha\beta}/f$ and, therefore, the composite strain approaches

$$\varepsilon_{ij} = \frac{1+\nu_f}{fE_f}\ \sigma_{ij} - \frac{\nu_f}{fE_f}\delta_{ij}\sigma_{\gamma\gamma} \tag{55}$$

which, of course, is restricted to plane stress. It can be seen that in uniaxial stress, the effective asymptotic modulus would now equal fE_f. A properly calculated solution for $\varepsilon_{kk}(t)$ is required to investigate whether this result holds true.

9.4 Uniaxial Reinforcement with Long Brittle Fibers

The reinforcement configuration of interest now is once more that depicted in Fig. 9.1 and the loading will be restricted to a longitudinal steady stress σ_{zz}. The possibility will be taken into account that the fibers might be overstressed and, therefore, could fail. Only elastic fibers which break in a brittle manner will be considered, although ceramic fibers are also known to creep and possibly rupture due to grain boundary damage. Only frictionally constrained fibers will be considered since well-bonded fibers will fail upon matrix cracking and vice versa. The case where the fibers have a deterministic strength S can be considered. In that situation, the fibers will remain intact when the fiber stress is below the deterministic strength level and they will break when the fiber stress exceeds the strength. The fracturing of the fibers could occur during the initial application of the load, in which case elastic analysis is appropriate. If the fibers survive the initial application of the load, then subsequent failure can occur as the matrix relaxes according to Eqns. (26) or (27) and the fiber stress increases. Thus, the time elapsed before first fiber failure can be estimated based on Eqns. (26) or (27) by setting the fiber stress equal to the deterministic strength. This predicts that failure of a fiber will occur when

$$\sigma_m = [\sigma - fS]/(1 - f) \tag{56}$$

from which the time to failure can be computed through Eqns. (26) or (27). The failure of one fiber in a homogeneous stress state will cause a neighboring fiber to fail nearby because of the fiber/matrix shear stress interaction and the resulting localized load sharing around the broken fiber. Thus, a single fiber failure will tend to cause a spreading of damage in the form of fiber breaks near a single plane across the section. This will lead to localized rapid creep and elastic strains in the matrix near the breaks, perhaps giving rise to matrix failure. It follows, therefore, that tertiary failure of the composite will tend to occur soon after the occurrence of one fiber failure when the fiber strength is deterministic.

Tertiary failure processes akin to this have been modeled by Phoenix and co-workers[13–15] in the context of epoxy matrix composites. Indeed, they show that such tertiary failure can occur even when the fiber strength is statistical in nature. This mechanism will not be pursued further in this chapter but some other basic results considered on the assumption are that when there is a sufficient spread in fiber strengths, such tertiary failures can be postponed well beyond the occurrence of first fiber failure or indeed eliminated completely. Thus, attention will be focused on fibers which obey the classical Weibull model that the probability of survival of a fiber of length L stressed to a level σ_f is given by

$$P_s = \exp\left[-\frac{L}{L_g}\left(\frac{\sigma_f}{S}\right)^m\right] \tag{57}$$

where L_g is a datum gauge length, S is a datum strength, and m is the Weibull modulus. Clearly the results given below can be generalized to account for variations on the statistical form which differ from Eqn. (57). However, the basic ideas will remain the same.

9.4.1 Long-Term Creep Threshold

Consider a specimen of length L_s containing a very large number of wholly intact fibers. A stress σ is suddenly applied to the specimen parallel to the fibers. The temperature has already been raised to the creep level and is now held fixed. Upon first application of the load, some of the fibers will break. The sudden application of the load means that the initial response is elastic. This elastic behavior has been modeled by Curtin,[16] among others, but details will not be given here. If the applied stress exceeds the ultimate strength of the composite in this elastic mode of response, then the composite will fail and long-term creep is obviously not an issue. However, it will be assumed that the applied stress is below the elastic ultimate strength and, therefore, creep can commence. It should be noted, however, that matrix cracking can occur in the ceramic matrix and the characteristics of creep relaxation would depend on the degree of matrix cracking. However, this aspect of the problem will not be considered in detail. For cases where there is matrix cracking and for which the specimen length L_s is sufficiently long, Curtin[16] has given the theoretical prediction that the ultimate elastic strength is

$$S_u = f[4L_g S^m \tau/D(m+2)]^{1/(m+1)} (m+1)/(m+2) \tag{58}$$

where τ is the interface shear strength between the fiber and the matrix, and D is the diameter of the fibers. The interface shear strength is usually controlled by friction. For specimens shorter than δ_c, the ultimate brittle strength exceeds S_u where δ_c is given by[16]

$$\delta_c = [SL_g^{1/m} D/2\tau]^{m/(m+1)} \tag{59}$$

This critical length is usually somewhat less than the datum gauge length.

When the applied stress σ is less than S_u, creep of the matrix will commence after application of the load. During this creep, the matrix will relax and the stress on the fibers will increase. Therefore, further fiber failure will occur. In addition, the process of matrix creep will depend on the extent of prior fiber failure and, as mentioned previously, on the amount of matrix cracking. The details will be rather complicated. However, the question of whether steady-state creep or, perhaps, rupture will occur, or whether sufficient fibers will survive to provide an intact elastic specimen, can be answered by consideration of the stress in the fibers after the matrix has been assumed to relax completely. Clearly, when the matrix carries no stress, the fibers will at least fail to the extent that they do in a dry bundle. It is possible that a greater degree of fiber failure will be caused by the transient stresses during creep relaxation, but this effect has not yet been modeled. Instead, the dry bundle behavior will be used to provide an initial estimate of fiber failure in these circumstances.

Given Eqn. (57), the elastic stress–strain curve for a fiber bundle is

$$\sigma = fE_f \varepsilon \exp\left[-\frac{L_s}{L_g}\left(\frac{E_f \varepsilon}{S} \right)^m \right] \tag{60}$$

Thus, when a stress σ is applied to the composite, creep will occur until the strain has the value consistent with Eqn. (60). Numerical inversion of Eqn. (60) can be used to establish this strain. The stress–strain curve in Eqn. (60) has a stress maximum when

$$\varepsilon = \frac{S}{E_f}\left(\frac{L_g}{mL_s} \right)^{1/m} \tag{61}$$

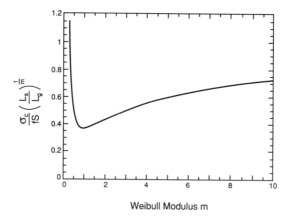

Fig. 9.2 Threshold for long-term creep of a uniaxially reinforced composite as a function of the Weibull modulus for the fiber strength distribution.

with a corresponding stress level given by

$$\sigma_c = fS\left(\frac{L_g}{mL_s}\right)^{1/m}\exp(-1/m) \tag{62}$$

This result is plotted as a function of m in Fig. 9.2. If $\sigma < \sigma_c$, the composite will creep until the strain is consistent with Eqn. (60) and thereafter no further creep strain will occur. Of course, the noncreeping state will be approached asymptotically. (It should be noted that due to possible fiber failure during the creep transient, the true value for σ_c may lie below the result given in Eqn. (62).) For an applied composite stress equal to, or exceeding, σ_c, creep will not disappear with time because all of the fibers will eventually fail and the strain will continue to accumulate.

The critical threshold stress for ongoing creep given by Eqn. (62) is specimen length dependent. For very long specimens, the threshold stress is low whereas short specimens will require a high stress for ongoing creep to continue without limit. On the other hand, the ultimate brittle strength as given by Eqn. (58) for a composite specimen longer than δ_c is specimen length independent. Thus, there are always specimens long enough so that σ_c is less than S_u. This means that the specimen can be loaded without failure initially and if σ exceeds σ_c, the specimen will go into a process of long-term creep. (It should be remembered, however, that this model is based on the assumption that tertiary failure is delayed and does not occur until a substantial amount of matrix creep has occurred.) For shorter specimens, the relationship between σ_c and S_u depends on the material parameters appearing in Eqns. (58) and (62). However, for typical values of the parameters, σ_c is less than S_u so that there is usually a window of stress capable of giving rise to long-term creep without specimen failure when the specimen length exceeds δ_c. Typical values for the parameters are given by, among others, Hild *et al.*[17] From these parameters, predictions for σ_c can be made. For example, a LAS matrix composite containing 46% of SiC (Nicalon) fibers ($m = 3$ or 4) is predicted to have a value for σ_c between 400 MPa and 440 MPa for a specimen length of 25 mm, whereas its measured ultimate brittle strength is between 660 MPa and 760 MPa. At 250 mm specimen length, the long-term creep threshold σ_c is predicted to fall to the range between 185 MPa to 250 MPa. Similarly, a CAS matrix composite with 37% SiC (Nicalon) fibers ($m = 3.6$) in a specimen length of 25 mm is predicted to suffer long-term creep if the stress exceeds 160 MPa, whereas the measured ultimate brittle strength is 430 MPa. For a 250 mm specimen length, this creep threshold is predicted to fall to 85 MPa. Thus, it is clear that in some practical cases, applied stresses which are modest fractions of the elastic ultimate strength will cause long-term creep.

9.4.2 Steady-State Creep

For specimens which have (1) previously experienced an applied stress exceeding the long-term creep threshold, or (2) had every fiber broken prior to testing or service (e.g., during processing) or (3) had few fibers intact to begin with so that initially the long-term creep threshold is much lower than σ_c as predicted by Eqn. (62), a prediction of the long-term creep behavior can be made. Prior to this state, there will, of course, be a transient which involves matrix creep and, perhaps, the fragmentation of fibers. This transient has not been fully modeled. Only a rudimentary assessment of the creep behavior has been made revealing the following features.

For those composites initially having some of the fibers intact, there will always be some which must be stretched elastically. This will require a stress which will tend towards the value given by Eqn. (60) with f replaced by f_i, the volume fraction of fibers initially intact. If a relaxation test were carried out, the stress would become asymptotic to the level predicted by Eqn. (60). The remaining broken fibers will interact with the matrix in a complex way, but at a given strain and strain rate, a characteristic stress contribution can be identified in principle. Details have not been worked out. However, the total stress would be the sum of the contribution from the broken and unbroken fibers. If the transient behavior is ignored (i.e., assumed to die away relatively fast compared to the strain rate), a basic model can be constructed.

9.4.3 Steady-State Creep with Broken Fibers

First, consider a composite with a volume fraction f of fibers, all of which are broken. There are two possible models for the steady-state creep behavior of such a material. In one, favored by Mileiko[18] and Lilholt,[19] among others, the matrix serves simply to transmit shear stress from one fiber to another and the longitudinal stress in the matrix is negligible. The kinematics of this model requires void space to increase in volume at the ends of the fibers. However, with broken fibers there is no inherent constraint on this occurring. Furthermore, if matrix cracking has occurred, the matrix will not be able to sustain large amounts of longitudinal tension and its main role will be to transmit shear from fiber to fiber. Indeed, matrix cracking will probably promote this mode of matrix flow since there will be no driving stress for other mechanisms of straining. The other model, favored by McLean,[20] and developed by Kelly and Street,[21] involves a stretching flow of the matrix between fibers at a rate equal to the macroscopic strain rate of the composite material. This requires substantial axial stress in the matrix. In addition, volume is preserved by the flow and there is no need for space to develop at the end of the fiber. The model requires a considerable matrix flow to occur, transporting material from the side of a given fiber to its end, and the injection of matrix in between adjacent ends of the broken fibers. There is good reason to believe that the Mileiko[18] pattern of flow prevails when there are broken fibers.

In a version of the Mileiko[18] model in which it is assumed that each of six neighboring fibers has a break somewhere within the span of the length of a given fiber, but that the location of those breaks is random within the span, the relationship between the steady-state creep rate and the composite stress is

$$\dot{\varepsilon} = g(n,f)(D/L)^{n+1} B\sigma^n \tag{63}$$

where L is the average length of the broken fiber segments and

$$g(n,f) = 2\sqrt{3}\left[\frac{\sqrt{3}(2n+1)}{2nf}\right]^n \left[\frac{(1-f^{(n-1)/2})}{(n-1)}\right] \tag{64}$$

when $n \neq 1$ and

$$g(1,f) = (9/f)\ln(1/\sqrt{f}) \tag{65}$$

These functions have been computed for uniform fiber length and based on a hexagonal shape for the fiber even though it is interpreted to be circular. That explains why creep strength goes to infinity at $f = 0$ rather than at $f < 1$. In this creep model, the influence of both volume fraction and the aspect ratio L/D on the strain rate is clear, with both having a strong effect. As noted, this model could serve as a constitutive law for the creep of a material in which all of the fibers are broken to fragments of average length L. In addition, it could be used for short fiber composites which have weak bonds between the fiber end and the matrix so that debonding can readily occur and void space can develop as a result. However, the aspect ratio L/D should be large so that the Mileiko[18] flow pattern will occur and end effects can be neglected when the composite creep law is computed.

The shear stress transmitted to a fiber is limited to the shear strength τ. As a result, the formula given in Eqn. (63) is valid only up to a composite macroscopic stress of

$$\sigma = \frac{2nf}{2n+1}\left(\frac{L}{D}\right)\tau \tag{66}$$

for both the linear and non-linear cases. According to the model, at this level of applied stress, the shear stress on the fiber interface will start to exceed τ. Therefore, at stresses higher than the value given in Eqn. (66), the strain rate will exceed the level predicted in Eqn. (63). This situation will persist in the presence of matrix cracks up to a composite macroscopic stress of

$$\sigma_{LIM} = f\tau L/D \tag{67}$$

at which stress the entire fiber surface is subject to a shear stress equal in magnitude to τ. Then, the mechanism represented by Eqn. (63) provides an indeterminate strain rate as in rate-independent plasticity. Thus σ_{LIM} can be thought of as a yield stress. This concept is probably satisfactory for materials with many matrix cracks so that there is no constraint on stretching the matrix. However, when there are no matrix cracks, the strain rate is probably controlled by the mechanism which generates void space at the fiber ends. This has been considered to require negligible stress in the version of the model leading to Eqn. (63). For a proper consideration of the limit behavior, the contribution to the stress arising from void development at the fiber ends should be taken into account.

9.4.4 The Effect of Fiber Fracture

If the stress applied to the composite is increased, the stress sustained by fibers will increase also. When the probability of survival of fibers obeys the statistical relationship given by Eqn. (57), the effect of a raised stress will be to fracture more fibers, with a preference for breaking long fibers. This will have the effect of reducing the average fiber length L and, therefore, raising the strain rate at a given applied stress as can be deduced from Eqn. (63). Therefore, the composite will no longer have a simple power-law behavior in steady-state creep since the fiber fragment length will depend on the largest stress which the composite material has previously experienced. In this regard, the elastic transients will play an important role in determining the fiber fragment length. However, the average fragment length in steady-state creep will generally be smaller than the average fragment length arising during initial elastic response. Therefore, some guidance can be obtained from a model designed to predict the steady-state creep response only.

For the Mileiko[18] model of composite creep leading to the steady-state creep rate for fixed fiber length given in Eqn. (63), a rudimentary fiber fragment length model gives

$$L = (m+1)\left[\frac{(n+1)fS}{(2n+1)\sigma}\right]^m L_g \tag{68}$$

subject to L being less than the specimen length. When a stress σ is applied to the composite material and steady state is allowed to develop, the average length for the fiber fragments is predicted by Eqn. (68). This model is by no means precise, based as it is on some approximations in the calculations, as well as the notion that all fibers can be treated as if they had the same length. However, the model conveys the important notion that the fiber fragment length will decrease as the applied stress is increased.

The fiber fragment average length during steady-state creep can be substituted into Eqn. (63), from which results

$$\dot{\varepsilon} = h(n, f, D/L_g, m, S) B \sigma^{n+m+nm} \tag{69}$$

where h is a rather complicated function of its arguments and can readily be calculated. A significant conclusion is that the creep index for the composite is no longer just n but is $n + m + nm$. Thus, a ceramic matrix material with a matrix creep index of 1 will have composite creep index of $2m + 1$. In the case of a fiber with a Weibull modulus of $m = 4$, the composite creep index will be 9. Similar effects will be apparent in composites with a non-linearly creeping ceramic matrix, say with $n = 2$. It has been observed that *metal* matrix composites with non-creeping reinforcements often have a creep index which differs from that of the matrix[5,22] and the effect is usually attributed to damage of the fibers or of the interface. It can be expected that ceramic matrix composites will exhibit a similar behavior.

It should be noted that the model leading to Eqn. (69) is incomplete since the stress required to cause the enlargement of void space at the fiber breaks is omitted from consideration. At high strain rates this contribution to stress can be expected to dominate other contributions. Therefore, at high stress or strain, the creep behavior will diverge from Eqn. (69) and perhaps exhibit the nth power dependence on stress as controlled by the matrix. The creep rate at these high stresses can be expected to *exceed* the creep rate of the matrix at the same applied stress since the void space at the fiber ends is a form of damage.

9.4.5 Creep of an Initially Undamaged Composite

The issue to be addressed in this section is the long-term behavior of a composite stressed above the threshold σ_c given by Eqn. (62) which means that the specimen will creep continuously. As in the immediately preceding sections, elastic transient effects will be omitted from the model of long-term creep of the initially undamaged composite. No model exists, as yet, for the transient behavior, but there is little doubt that the transient behavior is important. Many composite materials in service at creep temperatures will probably always respond in the transient stage since the time for that to die away will typically be rather long. However, a quasi-steady-state model, as before, will give some insight into the state towards which the transients will be taking the material. However, the model presented below is rather selective, since it includes some elastic effects and ignores others. It is not known how deficient this feature of the model is. Perhaps the material state will evolve rather rapidly towards the state predicted below and, therefore, the model may have some merit.

The specimen is composed of a mixture of matrix, unbroken fibers, and broken fibers. The volume fraction of intact fibers is given by Eqn. (57) with $L = L_s$, the specimen length. To the neglect of transients, the macroscopic stress supported by these intact fibers is given by Eqn. (60). The strain will now exceed the level of Eqn. (61) associated with the ultimate strength of the fiber bundle. Therefore, the stress supported by the intact fibers will be less than σ_c, which is the ultimate strength of the fiber bundle without matrix. The applied stress exceeds σ_c and the balance in excess of the amount borne by the intact fibers will cause the composite material to creep.

The steady-state result given in Eqn. (69) will be taken to express the creep behavior controlled by the broken fibers. The volume fraction of broken fibers is

$$f_b = 1 - \exp\left[-\frac{L_s}{L_g}\left(\frac{E_f \varepsilon}{S} \right)^m \right] \tag{70}$$

and a material with this volume fraction of broken fibers creeping at a rate $\dot{\varepsilon}$ will support a stress

$$\sigma_b = [\dot{\varepsilon}/Bh(n, f_b, D/L_g, m, S)]^{1/p} \tag{71}$$

where

$$p = n + m + nm \tag{72}$$

which comes directly from Eqn. (69). The total stress sustained by the composite material is, therefore

$$\sigma = f_b \sigma_b + \sigma_u \tag{73}$$

where σ_u is the contribution due to unbroken fibers. This leads to

$$\sigma = f_b[\dot{\varepsilon}/Bh(f_b)]^{1/p} + fE_f \varepsilon \exp\left[-\frac{L_s}{L_g}\left(\frac{E_f \varepsilon}{S} \right)^m \right] \tag{74}$$

which can be seen to be a rather nonlinear Kelvin–Voigt material in which the stress is the sum of a viscous element and an elastic element, both of which are nonlinear. As the strain increases, the second term on the right-hand side of Eqn. (74) (i.e., the term due to the intact fibers) will diminish and become rather small when only a few unbroken fibers are left. At the same time, f_b will approach f and so the strain rate will approach the steady-state rate for a material in which all of the fibers are broken. However, as long as a few fibers remain intact, the creep behavior will not precisely duplicate that for the fully broken material. This transient effect will be compounded by the redistribution of stress from the matrix to the fibers which will occur both after the first application of load to the composite material and after each fracture of a fiber, both effects having been omitted from this version of the model.

9.5 Creep of Materials with Strong Interfaces

It seems unlikely that long-fiber ceramic matrix composites with strong bonds will find application because of their low temperature brittleness. However, for completeness, a model which applies to the creep of such materials can be stated. It is that due to Kelly and Street.[21] It is possible also that the model applies to aligned whisker-reinforced composites since they may have strong bonds. In addition, the model has a wide currency since it is believed to apply to weakly bonded composites as well. However, the Mileiko[18] model predicts a lower creep strength for weakly bonded or unbonded composites and therefore is considered to apply in that case.

The Kelly and Street[21] model uses the notion that creep of the composite material can be modeled by the behavior of a unit cell. Each unit cell contains one fiber plus matrix around it so that the volume of the fiber divided by the volume of the unit cell equals the fiber volume fraction of the composite material. The perimeter of the unit cell is assumed to be deforming at a rate consistent with the macroscopic strain rate of the composite material. (It can be observed at this stage that this notion is inconsistent with the presence of transverse matrix cracks which would make it impossible to sustain the longitudinal stress necessary to stretch the matrix. This is an additional reason why the Kelly and Street[21] model is not likely to be applicable to unbonded ceramic matrix materials which are likely to have matrix cracks.) Only steady-state creep of materials with aligned reinforcements which are shorter than the specimen is considered. The unit cell is assumed to conserve volume. This means that material originally adjacent to the reinforcement must flow around the fiber and finish up at its end. This phenomenon has to occur when the end of the fiber or whisker is strongly bonded to the matrix. For this reason, the Kelly and Street[21] model is considered to be relevant to materials with strong bonds.

Kelly and Street[21] analyzed this model but their deductions were not consistent with the mechanics. McMeeking[23] has remedied this deficiency for nonlinear materials. His results for $n = 2$ are relevant to composite materials with nonlinearly creeping ceramic matrices which tend to have low creep indices. In that case, the steady-state creep rate is given by Eqn. (63) with $n = 2$ and

$$g(2,f) = \frac{75\sqrt{3}}{8f^2}\left(\frac{1}{2} - \frac{8}{5}\sqrt{f} + \frac{3}{2}f - \frac{1}{2}f^2 + \frac{1}{10}f^3\right) \tag{75}$$

which is invalid for $f = 0$. When f is close to zero a different form should be used which accounts for the matrix stress so that the matrix creep law is recovered smoothly as the volume fraction of fiber disappears. This result is developed below and is given in Eqn. (77). Comparison of Eqn. (75) with Eqn. (64) for $n = 2$ will show that the model of Kelly and Street[21] creeps more slowly than the Mileiko[18] law confirming that the Mileiko model is the preferred one when it is kinematically admissible.

It is thought that at higher temperatures, the interface between the fiber and the matrix becomes weak and sliding occurs according to the constitutive law given in Eqn. (3). In that case, creep of a composite with a well-bonded interface obeys Eqn. (63) with $n = 2$ and[23]

$$g(2,f) = \frac{25\sqrt{3}}{8f^2}\left[3\left(\frac{1}{2} - \frac{8}{5}\sqrt{f} + \frac{3}{2}f - \frac{1}{2}f^2 + \frac{1}{10}f^3\right) + \frac{(1-f)^3\bar{B}}{2DB}\right] \quad (76)$$

This form for g is identical with that in Eqn. (75) when $\bar{B} = 0$. Thus, sliding at the interface increases the creep rate at a given stress. If $\bar{B}/BD$ is very large, signifying a very weak interface, then the interface term will dominate the matrix term in Eqn. (76). It should be noted that there is a relative size effect, with large diameter fibers making sliding less important.

At large strain rates, stretching of the matrix as it slides past the fiber will contribute to the creep strength. Under those circumstances, the term $g(2,f)$ in Eqn. (63) should be replaced by[23]

$$\bar{g}(2,f) = \left[1/\sqrt{g(2,f)} + (1-f)\left(\frac{D}{L}\right)^{1.5}\right]^{-2} \quad (77)$$

where, in Eqn. (77), $g(2,f)$ is to be calculated according to Eqn. (76). Note that as $g(2,f)$ becomes large (i.e., the composite strain rate is large because either f is small or $\bar{B}$ is large), the composite strain rate will approach

$$\dot{\varepsilon} = B[\sigma/(1-f)]^2 \quad (78)$$

which is the rate that would prevail if the fibers were replaced by long cylindrical holes.

9.5.1 Creep of Materials with a Linear Rheology

The equivalent correction to the Kelly and Street[21] model for cases where the matrix creep obeys a linear rheology ($n = 1$) was not given by McMeeking.[23] However, consideration of this case can be included in a model with accounts for the ability of a well-bonded interface between the fiber and the matrix to sustain sliding according to Eqn. (3), and in which mass transport may cause the effect described by Eqn. (4). Kim and McMeeking[24] have given the steady-state creep law for the composite material in these circumstances to be

$$\sigma = \frac{\dot{\varepsilon}}{B}\left[h(f)\left(\frac{L}{D}\right)^2 + 1 - f\right] \quad (79)$$

where

$$1/h(f) = \frac{9}{8f}[4\ln(1/\sqrt{f}) - 3 + 4f - f^2]$$

$$+ \frac{3\bar{B}(1-f)^2}{fBD} + \frac{48f\mathcal{D}}{BD^3} \tag{80}$$

Recall that if sliding between the fiber and the matrix occurs readily, $\bar{B}$ will be large and also rapid mass transport is associated with a large value of $\mathcal{D}$.

It is thought that as the temperature is increased, the relative importance of sliding and mass transport is enhanced. Thus, at low creep temperatures, $\bar{B}/BD$ and $\mathcal{D}/BD^3$ would be small. Then only the first term on the right-hand side of Eqn. (80) will be important and when L/D is large, as required by this asymptotic model, the creep strength will be high. As the temperature is increased, either $\bar{B}/BD$ or $\mathcal{D}/BD^3$, or both, will increase in magnitude. When they become large, $h(f)$ will become small and the creep strength of the composite will fall, as can be seen in seen in Eqn. (79). However, if $h(f)$ becomes negligible, the steady-state creep law for the composite will be approximately

$$\dot{\varepsilon} = B\sigma/(1-f) \tag{81}$$

As in the case of the quadratic matrix rheology, the creep behavior when sliding dominates (or, as in this new case, mass transport is significant) is the same as for a material containing cylindrical holes instead of fibers even if the interface is nominally well bonded. This behavior will occur when $h(f)$ is much smaller than $(D/L)^2$ so that the relevant term containing $h(f)$ in Eqn. (79) is negligible.

It should be noted that the creep behavior is affected in the way predicted by Eqns. (79) and (80) whether interface sliding occurs readily or mass transport occurs rapidly at the interface between the fiber and the matrix. It follows that rapid sliding by itself is sufficient to diminish the creep strength of the composite material and long range mass transport at the interface is not necessary. Note also that if the matrix does not creep (i.e., $B = 0$) neither sliding nor mass transport will have any effect on creep and the composite will be rigid. This feature arises because the matrix must deform when any sliding or mass transport occurs at the interface.

An additional feature is a size effect in the creep law when sliding or mass transport at the interface are significant enough to affect the composite behavior. A small diameter fiber (i.e., small D) will tend to enhance the effect of sliding or mass transport on the creep rate of the composite, and the composite will creep faster. Similarly, a large diameter fiber will tend to suppress the effect of sliding or mass transport, and the creep strength of the composite will correspondingly be increased. Similar effects tied to grain size are known to occur in the creep of ceramics and metals controlled by mass transport on the grain boundaries.[1] Note that the mass transport term in Eqn. (80) is much more sensitive to fiber diameter than the sliding term. The cubic dependence on fiber diameter than the sliding term. The cubic dependence on fiber diameter in the mass transport controlled term will cause it to disappear rapidly as D is increased. However, if both $\mathscr{D}$ and $\bar{B}$ are substantial, the creep strength of a composite will not be improved substantially by increase of fiber diameter until both the effects of sliding and mass transport are suppressed. It seems likely that in practice this will mean that mass transport will be relatively easy to eliminate as a contributor to rapid creep strain of the composite by increase of the fiber diameter, whereas the effect of sliding at a given temperature will be more persistent. Furthermore, there is also an interplay with volume fraction, with the importance of interface sliding being greater at low volume fractions of fibers and mass transport being more significant at higher volume fractions.

9.6 Discussion

As previously noted, this chapter has been concerned mainly with those models for the creep of ceramic matrix composite materials which feature some novelty that cannot be represented simply by taking models for the linear elastic properties of a composite and, through transformation, turning the model into a linear viscoelastic one. If this were done, the coverage of models would be much more comprehensive since elastic models for composites abound. Instead, it was decided to concentrate mainly on phenomena which cannot be treated in this manner. However, it was necessary to introduce a few models for materials with linear matrices which could have been developed by the transformation route. Otherwise, the discussion of some novel aspects such as fiber brittle failure or the comparison of non-linear materials with linear ones would have been incomprehensible. To summarize those models which could have been introduced by the transformation route, it can be stated that the inverse of the composite linear elastic modulus can be used to represent a linear steady-state creep coefficient when the kinematics are switched from strain to strain rate in the relevant model.

No attempt has been made to discuss, in a comprehensive manner, models which are based on finite element calculations or other numerical analyses. Only some results of Schmauder and McMeeking[10] for transverse creep of power-law materials were discussed. The main reason that such analyses were, in general, omitted, is that they tend to be in the literature for a small number of specific problems and little has been done to provide comprehensive results for the range of parameters which would be technologically interesting, i.e., volume fractions of reinforcements from zero to 60%, reinforcement aspect ratios from 1 to 10^6, etc. Attention in this chapter was restricted to cases where comprehensive results could be stated. In almost all cases, this means that only approximate models were available for use.

Furthermore, numerical analyses for creep in the literature tend to be for metal matrix composites and so use creep indices which are rather high for ceramic matrices. Indeed, this latter fault applies to the finite element calculations so far performed by Schmauder and McMeeking,[10] even though there was an attempt to be comprehensive. Those finite element results which are available in the literature, such as the work by Dragone and Nix,[25] are very valuable and provide accurate results for a number of specific cases against which the more approximate models discussed in this chapter can be checked. A limited amount of this checking for a single model has been done by McMeeking[23] in comparison with the Dragone and Nix[25] calculations. The results show that the approximate model is reasonably accurate. However, more extensive checking of the approximate models is required and to do this, in many cases, it will be necessary to create the finite element analyses.

Also omitted from this chapter was any attempt to compare the models with experiments. This would require a lengthy chapter by itself and some comparisons are given elsewhere in this book. In addition, limited data are available for such comparisons, in general. For metals, there are some successful comparisons[5] and some unsuccessful ones.[22] It seems that when there is good knowledge of the material properties and the operating mechanisms, the right model can be chosen, but lack of such knowledge makes it virtually impossible to identify which features must be present in the model. Thus, multidisciplinary work is necessary to understand the microstructure, to identify the mechanisms, and to select and develop the appropriate model. An example of such an effort, although for the closely related subject of the plastic yielding of a metal matrix composite, is the work of Evans et al.,[26] where careful control of the metallurgy and the experiments was used to confirm the validity of the models.

Acknowledgment

This research was performed while the author was supported by the DARPA University Research Initiative at the University of California, Santa Barbara: contract ONR N00014-86-K0753.

References

1. M. F. Ashby and H. Frost, *Deformation Maps*, Pergamon Press, Oxford, U.K., 1982.
2. R. M. Christensen, *Theory of Viscoelasticity: An Introduction*, Academic Press, New York, NY, 1982.
3. R. A. Schapery, "On the Characterization of Nonlinear Viscoelastic Materials," *Polymer Engineering and Science*, **9**, 295–310 (1969).
4. R. Hill, "Elastic Properties of Reinforced Solids: Some Theoretical Principles," *Journal of the Mechanics and Physics of Solids*, **11**, 357–372 (1963).

5. M. McLean, "Creep Deformation of Metal-Matrix Composites," *Composites Science and Technology*, **23**, 37–52 (1985).

6. D. McDanels, R. A. Signorelli, and J. W. Weeton, NASA Report No. TND-4173, NASA Lewis Research Center, Cleveland, OH, 1967.

7. R. M. Christensen, *Mechanics of Composite Materials*, Wiley-Interscience, New York, NY, 1979.

8. S. Jannson and F. A. Leckie, "Mechanical Behavior of a Continuous Fiber-Reinforced Aluminum Matrix Composite Subjected to Transverse and Thermal Loading," *Journal of the Mechanics and Physics of Solids*, **40**, 593–612 (1992).

9. G. Bao, J. W. Hutchinson and R. M. McMeeking, "Particle Reinforcement of Ductile Matrices against Plastic Flow and Creep," *Acta Metallurgica et Materialia*, **39**, 1871–1882 (1991).

10. D. B. Zahl, S. Schmauder, and R. M. McMeeking, "Transverse Strength of Metal Matrix Composites Reinforced with Strongly Bonded Continuous Fibers in Regular Arrangements," *Acta Metallurgica et Materialia*, **42**, 2983–2997 (1994).

11. J. E. Ashton, J. C. Halpin, and P. H. Petit, *Primer on Composite Materials Analysis*, Technomic Publishing, Stanford, CT, 1969.

12. J. M. Whitney and R. L. McCullough, *Micromechanical Materials Modeling*, Delaware Composites Design Encyclopedia, Vol. 2, Technomic Publishing, Lancaster, PA, 1989.

13. S. L. Phoenix, P. Schwartz, and H. H. Robinson, IV, "Statistics for the Strength and Lifetime in Creep-Rupture of Model Carbon/Epoxy Composites," *Composites Science and Technology*, **32**, 81–120 (1988).

14. D. C. Lagoudas, C.-Y. Hui, and S. L. Phoenix, "Time Evolution of Overstress Profiles Near Broken Fibers in a Composite with a Viscoelastic Matrix," *International Journal of Solids and Structures*, **25**, 45–66 (1989).

15. H. Otani, S. L. Phoenix, and P. Petrina, "Matrix Effects on Lifetime Statistics for Carbon Fibre-Epoxy Microcomposites in Creep Rupture," *Journal of Materials Science*, **26**, 1955–1970 (1991).

16. W. A. Curtin, "Theory of Mechanical Properties of Ceramic Matrix Composites," *Journal of the American Ceramic Society*, **74**, 2837–2845 (1991).

17. F. Hild, J.-M. Domergue, F. A. Leckie, and A. G. Evans, "Tensile and Flexural Ultimate Strength of Fiber-Reinforced Ceramic-Matrix Composites," to be published.

18. S. T. Mileiko, "Steady State Creep of a Composite Material with Short Fibres," *Journal of Materials Science*, **5**, 254–261 (1970).

19. H. Lilholt, "Creep of Fibrous Composite Materials," *Composites Science and Technology*, **22**, 277–294 (1985).

20. D. McLean, "Viscous Flow of Aligned Composites," *Journal of Materials Science*, **7**, 98–104 (1972).

21. A. Kelly and K. N. Street, "Creep of Discontinuous Fibre Composites II, Theory for the Steady State," *Proceedings of the Royal Society, London*, **A328**, 283–293 (1972).

22. T. G. Nieh, "Creep Rupture of a Silicon Carbide Reinforced Aluminum Composite," *Metallurgical Transactions A*, **15A**, 139–146 (1984).

23. R. M. McMeeking, "Power Law Creep of a Composite Material Containing Discontinuous Rigid Aligned Fibers," *International Journal of Solids and Structures*, **30**, 1807–1823 (1993).

24. T. K. Kim and R. M. McMeeking, "Power Law Creep with Interface Slip and Diffusion in a Composite Material," *Mechanics of Materials*, to be published.

25. T. L. Dragone and W. D. Nix, "Geometric Factors Affecting the Internal

Stress Distribution and High Temperature Creep Rate of Discontinuous Fiber Reinforced Metals," *Acta Metallurgica et Materialia*, **38**, 1941–1953 (1990).

26. A. G. Evans, J. W. Hutchinson, and R. M. McMeeking, "Stress-Strain Behavior of Metal Matrix Composites with Discontinuous Reinforcements," *Scripta Metallurgica et Materialia*, **25**, 3–8 (1991).

CHAPTER **10**

Macro- and Micromechanics of Elevated Temperature Crack Growth in Ceramic Composites

S. V. Nair and J. L. Bassani

10.1 Introduction

In metallic alloys[1–3] and structural ceramics,[4–11] cracks have been well known, under elevated temperature creep conditions, to exhibit time-dependent stable crack growth termed creep crack growth. Bulk creep failure mechanisms in both metals and ceramics have been associated with stress-induced cavity nucleation, growth and coalescence, usually at grain boundaries.[1–11] When such failure mechanisms are localized within a crack tip process zone, stable crack growth can result by linkage of process zone cavities with each other and with the crack tip itself. The rate of crack growth is then controlled by the local, or crack tip, creep deformation rates which govern the rate at which cavities nucleate and grow ahead of the primary crack.

This chapter is concerned with a fracture mechanics description of creep crack growth in ceramic composites, namely, in monolithic ceramics reinforced with fibers or whiskers or a second ceramic phase. Current structural ceramic composites also include particulate-reinforced ceramic matrices; however, this category of ceramic composite is not specifically considered in this chapter. The emphasis on whisker- and fiber-reinforced composites reflects the generally much larger ambient temperature toughening obtained in these systems when compared to the particulate-reinforced case.[12] There are two components to the development of a satisfactory description of crack growth in any material. One is the macromechanics component which is concerned with the applicable bulk fracture mechanics parameter needed to correlate creep crack growth rates. For example, at ambient temperature, fracture toughness or R-curve behavior can be correlated with either the linear elastic stress intensity factor,

437

K, or the elastic-plastic J-integral fracture parameter, J. At elevated temperature, in addition to K or J, the appropriate fracture parameter can be $C(t)$, C_t or C^* which are power release rate parameters, rather than energy release rate parameters. Recently, experimental work[8–10] points to the importance of considering power release rate parameters for the case of structural ceramics at elevated temperature. The second component is the micromechanics component which relates the bulk fracture parameters to the local, or crack tip, creep strain rates responsible for local damage accumulation and crack growth. Also included in this category is the micromechanics treatment of reinforcing phases and their influence on local crack tip fields.

In fiber- or whisker-reinforced ceramics, the primary toughening mechanism at ambient temperature has been observed to be the bridging of cracks by unbroken fibers or whiskers[13–28] resulting in a zone of crack-cohesive forces. In other words, crack surfaces can now no longer be considered traction free along their entire length as is generally assumed in monolithic materials. In order for bridging to occur, a propagating matrix crack must get past fibrous reinforcements without fracturing them. In ceramic composites at ambient temperature, this has been achieved by designing a sufficiently weak interface between the fiber and matrix so that an approaching matrix crack is deflected along the fiber–matrix interface and then subsequently continues past the fiber as shown in Fig. 10.1.[27,28] From a theoretical standpoint such bridging action results in a "shielding" of the crack tip due to the presence of crack-face closure tractions exerted by the bridging or cohesive forces. This shielding is associated with a reduction of the effective value of the applied crack driving force, hence giving rise to an apparent toughening effect.[13–33] Several fracture mechanics models based on the presence of a bridging or cohesive zone[13,23–26,29–33] have predicted the magnitude of the apparent toughening expected in terms of fiber, matrix and fiber–matrix interface properties.

Recent studies[14,22,29–33] have shown that at elevated temperatures the bridging action of fibers also occurs, and cracks must be considered to exist with cohesive zones. However, the bridging or cohesive action is, in general, time dependent due, for instance, to the presence of viscous phases at fiber–matrix interfaces, fiber degradation, fiber creep, or matrix creep deformation. Secondly, there is the presence of a creep-plastic crack tip zone within which failure criteria determining crack growth apply. Creep and damage within this creep zone would be influenced by the fiber bridging action. This elevated temperature situation has received very little attention, for example, in terms of fracture mechanics models, with the result that there is little guidance for microstructural or interface design for elevated temperature use of ceramic composites.

This chapter first provides a tutorial-type presentation of fracture macromechanics and micromechanics related to cracks in creeping solids, including models for creep crack growth in monolithic solids. Next, the topic of elevated temperature crack growth in ceramic composites is considered. General aspects of the modeling of stationary cracks containing cohesive or

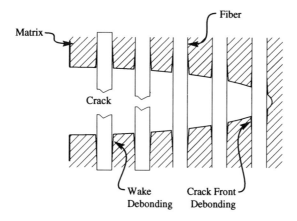

Fig. 10.1 Schematic of the development of a bridging zone due to the presence of a weak fiber/matrix interface.

bridging zones is reviewed. We then present very recent models that have attempted to describe the case of growing creep cracks containing cohesive zones. All these models are preliminary and more development is necessary before specific microstructural guidelines begin to emerge for the use of ceramic composites at elevated temperature.

10.2 Crack Tip Fields Under Nonlinear Creep Conditions

Analyses of crack tip stress and strain fields are central both in understanding mechanisms of failure in the crack tip region and in identifying the macroscopic load and geometry-dependent fracture parameter that correlates cracking behavior in a given material under given conditions. A wide range of asymptotic analyses have been developed which, in many cases, were derived from close analogies with problems in elastic-plastic fracture. Nevertheless, these analogies generally involve the mathematical structure and do not directly lead to analogies involving the appropriate fracture parameter, particularly in the case of crack growth under small-scale creep. Important aspects of the latter problem are discussed in this chapter. One of the key issues concerns the identification of macroscopic parameters (e.g., K_I, C^* or C_t) that correlate the crack velocity in terms of loading conditions (e.g., magnitude, time dependence, temperature), specimen or component geometry, crack length and shape, and material properties.

In the 1980s there was significant progress in the understanding of crack tip stresses and strains under creep conditions; an excellent review of research up to 1986 is found in Riedel's book.[3] This relatively rapid progress has possibly been due to the developments of previous decade in time-dependent elastic-plastic fracture. The cornerstone analysis in creep fracture of Riedel and Rice[34] is an example of this. For example, under extensive creep conditions the fracture mechanics parameter C^* is analogous to the fully plastic J. Under extensive creep, the crack tip stresses are given by the HRR fields (as in J-controlled plastic solids) according to

$$\sigma_{ij} = \left(\frac{C^*}{BI_n r} \right)^{1/(n+1)} \tilde{\sigma}_{ij}(\theta, n) \tag{1}$$

where σ_{ij} is the stress distribution in front of the crack tip in tensor notation, C^* is the power release rate, r is the radial distance from the crack tip, I_n is an integration constant that depends on stress state and secondary creep stress exponent, n, and $\tilde{\sigma}_{ij}$ is a dimensionless function of θ and n that depends on the stress state, the angle θ from the crack plane and n. The material constants B and n in Eqn. (1) are defined in the uniaxial elastic-creep constitutive law

$$\dot{\varepsilon} = \frac{\dot{\sigma}}{E} + B\sigma^n \tag{2}$$

where E is the Young's modulus. There are limitations to the analogy between time-independent and time-dependent fracture mechanics. One must not lose sight of the fact that although J can describe continuously the range in behavior from elastic to small-scale yielding (SSY) to fully plastic, C^* is only valid in extensive creep, where the creep strains and strain rates are everywhere much larger than the elastic strains and strain rates, respectively. This point has been overlooked in several investigations where a quantity that only equals C^* in extensive creep, i.e., a constant times load times load-line deflection rate, is measured and labeled C^* even when extensive creep conditions are not met.

In ceramics and ceramic composites wherein creep deformation rates are expected to be much smaller than in alloys at equivalent temperature, the regime of small-scale rather than extensive creep deformation would be most important. There is much debate about parameters that can correlate small-scale creep (SSC) behavior. SSC is defined by the condition that the region surrounding the crack tip where the creep strains are larger than the elastic strains is small compared to all relevant geometric length scales, e.g., crack length and remaining ligament length.[1–3,35] When elastic strains cannot be neglected everywhere in the specimen (then a single potential for the total strain rate in terms of stress does not exist), an integral that is path-independent everywhere does not exist. This is why C^* is only valid under extensive creep. The problem we will focus on is closely related to laboratory situations where a cracked specimen is suddenly loaded, e.g., in Mode I, at high temperature and, possibly after some incubation time, the crack begins to propagate. Accordingly, it is necessary, in general, to consider crack tip fields ahead of both stationary *and* growing cracks. This SSC problem has been considered in detail by Hawk and Bassani,[36–38] and Hui,[39] whose analyses extended the analyses of Riedel and Rice,[34] and Hui and Riedel.[40] For brevity, and because the essential mechanics is controlled by relative effects of time-independent elastic deformation, time-dependent creep deformation, and crack growth, the primary creep will be neglected; see Refs. 3 and 41. The effects of (continuum) damage on the crack tip fields can also be included. One approach is to use Kachanov-type creep damage equations where, for example, under a constant strain rate, the stress at a material point vanishes as the point becomes fully damaged.[38] In the case of small-scale damage, there is a zone surrounding the crack tip where the stresses more or less monotonically increase from zero at the crack tip to match up with the (singular) asymptotic crack tip fields summarized below.

10.2.1 *Stationary Cracks*

Consider a sharp, Mode I crack in an elastic-power-law creeping solid (Eqn. (2)). The isotropic multiaxial generalization of Eqn. (2) is in terms of the Von Mises effective stress $\sigma_e = (\frac{3}{2}s_{ij}s_{ij})^{1/2}$ where $s_{ij} = \sigma_{ij} - \frac{1}{3}\sigma_{kk}\delta_{ij}$, Young's modulus E, and Poisson's ratio ν. Upon sudden loading at $t = 0$ the instantaneous response is purely elastic since creep deformation takes time, and the crack tip stress field (referred to as the K-field) as $r \to 0$ is

$$\sigma_{ij} = \frac{K_I}{\sqrt{2\pi r}} f_{ij}(\theta) \quad (K\text{-field}) \tag{3}$$

where K_I is the stress intensity factor, and f_{ij} depends on the stress state and is a function of angle θ from the crack plane.

High crack tip stresses cause fast crack tip creep in the creep zone surrounding the crack tip. As $r \to 0$, $\dot{\varepsilon}^c \gg \dot{\varepsilon}^e$, i.e., the creep strain rates are much greater than the elastic strain rates, but this intense creep is constrained by the surrounding elastic material which causes *crack tip stress relaxation*. This field is of the HRR type, where as $r \to 0$ (referred to as the *RR*-field for Riedel and Rice[3,34]

$$\sigma_{ij} = \left(\frac{C(t)}{BI_n r} \right)^{1/(n+1)} \tilde{\sigma}_{ij}(\theta, n) \quad (RR\text{-field}) \tag{4}$$

where the amplitude $C(t)$ is given by

$$C(t) = \frac{(1 - \nu^2) K_I^2}{(n + 1) Et} \tag{5}$$

For long times, extensive creep conditions are reached, in which case the amplitude $C(t)$ is replaced by C^*. The transition from SSC to extensive creep under constant load condition is

$$t_T = \frac{(1 - \nu^2) K_I^2}{(n + 1) EC^*} \tag{6}$$

It is important to keep in mind that this transition time is an estimate based on a stationary crack analysis and that only when crack growth behavior is measured at times very much larger than t_T can extensive creep conditions be reasonably assumed. In many cases, Eqn. (6) underestimates the transition time; a better estimate may be the time when the creep strain equals the elastic strain in the far field.[35,42]

10.2.2 Growing Cracks

For $n < 3$, as $r \to 0$, $O(\dot{\varepsilon}^e) > O(\dot{\varepsilon}^c)$ and, therefore, the elastic field, Eqn. (3), dominates at the crack tip. In this case, the near-tip and far-fields are both of the type $\sigma \propto K/\sqrt{r}$ where K_{tip} generally differs from $K_{applied}$.[43,44] This leads to a shielding mechanism. Nevertheless, an intermediate field of the HRR type can also exist for $n > 1$, but this case $(1 < n < 3)$ which is important for ceramics has not received much attention. The case of $n > 3$ has been widely investigated.

For $n > 3$, as the crack begins to grow the elastic and the creep strain rates are of the same order of magnitude near the crack tip, i.e., $O(\dot{\varepsilon}^e) = O(\dot{\varepsilon}^c)$ as $r \to 0$, and the asymptotic stress field has been the form (referred to as the *HR*-field after Hui and Riedel[40])

$$\sigma_{ij} = \alpha_n \left(\frac{\dot{a}}{EBr} \right)^{1/(n-1)} \hat{\sigma}_{ij}(\theta, n) \quad (HR\text{-field}) \tag{7}$$

where $\dot{a}$ is the crack velocity, $\hat{\sigma}_{ij}$ is a nondimensional function of θ and n, and α_n is a numerical constant depending on n. Accordingly, the local stresses depend only upon the current crack growth rate and are not directly related to the loading parameters. Since an increasing crack velocity leads to higher stresses very near the crack tip, as seen in Eqn. (7), unstable crack growth is expected if the cracking process is dominated by this stress field. This has been predicted at low load levels from the model for crack growth discussed below.[45]

10.2.3 Nesting of SSC Fields

We continue to consider the problem where the load is suddenly applied at time $t = 0$ and either instantaneously, or after an incubation period, the crack begins to propagate. While the crack is stationary the *HR*-field of Eqn. (7) is absent. The appropriate crack tip field is the *RR*-field, Eqn. (4), which joins the K-field at the creep zone boundary, $R_{RR,K}$. Once the crack begins to grow, say, at time t_2 in Fig. 10.2, an *HR*-field (Eqn. (7)) develops at the crack tip which is enveloped by the *RR*-field at the boundary defined by $R_{HR,RR}$. At time t_3 in Fig. 10.2, the *HR*-field zone has grown beyond the original *RR*-field zone so that the *RR*-field now no longer defines the crack tip fields and the only boundary is $R_{HR,K}$. Estimates for the size of the region where each of these

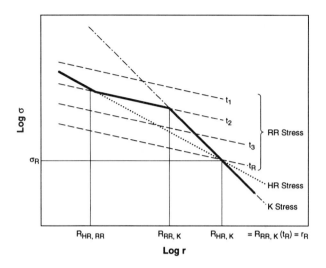

Fig. 10.2   Effective stress versus distance from crack tip based on matching of the singular crack tip stress fields of Eqs. (3), (4), and (7).

three nested fields dominates have been derived by Bassani *et al.*[37] and are depicted in Fig. 10.2,

$$R_{HR,RR} \propto \frac{(\dot{a}^{n+1} t^{n-1})^{1/2}}{K_I^{n-1}}$$

$$R_{RR,K} \propto K_I^2 t^{2/(n-1)}$$

$$R_{HR,K} \propto \left(\frac{K_I^{n-1}}{\dot{a}} \right)^{2/(n-3)} \tag{8}$$

The size of these regions varies with time, with far-field load level K_I, and with crack velocity. Most significantly, since the evolution of damage ahead of the propagating crack tip is directly influenced by the strength of the crack tip fields (and obviously their regions of dominance), in SSC the crack velocity is strongly *history dependent*; this has been demonstrated using a simplified crack growth model[45] as summarized in Section 10.3.2. Consequently, a simple correlation between crack velocity and a fracture parameter is rather elusive in SSC.

Recognizing these difficulties and motivated to identify a parameter that could readily be measured, Saxena[1,2] proposed the C_t parameter, where

$$C_t \propto P\dot{V} \tag{9}$$

where P is the magnitude of load on the cracked specimen, and $\dot{V}$ is the load-line displacement rate. The proportionality factor depends on specimen geometry and material properties; explicit expressions and a detailed discussion are found in Ref. 42.

Here, we note that under extensive creep conditions, C^* is also proportional to $P\dot{V}$, but the proportionality factor is different from that which applies in Eqn. (9). In some specimens, the proportionality factor for C_t in SSC is not appreciably different than that for the extensive creep parameter C^*, so that apparent correlations of experimental data in terms of C^* are actually correlations in terms of C_t. The important distinction arises when an attempt is made to calculate the value of the fracture parameter for use in design or life estimation. Under SSC conditions, C_t can be much greater than C^* so that predictions based upon C^* in SSC will be nonconservative.

Under SSC, based on the Riedel–Rice analysis, it can be shown from compliance analysis that (see Ref. 42)

$$C_t \propto K_I^2 \dot{r}_c \tag{10}$$

where r_c is the radius of the creep zone and, again, the proportionality factor depends on specimen geometry and material properties. Furthermore, C_t is defined so that in the transition from SSC to extensive creep, $C_t \rightarrow C^*$.

One interpretation of C_t is that it reflects the history of crack tip deformation through the rate of growth of the creep zone. From the Riedel–Rice SSC analysis of stationary cracks

$$\dot{r}_c \propto \frac{K_I^2}{t^{(n-3)/(n-1)}} \tag{11}$$

With Eqn. (11) substituted into Eqn. (10)

$$C_t \propto \frac{K_I^4}{t^{(n-3)/(n-1)}} \tag{12}$$

This dependence is certainly different from the amplitude of the RR stress and strain-rate fields which is K_I^2/t. This is an illustration of why the amplitude of the RR-field, $C(t)$, is not necessarily the crack driving force parameter. This is in contrast to the ambient temperature situation wherein the strain energy release rate correlates exactly with either $G \, (= K^2/E)$ or J, both of which also govern the amplitude of the appropriate elastic or elastic-plastic stress fields.

Numerous investigations that support the use of the C_t parameter have been carried out by Saxena and co-workers,[1,2,42] although there are many fundamental issues to be addressed. The latter essentially are related to the nonexistence of a strain-rate potential for simultaneous elastic and creep deformations. Nevertheless, based upon rigorous mechanics arguments, it is certain that neither C^* nor K_I alone can adequately correlate crack growth in SSC. See the next section for more details on crack growth models.

10.3 Creep Crack Growth Models in Monolithic Solids

Crack growth models in monolithic solids have been well document-ed.[1-3,36-45] These have been derived from the crack tip fields by the application of suitable fracture criteria within a creep process zone in advance of the crack tip. Generally, it is assumed that secondary failure in the crack tip process zone is initiated by a creep plastic deformation mechanism and that advance of the primary crack is controlled by such secondary fracture initiation inside the creep plastic zone. An example of such a fracture mechanism is the well-known creep-induced grain boundary void initiation, growth and coalescence inside the creep zone observed both in metals[1-3] and ceramics.[4-10] Such creep plastic-zone-induced failure can be described by a criterion involving both a critical plastic strain as well as a critical microstructure-dependent distance. The criterion states that advance of the primary creep crack can occur when a critical strain, ε_c, is exceeded over a critical distance, l_c in front of the crack tip. In other words

$$\varepsilon = \varepsilon_c \quad \text{at} \quad r = l_c \tag{13}$$

Two types of crack growth have been considered. One is intermittent, or stop–start crack growth, for which case the crack tip fields applicable to the static crack have been used to develop crack growth models.[32] An example of intermittent crack growth in an alumina ceramic is shown in Fig. 10.3,[46] and these results are supported also by work in glass-ceramics.[8,47] The other mode of crack growth advance is continuous crack growth for which case the *HR*-fields are taken into account.

For both intermittent and continuous crack growth, the fracture criterion $\varepsilon = \varepsilon_c$ is taken to be satisfied at a distance l_c from either the stationary or growing crack tip. Earlier models[3,34,40] for crack growth are generally based on this fracture criterion alone. A more comprehensive mechanistic model may need to take into account the additional criterion based on the earlier work of Purushothaman and Tien,[48] namely, that *rate* of crack growth must also be related to the *rate* at which tensile ligaments of width, dr, in front of the crack tip fail. If these ligaments fail when their rupture life reaches $t_f(r)$ then, as proposed by Nair and Gwo,[32]

$$\dot{a} = \frac{dr}{dt_f}\Big|_{r=l_c} \tag{14}$$

Thus Eqn. (14) needs to be satisfied in addition to the condition $\varepsilon = \varepsilon_c$ at $r = l_c$.

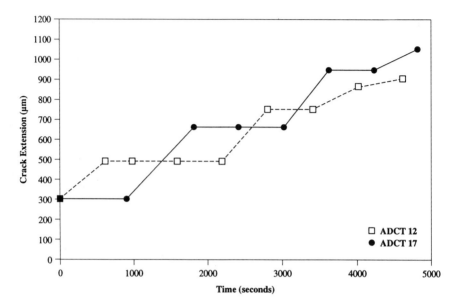

Fig. 10.3 Curves showing the crack extension as a function of time at temperature 1200°C.[46] The material used in the study was Coors AD998 polycrystalline alumina with grain size ranging from 2 μm to 10 μm. The data was obtained by testing two compact tension samples, ADCT12 and ADCT17.

10.3.1 Intermittent Crack Growth

10.3.1.1 Extensive Creep

Under extensive creep conditions, the crack is engulfed by the creeping solid and the appropriate fracture parameter is C^*. In order to calculate the strain, $\varepsilon(r)$, ahead of the crack tip, one starts with the *HRR*-field pertinent to a static crack, Eqn. (4), and calculates the strain rate from the material law, Eqn. (2). The strain then follows from time integration of the strain rate. Under these assumptions, Riedel[3] calculated the crack growth rate by taking into account the contribution to the strain at distance, r, from the current crack tip due to prior periods of crack extension, and showed that for sufficiently large crack extensions

$$\dot{a} \propto (C^*)^{n/(n+1)} \qquad (15)$$

This is the case wherein the stress amplitude factor C^* in large-scale creep also governs the crack growth rate. This analogy occurs because C^*, as for the case of G or J at ambient temperature, is a path-independent integral.

10.3.1.2 Small-Scale Creep

Under SSC the time-dependent fracture parameter $C(t) \propto K_I^2/t$ and the RR-field for a static crack is given by Eqn. (4). Nair and Gwo[32] showed that when the fracture criteria shown in Eqns. (13) and (14) are both satisfied

$$\dot{a} = \frac{(n+1)B}{nl_c^{n-1}\varepsilon_c^{n+1}}\left[\frac{K_I^2(1-\nu^2)}{I_n E}\right]^n \tag{16}$$

In other words, $\dot{a} \propto K_I^{2n}$ and cannot be directly expressed in terms of $C(t)$. In Eqn. (16) the crack growth history is neglected, that is, the contribution to the strain at l_c from the prior periods of crack extension is not accounted for. Thus $\dot{a}$ from Eqn. (16) may be strictly viewed as the *initial* crack growth rate, $\dot{a}_0$, associated with a crack of some initial size, a_0. Once crack growth begins and $a > a_0$, then Eqn. (16) may not correctly predict the growth rate. From an experimental standpoint, this pertains to the case where $\dot{a}_0$ is measured in multiple specimens of various a_0 values, rather than to the case where $\dot{a}$ is measured *in situ* as a function of the growing crack length, a, from a single specimen.

10.3.2 Continuous Crack Growth

As mentioned above, for a creep stress exponent, $n < 3$, the crack tip stress field for a continuously growing crack is the applied K_I-field. For this case, based on the model of Purushothaman and Tien[48] the crack growth rate is, for strain-controlled failure,

$$\dot{a} = \frac{2Bl_c}{\varepsilon_c n}\left(\frac{K_I}{\sqrt{2\pi l_c}}\right)^n \tag{17}$$

In other words, $\dot{a} \propto K_I^n$. Equation (17) was derived by the elastic stress field, Eqn. (3), and the fracture criteria, Eqns. (13) and (14). This analogy between $\dot{a}$ and K_I is similar to the ambient temperature stress–corrosion cracking situation wherein the elastic K-field governs the crack tip stresses.

For $n > 3$, the asymptotic HR-field is, as mentioned, independent of applied load which would predict a creep crack growth rate that is independent of applied load, contrary to experimental data. This further illustrates the complexity of relating crack growth to the crack tip stress fields under creep conditions. Hui and Riedel[40] show that the load dependence arises because of the contribution to crack growth of load-dependent nonsingular terms of the strain field. This occurs because the elastic field sets the remote boundary condition to the HR-field. The results from their work are shown in Fig. 10.4. Unstable crack growth is predicted as expected when the HR-field controls the strain contribution at $r = l_c$. Stable crack growth is predicted at larger values of

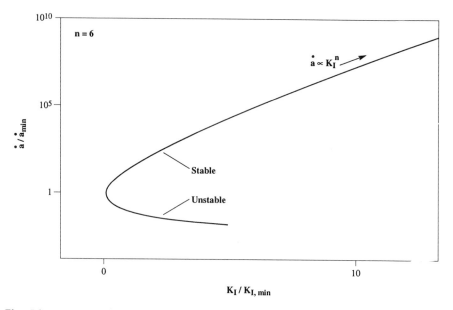

Fig. 10.4 Curve showing the predicted creep crack growth rate versus stress intensity factor based on the model of Hui and Riedel.[40]

the normalized crack growth rates when the elastic singular field governs the creep strain at $r = l_c$. In this limit of elastic field dominance

$$\dot{a} = \frac{2Bl_c}{(n-2)\,\varepsilon_c}\left(\frac{K_I}{\sqrt{2\pi l_c}}\right)^n \tag{18}$$

Equation (18) is very similar to the result of Eqn. (17) which was derived for crack growth under elastic field control. From the result of Fig. 10.4, it is clear that K_{min} represents the threshold stress-intensity factor for creep crack growth. K_{min} is proportional to $\varepsilon_c E\sqrt{l_c}$ so that an increase in ε_c and l_c leads to an increase in the threshold stress-intensity factor. The creep crack growth rate, on the other hand, is decreased in the asymptotic limit by an increase in l_c and ε_c, as would be expected.

The results summarized above in Eqns. (15)–(18) are for steady-state conditions and, therefore, neglect transient effects due to crack tip damage evolution from some initial state, and stress relaxation from constrained creep under small-scale yielding conditions. For a mechanism of damage and crack growth that depends strongly on the history of crack tip stressing and straining (e.g., grain boundary cavitation), in the presence of the nested crack tip fields discussed in Section 10.2 and depicted in Fig. 10.2 (even when the dominance of the *HR*-field is so small that it is not important), one can anticipate complexity in the crack growth histories. This certainly was the conclusion from the analyses of Wu *et al.*[45] and Hui and Wu.[49] Examples of predicted histories in SSC at constant values of K_I are shown in Fig. 10.5, where $A = \Delta a/l$ is the crack extension, which monotonically increases with time, divided by a microstructural dimension l such as grain size, and $\dot{A} \propto \dot{a}/K_I^{n-1}$ is the normalized crack velocity. With strong history dependence in Δa and $\dot{a}$, one does not find simple correlation of the transient crack growth history in Fig. 10.5, either in terms of K_I or the amplitude of the *RR*-field, K_I^2/t, except at high K_I or fast crack growth rates when $\dot{a} \propto K_I^n$ in the steady-state limit (as in Eqn. (18)). At lower values of K_I, $\dot{a}$ severely decreases with time (or crack length). The sudden steep transients in the predicted crack velocity arise due to the strong influence of the *HR*-field, Eqn. (7), on the crack tip damage. Similar transients have been seen in recent experiments on a diffusion-controlled crack growth in a model Cu–Sn alloy.[50] In this transient regime of crack growth, the C_t parameter (see Section 10.2.3) may serve as a better parameter to characterize $\dot{a}$ than K_I. For example, in the SSC regime, the decrease of $\dot{a}$ with time (or crack length) is consistent with the diminishing value of C_t with time, as suggested by Eqn. (12), although the applied K_I is increasing.

10.4 Modeling of Creep Crack Growth in Ceramic Composites

The focus of this section is on crack bridging and its role in creep crack growth. A general picture of a crack in a ceramic composite may be viewed as containing three zones as shown schematically in Fig. 10.6. One, a zone wherein the crack surface is traction-free, is similar to that of a crack in a monolithic ceramic. In the case of cracks growing from notches, this zone would include the notched portion of the crack. The second zone is composed of a bridging or cohesive zone. These bridges may be unbroken fibers as in ceramic composites, or grain bridges as in large-grained alumina.[4–7,9,27,28] However, it is not strictly necessary for there to be any physical bridges. The general feature of this zone is the existence of a cohesive traction force displacement relationship, $p(u)$, relating the crack face traction to the crack opening for time-independent bridging, or a traction force displacement time, $p(u,t)$, for time-dependent bridging. By this general definition, a transformation[27] or microcracking[11,27,28] zone in the wake of the crack can also provide a $p(u)$ characteristic. For a time-dependent bridging that involves viscous bridges or perhaps creep deformation of bridging fibers, $p(u,t)$ can reduce to $p(\dot{u})$, that is, the traction forces are related to the crack opening rate.[31,33] The last zone is a creep deformation zone exhibiting a strain rate–stress relationship, $\dot{\varepsilon}(\sigma)$, within the creep zone boundary. From a mathematical standpoint, the creep zone may be viewed as part of the cohesive or bridging zone itself so that the effective crack extends to the end of the creep zone boundary as in the Dugdale–Barenblatt model for cracks in elastic-plastic materials. Specific cases of cracks in ceramic composites (as shown in Fig. 10.6) can include a fully bridged crack, in which the first zone is

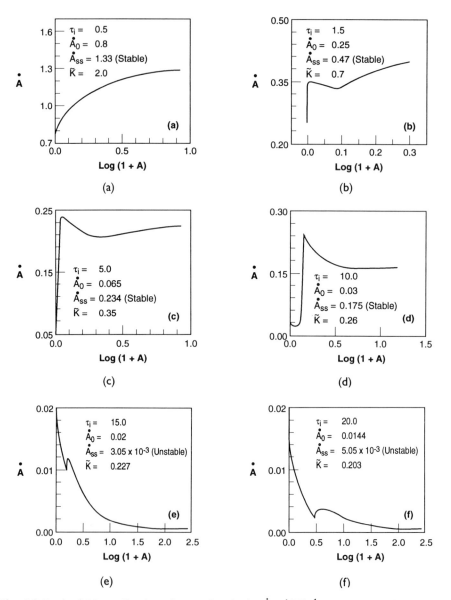

Fig. 10.5 (a–f) Normalized crack growth velocity $\dot{A} \propto \dot{a}/K_I^{n-1}$ versus normalized crack extension $A = \Delta a/l$ for transient SSC crack growth under constant K_I loading based on a model for crack growth by grain boundary cavitation (taken from Ref. 45). The parameters $\tilde{K}$, τ_i, $\dot{A}_0$, and $\dot{A}_{ss}$ are normalized values of, respectively, the stress intensity factor (which is held constant), the crack growth initiation time, the initial crack velocity, and the steady-state crack velocity.

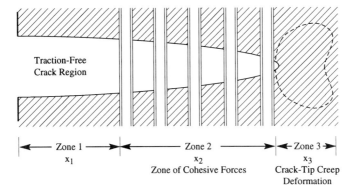

Fig. 10.6 General picture of a crack in a ceramic composite at elevated temperature showing the existence of three zones.

absent, or the case of crack growth into an elastic body with negligible creep, where the last zone is absent.

Currently, mathematical tools are available only for the modeling of cohesive or bridging zones for cracks in linear-elastic solids, although the closure pressure function $p(u)$ or $p(u,t)$ can itself be nonlinear. We first review some basic approaches for the modeling of cohesive zones, beginning with time-independent bridging and then discuss the relationship between cohesive zones and crack growth at elevated temperature primarily based on some recent or just-completed studies.[29,30,32,33]

10.4.1 Time-Independent Bridging

If the crack tip is considered to be located at the end of the zone of bridging or cohesion (see Fig. 10.6), then the cohesive forces exist in the wake of the crack. Under these circumstances, the cohesive forces essentially reduce the effective value of the stress intensity factor at the crack tip, K_{eff}. Accordingly, the crack tip in this case is shielded. When the shielding is only partial, K_{eff} is finite, whereas for complete shielding, $K_{eff} = 0$. Both cases are valid fracture mechanics representations for cracks that include cohesive zones. We consider partial shielding first.

For $K_{eff} > 0$, depending on the loading and crack/specimen geometry conditions, one can draw on known[32,33] solutions for the crack shape and K_{eff}. If $u(x)$ is the crack opening at some location, x (see Fig. 10.6), then, in the presence of a remote applied stress, σ_a,

$$u(x) = \frac{4}{E'} \int_x^a \left\{ \int_0^{a'} G(x',a',w)[\sigma_a - H(x'-x_1)p(u(x'))]\,dx' \right\} G(x,a',w)\,da'$$

$$(19)$$

where a is the crack length, w is the specimen width, G is a weight function, and H is the unit step function; $H = 0$ in Zone 1 of the crack: Fig. 10.6. $E' = E$ for plane stress and $E' = E/(1 - \nu^2)$ in plane strain. For a straight crack embedded in an infinite specimen

$$G(x,a,w) = \frac{1}{\sqrt{\pi a}}\left(1 - \frac{x^2}{a^2}\right)^{-1/2} \tag{20}$$

The corresponding K_{eff} is given by

$$
\begin{aligned}
K_{eff} &= K_{applied} - K_{bridging} \\
&= K_I - K_b \\
&= 2\int_0^a G(x,a,w)[\sigma_a - H(x - x_1)p(u(x))]\,dx
\end{aligned}
\tag{21}
$$

Thus, the magnitude of the reduction in the applied stress intensity factor, K_I, due to the cohesive zone is

$$K_b = 2\int_0^a G(x,a,w)H(x - x_1)p(u(x))\,dx \tag{22}$$

K_b accordingly represents the toughening provided by the bridging zone. If failure of the composite occurs at $K_{eff} = K_c$ (the fracture toughness), then the fracture resistance, K_R is given by

$$K_R(a) = K_c + K_b(a) \tag{23}$$

so that R-curve, or fracture resistance curve defined as the increase in fracture resistance with crack size, also arises from the bridging zone contribution, $K_b(a)$. In this approach the size of the bridging zone, x_2, depends on the cohesive mechanism. For physical bridges which fracture above a threshold stress, x_2 is determined based on the fracture strength of bridges. In other cases, there may be no unique demarcation between Zone 1 and Zone 2 except for crack growth from notches. Such is the case where $p(u)$ at first increases with u and then gradually reduces to zero as u tends to infinity, as in the case of cohesive forces between atoms in solids. Specific solutions for K_b have been obtained for different cohesive zone characteristics. For this the reader is referred to Refs. 31, 33 and 51.

Another mathematical approach to modeling cohesive zones is to consider the crack tip fully shielded, that is $K_{eff} = 0$, rather than partially shielded as in the case considered above. In this case, a cohesive zone lies in front of a traction free crack (Zone 1). This is the classical Barenblatt–Dugdale model in which the stress-intensity factor at the end of the cohesive zone is now zero; that is, stress singularities are completely removed by the cohesive forces.[29] The requirement of complete shielding results in a cusp-shaped cohesive zone or bridging zone profile. This approach has advantages, particularly for the elevated temperature case, in that the cohesive zone can include both Zone 2 and Zone 3 (the creep deformation zone) in Fig. 10.6 with different physical properties in the two cohesive regimes. The equivalence between the two approaches for the case of time-independent bridging can be noted by using a J-integral approach and performing the J-integral over the cohesive zone boundary. In this case, the steady-state value of the toughening is simply[27]

$$\Delta J_c = 2 \int_0^{u_c} p(u)\, du \tag{24}$$

where u_c here is the critical value of u corresponding to ligament failure. Equation (24) is valid for large cracks with a small-scale bridging zone as long as the crack size is much smaller than the specimen dimensions.[52] In this case, ΔJ_c is essentially equivalent to $K_R^2/E - K_c^2/E$ where K_R is defined in Eqn. (23).

The mathematical approaches pertaining to the full-shielded case ($K_{eff} = 0$) for time-dependent bridging are very similar. In this case, Eqn. (19) still holds, but for a particular $p(u)$ relation holding over a portion of the crack, the extent of that zone is determined by the $K_{eff} = 0$ condition. For example, if, as in Eqn. (19), the cohesive tractions act over the entire crack, then the crack length itself is determined by that condition.

10.4.2 Time-Dependent Bridging

If the crack is considered to be a wake zone and $p = p(u,t)$ in Eqns. (19) and (21) then, for a stationary crack, both the crack shape and K_{eff} are time dependent, given respectively by $u(x,t)$ and $K_{eff}(t)$ from Eqns. (19) and (21) for the case of partial shielding. Cox and Rose[33] recently considered an elastic time-dependent bridging law of the form

$$p = \beta(t)\, u \tag{25}$$

where, above a threshold value u_0,

$$\frac{d\beta}{dt} = -r_1 \beta(u - u_0) \tag{26}$$

where r_1 is a rate constant of the bridging process. For $u < u_0$, $\beta = $ constant (β_0) independent of t. The selection of Eqns. (25) and (26) was not based on a physical bridging mechanism at elevated temperature but, in part, on its analogy to elastic-plastic time-independent bridging when $r_1 \to \infty$. Also, in Eqn. (25) time dependence does not arise until above a threshold bridging stress $p_0 = \beta_0 u_0$, a situation corresponding to damage of bridging ligaments in patched cracks.

A specific physical time-dependent bridging mechanism relevant to ceramic matrix composites at elevated temperature was modeled recently by Nair *et al.*[31] They considered the case wherein the fiber–matrix interface was viscous in character, as can be the case at elevated temperatures in many ceramic matrix composites which are known[14–18] to contain glassy, or viscous, interfacial layers. The thermally activated behavior was confined to the interface with the fibers allowed to deform elastically within a perfectly rigid matrix, whereas the interface was allowed to deform in a Newtonian viscous fashion with viscosity, μ. Also, in this model the crack was considered to be fully bridged, that is, Zone 1 is absent.

A treatment of the mechanics of this time-dependent fiber pull-out by interfacial viscous flow showed that the displacements, u', along the fiber axis, y, satisfied the differential equation

$$\frac{\partial u'^2}{\partial y^2} = \frac{\partial u'}{\partial t} \tag{27}$$

Equation (27) had to be solved subject to the boundary condition that $u'(x, y = 0, t) = u(x, t)$ given by Eqn. (19) with $p(u, t)$ in Eqn. (19) given by

$$p(u, t) = \alpha \frac{\partial u'}{\partial y}\Big|_{y=0} \tag{28}$$

where $\alpha = -(V_f E_f / \sigma_a)$ and V_f is the fiber volume fraction, E_f is the fiber Young's modulus, and σ_a is the applied far field stress. Equation (28) illustrates that the fundamental bridging law cannot always be assumed *a priori* and can depend on the crack shape itself. For the case where the crack shape is parabolic, simple closed-form solutions result. The time-dependent crack shape can be given by

$$u(x, t) = \frac{2(1 - \nu^2)\sigma_a(a^2 - x^2)^{1/2}}{E_c} \left[1 - \exp\left(\frac{\phi t}{a^2}\right) erfc \left(\frac{\phi t}{a^2}\right)^{1/2} \right] \tag{29}$$

and effective partially shielded crack tip stress intensity is

$$K_{eff} = \sigma_a(\pi a)^{1/2} \left[1 - \exp\left(\frac{\phi t}{a^2}\right) erfc \left(\frac{\phi t}{a^2}\right)^{1/2} \right] \tag{30}$$

where ϕ is a function of material properties given by

$$\phi = \frac{\pi^2 E_c^2 \Delta R}{32(1 - \nu^2)^2 E_f V_f^2 \mu} \tag{31}$$

In the above equation, R is the fiber radius, Δ is the interface thickness, and E_c is the composite modulus. The result of the opening of the matrix crack is to increase K_{eff} as shown in Fig. 10.7. The rate at which K_{eff} increases is shown to depend on the size of the crack. As can be seen from Eqn. (30), as $t \to \infty$, K_{eff} approaches $\sigma_a(\pi a)^{1/2}$ which is the applied stress intensity value for an unbridged crack. The quantity a^2/ϕ is a time constant of the relaxation process. For large values of the time constant, the crack opening by viscous interface relaxation is extremely sluggish, so that wake effects are predominant over long periods of time. Thus, large crack sizes and/or large values of the interface

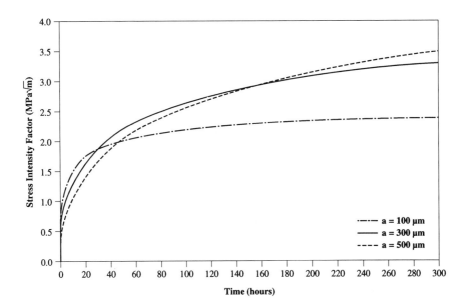

Fig. 10.7 Curves showing the stress intensity value as a function of time for different values of the crack size. The stress intensity value increases more sluggishly for larger crack sizes. On the other hand, the plateau value of K increases as the crack size increases. This plateau value of K is exactly equal to the stress intensity value for the unbridged crack.

viscosity and fiber volume fraction provide wake-dominant behavior. Wake-dominant behavior is also provided by a small interface thickness and small fiber diameter. The effect of fiber diameter is in contrast to ambient temperature wake-toughening behavior, wherein larger fiber diameters provide higher value of steady-state toughening for the wake. This aspect of the elevated temperature wake effect also plays a role in the creep failure behavior of the composite, as will be seen in the next section. The closure pressure was found to be linearly related to u and given by

$$p(x,t) = \frac{\pi \sigma_a}{2}\left(1 - \frac{x^2}{a^2}\right)^{1/2} \exp\left(\frac{\phi t}{a^2}\right) erfc \left(\frac{\phi t}{a^2}\right)^{1/2} \qquad (32)$$

so that,

$$p(x,t) = \frac{\pi \sigma_a}{2a}(a^2 - x^2)^{1/2} - \frac{\pi E_c}{4(1 - v^2)a}u(x,t) \qquad (33)$$

Clearly, the function $p(u,t)$ for a stationary crack depends on the crack shape as well as on crack/specimen geometry.

For a fully shielded crack tip ($K_{eff} = 0$) time-dependent bridging is modeled in close conjunction with the crack growth rate itself, since da/dt

appears as a natural consequence of the calculation of the cohesive stresses themselves. The $K_{eff} = 0$ approach is summarized in the next section.

10.4.3 Creep Crack Growth in Fiber Composites

In this section we discuss recent models that have attempted to combine crack growth with the presence of cohesive zones. Three classes of bridging or cohesive zone behavior are presented: (1) an elastic bridging law as given by Eqn. (25) together with Eqn. (26) (this provides some general insights into the type of behavior that can be anticipated when crack growth is combined with crack bridging; (2) a specific cohesive zone relevant to ceramic composites when unbroken fibers are linked to the matrix through a viscous interface, and (3) power-law creep within the cohesive zone as when the bridging ligaments themselves undergo creep deformation. The latter can apply, for example, to the case of ceramics toughened by ductile phases[4-7,9] or to glass-ceramics wherein the glassy phase can form viscous bridges connecting crack faces.[8,10]

10.4.3.1 Crack Growth with Elastic Bridging

The approach here is that of partial shielding with a finite value of the effective crack tip stress intensity factor, K_{eff}. The crack growth rate law assumed was of the form

$$\frac{da}{dt} = r_2 K_{eff}^m \qquad (34)$$

as long as $K_{eff} > K_0$, the threshold value for crack growth. Here r_2 is a rate parameter associated with crack growth in contrast with r_1 (Eqn. (25)) which is a rate parameter associated with bridging. Fundamental models of creep crack growth (see Section 10.3) do provide a power-law relationship of da/dt to applied K, but these models are for traction-free crack surfaces. More work is needed to verify the rigorous applicability of Eqn. (34) to cracks containing cohesive zones. Combining Eqn. (34) with Eqns. (19), (21), (25), and (26) it is possible to derive K_{eff} and hence $\dot{a}$ as a function of crack length for the case where cohesive forces are time dependent.

The results are shown here only for the case of a fully bridged crack (Zone 1 in Fig. 10.6 absent) and for a ratio of the rate constants r_1/r_2 of unity. For a discussion of other cases, the reader is referred to Ref. 33. Figure 10.8 shows results of the normalized value of K_{eff} with normalized crack length, A, for three initial crack length values, $A_i \leq 1.2$, $A_i = 4$, and $A_i = 8$. The initial value of K_{eff} at each A_i value falls on the K_{eff}^E curve which is the lower bound for K_{eff} assuming that the bridging ligaments remain elastic.

The results indicate a history dependence on K_{eff} values, that is, K_{eff} depends both on a and a_0. However, after sufficient crack extension, knowledge of initial conditions is lost and a quasi-steady state is achieved.

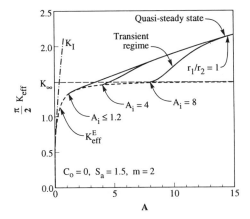

Fig. 10.8 Results of the development of K_{eff} with crack size, A, from the model of Cox and Rose[33] for the case of time-dependent elastic bridging of matrix cracks.

Once quasi-steady state has been achieved, K_{eff} depends only on a and on the ratio r_1/r_2.

10.4.3.2 Creep Crack Growth with Fiber Bridging Through a Viscous Fiber–Matrix Interface

Nair and co-workers[31,32,53] considered the growth rate of cracks with bridging zones having a time dependence defined by Eqn. (27). Both presence[32,53] and absence[31] of a crack tip creep zone were considered. In these studies, fundamental physical models were used for the cohesive zone and, in cases where a crack tip creep zone was present, the creep tip fields were used to derive creep crack growth rates as discussed in Section 10.2. This allowed for the relation of crack growth behavior to actual material and microstructural factors.

In the first case considered by Nair *et al.*,[31] a fully bridged crack, with the crack tip creep zone neglected, grows by the failure criterion $K_{eff} = K_c$, the fracture toughness at elevated temperature. This corresponds to neglecting Zone 1 and Zone 3 in Fig. 10.6. We identify this failure behavior as relatively brittle matrix toughness-controlled failure because the crack tip toughness is exceeded prior to a significant crack front creep deformation. Figure 10.9, from their result, shows the time needed to achieve $K_{eff} = K_c$ at the crack tip for various initial crack lengths. It is evident that the C-curve behavior of the curve in Fig. 10.9 exhibits a region of positive slope wherein stable crack growth can occur, similar to the Hui and Riedel[40] result of Fig. 10.4. In the figure, a_0 is defined by $K_c = \sigma_a(\pi a_0)^{1/2}$. Accordingly, cracks of size a less than a_0 would never attain a stress intensity value of K_c, and consequently cannot extend. For crack sizes a greater than a_0, the crack tip stress intensity value will attain to K_c in finite times, so that crack extension is predicted to occur. If a_n is the size of

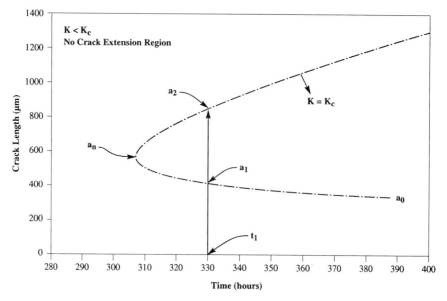

Fig. 10.9   Curve showing the time needed for the stress intensity factor to attain K_c, the matrix toughness, as a function of the initial crack length, a. The curve has a C-curve appearance. The stable portion of the curve has a positive slope, and the unstable portion has a negative slope.

the crack for which K_c is reached at the crack tip in the shortest time, i.e., the nose of the C-curve in Fig. 10.9, then two crack size regions may be identified. One, where $a_0 \leqslant a \leqslant a_n$ and the other, where $a \geqslant a_n$. In the first region, larger cracks begin to extend in smaller times. Clearly this represents an unstable condition, and unstable crack extension is predicted. On the other hand, in the second region for which a is greater than a_n, larger cracks take a longer time to extend, thereby representing a condition wherein stable crack extension is predicted. Consequently, the positive slope region of the C-curve represents the region of stable crack growth and the negative slope region represents the region of unstable crack extension. Because of the existence of the stable crack growth region on the C-curve, unstable crack extension, for crack sizes lying below a_n, does not lead to catastrophic failure. This is illustrated in Fig. 10.9. At time $t = t_1$, a crack of size $a = a_1$ will begin to extend in an unstable fashion in the direction of the vertical arrow until the crack length $a = a_2$ is reached, namely where the vertical arrow meets the stable portion of the C-curve. Continued extension of the crack then occurs in a stable fashion. Thus, at any given time, all cracks in the composite have sizes either greater than, or equal to, the corresponding crack size on the stable portion of the curve (a_2), or less than, or equal to, the corresponding size on the unstable portion of the curve (a_1). In other words, all growing cracks have the *same* size after a given time at elevated temperature irrespective of the initial crack size distribution. Cracks having sizes larger than those of the growing cracks are stationary. Consequently, the distribution of crack sizes is not critical to the overall failure behavior. Further, as shown below, the crack growth rates are also relatively independent of crack size in the steady-state regime. This is analogous to the flaw size independence of failure stresses for bridged cracks at room temperature.[26]

In this equilibrium toughness-controlled (no creep process zone) model, for $a \gg a_0$, the crack growth rate was derived to be[31]

$$\frac{da}{dt} = \frac{\pi^2 R \Delta E_c^2 \sigma_a^2}{8(1 - \nu^2)^2 V_f^2 E_f K_c^2 \mu} \tag{35}$$

As mentioned da/dt was found to be independent of a for large crack sizes. This is in sharp contrast to the case of monolithic materials reviewed in Section 10.3. Equation (35) indicates that crack growth rates are decreased by increasing the matrix toughness, fiber volume fraction and the fiber–matrix interface viscosity, or by decreasing the fiber diameter. Equation (35) summarizes the role of crack wake toughening on elevated temperature crack growth wherein crack front creep effects can be neglected. In the remainder of this section, we consider the inclusion of a crack front creep zone and a crack front creep controlled failure criterion in combination with crack wake toughening.

The second case in this category of creep crack growth with a cohesive zone associated with viscous interfaces is the work of Nair and Gwo.[32,53] Nair and Gwo[32] considered intermittent crack growth whereas continuous crack growth was considered in the latter study. The principal assumption in these studies was that the stress field in the creep zone follows the *HR*-field or *RR*-field (see Section 10.2) with applied K replaced by K_{eff}. This is only a first approximation, because, as mentioned earlier, these crack tip fields were originally derived for traction-free crack surfaces.

For various initial crack sizes, growth can occur only after a certain delay period, t_0, corresponding to the time needed to establish $\varepsilon = \varepsilon_f$ at $l = l_c$ in front of the crack. Crack growth begins only when $t = t_0$ for a given $a = a_0$. That dependence of this delay period on the magnitude of the initial crack size, a_0, is shown in Fig. 10.10, indicating a C-curve behavior as for the case when the creep zone was neglected, shown in Fig. 10.9. For intermittent crack growth, the *initial* crack growth rate, namely $\dot{a}_0$, was computed in terms of the initial crack size, a_0. Thus, the history dependence of crack growth as shown in Fig. 10.8 from the Cox and Rose[33] work is not available for this cohesive zone mechanism.

The crack growth rate is given by

$$\dot{a}_0 = \frac{n+1}{n\varepsilon_c} (l_c B)^{1/(n+1)} \left[\frac{(1 - \nu^2) \sigma_a^2 \pi a_0}{I_n(n+1) E_c t_f} \right]^{n/(n+1)}$$

$$\times \left[1 - \exp\left(\frac{\phi t_f}{a_0^2} \right) erfc \left(\frac{\phi t_f}{a_0^2} \right)^{1/2} \right]^{2n/(n+1)} \tag{36}$$

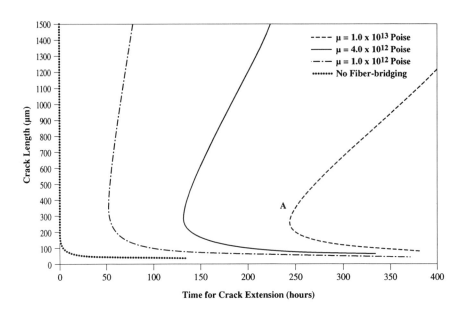

Fig. 10.10 Curves showing the time needed for crack extension by matrix creep as a function of crack length for various values of the interface viscosity, μ. Also shown in the figure is the case wherein fiber bridging is absent. See text for discussion.

where t_f is obtained from a solution of

$$
\varepsilon(r, t_f) = \varepsilon_c
$$

$$
= B^{1/(n+1)} \left(\frac{\sigma_a^2 \pi a_0 (1 - \nu^2)}{r E_c (n+1) I_n} \right)^{n/(n+1)} \left[2 \left(\frac{a_0^2}{\phi} \right)^{1/(n+1)} \right]
$$

$$
\int_0^{\sqrt{(\phi t_f)/(a_0^2)}} [1 - \exp(x^2) \, erfc(x)]^{2n/(n+1)} x^{(1-n)/(n+1)} \, dx \qquad (37)
$$

The crack growth rate results from a numerical computation of Eqn. (36) are shown in Fig. 10.11 for different values of the viscosity, μ, and compared with the case wherein fiber bridging is absent. As shown in Fig. 10.11, $\dot{a}_0$ at first increased and then decreased with a_0. The decrease in $\dot{a}_0$ with a_0 is again directly related to the dominance of fiber bridging effects. As mentioned, larger crack sizes provide wake-dominant behavior because of the time constant factor a_0^2/ϕ in Eqn. (30). Figure 10.11 also illustrates that a decrease in the viscosity also results in behavior that correctly approaches the no fiber bridging case. This also applied to the case when the fiber volume fraction, V_f, is reduced to zero. The decrease in $\dot{a}_0$ is again in dramatic contrast to the monolithic results wherein the crack growth rate always increases with crack length (Section 10.3).

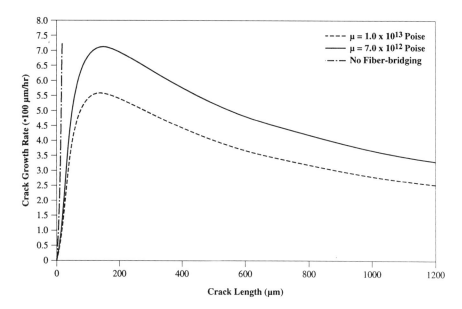

Fig. 10.11 Curves showing the creep crack growth rate as a function of crack length for different values of the interface viscosity, μ. Also shown in the figure is the case wherein fiber bridging is absent. See text for discussion.

The role of microstructure-dependent fracture process zone size, l_c, on creep crack growth rates is shown in Fig. 10.12. The abscissa is plotted as a percentage of frontal creep fracture process zone size, l_c, to the crack length. Figure 10.12 indicates that the creep crack growth rate first increased and then decreased with increase in l_c. This result suggests that the fracture frontal process zone plays the dual role of (1) decreasing the effectiveness of bridging, and (2) enhancing frontal zone creep resistance. The former appears to apply in the bridging-dominated case, that is, large a (or small l_c/a), hence an increase in $\dot{a}$ with increasing l_c. However, the dominant effect of l_c appears to be that of increasing creep resistance in the frontal zone through its effect on decreasing creep crack growth rates, as is established to be the case in monolithic alloys. These results suggest the complexities associated with an interaction between bridging zone and the creep zone. The creep zone does not merely have an additive effect but exists in a synergistic relation with the bridging zone.

10.4.3.3 Cohesive Zone with Power-Law Creep and Damage
In this section we present a model that applies to a fully shielded crack tip model (Dugdale–Barenblatt-type cohesive zone). We consider the problem of a fully shielded crack ($K_{eff} = 0$) under small-scale creep/bridging and

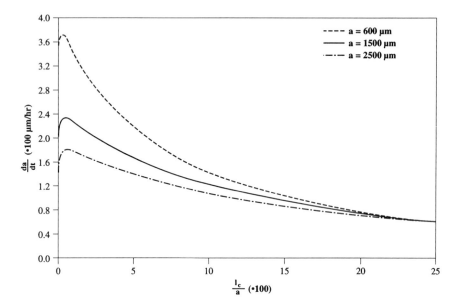

Fig. 10.12 Curves showing the creep crack growth rate as a function of crack tip fracture process zone size for different values of crack length. The abscissa is plotted as a percentage of frontal creep fracture process zone size to the crack length. See text for discussion.

investigate the effects of damage on crack growth. Crack growth in this approach occurs as a natural consequence of damage evolution in the cohesive zone. Damage is described by a $\dot{\delta}(\delta, \sigma^{cz})$ (traction–displacement) relationship, where σ^{cz} is the stress in the nonlinear creep/bridging/damage zone (the cohesive zone) and $\sigma^{cz} \rightarrow 0$ as $\delta \rightarrow \delta_c$, i.e., at fracture. With fracture at a material point arising naturally from the evolution of damage, the stress at the boundary of the open crack and the bridging zone vanishes during crack growth. Therefore, since the stress at this point is bounded (in fact zero), a singular crack tip stress field is precluded (a Dugdale-type solution). Under conditions of small-scale creep and damage, complete shielding in the sense that $K_{eff} = 0$ (the Dugdale condition) is considered in this section following the analysis of Fager et al.[29] and Fager and Bassani.[30]

The cohesive zone material is described by a Kachanov-like power-law creep/damage constitutive equation which relates the opening displacement, the rate of opening, and stress. The measure of damage is taken to be the ratio of the local zone opening and its critical value, δ/δ_c. At the fully damaged state, i.e., the fracture state, the stress at the material point vanishes. As noted, $\delta/\delta_c = 1$ with $\sigma \rightarrow 0$ at the boundary of the open crack and the bridging zone requires that $K_{eff} = 0$. The reader is referred to Fager and Bassani[30] for details of the crack growth model.

The rate of opening in the creep/bridging/damage zone, $\dot{\delta}$, is taken to be

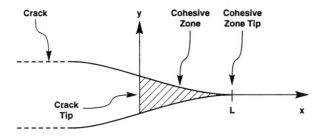

Fig. 10.13 Schematic of a cohesive zone ahead of a crack. Note the origin of the coordinate system.

a function of the stress on the zone, σ^{cz}, and the opening itself. With δ/δ_c denoting the damage measure, the stress that the material in the damage zone feels (in the sense of Kachanov) is

$$\sigma = \frac{\sigma^{cz}}{\left(1 - \dfrac{\delta}{\delta_c}\right)^{\kappa}} \tag{38}$$

where κ is a material parameter. The creep, i.e., the rate of opening in the zone is taken to be a power-law relation

$$\frac{\dot{\delta}}{\dot{\delta}_0} = \left(\frac{\sigma}{\sigma_0} - 1\right)^n = \left[\frac{\sigma^{cz}}{\sigma_0\left(1 - \dfrac{\delta}{\delta_c}\right)^{\kappa}} - 1\right]^n \tag{39}$$

where $\dot{\delta}_0$ and σ_0 are reference material parameters associated with the undamaged state. This constitutive law is taken to describe the material in the "cohesive" zone, depicted in Fig. 10.13, which extends a distance L from the crack tip. The limiting case of no damage, $\kappa = 0$, has been treated in detail by Fager *et al.*[29] In this case the steady-state crack growth model predicts a minimum value of applied K in the small-velocity limit and an applied K dependence in the high-velocity limit of $\dot{a} \propto K_I^{2(n-1)}$ for $n > 2$. Note the difference in K_I dependence predicted in Eqns. (17) and (18).

The model is formulated in terms of an integral equation which is solved with the condition at the boundary of the open crack and the bridging zone ($x = 0$) that $\delta(0) = \delta_c$. It is interesting to note that the structure of the rate-dependent problem is such that, aside from the material parameters, the solution is completely determined for a given crack velocity. For a given velocity, the value of the applied stress intensity factor, K_I, and the length of the cohesive zone, L, that maintains this condition is determined. Selected results are presented below.

For the case of damage with $\kappa = 0.4$ and $n = 4$ in Eqn. (38), the

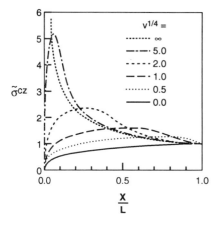 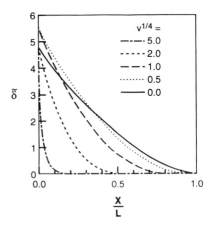

Fig. 10.14 The normalized stress in the cohesive zone, $\bar{\sigma}^{cz}$, and the normalized opening, $\tilde{\delta}$, versus distance ahead of the crack tip for $n = 4$ and $\kappa = 0.4$ at various normalized velocities, ν.

normalized stress, $\bar{\sigma}^{cz} = \sigma^{cz}/\sigma$, and the normalized opening, $\tilde{\delta} = E\delta/\sigma_0 L$ (E is the plane strain Young's modulus), ahead of the crack in the creep/bridging zone are plotted in Fig. 10.14, where the distance x ahead of the boundary of the open crack and the bridging zone is normalized by the zone length L, for various values of normalized crack velocity $\nu = 4\pi\sigma_0\dot{a}/E\dot{\delta}_0$. At low crack velocity the stress in the zone is at most in the order of σ_0. At the high velocity limit the stress approaches the elastic field of Eqn. (3) and only in a very small region at the crack tip does the stress drop rapidly to zero. The significant opening is concentrated in that region. The effects of n on these results are seen in Fig. 10.15 for $\kappa = 0.4$ and $\nu = 1$ while the effect of κ is seen in Fig. 10.16 for $n = 4$ and $\nu^{1/4} = 5$ (or $\nu = 625$).

The effects of n and κ on velocity dependence of the ratio of the crack tip opening to the zone length, $\tilde{\delta}_c \propto \delta(0)/L = \delta_c/L$, which is a measure of the cohesive zone opening angle, are seen in Fig. 10.17. The relationship between the normalized applied stress intensity factor, $K^* = K_I/E\sigma_0\delta_c$, and the crack velocity is plotted in Fig. 10.18. Both of these figures include the results that neglect the effects of damage on the cohesive zone stress ($\kappa = 0$). As noted above, when neglecting damage ν (or $\dot{a}$) $\propto K_I^{2(n-1)}$ in the high velocity limit. With the effects of damage included, the numerical results of Fager and Bassani[30] suggest that $\nu \propto K_I^{2(n-1)}/\sqrt{\kappa + 1}$.

10.5 Concluding Remarks

Because of the presence of a creep zone at the tip of a growing creep crack, the crack tip fields can be expressed in terms of stress–strain rate fields as a function of the distance from the crack tip within the creep zone. These

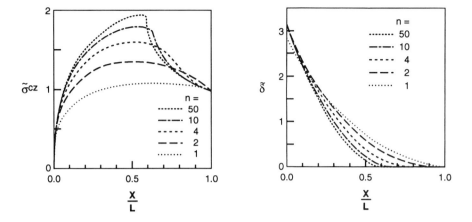

Fig. 10.15 The normalized stress in the cohesive zone, $\bar{\sigma}^{cz}$, and the normalized opening, $\bar{\delta}$, versus distance ahead of the crack tip for $\kappa = 0.4$ and normalized velocities $\nu = 1$ for various values of n.

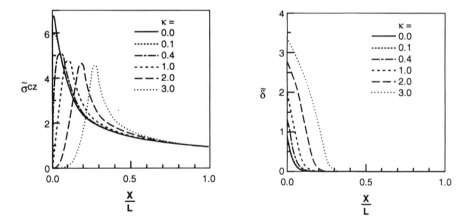

Fig. 10.16 The normalized stress in the cohesive zone, $\bar{\sigma}^{cz}$, and the normalized opening, $\bar{\delta}$, versus distance ahead of the crack tip for $n = 4$ and normalized velocities $\nu^{1/4} = 5$ for various values of κ.

fields are strongly time and history dependent and, consequently, unlike in the ambient temperature case, a path-independent integral cannot be defined around the crack tip except under extensive steady–state creep conditions at the crack tip. This complexity gives rise to the existence of several possible controlling fracture mechanics parameters for creep crack growth depending on the extent of crack tip creep plasticity.

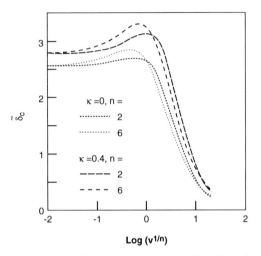

Fig. 10.17 The ratio of the crack tip opening to zone length as function of normalized crack velocity.

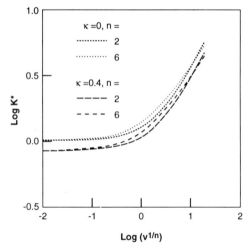

Fig. 10.18 The normalized stress intensity factor, $K^* = K_I/E\sigma_0\delta_c$, as a function of the normaized crack velocity.

In structural ceramic composites, the principal effect considered was one of crack-face closure tractions, or cohesive forces, brought about, for instance, by bridging fibers. A rigorous evaluation of the crack tip fields where the crack faces are not traction free has not yet been attempted. However, an approximate approach for the small-scale creep case is to assume that the crack tip fields are not functionally altered by crack-face tractions, with the effect of the traction being only to introduce a zone of crack tip shielding. This allows for the development of preliminary models for creep crack growth which is inclusive of the role of crack bridging. These preliminary models predict that, depending on the nature of the bridging mechanism, creep crack growth rate can be retarded in ceramic composites and that larger creep cracks may not necessarily be more damaging. In effect, this is a prediction of improved elevated temperature reliability for the structural ceramic.

Different specific mechanisms of elevated temperature bridging, which provides the crack cohesive traction, were considered and the predictions of each of these mechanisms with respect to creep crack growth were presented. All of these bridging models emphasize the fundamental fact that elevated temperature bridging is time dependent, that is, the crack shape changes as a function of time for a fixed crack length when subjected to some applied load. This is in contrast to time-independent bridging, extensively considered for the understanding of ambient temperature behavior of ceramic composites.

Further development of theoretical models is limited by the lack of availability of information in ceramic composites pertaining to crack shape, crack growth rates, bridging, and process zone sizes associated with growing creep cracks in ceramic composites.

Acknowledgments

S. V. Nair acknowledges NSF Grant MSS-9201625 and also the support of Tsung-Ju Gwo who helped with the preparation of this manuscript. Many portions of this chapter are also based on Gwo's doctoral thesis research at the University of Massachusetts, Amherst. Appropriate citations of his work appear in the text. J. L. Bassani acknowledges the support of NSF under grant DMR 88-06966 and the NSF/MRL program at the University of Pennsylvania under grant DMR 88-19885.

References

1. A. Saxena, "Creep Crack Growth under Non-Steady-State Conditions," in *Fracture Mechanics: Seventeenth Volume*, eds. J. H. Underwood, R. Chait, C. W. Smith, D. P. Wilhem, W. A. Andrews, and J. C. Newman, ASTM STP 905, American Society for Testing and Materials, Philadelphia, PA, 1986, pp. 185–201.
2. A. Saxena, "Mechanics and Mechanisms of Creep Crack Growth," in *Fracture Mechanics: Microstructure and Micromechanisms*, eds. S. V. Nair, J. K. Tien, R. C. Bates, and O. Buck, ASM Materials Science Seminar, ASM International, OH, 1987, pp. 283–334.
3. H. Riedel, *Fracture at High Temperatures*," Springer-Verlag, New York, NY, 1987.
4. B. J. Dalgleish, S. M. Johnson, and A. G. Evans, "High-Temperature Failure of Polycrystalline Alumina: I, Crack Nucleation," *J. Am. Ceram. Soc.*, **67**[11], 741–750 (1984).
5. W. Blumenthal and A. G. Evans, "High-Temperature Failure of Polycrystalline Alumina: II, Creep Crack Growth and Blunting," *J. Am. Ceram. Soc.*, **67**[11], 751–758 (1984).

6. S. M. Johnson, B. J. Dalgleish, and A. G. Evans, "High-Temperature Failure of Polycrystalline Alumina: III, Failure Times," *J. Am. Ceram. Soc.*, **67**[11], 759–763 (1984).

7. A. G. Robertson, D. S. Wilkinson, and C. H. Cáceres, "Creep and Creep Fracture in Hot-Pressed Alumina," *J. Am. Ceram. Soc.*, **74**[5], 915–921 (1991).

8. R. A. Page, K. S. Chan, D. L. Davidson, and J. Lankford, "Micromechanics of Creep-Crack Growth in a Glass-Ceramic," *J. Am. Ceram. Soc.*, **73**[10], 2977–2986 (1990).

9. D. S. Wilkinson, C. H. Cáceres, and A. G. Robertson, "Damage and Fracture Mechanisms During High-Temperature Creep in Hot-Pressed Alumina," *J. Am. Ceram. Soc.*, **74**[5], 922–933 (1991).

10. K. Kromp, T. Haug, R. F. Pabst, and V. Gerold, "C^* for Ceramic Materials? Creep Crack Growth at Extremely Low Loading Rates at High Temperatures Using Two-Phase Ceramic Materials," in *Proceedings of the Third International Conference on Creep and Fracture of Engineering Materials and Structures* (Swansea 1987), Eds. B. Wilshire and R. H. Evans, The Institute of Metals, London, U.K., 1987, pp. 1021–1032.

11. K. S. Chan and R. A. Page, "Creep Damage Development in Structural Ceramics," *J. Am. Ceram. Soc.*, **76**[4], 803–826 (1993).

12. A. Lightfoot, L. Ewart, J. S. Haggerty, Z.-Q. Cai, J. E. Ritter, and S. V. Nair, "Processing and Properties of SiC Whisker and Particulate Reinforced Reaction Bonded Si_3N_4," *Ceram. Eng. Sci. Proc.*, **12**[7–8], 1265–1291 (1991).

13. S. V. Nair, "Crack Wake Debonding and Toughness in Fiber or Whisker Reinforced Brittle-Matrix Composites," *J. Am. Ceram. Soc.*, **73**[10], 2839–2847 (1990).

14. S. V. Nair, K. Jakus, and C. Ostertag, "Role of Glassy Interface in High Temperature Crack Growth in SiC Fiber Reinforced Alumina," *Ceram. Eng. Sci. Proc.*, **9**[7–8], 681 (1988).

15. N. D. Corbin, G. A. Rossetti, Jr., and S. D. Hartline, "Microstructure/Property Relationships for SiC Filament Reinforced RBSN," *Ceram. Eng. Sci. Proc.*, **7**[7–12], 958–968 (1986).

16. T. F. Foltz, "SiC Fibers for Advanced Ceramic Composites," *Ceram. Eng. Sci. Proc.*, **6**[9–10], 1206–1220 (1985).

17. H. Kodama, H. Sakamoto, and T. Miyoshi, "Silicon Carbide Monofilament-Reinforced Silicon Nitride or Silicon Carbide Matrix Composites," *J. Am. Ceram. Soc.*, **72**[4], 551–558 (1989).

18. R. J. Kerans, R. S. Hay, N. J. Pagano, and T. A. Parthasarathy, "The Role of Fiber-Matrix Interface in Ceramic Composites," *Ceram. Bull.*, **68**[2], 429–442 (1989).

19. R. Chaim, L. Baum, and D. G. Brandon, "Mechanical Properties and Microstructure of Whisker-Reinforced Alumina-30 vol% Glass Matrix Composite," *J. Am. Ceram. Soc.*, **72**[9], 1636–1642 (1989).

20. M. H. Rawlins, T. A. Nolan, D. P. Stinton, and R. A. Lowden, "Interfacial Characterizations of Fiber Reinforced SiC Composites Exhibiting Brittle and Toughened Fracture Behavior," in *Advanced Structural Ceramics, Materials Research Society Symposia Proceedings*, Vol. 78, MRS, Pittsburgh, PA, 1987, p. 223.

21. G. H. Campbell, M. Rühle, B. J. Dalgleish, and A. G. Evans, "Whisker Toughening: A Comparison Between Al_2O_3 and Si_3N_4 Toughened with SiC," *J. Am. Ceram. Soc.*, **73**[3], 521–530 (1990).

22. S. V. Nair, T.-J. Gwo, N. Narbut, J. G. Kohl, and G. J. Sundberg, "Mechanical Behavior of a Continuous SiC Fiber Reinforced RBSN Matrix Composite," *J. Am. Ceram. Soc.*, **74**[10], 2551–2558 (1991).

23. B. Budiansky, J. W. Hutchinson, and A. G. Evans, "Matrix Fracture in Fiber Reinforced Ceramics," *J. Mech. Phys. Solids*, **34**[2], 167–189 (1986).
24. A. G. Evans and R. M. McMeeking, "On the Toughening of Ceramics by Strong Reinforcements," *Acta Metall.*, **34**[12], 2435–2441 (1986).
25. D. B. Marshall and B. N. Cox, "Tensile Fracture of Brittle-Matrix Composites: Influence of Fiber Strength," *Acta Metall.*, **35**[11], 2607–2619 (1987).
26. D. B. Marshall, B. N. Cox, and A. G. Evans, "The Mechanics of Matrix Cracking in Brittle-Matrix Fiber Composites," *Acta Metall.*, **33**[11], 2013–2021 (1985).
27. A. G. Evans, "Perspective on the Development of High-Toughness Ceramics," *J. Am. Ceram. Soc.*, **73**[2], 187–206 (1990).
28. P. F. Becher, "Microstructural Design of Toughened Ceramics," *J. Am. Ceram. Soc.*, **74**[2], 255–269 (1991).
29. L.-O. Fager, J. L. Bassani, C.-Y. Hui, and D.-B. Xu, "Aspects of Cohesive Zone Models and Crack Growth in Rate-Dependent Materials," *Int. J. Fract.*, **52**, 119–144 (1991).
30. L.-O. Fager and J. L. Bassani, "Stable Crack Growth in Rate-Dependent Materials with Damage," *Journal of Engineering Materials and Technology*, **115**, 252–261 (1993).
31. S. V. Nair, K. Jakus, and T. Lardner, "The Mechanics of Matrix Cracking in Fiber-Reinforced Ceramic Composites Containing a Viscous Interface," *Mech. Mater.*, **12**, 229–244 (1991).
32. S. V. Nair and T.-J. Gwo, "Role of Crack Wake Toughening on Elevated Temperature Crack Growth in a Fiber Reinforced Ceramic Composite," *Journal of Engineering Materials and Technology*, **115**, 273–280 (1993).
33. B. N. Cox and L. R. F. Rose, "Time or Cycle Dependent Crack Bridging," *Mechanics of Material*, submitted.
34. H. Riedel and J. R. Rice, "Tensile Cracks in Creeping Solids," in *Fracture Mechanics: Twelfth Conference*, ASTM STP 700, American Society for Testing and Materials, Philadelphia, PA, 1980, pp. 112–130.
35. J. L. Bassani and F. A. McClintock, "Creep Relaxation of Stress around a Crack Tip," *Int. J. Solids Struct.*, **17**, 479–492 (1981).
36. D. E. Hawk and J. L. Bassani, "Transient Crack Growth under Creep Conditions," *J. Mech. Phys. Solids*, **34**[3], 191–212 (1986).
37. J. L. Bassani, D. E. Hawk, and F.-H. Wu, "Crack Growth in Small-Scale Creep," in *Nonlinear Fracture Mechanics: Volume I—Time Dependent Fracture*, eds. A. Saxena, J. D. Landes, and J. L. Bassani, ASTM STP 995, American Society for Testing and Materials, Philadelphia, PA, 1988, pp. 68–95.
38. J. L. Bassani and D. E. Hawk, "Influence of Damage on Crack-Tip Fields under Small-Scale-Creep Conditions," *Int. J. Fract.*, **42**, 157–172 (1990).
39. C.-Y. Hui, "The Mechanics of Self-Similar Crack Growth in an Elastic Power-Law Creeping Material," *Int. J. Solids Struct.*, **22**[4], 357–372 (1986).
40. C.-Y. Hui and H. Riedel, "The Asymptotic Stress and Strain Field near the Tip of a Growing Crack under Creep Conditions," *Int. J. Fracture*, **17**[4], 409–425 (1981).
41. H. Riedel, "Creep Crack Growth," in *Fracture Mechanics: Perspectives and Directions (Twentieth Symposium)*, ASTM STP 1020, American Society for Testing and Materials, Philadelphia, PA, 1989, pp. 101–126.
42. J. L. Bassani, D. E. Hawk, and A. Saxena, "Evaluation of the C_t Parameter for Characterizing Creep Crack Growth Rate in the Transient Regime," in *Nonlinear Fracture Mechanics: Volume I—Time Dependent Fracture*, eds. A. Saxena, J. D. Landes, and J. L. Bassani, ASTM STP 995, American Society for Testing and Materials, Philadelphia, PA, 1988, pp. 7–26.

43. E. W. Hart, "A Theory for Stable Crack Extension Rates in Ductile Materials," *Int. J. Solids Struct.*, **16**, 807–823 (1980).

44. E. W. Hart, "Stable Crack Extension Rates in Ductile Materials: Characterization by a Local Stress-Intensity Factor," in *Elastic-Plastic Fracture: Second Symposium, Volume I—Inelastic Crack Analysis*, eds. C. F. Shih and J. P. Gudas, ASTM STP 803, American Society for Testing and Materials, Philadelphia, PA, 1983, pp. I-521–I-531.

45. F.-H. Wu, J. L. Bassani, and V. Vitek, "Transient Crack Growth under Creep Conditions due to Grain-Boundary Cavitation," *J. Mech. Phys. Solids*, **34**, 455–475 (1986).

46. S. V. Nair and T.-J. Gwo, "An Experimental Method for Fracture Mechanics Characterization of Elevated Temperature R-Curve Using the C-Integral Approach," Paper presented at the 93rd Annual Meeting and Exposition of the American Ceramic Society, Cincinnati, OH, April, 1991.

47. K. S. Chan and R. A. Page, "Creep-Crack Growth by Damage Accumulation in a Glass-Ceramic," *J. Am. Ceram. Soc.*, **74**[7], 1605–1613 (1991).

48. S. Purushothaman and J. K. Tien, "A Theory for Creep Crack Growth," *Scripta Metall.*, **10**, 663–666 (1976).

49. C.-Y. Hui and K.-C. Wu, "The Mechanics of a Constantly Growing Crack in an Elastic Power-Law Creeping Material," *Int. J. Fracture*, 3–16 (1986).

50. E. V. Barrera, M. Menyhard, D. Bika, B. Rothman, and C. J. McMahon, "Quasi-static Intergranular Cracking in a Cu-Sn Alloy; an Analog of Stress Relief Cracking of Steels," *Met. Trans. A*, to be published.

51. B. N. Cox and D. B. Marshall, "Stable and Unstable Solutions for Bridged Cracks in Various Specimens," *Acta Metall.*, **39**[4], 579–589 (1991).

52. B. N. Cox, "Extrinsic Factors in the Mechanics of Bridged Cracks in Various Specimens," *Acta Metall.*, **39**, 1189–1201 (1991).

53. S. V. Nair and T.-J. Gwo, "Role of Interface in Creep Crack Growth of a Fiber Reinforced Composite at Elevated Temperature," Paper presented at the 95th Annual Meeting of the American Ceramic Society, Cincinnati, OH, April, 1993.

Reliability and Life Prediction of Ceramic Composite Structures at Elevated Temperatures

S. F. Duffy and J. P. Gyekenyesi

11.1 Introduction

Engineered materials are poised to supplant conventional metal alloys in numerous applications. A variety of engineered materials have been proposed that increase either strength, fracture toughness, or both. For the most part, successful engineered materials have exploited transformation toughening caused by constituents undergoing a stress-induced martensitic transformation, or have embedded long fibers and particulate reinforcements in a matrix material to enhance the fracture or deformation behavior in a particular direction. It is anticipated that the availability of engineered materials in large quantities will herald the advent of new products that have been conceptualized, but are awaiting the right material with an optimum set of properties. Crucial material properties may vary for many industrial and aerospace applications; however, most applications will require high specific stiffness, strength, and toughness (which must be maintained at elevated temperatures in aggressive environments), and low density. A breakthrough material will possess the right property mix at a price that allows components to be fabricated economically.

The engineered material systems focused on here are composites with ceramic matrices reinforced with ceramic fibers. Monolithic ceramics exhibit useful properties such as retention of strength at high temperatures, chemical inertness, and low density. However, the use of monolithic ceramics has been limited by their inherent brittleness, susceptibility to damage from thermal shock, and a large variation in strength which is reflected in diminished component reliabilities. Adding a second ceramic phase with an optimized interface to a brittle matrix improves fracture toughness, decreases the sensitivity to microscopic flaws and, depending on the constituents and direction, may also increase strength. The presence of fibers in the vicinity of a critical defect modifies fracture behavior by increasing the required crack driving force through several mechanisms. These mechanisms include crack pinning, fiber bridging, fiber debonding, and fiber pull-out. This increase in fracture toughness allows for graceful rather than catastrophic failure.

Ceramic composite material systems have been produced using a variety of reinforcing schemes. These include whiskers, chopped fibers, particulates, long fibers, and woven fibers as reinforcements. In some instances, two types of reinforcements have been commingled to produce a hybrid ceramic matrix composite (CMC). These emerging composite materials can compete with metals in many demanding applications. Prototypes fabricated from the materials just mentioned have already demonstrated functional capabilities at temperatures approaching 1400°C, which is well beyond the operational limit of most metallic material systems, including composites manufactured from superalloys. Indeed the focus of this text and the issues concentrated on in this chapter primarily relate to the use of ceramic matrix composites at elevated service temperatures.

The authors anticipate that the majority of near-term high temperature applications of components fabricated from ceramic matrix composites will be found in the power generation and the aerospace industries. The use of gas turbines is being advocated in meeting peak service demands in the electric power industry. In addition, progress in utilizing aero-derivative gas turbines has supported cogeneration technologies. Cogeneration (combined cycle) power plants are able to utilize the exhaust gases from gas turbines to generate steam. Due to low costs, improved efficiency ratings, and plentiful supply of natural gas, cogeneration power plants will play a significant role in future electric power generation. Use of CMC material systems in power generation applications (as well as marine propulsion systems) has given rise to expectations of increased fuel efficiency, multifuel capability, and meeting stringent emission standards.

There are many potential aerospace applications of CMC material systems currently under consideration. Examples include propulsion subsystems in High Speed Civil Transport (HSCT). In this program, ceramic composites are being proposed for use as segmented combustor liners and exhaust nozzles. CMC materials are also under consideration in the fabrication of components for advanced gas-turbine engines, including small missile turbine rotors. Ceramic matrix composites offer a significant potential for raising the thrust-to-weight ratio of gas-turbine engines by tailoring directions of high specific reliability. Improvements in fuel efficiency are anticipated due to increased engine temperatures and pressures, which, in turn, generate more power and thrust. The anticipated goals again include developing multifuel capability, and reducing NO_x (oxides of nitrogen) emissions. For these reasons, programs such as HSCT and the National Aero-Space Plane (NASP) advocate the use of CMC materials in critical hot sections of advanced propulsion systems.

The use of ceramic matrix composites is not limited to the two fields cited above. Other potential industrial applications include various components of automotive engines, heat exchangers, waste incinerators, and membrane filtration systems. These include applications with ambient and elevated service temperatures. CMC materials should be considered in any application where toughness and wear resistance are key design factors, or where components are subjected to high service temperature in corrosive environments. In the near term, aqueous waste processing and diesel engine exhaust filters are examples of separation and purification processes that represent potential markets for this material system. In addition, components fabricated from CMC materials are currently employed in heat exchanger design. Examples include recuperators, bayonet and plate-fin heat exchangers. Wherever stringent performance criteria are specified in operating environments that challenge the limits of metal-based material systems, CMC materials will provide viable alternatives.

Yet the impact of CMC materials will be felt only as engineers become comfortable designing components fabricated from these materials. As noted in the following sections, a number of aspects relating to the mechanical behavior of ceramic matrix composites necessitates reconsidering traditional design methods. To meet this challenge, this chapter can be used as an aid for those individuals willing to pursue the probabilistic methods that address the mechanical response of ceramic matrix composites. Many references are provided to enable the reader to pursue a topic in greater detail than is allowed for here. It was noted earlier that the focus of this text is the use of ceramic matrix composites in applications with elevated service temperatures. To this end, the authors have highlighted methods to ascertain the structural reliability of components subject to quasi-static load conditions. Each method focuses on a particular composite microstructure. In addition, since elevated service temperatures usually involve time-dependent effects, a section dealing with reliability degradation as a function of load history has been included. However, this field of research is not as mature as the other issues covered here, and it is treated in a preliminary fashion. The authors believe that as these new design concepts are embraced, fabricating components from ceramic matrix composites will take place in large volumes. As a result, manufacturing economies of scale will quickly bring fabrication costs in line with the more conventional material systems.

In addressing issues relating to component failure, the reader will note a recurring theme throughout this chapter. Even though component failure is controlled by a sequence of many microfailure events, failure of ceramic composites will be modeled using macrovariables. The sheer number and order of microfailure events will preclude the design engineer from predicting analytically the sequence of events leading to component failure (in whatever manner this macro-level event is defined). This does not imply that the engineer can design components fabricated from CMC material systems without a working knowledge of the physics of failure taking place within the microstructure of a ceramic composite. However, it is the authors' viewpoint that failure at the component level can best be described with phenomenological failure models using macrovariables. At this point micromechanics is ill-suited for component failure analyses, yet it can be utilized quite effectively in predicting global stiffness properties. Although global stiffness properties are extremely important when designing components fabricated from composite materials, with the exception of woven ceramic composites, this subject is not discussed here. Otherwise, the reader is directed to the wealth of literature in this field (e.g., Rosen[1] and McCullough[2]). In the case of woven ceramic composites the authors provide a focused literature survey regarding the prediction of stiffness properties for this emerging material system.

11.2 Whisker-Toughened Ceramics

11.2.1 Introduction

The reason behind adding a second ceramic phase is that through proper tailoring of the composite microstructure one can dramatically improve the mechanical properties (fracture toughness and creep resistance) of monolithic ceramics. For ceramic matrix composites this effort began with attempts to uniformly disperse whiskers and particle reinforcements with tailored interfaces throughout a ceramic matrix. Conceptually, one hopes to perturb the stress field in the vicinity of the tip of a critical matrix crack with the inclusion of a second phase. This should cause a reduction in the stress intensity. Lange[3] originally proposed a method to compute the stress necessary to propagate a crack front that bows between two obstacles, and developed a modified Griffith equation where the increase in fracture surface energy is directly related to a line tension effect. The approach assumes that the whiskers have a higher fracture toughness than the matrix, and that the crack front penetrates the matrix material in a nonlinear fashion (i.e., "bows out"). The postulated behavior is analogous to the motion of dislocations through a precipitate-hardened material. Evans[4] and Green[5] have suggested modifications to this model that account for whisker morphologies. Toughening by crack deflection arises from the crack front tilting and/or twisting as it encounters the second phase. This produces noncoplanar crack extension and a decreased stress intensity at the crack tip. The direction taken by the crack front after deflection is controlled by whisker morphology and the residual stress field (e.g., tensile or compressive) arising from high temperature processing and subsequent cooling. A fracture mechanics model for crack deflection has been proposed by Faber and Evans[6] that accounts for mixed-mode behavior. Angelini and Becher[7] have presented evidence from electron microscopy observations that indicates whiskers may bridge the crack as it propagates. Based on these data, Becher and co-workers[8] developed a model to predict the increase in fracture toughness due to the closure stresses imposed on the crack by the bridging whiskers. Finally, Wetherhold[9] developed a model based on probabilistic principles that enables the computation of increased energy absorption during fracture due to whisker pull-out. However, compared to continuous fiber-reinforced ceramics (discussed in Section 11.3), whisker pull-out is limited due to the short lengths of the whiskers (typically less than 100 μm).

The primary intent of the models mentioned above is to consider the composite as a structure and develop predictive methods to optimize the microstructure. Ultimately, one would then make use of these concepts to similarly refine the design of structural components. Whenever practical, elegant design methods incorporate the relevant physics of failure into the analysis of the macroscopic response of a component. However, at this point difficulties arise that prevent taking this approach in the analysis of components manufactured from whisker-toughened ceramics. First, the crack mitigation processes strongly interact, and it is difficult to experimentally detect or analytically predict the sequence of mechanisms leading to failure (e.g., crack deflection, then whisker pull-out, then crack bridging, etc.). Furthermore, with the exception of the model by Faber and Evans,[6] the methods mentioned previously only consider Mode I failure. This precludes conducting a structural analysis on a component subject to multiaxial states of stress that vary from point to point in the component. Finally, these initial analytical techniques have focused primarily on predicting behavior using deterministic approaches. Even though improved processing techniques have resulted in the reduction of inhomogeneities, uniform whisker distributions, and dense matrices, failure remains a stochastic process for discrete particle-toughened materials.

An alternate approach that conveniently sidesteps the aforementioned obstacles is to compute reliability in terms of macrovariables using phenomenological criteria. Focusing a design approach at the macroscopic level represents a minor philosophical drawback, for it excludes any consideration of the microstructural events that involve interactions between individual whiskers and the matrix. The phenomenological point of view implies that the material element under consideration is small enough to be homogeneous in stress and temperature, yet large enough to contain a sufficient number of whiskers such that the element is a statistically homogeneous continuum. Obviously these conditions cannot always be met for they depend on characteristic component dimensions, the severity of gradients within the component, and the geometry of the microstructure. When these conditions are satisfied, one can systematically formulate multiaxial reliability models under the assumption that the material is homogeneous, with stochastic properties and behavior that can be deduced from well-chosen phenomenological experiments. From a historical perspective, phenomenological approaches are usually (but not always) supplanted by techniques that accurately reflect the physics of the microstructure. Precedence for this can be found in modeling monolithic ceramics where the detailed concepts of Batdorf and Crose[10] have, for the most part, superseded phenomenological approaches such as the reliability model based on the principle of independent action (PIA: see Shih[11]).

11.2.2 *Noninteractive Reliability Models*

Duffy *et al.*[12] presented the details of using phenomenological approaches in designing whisker-toughened ceramic components. Depending on fabrication, a whisker-toughened composite may have an isotropic, transversely isotropic, or orthotropic material symmetry. One can postulate that a model based on the principle of independent action would be an appropriate first approximation phenomenological theory for isotropic whisker composites. However, an interactive reliability model presented in the next section more accurately reflects anticipated phenomenological behavior such as reduced scatter in compression, and sensitivity to the hydrostatic component of the stress state. Duffy and Arnold[13] presented a noninteractive phenomenological model for whisker-toughened composites with a transversely isotropic material symmetry often encountered in hot-pressed and injection-molded whisker-toughened ceramics. In their discussion, reliability is governed by weakest link theory. The existence of a failure function per unit volume was assumed, where the unit volume was considered as an individual link. The failures of the individual links are considered as statistical events, which are assumed to be independent. Defining f as the failure of an individual link, then

$$f = \psi \Delta V \tag{1}$$

where V denotes volume and ψ is the failure function. Taking r as the reliability of an individual link, then

$$r = 1 - \psi \Delta V \tag{2}$$

The reliability of the continuum, denoted as R, is

$$R = \lim_{N \to \infty} \left[\prod_{\xi=1}^{N} r_\xi \right]$$
$$= \lim_{N \to \infty} \left[\prod_{\xi=1}^{N} [1 - \psi(x_i) \Delta V]_\xi \right] \tag{3}$$

Here $\psi(x_i)$ is the failure function per unit volume at position x_i within the continuum, and N is the number of links in the continuum. Unless noted otherwise, roman letter subscripts denote tensor indices with an implied range from 1 to 3. Greek letter subscripts are associated with products or summations with ranges that are explicit in each expression. Adopting an argument used by Cassenti,[14] the reliability of the continuum is given by the following expression

$$R = \exp \left[- \int_V \psi dV \right] \tag{4}$$

As was noted earlier, a reliability model based on the principle of independent action would be a candidate for a first approximation macroscopic

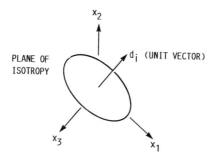

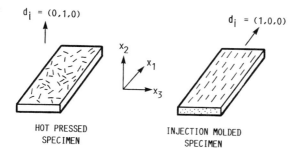

Fig. 11.1   Examples of transversely isotropic whisker-reinforced ceramic composites.

theory for isotropic whisker composites. In this instance, the failure function ψ would depend only upon stress, or the principal invariants of stress, i.e.,

$$\psi = \psi(\sigma_{ij})$$
$$= \psi(\sigma_1, \sigma_2, \sigma_3) \qquad (5)$$

where σ_{ij} is the Cauchy stress tensor and σ_1, σ_2 and σ_3 are the associated principal stresses. However, a more comprehensive reliability model for isotropic whisker-toughened ceramics is pressented in the next section. For transversely isotropic whisker composites, the failure function must also reflect material symmetry. This requires

$$\psi = \psi(\sigma_{ij}, d_i) \qquad (6)$$

where d_i is a unit vector that identifies a local material orientation. This orientation, depicted in Fig. 11.1, is defined as the normal to the plane of isotropy.

The sense of d_i is immaterial, thus its influence is taken into account through the product $d_i d_j$, i.e.

$$\psi = \psi(\sigma_{ij}, d_i d_j) \qquad (7)$$

Note that $d_i d_j$ is a symmetric second-order tensor whose trace satisfies the identity

$$d_i d_i = 1 \tag{8}$$

Furthermore, the stress and local preferred direction may vary from point to point in the continuum. Thus Eqn. (7) implies that the stress field and unit vector field, i.e. $\sigma_{ij}(x_k)$ and $d_i(x_k)$, must be specified to define ψ. Current state of the art in designing structural components uses finite element methods to characterize the stress field. The reader will see shortly that this is conveniently utilized in analyzing component reliability.

Since ψ is a scalar valued function, it must remain form invariant under arbitrary proper orthogonal transformations. Work by Reiner,[15] Rivlin and Smith,[16] Spencer,[17] and others, demonstrates that through the application of the Cayley–Hamilton theorem and elementary properties of tensors, a finite set of invariants known as an integrity basis can be developed. Form invariance of ψ is ensured if dependence is taken on invariants that constitute the integrity basis, or any subset thereof. Adapting the above-mentioned work to ψ results in an integrity basis composed of ten tensor products. Following arguments similar to Spencer,[18] several of these tensor products are equal and others are trivial identities such that the final integrity basis for ψ contains only the invariants

$$I_1 = \sigma_{ii} \tag{9}$$

$$I_2 = \sigma_{ij} \sigma_{ji} \tag{10}$$

$$I_3 = \sigma_{ij} \sigma_{jk} \sigma_{ki} \tag{11}$$

$$I_4 = d_i d_j \sigma_{ji} \tag{12}$$

and

$$I_5 = d_i d_j \sigma_{jk} \sigma_{ki} \tag{13}$$

A slightly different set of invariants that corresponds to components of the stress tensor oriented to the material direction can be constructed from the above integrity basis. This new set of invariants includes

$$\hat{I}_1 = I_4 \tag{14}$$

$$\hat{I}_2 = [I_5 - (I_4)^2]^{1/2} \tag{15}$$

$$\hat{I}_3 = (\tfrac{1}{2})(I_1 - I_4) + [(\tfrac{1}{2})I_2 - I_5 + \tfrac{1}{4}(I_4)^2 - \tfrac{1}{4}(I_1)^2 + (\tfrac{1}{2})I_1 I_4]^{1/2} \tag{16}$$

and

$$\hat{I}_4 = (\tfrac{1}{2})(I_1 - I_4) - [(\tfrac{1}{2})I_2 - I_5 + \tfrac{1}{4}(I_4)^2 - \tfrac{1}{4}(I_1)^2 + (\tfrac{1}{2})I_1 I_4]^{1/2} \tag{17}$$

Considering a uniformly stressed volume, or, in the context of Weibull analysis, a single link, the invariant $\hat{I}_2$ corresponds to the magnitude of the stress component in the direction of d_i, as shown in Fig. 11.1. $\hat{I}_2$ corresponds to the shear stress on the face normal to d_i. The invariants $\hat{I}_3$ and $\hat{I}_4$ are the maximum and minimum normal stresses in the plane of isotropy. Note that these physical mechanisms are independent of I_3, and the implication of this is discussed in the following section. Taking

$$\psi = \psi(\hat{I}_1, \hat{I}_2, \hat{I}_3, \hat{I}_4) \tag{18}$$

ensures ψ is form invariant.

At this point it is assumed that the stress components identified by the invariants above act independently in producing failure. Following reasoning similar to Wetherhold[19]

$$\psi = \left[\frac{<\hat{I}_1>}{\beta_1}\right]^{\alpha_1} + \left[\frac{<\hat{I}_2>}{\beta_2}\right]^{\alpha_2} + \left[\frac{<\hat{I}_3>}{\beta_3}\right]^{\alpha_3} + \left[\frac{<\hat{I}_4>}{\beta_3}\right]^{\alpha_3} \tag{19}$$

It is further assumed that compressive stresses associated with $\hat{I}_1$, $\hat{I}_3$ and $\hat{I}_4$ do not contribute to failure so that

$$<\hat{I}_1> = \begin{cases} \hat{I}_1 & \hat{I}_1 > 0 \\ 0 & \hat{I}_1 \leq 0 \end{cases} \tag{20}$$

$$<\hat{I}_3> = \begin{cases} \hat{I}_3 & \hat{I}_3 > 0 \\ 0 & \hat{I}_3 \leq 0 \end{cases} \tag{21}$$

and

$$<\hat{I}_4> = \begin{cases} \hat{I}_4 & \hat{I}_4 > 0 \\ 0 & \hat{I}_4 \leq 0 \end{cases} \tag{22}$$

Again, the implications of this assumption are discussed in the next section. In addition,

$$<\hat{I}_2> = |\hat{I}_2| \tag{23}$$

for all values of $\hat{I}_2$.

In association with each invariant, the α values correspond to the Weibull shape parameters and the β values correspond to Weibull scale parameters. A two-parameter Weibull probability density function has the following form

$$p(x) = \left[\frac{\alpha}{\beta}\right]\left[\frac{x}{\beta}\right]^{(\alpha-1)} \exp\left[-\left[\frac{x}{\beta}\right]^{\alpha}\right] \quad x > 0 \tag{24}$$

and

$$p(x) = 0 \quad x \leq 0 \tag{25}$$

The cumulative distribution function is given by

$$P(x) = 1 - \exp\left[-\left[\frac{x}{\beta}\right]^{\alpha}\right] \quad x > 0 \tag{26}$$

and

$$P(x) = 0 \quad x \leqslant 0 \tag{27}$$

where $\alpha(>0)$ is the Weibull modulus (or the shape parameter), and $\beta(>0)$ is the Weibull scale parameter.

Note that $\hat{I}_3$ and $\hat{I}_4$ are stress components in the plane of isotropy and, therefore, have the same Weibull parameters. The parameters α_1 and β_1 would be obtained from uniaxial tensile experiments along the material orientation direction, d_i. The parameters α_2 and β_2 would be obtained from torsional experiments of thin-walled tubular specimens where the shear stress is applied across the material orientation direction. The final two parameters, α_3 and β_3, would be obtained from uniaxial tensile experiments transverse to the material orientation direction.

Insertion of Eqn. (19) into Eqn. (4) along with Eqn. (20) through Eqn. (23) yields a noninteractive reliability model for a three-dimensional state of stress in a transversely isotropic whisker-reinforced ceramic composite. The noninteractive representation of reliability for orthotropic ceramic composites would follow a similar development. The analytical details for this can be found in Duffy and Manderscheid.[20]

Figure 11.2A depicts level surfaces of R (for a uniformly stressed element of unit volume) projected onto the σ_{11}–σ_{33} stress plane for the special case where $d_i = (1,0,0)$, It is assumed that the whiskers are highly aligned along the 1-axis, and the 2–3 plane is the plane of isotropy as shown in Fig. 11.1. Here $\alpha_1 = 15$, $\beta_1 = 700$, $\alpha_2 = 12$, $\beta_2 = 600$, $\alpha_3 = 10$, and $\beta_3 = 500$. The three surfaces correspond to $R = 0.95$, 0.5 and 0.05. Note that the contours of reliability are not symmetric about the line $\sigma_{11} = \sigma_{33}$. The intercepts for the various contours are larger in value along the σ_{11} axis. This increase in reliability results from the alignment of whiskers relative to the load direction (i.e., the direction of σ_{11} and d_i are coincident). Contrast this with the contours of R depicted in Fig. 11.2B. Here the Weibull parameters are unchanged but the orientation of d_i is rotated 45° in the 1–3 plane, thus $d_i = (1/\sqrt{2}, 0, 1/\sqrt{2})$. A comparison of Fig. 11.2A and Fig. 11.2B reveals that the tensile intercepts along both axes increase. An increase in the tensile intercepts along the σ_{33} axis is expected. However, an increase of the tensile intercepts along the σ_{11} axis needs explanation. Consider a flaw located in the material that is initially oriented such that the application of σ_{11} would cause Mode I crack propagation. Upon encountering the aligned whiskers, the flaw deflects in the direction of the whiskers (assuming weak bonding between whisker and matrix) or away from the whiskers (see Faber and Evans[6]). In either case, further propagation would proceed under mixed mode conditions and requires an increase in σ_{11}. This is reflected in an increase of the intercepts along the axes. A further consideration of Fig. 11.2B shows that for equibiaxial compression, reliability is unity. Expansion of the invariants under these conditions would result in all four invariants being equal to zero.

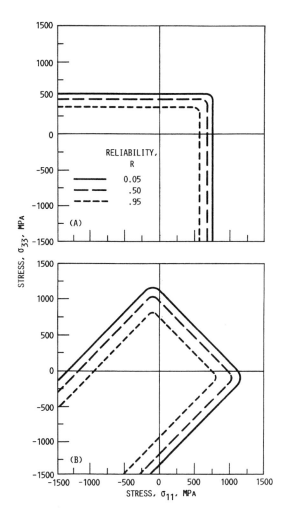

Fig. 11.2 Typical family of reliability contours. Weibull parameters, $\alpha_1 = 15$, $\beta_1 = 700$, $\alpha_2 = 12$, $\beta_2 = 600$, $\alpha_3 = 10$, and $\beta_3 = 500$. Contours shown with (A) local material direction coincident with σ_{11}, and (B) effect of rotating the local material direction 45° in the 1–3 plane.

11.2.3 Interactive Reliability Model

An essential step in developing an interactive reliability model requires formulating a deterministic failure criterion that reflects the limit state behavior of the material. Miki *et al.*,[21] and de Roo and Paluch[22] have adopted this approach in computing the reliability of unidirectional composites. In both articles, the Tsai–Wu failure criterion is adopted, where different failure behavior is allowed in tension and compression, both in the fiber direction and transverse to the fiber direction. Here, the authors assume that failure behavior for isotropic whisker-toughened ceramics will exhibit a reduced tensile strength in comparison to the material's compressive strength, and exhibit a dependence on the hydrostatic component of the stress state. This behavior has not been documented experimentally in the open literature, but this type of response has been exhibited by monolithic ceramics, and we have similar expectations for the whisker-toughened ceramics.

In general, a failure criterion (or limit state) defines the conditions under which a structural component can no longer fulfill its design function. Let

$$\tilde{Y} = (Y_1, Y_2, \ldots, Y_n) \tag{28}$$

denote a vector of design variables (e.g., strength parameters, cyclic load limits, allowable deformation at critical locations of a component, etc.). A limit state function $f(\tilde{Y})$, which stipulates how the design variables interact to produce failure, defines a surface in the n-dimensional design variable space. Typically, the following simple expression

$$f(\tilde{Y}) = 0 \tag{29}$$

is used to define the failure surface. A design point for a structural component that falls within the surface defines a successful operational state. This region will be denoted ω_s. If the design point lies on the surface (denoted ω_f), the component fails. Note that points outside the failure surface are inaccessible limit states. Here, we restrict the design variable space to material strength parameters. These material strength parameters will be closely associated with components of the stress tensor, hence, in general

$$f = f(\tilde{Y}, \sigma_{ij}) \tag{30}$$

where σ_{ij} represents the Cauchy stress tensor. As was discussed in the previous section, due to the dependence upon the stress tensor, phenomenological failure criteria must be formulated as invariant functions of σ_{ij}. Using this approach makes a criterion independent of the coordinate system used to define the stress tensor (i.e., the criterion exhibits frame indifference). In general, for isotropic materials the failure function can be expressed as

$$f(\tilde{Y}, I_1, J_2, J_3) = 0 \tag{31}$$

which guarantees that the function is form invariant under all proper orthogonal transformations. Here I_1 is the first invariant of the Cauchy stress defined by Eqn. (9), J_2 is the second invariant of the deviatoric stress (S_{ij}), and J_3 is the third invariant of the deviatoric stress. These quantities are defined in the following manner

$$S_{ij} = \sigma_{ij} - (1/3)\delta_{ij}\sigma_{kk} \tag{32}$$

$$J_2 = (1/2)S_{ij}S_{ji} \tag{33}$$

$$J_3 = (1/3)S_{ij}S_{jk}S_{ki} \tag{34}$$

where δ_{ij} is the identity tensor. Admitting I_1 to the functional dependence allows for a dependence on hydrostatic stress. Admitting J_3 (which changes sign when the direction of a stress component is reversed) allows different behavior in tension and compression to be modeled with a single unified failure function.

Two representative failure criteria formulated by Willam and Warnke,[23] and Ottosen[24] satisfy the requisite failure behavior of reduced tensile strength, and sensitivity to hydrostatic stress. For the sake of brevity, the discussion here is limited to the Willam–Warnke criterion. The Willam–Warnke criterion can be expressed as

$$f = \lambda \left[\frac{\sqrt{J_2}}{Y_c} \right] + B \left[\frac{I_1}{Y_c} \right] - 1 \tag{35}$$

where

$$B = B(Y_t, Y_c, Y_{bc}) \tag{36}$$

and

$$\lambda = \lambda(Y_t, Y_c, Y_{bc}, J_3) \tag{37}$$

Here Y_t denotes the tensile strength of the material, Y_c is the compressive strength, and Y_{bc} represents equal biaxial compression strength of the material. The reader is directed to Palko[25] where the specific forms of B and λ are presented. This model is termed a three-parameter model, referring to the three material strength parameters (Y_t, Y_c, and Y_{bc}) used to characterize the model. Failure is defined when $f = 0$, and the multiaxial criterion is completely defined in the six-dimensional stress space.

The general nature of the criterion can be examined by projecting the failure function into various coordinate systems associated with the applied stress state. In Fig. 11.3 the function has been projected onto the principal stress plane, i.e., the σ_1–σ_2 stress plane. Note that this plane coincides with the σ_{11}–σ_{22} stress plane. Here the model parameters have been arbitrarily chosen ($Y_t = 0.2$, $Y_c = 2.0$, and $Y_{bc} = 2.32$). Note that the reduction in tensile strength, and the ratio of intercepts along the tensile and compressive axes, is equal to the ratio of Y_t to Y_c. The functional dependence of the limit state function governed by the Willam–Warnke criterion is given by

$$f(\tilde{Y}, \sigma_{ij}) = f(Y_t, Y_c, Y_{bc}, I_1, J_2, J_3) \tag{38}$$

where the specific form of $f(\tilde{Y}, \sigma_{ij})$ is given by Eqn. (35). It is assumed that the material strength parameters are independent random variables. The objective is to compute the reliability (denoted $\mathcal{R}$) of a material element of unit volume, given a stress state, where the components of the stress tensor are considered deterministic parameters. This philosophy, i.e., assuming that material strength parameters are independent random variables and load parameters are deterministic, is not without precedent. The reader is directed to the overview

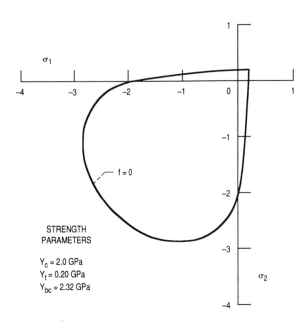

Fig. 11.3   Level failure surface ($f = 0$) of the Willam and Warnke failure criterion projected onto the σ_1–σ_2 stress plane.

by Duffy et al.[26] for references. The reliability of the ith element of a structural component is given by the following expression

$$\mathcal{R}_i = \text{Probability}[f(\tilde{Y}) < 0] \tag{39}$$

Under the assumption that the random variables are independent, this probability is given by the product of the marginal probability density functions integrated over the region ω_s, i.e.,

$$\mathcal{R}_i = \int\int_{\omega_s}\int p_1(y_c)p_2(y_t)p_3(y_{bc})\,dy_c\,dy_t\,dy_{bc} \tag{40}$$

where $p_1(y_c)$, $p_2(y_t)$, and $p_3(y_{bc})$ are the marginal probability density functions of the random variables representing the material strength parameters. Here, the two-parameter Weibull formulation, defined by Eqn. (24) and Eqn. (25), is used for the probability density function. A feature of this analytical approach is the flexibility of using other formulations of the marginal probability density function, such as a three-parameter Weibull distribution or a log–normal distribution. The appropriate distribution function would be dictated by the experimental failure data for a particular strength parameter. Here, for the purposes of illustration, the two-parameter Weibull distribution is adopted for simplicity.

Explicit integration of Eqn. (40) is intractable due to the form of the limit state function which defines the integration domain (ω_s). The reader is referred

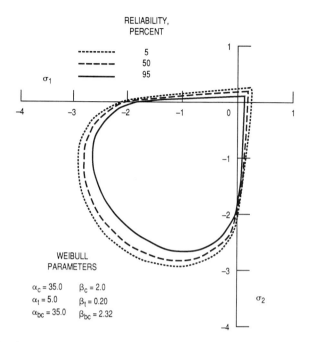

Fig. 11.4 Family of reliability contours associated with an interactive reliability model projected onto the $\sigma_1 - \sigma_2$ stress plane.

to Sun and Yamada,[27] and Wetherhold[19] for the details of this type of integration for simpler interactive failure criteria. Here, Monte Carlo simulation is used to numerically evaluate the triple integral. This technique involves generating a uniform random sample of size k for each of the random variables. For a given stress state, the failure function is computed for each trial sample of random variables. If $f(\tilde{Y}) < 0$ for a given trial, then that trial is recorded as a success. By repeating this process a suitable number of times for a given state of stress, a probability distribution for the element reliability is generated. For a sufficiently large sample size, reliability can be computed as

$$\mathscr{R}_i = n/k \tag{41}$$

where n is the number of trials where $f(\tilde{Y}) < 0$.

This technique was employed in calculating the reliability contours depicted in Fig. 11.4. The reliability contours represent a homogeneously stressed material element, and for dimensionless $\mathscr{R}_i$, the Weibull parameter β has units of stress $\cdot$ (volume)$^{1/\alpha}$. Here $\alpha_t = 5$, $\beta_t = 0.2$, $\alpha_c = 35$, $\beta_c = 2$, $\alpha_{bc} = 35$, and $\beta_{bc} = 2.32$. The three surfaces correspond to $\mathscr{R}_i = 0.95$, 0.5, and 0.05. Note that the reliability contours retain the general behavior of the deterministic failure surface from which they were generated. In general, as the α values increase, the spacing between contours diminishes. Eventually the contours would not be distinct and they would effectively map out a deterministic failure surface. An increase in the β values shifts the relative position of the contours in an outward direction.

The noninteractive and interactive reliability models discussed above have been incorporated into a public domain computer algorithm given the acronym T/CARES (Toughened Ceramics Analysis and Reliability Evaluation of Structures). The reliability analysis of a structural component requires that the stress field must be characterized. Commercial finite element programs allow the design engineer to determine the structural response of composite components subjected to thermomechanical loads. Currently, this algorithm is coupled to the MSC/NASTRAN finite element code. For a complete description of the algorithm, see Duffy *et al.*[12] Before utilizing T/CARES, the reliability models presented here (and later in this chapter) must be characterized using an extensive database that includes multiaxial experiments. It is not sufficient to simply characterize the Weibull parameters for each strength parameter. Multiaxial experiments should be conducted to assess the accuracy of the interactive modeling approach. However, once the Weibull parameters have been characterized for each strength variable, the computer algorithm allows a design engineer to predict the reliability of a structural component subject to quasi-static multiaxial loads. Note that the algorithm is capable of nonisothermal analyses if the Weibull parameters are specified at a sufficient and appropriate number of temperature values.

11.3 Laminated Ceramic Matrix Composites

11.3.1 Introduction

Although whisker-reinforced ceramics have enhanced toughness and reliability, they do not substantially lessen the possibility of catastrophic failure, a problem that restricts their use in certain applications. Continuous fiber-reinforced ceramic composites, however, can provide significant increases in fracture toughness along with ability to fail in a noncatastrophic manner (often referred to as "graceful" failure). Prewo and Brennan[28–30] have demonstrated that incorporating fibers with high strength and stiffness into brittle matrices with similar coefficients of thermal expansion and optimized interfaces yields ceramic composites with the potential of meeting high temperature performance requirements. Typical stress–strain curves of unidirectional systems (see Fig. 11.5) tend to be bilinear when loaded along the fiber direction, with a distinct breakpoint that usually corresponds to the formation of an initial transverse matrix crack. From Fig. 11.5 it is somewhat apparent that an engineer may focus attention on three factors when designing components fabricated from ceramic matrix composites. They are:

(1) the stress (or strain associated with the formation of the first transverse matrix crack (often referred to as the microcrack yield strength);

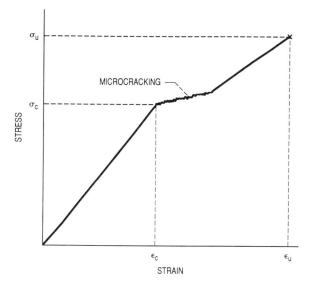

Fig. 11.5 Generic stress–strain curve for a unidirectional ceramic matrix composite with a tensile load imposed in the fiber direction.

(2) the stress (or strain) associated with the ultimate strength of the material; and

(3) the work of fracture, associated with the total area under the stress–strain curve.

Which design factor is of primary importance will depend on the application of the structural component. If limiting strength degradation during the life of a component is important, then the transverse matrix cracking stress becomes the design focus. This restricted focus limits full exploitation of the ultimate load carrying capacity of the material. However, adopting this viewpoint should extend the life of a component since the matrix serves as protection for the fibers and the interface, shielding them from high temperature service environments that cause oxidation and rapid degradation in composite strength. If life is not an important design issue (e.g., the turbine components used in a single-mission cruise missile) then the engineer may extend the design envelope past first matrix cracking and utilize more of the ultimate strength of the composite. In either case, a component benefits from the increase in work of fracture which is a measure of the graceful failure behavior associated with unidirectional reinforced ceramic composites.

Theoretically, the addition of a ceramic fiber to a ceramic matrix should increase strength (both microcrack yield and the ultimate) *and* fracture toughness. Traditionally, fiber reinforcement has been employed to strengthen a matrix with a low modulus and low strength, by transferring load to high modulus, high strength fibers. This type of transfer mechanism demands a microstructure with reasonably strong bonding at the interface between the matrix and the fiber. The result is a composite with fiber-dominated failure behavior. To date, ceramic composite microstructures that promote increased toughness (mainly due to fiber pull-out) have usually done so at the expense of the microcrack yield strength. In the presence of an interface that is too strongly bonded, a transverse matrix crack merely propagates through both the fiber and the matrix in a self-similar fashion. In essence, monolithic failure behavior is obtained. Since catastrophic fracture is the overriding concern with monolithic ceramics, fibers are added to mitigate tensile failure by bridging flaws, and this requires a weakly bonded interface. As a consequence of a weak interface, a transverse matrix crack encounters the interface, changes direction and debonds along the fiber. The applied far-field stress must then be increased for further self-similar crack extension, thus the toughness of the material is increased. The lack of improvement in the first matrix cracking strength of continuous fiber-reinforced ceramic matrix composites is a direct result of current fabrication processes that do not yield optimized interfaces and uniform matrix densities. As fabrication technologies evolve and fully dense matrices with high fiber volumes become a reality, composite matrix cracking strength should also improve.

11.3.2 Reliability Issues

Relative to component reliability, current state of the art focuses on composite tensile strength in the fiber direction, which addresses the upper bound problem. Conversely, a tensile load applied transverse to the fiber direction results in failure behavior similar to, or worse than, a monolithic ceramic. This represents the lower bound of composite strength. The transverse direction also does not possess the graceful failure behavior, and for this reason (and since the majority of structural applications involve multiaxial states of stress) practical continuous fiber composites are reinforced in two or three directions using laminated, woven, or braided architectures. Two-dimensional laminate construction is the focus in this section.

As is indicated throughout this chapter, there is a philosophical division that separates analytical schools of thought for predicting failure behavior into micro- and macro-level methods. Blass and Ruggles[31] point out that analysts from the first school would design the material in the sense that the constituents are distinct structural components, and the composite ply (or lamina) is considered a structure in its own right. Analysts from the latter school of thought would design with the material, i.e., analyze structural components fabricated from the material. Since failure (assuming that deformation or stability does not control component design) usually emanates from crack-like defects in most brittle materials, it is not unreasonable to apply the principles of fracture mechanics to ceramic composite systems. There is a wealth of literature concerning the application of linear elastic fracture mechanics (LEFM) to many types of composite materials. The reader is directed to the volume edited by Friedrich[32] for a state-of-the-art review on this topic. If a unidirectional composite is treated as a homogenized material using fracture mechanics, then the details of the microstructure are ignored. Usually the ply is treated as an anisotropic plate and the approach is used to predict failure loads when notches or holes are present in a structure. However, problems arise with this approach when obtaining critical stress intensity factors and/or critical strain energy release rates in composites where self-similar crack growth is an experimental myth. In addition, since failure in ceramic composites is sensitive to flaws and geometric discontinuities at a much smaller scale, this approach has not found many proponents in the past.

The alternative is to account for the details of the microstructure, since it is typical to assume that the size of critical flaws is of the same order of magnitude as the characteristic dimensions of the microstructure (e.g., fiber diameter, crack spacing). Rigorous fracture mechanics criteria that adopt the microstructural viewpoint have been proposed to predict first matrix cracking (e.g., Budiansky *et al.*,[33] and Marshall *et al.*[34]), yet these approaches make the rather strong simplifying assumption that the analyst has a working knowledge of the constituent properties, constituent geometries, and arrangements. This is simplistic, especially in light of the mechanical properties of the interface. The authors wish to point out that different types of pull-out and push-out tests have been advocated to characterize interfacial shear properties. With these experiments, assumptions must be made concerning the stress distribution in the interface along the fiber length. A uniform interface is universally assumed (along with the various postulated stress distributions) which is rather remarkable since post-failure investigations of typical CMC microstructures usually indicate that the interface is highly nonuniform, or even totally absent. Current fabrication technology does not produce a consistent interface in most ceramic composites. This strongly indicates that stochastic approaches should be considered in characterizing the strength of unidirectional ceramic composites. If probabilistic methods are utilized, then the reader will notice in the previous discussion, and in what follows, that employing reliability analyses requires the design engineer to characterize the state of stress throughout the component. Local–global techniques will not work since component probability of failure is computed based on the reliability of each discrete element. Using finite element techniques at the constituent level to characterize the stress state of an entire component would quickly exhaust the capacity of a supercomputer. Thus, conducting component failure analyses at the microstructural level using fracture mechanics is not a viable design alternative.

As a final note regarding the use of fracture mechanics, several authors have combined fracture mechanics with a probabilistic Weibull analysis of fiber failure to determine the stress–strain behavior and subsequent work of fracture for unidirectional composites (e.g., Thouless and Evans,[35] and Sutcu[36]). However, the focus here is on first matrix cracking, i.e., designing composite structures for extended service life. We note that mature reliability-based design methods using fracture mechanics concepts will not surface until a coherent mixed mode fracture criterion has been proposed that accounts for complex crack geometries, fiber/matrix debonding, and closing pressures resulting from fibers that bridge highly irregular matrix cracks. Even when the state of the art reaches this point, fracture mechanics may have a practical utility in designing the material, but not a component.

If component failure analyses based on fracture mechanics are not the answer, the engineer can turn to design methods based on phenomenological strength models. As in fracture mechanics, phenomenological models can be proposed at the macrostructural and microstructural levels. The authors advocate the school of thought mentioned above which idealizes the ply (or lamina) as a homogenized material with strength properties that are determined from phenomenological experiments. There are practical reasons for embracing this viewpoint. The authors fully recognize that the failure characteristics of these composites are controlled by a number of local phenomena including matrix cracking, debonding and slipping between matrix and fibers, and fiber breakage, all of which strongly interact. Understanding the underlying analytical concepts associated with the microstructural viewpoint allows one to gain insight and intuition prior to constructing multiaxial failure theories that, in some respect, reflect the local behavior. However, proposing phenomenological models at the microstructural level entails developing an extensive property database that must include all the constituents. As noted earlier, constituent properties are not often uniform throughout the microstructure. A top-down approach, that is first proposing analytical models at the ply level, will establish viable and working design protocols. Initially adopting the bottom-up approach allows for the possibility of becoming mired in detail (experimental and analytical) when multiaxial reliability analyses are conducted at the constituent level.

Size effect (i.e., decreasing bulk strength with increasing component size) is an important feature of ceramic composites that must be addressed in failure analysis. How it is addressed depends on whether the material is modeled as a series system, a parallel system, or a combination. Current analytical practice uses finite element methods to determine the state of stress throughout the component. It is assumed that failure is dependent upon the stress state in a component, such that deformations are not controlling design. Since failure may initiate in any of the discrete volumes (elements), it is useful to consider a component as a system. A component comprised of discrete volumes is a series system if it fails when one of the discrete volumes fails. This approach gives rise to weakest-link theories. In a parallel system, failure of a single element does not dictate that the component has failed, since the remaining elements may be able to sustain the load through redistribution. Adopting a parallel system approach leads to what has been referred to in the literature as bundle theories.

The basic principles of bundle theory were originally discussed by Daniels[37] and Coleman.[38] Their work was extended to polymer matrix composites by Rosen[39] and Zweben.[40] Here, a relatively soft matrix serves to transfer stress between fibers, and contributes little to the composite tensile strength. Hence, when a fiber breaks, the load is transferred only to neighboring fibers. Their analysis is rather complex and limited to establishing bounds on the stress at which the first fiber breaks and the stress at which all the fibers are broken. Harlow and Phoenix[41] proposed a rather abstract approach that established a closed-form solution for all the intermediate stress levels in a two-dimensional problem, and Batdorf[42] used an approximate solution to establish the solutions for the three-dimensional problem. Batdorf's model includes the two-dimensional model as a special case. In both of the latter two models, the concept of an effective Weibull modulus that increases with increasing component volume is proposed. This implies a diminished size effect. However, these current bundle theories are predicated on the fact that fibers are inherently much stronger and stiffer than the matrix. In laminated CMC materials this is not always the case. The magnitudes of the fiber and the matrix stiffness are usually close in value while fiber strength is usually much higher due to the small fiber size effect. Since fiber bundle ultimate strength theories are discussed elsewhere in this book, they will not be considered in this chapter.

The authors advocate the use of a weakest link reliability theory in designing components manufactured from laminated CMC materials that exhibit a limited size effect in specific directions. Assuming that a laminated structure behaves in a weakest link manner allows the calculation of a conservative estimate of structural reliability. Provided that appropriate failure data are used, Thomas and Wetherhold[43] point out that this is also consistent with predicting the probability of the first matrix crack occurring in an individual ply, either in the longitudinal or transverse directions. For most applications, the design failure stress for a laminated material will be taken to coincide with this event (i.e., first ply matrix cracking). The reason for this is that matrix cracking usually leads to high temperature oxidation of the interface and consequent embrittlement of the composite.

11.3.3 Laminate Reliability Model

There is a great deal of intrinsic variability in the strength of each brittle constituent of a ceramic matrix composite, but depending on the composite system, the first matrix cracking strength and the ultimate strength may either be deterministic or probabilistic. Statistical models are a necessity for those composite systems which exhibit appreciable scatter in either of these. Here, we focus on first matrix cracking and treat it in a probabilistic fashion, requiring that deterministic strength be a limiting case that is readily obtainable from the proposed reliability model. With regard to high temperature service conditions, the reader should keep in mind that the parameters associated with the reliability models presented in this section are temperature dependent. In a sense, this is analogous to taking the stiffness parameters in linear elasticity to be temperature dependent. The structure of the reliability models does not change with service temperature, but the model parameters and component reliability do.

Predicting the reliability due to loads in the fiber direction addresses an upper bound for ply reliability in a structural design problem. Conversely, a tensile load applied transverse to the fiber direction results in failure behavior similar to a monolithic ceramic, which corresponds to the lower bound of ply reliability. Thus, multiaxial design methods must be capable of predicting these two bounds as well as accounting for the reduction in reliability due to an in-plane shear stress, and compressive stresses in the fiber direction and transverse to the fiber direction. The reader should also be cognizant of the delamination failure mode, and for relatively thick laminates, failure may emanate from out-of-plane stresses; however, neither will be addressed in what follows. A number of macroscopic theories exist that treat unidirectional composites as homogenized, anisotropic materials, assuming a plane state of stress. These methods use phenomenological strength data directly without hypothesizing specific crack shapes or distributions. Theories of this genre are generally termed noninteractive if individual stress components are compared to their strengths separately. In essence, failure modes are assumed not to interact and this results in component reliability computations that are quite tractable. Noninteractive models mentioned in the previous section dealing with whisker-toughened composites, as well as work by Thomas and Wetherhold,[43] are representative of multiaxial noninteractive reliability models for anisotropic materials. In addition Wu,[44] and Hu and Goetschel[45] have proposed simpler unidirectional reliability models for laminated composites that can be classified as noninteractive. Alternatively, one can assume that for multiaxial states of stress, failure modes interact and depend on specific stochastic combinations of material strengths. The reader is directed to an excellent generic treatment of this approach by McKernan.[46] Failure modes can be either identified and characterized by micromechanics (see McKernan[46]), or a phenomenological failure criterion is adapted from existing polymer matrix design technologies. The probability that the criterion has been violated for a given stress state is computed using Monte Carlo methods,[21] or first-order–second-moment (FOSM) methods.[22] For the sake of brevity only the noninteractive approach is presented here.

A noninteractive phenomenological approach is presented where a unidirectional ply is considered a two-dimensional structure in a state of plane stress which is assumed to have five basic strengths (or failure modes). They include a tensile and compressive strength in the fiber direction, a tensile and compressive strength in the direction transverse to the fiber direction, and an in-plane shear strength. In addition, each ply is discretized into individual sub-ply volumes. For reasons discussed in the previous section, we assume that failure of a ply is governed by its weakest link (or sub-ply volume). Under this assumption, events leading to failure of a given link do not affect other links (see, for example, Batdorf and Heinisch,[47] Wetherhold,[19] and Cassenti[14]);

thus, the reliability of the ith ply is given by the following expression

$$R_i = \exp\left[-\int_V \psi_i \, dV\right] \tag{42}$$

Here, $\psi_i(x_j)$ is the failure function per unit volume at position x_j within the ply, given by

$$\psi_i = \left[\frac{<\sigma_1 - \gamma_1>}{\beta_1}\right]^{\alpha_1} + \left[\frac{|\tau_{12} - \gamma_2|}{\beta_2}\right]^{\alpha_2} + \left[\frac{<\sigma_2 - \gamma_3>}{\beta_3}\right]^{\alpha_3} +$$

$$+ \left[\frac{<(-1)(\sigma_1 + \gamma_4)>}{\beta_4}\right]^{\alpha_4} + \left[\frac{<(-1)(\sigma_2 + \gamma_5)>}{\beta_5}\right]^{\alpha_5} \tag{43}$$

where V is the component volume. The α values associated with each term in Eqn. (43) correspond to the Weibull shape parameters, the β values correspond to Weibull scale parameters, and the γ values correspond to the Weibull threshold stresses. In addition, σ_1 and σ_2 represent the in-plane normal stresses that are aligned with, and transverse to, the fiber direction, respectively. Also, τ_{12} is the in-plane shear stress. Normal stresses appear twice and this allows for different failure strengths to emerge in tension and compression. Note that the brackets indicate a unit step function, i.e.,

$$<x> = x \cdot u[x] = \begin{cases} x & x > 0 \\ 0 & x \leq 0 \end{cases} \tag{44}$$

The reader should note that Weibull threshold parameters were employed in Eqn. (43). Thus, it is assumed that each random strength variable is characterized by a three-parameter Weibull distribution. This is a broad (but a less restrictive) assumption that necessitates some reflection. First, not enough failure data exist to completely characterize an existing CMC material system. However, the experience that monolithic ceramics occasionally exhibit threshold behavior (see Duffy *et al.*,[48] and Margetson and Cooper[49]), combined with the fact that one would intuitively assume that threshold behavior would occur in the fiber direction, indicates that this issue should be approached with an open mind. Estimation techniques have been proposed by numerous authors, and the reader is directed to Duffy *et al.*[50] for an overview.

Inserting Eqn. (43) into the volume integration given by Eqn. (42) yields the reliability of the ith ply, and the probability of first ply failure for the laminate is given by the expression

$$P_{fpf} = 1 - \prod_{i=1}^{N} R_i \tag{45}$$

where N is the number of plies. The computation given in Eqn. (45) represents an upper bound on component probability of failure. Alternatively, Thomas

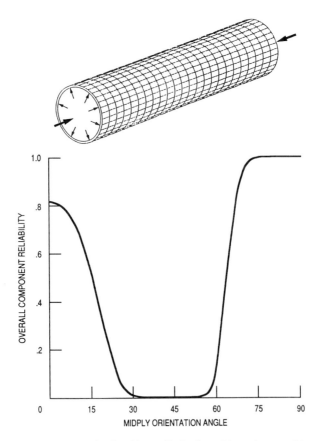

Fig. 11.6 Finite element mesh of a thin-walled tube with an imposed internal pressure and axial compressive stress. Component reliability is plotted as a function of the mid-ply orientation angle.

and Wetherhold[43] proposed a lower bound on component probability of failure by assuming the failure behavior of a component is dominated by the strongest link in the chain. Failure does not occur until the strongest link has failed. Component probability of failure is obtained by subtracting the product of probabilities of failure of each ply from unity, i.e.,

$$(P_f)_{lower} = \prod_{i=1}^{N} (1 - R_i) \tag{46}$$

Either reliability model can be readily integrated with laminate analysis options available in several commercial finite element codes. A preliminary version of a public domain computer algorithm (C/CARES)[48] has been coupled with MSC/NASTRAN to perform this analysis. A simple benchmark application illustrates the approach. A thin-walled tube is subjected to an internal pressure and an axial compressive load. The component is fabricated from a

Table 11.1 Composite Weibull parameters for thin-walled tube (Weibull threshold stress, $\gamma_i \equiv 0$)

Index[a]	Type and direction of stress	Weibull parameters	
		Shape (α_i)	Scale (β_i)
1	Normal tensile stress in fiber direction	25	450
2	In-plane shear stress	22	420
3	Normal tensile stress transverse to fiber direction	10	350
4	Normal compressive stress in fiber direction	35	4500
5	Normal compressive stress transverse to fiber direction	30	3500

[a] Indices correspond to subscripts in Eqn. (43).

three-ply laminate, with a 90°/θ/90° lay-up. Here, angle θ is measured relative to the longitudinal axis of the tube (see Fig. 11.6). An arbitrary internal pressure of 4.25 MPa and axial compressive stress of 87.5 MPa were applied to the tube. The Weibull parameters were also arbitrarily chosen (see Table 11.1). Indices 1 and 4 from Table 11.1 are associated with the normal stress (tensile and compressive, respectively) in the fiber direction. Similarly, indices 3 and 5 are associated with the normal stress (tensile and compressive, respectively) transverse to the fiber direction. Finally, index 2 is associated with the in-plane shear stress. Note that the threshold stresses are taken equal to zero for simplicity. In design, setting the threshold stresses equal to zero would represent a conservative assumption. The overall component reliability is depicted as a function of the mid-ply orientation angle (θ) in Fig. 11.6. This illustrates that ply orientation has a decided effect on component reliability, as expected. Similar studies can be conducted that demonstrate the effect that component geometry, ply thickness, load, and/or Weibull parameters have on component reliability. Hence, the C/CARES code allows the design engineer a wide latitude to optimize a component relative to a number of design parameters.

11.4 Fabric-Reinforced Ceramic Matrix Composites

Advancements in textile weaving technology have resulted in significant new opportunities for utilizing high performance two- and three-dimensional fabric-reinforced ceramic matrix composites in high temperature structural applications (see Fareed et al.,[51] and Ko et al.[52]). Attractive features include improvements in damage tolerance and reliability, flexibility in fiber placement and fabric architecture, and the capability of near-net-shape fabrication. This latter feature is of particular interest since applications where these materials can make a significant impact often require complex geometric shapes. However, designing structural components that are fabricated from materials incorporating ceramic fiber architectures also represents new and distinct challenges in analysis and characterization. Preforms, which serve as the composite skeleton, are produced by weaving, knitting and braiding techniques (see Ko[53]). Woven fabrics (i.e., two-dimensional configurations) exhibit good stability in the mutually orthogonal warp and weft directions. Triaxially woven fabrics, made from three sets of yarns which interlace at 60° angles, offer nearly isotropic behavior and higher out-of-plane shearing stiffness. A three-dimensional fabric, consisting of three or more yarn diameters in the thickness direction, is a network where yarns pass from fabric surface to fabric surface. These three-dimensional systems can assume complex shapes and provide good transverse shear strength, impact resistance, and through-the-thickness tensile strength. Furthermore, the problem of interlaminar failure is totally eliminated. However, due to crimping of fibers and less than optimized fiber fractions in the primary load direction, some tailoring and optimization is usually sacrificed when compared to traditional laminated composites.

Complex textile configurations and complicated yarn–matrix interface behavior represent a challenge in determining the mechanical properties of these composites. Considerable effort has been devoted to evaluating the effectiveness of various reinforcement architectures based upon approximate geometrical idealizations. Chou and Yang[54] have summarized the results of extensive studies in modeling thermoelastic behavior of woven two-dimensional fabrics, and braided three-dimensional configurations. They proposed a unique method of constructing structure–performance maps, which show anisotropic stiffness variations of angle-ply, cross-ply, fabric and braided-fiber architectures. The relative effectiveness of various textile-reinforcing schemes can be assessed using these maps. In the future, as these materials emerge, the maps should serve as a guide for design engineers wishing to specify these materials.

Several analytical models have been developed to predict the mechanical properties and structural behavior of these composites. The approaches are based mainly on modified laminate theory, and/or a geometric unit cell concept. For two-dimensional woven fabric reinforced composites, Chou and Ishikawa[55] have proposed three models using laminate theory. These models are known as the mosaic, crimp, and bridging models. The mosaic model ignores fiber continuity and treats a fabric composite as an assemblage of cross-ply laminates. The crimp model takes into account the continuity and undulation of the yarns; however, it is only suitable for plain weaves. The bridging model, developed for satin weaves, is essentially an extension of the crimp model. This model takes into consideration the contribution to the total stiffness of the linear and nonlinear yarn segments.

The concept of a geometric unit cell has been widely used to characterize the complex structure of three-dimensional fiber-reinforced composites and to establish constitutive relations. In general, this approach assumes that the thermoelastic properties are functions of fiber spatial orientation, fiber volume fraction, and braiding parameters. Ma *et al.*[56] developed a fiber interlock model which assumes that the yarns in a unit cell of a three-dimensional braided composite consist of rods which form a parallelepiped. Contribution to the overall strain energy from yarn axial tension, bending, and lateral compression are considered, and formulated within the unit cell. Yang *et al.*[57] proposed a fiber inclination model that assumes an inclined lamina is represented by a single set of diagonal yarns within a unit cell. Ko *et al.*[58] have also proposed a unit cell model where the stiffness of a three-dimensional braided composite is considered to be the sum of the stiffnesses of all its laminae. The unique feature of this model is that it admits a three-dimensional braid as well as the other multidirectional reinforcements, including five-, six-, and seven-directional yarns that are either straight or curvilinear. Finally, Crane and Camponeschi[59] have modeled a multidimensional braided composite using modified laminate theory. Their approach determines the extensional stiffness in the three principal geometric directions of a braided composite. We note that these analytical models are based on previous research related to polymeric and metal matrix composites. Though much of the current stiffness modeling effort is in its infancy, the approaches mentioned above have demonstrated merit relative to experimental data. However, caution is advised since agreement between model predictions and experimental data is subject to interpretation. Yet accurate predictions of mechanical properties are a necessity when conducting stress–strain analyses which logically precede all failure studies.

Recently, characterization studies that focus on the failure behavior of woven ceramic composites have been appearing in the open literature. The work of Wang *et al.*[60] and Chulya *et al.*[61] represents the fundamental type of exploratory test programs necessary in the study of constitutive response and damage mechanisms. It is apparent from these studies that material behavior is dominated by voids and gaps located in the interstices of the weave. Behavior in tension and compression is different. Tensile behavior is influenced by microcracking which is the direct result of residual stresses developed during fabrication. The nonlinear tensile stress–strain curve exhibits a continuous reduction in stiffness in the initial load cycle. However, subsequent load cycles yield relatively linear stress–strain behavior. This indicates that most of the damage accumulation takes place during the initial load cycle. Wang *et al.*[60] postulated that the ultimate tensile strength of woven ceramic composites is governed by the formation of distributed transverse cracks that terminate at voids or fibers in the materials. These transverse cracks eventually link up, causing ultimate failure. Yen *et al.*[62] have proposed a model that accounts for this type of behavior by assuming the formation of H-shaped cracks which eventually merge, leading to ultimate failure. The mechanical response of this material under compressive loads yields nearly linear behavior up to the ultimate strength. The ultimate compressive strength was associated with buckling of fiber bundles and matrix fragmentation.

Predicting the onset of failure represents a complex task due to the numerous failure modes encountered in this class of materials. As research progresses, deterministic and probabilistic schools of thought will emerge. For either case, models based upon the principles of fracture mechanics, as well as phenomenological models, will be proposed. It is expected that deterministic approaches will precede the development of probabilistic concepts. Initial research regarding failure analysis for whisker-toughened and long fiber ceramic composites typically borrowed concepts from polymer and metal matrix composite research. We anticipate a similar trend with woven composites.[63] For example, Yang,[64] and the previously cited work of Chou and Ishikawa[55] have proposed techniques for predicting the onset of short-term failure in fabric-reinforced polymer composites that make use of the maximum stress and maximum strain failure criteria. The work by Walker et al.[65] represents another potential concept that could be utilized. Here, mesomechanics is employed to predict the behavior of metal matrix composites. This approach takes advantage of the periodic nature of the microstructure, and homogenizes the material through volume averaging techniques. In Walker's work, a damage parameter can be included as a state variable in formulating constitutive relationships. Other suitable failure concepts may emerge from the field of continuum damage mechanics as well.

Currently, phenomenological approaches may be the logical first choice due to the complexity of the microstructure. Yet one should be cognizant of the fact that random microcracks and voids are present in woven ceramics. These microdefects may lead to statistically distributed failures and failure modes. However, at this time, failure models based upon probabilistic concepts have not been proposed. A more extensive database is needed to identify the underlying mechanisms of failure, and innovative experimental techniques must accompany the development of constitutive models that incorporate damage evolution at the microstructural level. The task will be iterative, requiring constant refinement of the description of the physical mechanisms causing failure and their mathematical description. The authors wish to point out that fiber preform design, constitutive modeling, and material processing, will require the combined talents and efforts of material scientists and structural engineers. As processing methods improve, the popularity of fabric-reinforced ceramic matrix composites will increase.

11.5 Time-Dependent Reliability

11.5.1 Introduction

The utilization of ceramic composites in fabricating structural components used in high temperature environments requires thoughtful consideration of fast fracture as well as strength degradation due to time-dependent phenomena such as subcritical crack growth, creep rupture, and stress corrosion. In all cases, this can be accomplished by specifying an acceptable reliability level for a component. Methods of analysis exist that capture the variability in strength of ceramic composites as it relates to fast fracture (see Sections 11.2 through 11.4). However, the calculation of an expected lifetime of a ceramic component has been limited to a statistical analysis based on subcritical crack growth in monolithic materials (see Wiederhorn and Fuller[66] for a detailed development). The subcritical crack growth approach establishes relationships among reliability, stress, and time-to-failure based on principles of fracture mechanics. The analysis combines the Griffith[67] equation and an empirical crack velocity equation with the underlying assumption that steady growth of a preexisting flaw is the driving failure mechanism.

Several authors including Quinn,[68] Ritter *et al.*, [69] and Dalgleish *et al.*[70] have emphasized that time-dependent failure of monolithic ceramics is not limited to subcritical crack growth and may also occur by either stress corrosion or creep rupture. Stress corrosion involves nucleation and growth of flaws by environmental/oxidation attack. Creep rupture typically entails the nucleation, growth, and coalescence of voids dispersed along grain boundaries. The authors have similar expectations that these types of mechanisms will exist within the microstructure of ceramic composite constituents. However, the time-dependent mechanical response of a ceramic matrix composite will depend on how the material is loaded relative to the principal material direction. In this section, creep rupture is highlighted, and the intent is to outline a method that determines an allowable stress for a given component lifetime and reliability. One can assume that voids will nucleate, grow, and coalesce within the matrix first, and these voids will eventually produce a macrocrack within the composite, which is bridged by the fibers. Thus, one definition of component life is determined by the onset of the first transverse matrix crack. This is accomplished by combining Weibull analysis with the principles of continuum damage mechanics. Continuum damage mechanics was originally developed by Kachanov[71] to account for tertiary creep and creep fracture of ductile metal alloys. For this reason, many have wrongly assumed that the principles of continuum damage mechanics are only applicable to materials exhibiting ductile behavior. There are numerous articles (see, for example, Krajcinovic[72] and Cassenti[73]) that have advocated the use of continuum damage mechanics to model materials with brittle failure behavior.

Ideally, any theory that predicts the behavior of a material should incorporate parameters that are relevant to its microstructure (grain size, void spacing, etc.). However, this would require a determination of volume averaged effects of microstructural phenomena reflecting nucleation, growth, and coalescence of microdefects, that in many instances interact. This approach (see, for example, Ju[74]) is difficult, even under strongly simplifying assumptions. In this respect, Leckie[75] points out that the difference between the materials scientist and engineer is one of scale. He notes that the materials scientist is interested in mechanisms of deformation and failure at the microstructural level, and the engineer focuses on these issues at the component level. Thus, the former designs the material and the latter designs the component. Here, the engineer's viewpoint is taken and it is noted from the outset that continuum damage mechanics does not focus attention on microstructural events, yet it does provide a practical model which macroscopically captures the changes induced by the evolution of voids and defects. As pointed out by Duffy and Gyekenyesi,[76] a comparison of the continuum and microstructural kinetic equations for monolithic ceramics yields strong resemblances to one another. Thus, the persistent opinion that phenomenological models have little in common with the underlying physics is sometimes short-sighted, and most times self-serving. Adopting a continuum theory of damage with its attendant phenomenological view would appear to be a logical first approach, and as Krajcinovic[77] notes, probably has the best chance of being utilized by the engineering community.

11.5.2 Damage State Variable

The evolution of the microdefects represents an irreversible thermodynamic process. At the continuum level, this requires the introduction of an internal state variable that serves as a measure of accumulated damage. Consider a uniaxial test specimen and let A_0 represent the cross-sectional area in an undamaged (or reference) state. Denote A as the current cross-sectional area in a damaged state where material defects exist in the cross-section (i.e., $A < A_0$). Microstructurally, this can be represented by Fig. 11.7. The macroscopic damage associated with this specimen is represented by the scalar

$$\omega = (A_0 - A)/A_0 \tag{47}$$

or alternatively by $\eta = 1 - \omega$, which is referred to as "continuity." This is the simplest manner in which to represent damage. The variable η represents the fraction of cross-sectional area not occupied by voids. A material is undamaged if $\omega = 0$ or $\eta = 1$. It is quite typical (and somewhat natural) to associate the thermodynamic state variable representing damage with a decrease in material integrity.

Note that, as it appears above, ω is a scalar quantity (i.e., a zero-order tensor). This is appropriate for the limited case where components are subjected to uniaxial loads and undergo isotropic damage. Here, isotropic damage is defined as a microstructure with microdefects that are randomly oriented and randomly distributed. A number of authors have alternatively treated the damage state variable as a vector (a first-order tensor), a second-order tensor, a fourth-order tensor, or an eighth-order tensor. Representing the damage state variable as a vector admits information relative to microcrack area and orientation, but anomalies arise in transforming this first-order tensor relative to a general multiaxial state of stress. Other authors have proposed using second-order tensors to model damage, but problems occur in representing anisotropic damage in general. Using fourth-order

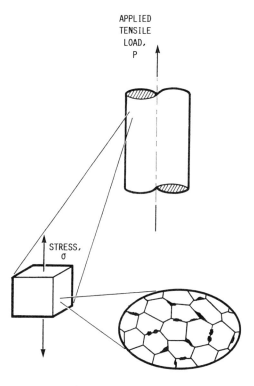

Fig. 11.7 Uniaxial tensile test specimen with distributed microdefects along the grain boundary.

tensors is appealing since an analogy can be made to the reduction (evolution) in stiffness properties that occurs in a damaged material. This analogy is illustrated using the concept of an effective stress.

The concept of an effective stress is based, in part, on the assumed existence of an elastic potential energy function that admits damage state variables. This function is a scalar valued function, and is defined by the expression

$$\Omega^e = (1/2)\,\eta\varepsilon_{ij}C^0_{ijkl}\varepsilon_{kl} \qquad (48)$$

where C^0_{ijkl} is the undamaged (or reference) fourth-order elastic stiffness tensor, ε_{ij} is the second-order elastic strain tensor, and $\eta(= 1 - \omega)$ is initially taken as a scalar quantity to illustrate a deficiency associated with the scalar representation. Based on the assumed existence of Ω^e, and the Clausius–Duhem inequality, a homogenized stress tensor can be derived such that

$$\sigma_{ij} = \eta C^0_{ijkl}\varepsilon_{kl} \qquad (49)$$

Thus, a damaged stiffness tensor can be easily obtained from Eqn. (49) and is given by the expression

$$C_{ijkl}^d = \eta C_{ijkl}^0 \tag{50}$$

The result is that the elements of the damaged stiffness tensor are equal to the elements of the reference stiffness tensor reduced by a common factor η. This indicates that the amount of damage induced in a material can be quantified (or monitored experimentally) by the change in elastic stiffness properties. Note that these stiffness properties are macrovariables, and no assumption was imposed at this point regarding the original material symmetry (i.e., the undamaged material can be elastically isotropic or elastically anisotropic, the method admits either).

At this point, an effective stress can be defined by the expression

$$\bar{\sigma}_{ij} = \sigma_{ij}/\eta \tag{51}$$

Here, the principal orientations of the effective stress will coincide with principal orientations of the Cauchy stress (σ_{ij}) when the damage variable is a scalar. Through the use of an effective stress, the damage state variable can be incorporated directly into the Weibull equation (see Section 11.5.3). However, there are deficiencies in associating a scalar damage variable with a three-dimensional stress state. Specifically, Ju[74] has pointed out that a scalar damage variable will reduce the bulk and the shear moduli by an equal amount, that is

$$G^d = \eta G^0 \tag{52}$$

and

$$K^d = \eta K^0 \tag{53}$$

where G^d and K^d are damage shear and bulk moduli. This leads to

$$\frac{G^d}{K^d} = \frac{G^0}{K^0} \tag{54}$$

which implies that Poisson's ratio does not evolve (i.e., $\nu^d = \nu^0$). This is a rather strong restriction that has limited experimental support.[78] The physical implication of this is depicted in Fig. 11.8. Here, both the uniaxial stress–strain curve and the transverse strain are graphed. If the material damages in the load direction then the stress–strain curve changes slope. Even if the material damages isotropically, it would not seem reasonable that the transverse strain curve would be unaffected.

To circumvent this difficulty (and other difficulties associated with first- and second-rank damage tensors) it is more likely that a fourth-order tensor would better represent the damaged state of a material, whether the material is isotropic or anisotropic. Keep in mind that materials that are originally isotropic can damage in an anisotropic fashion. The result is an anisotropic

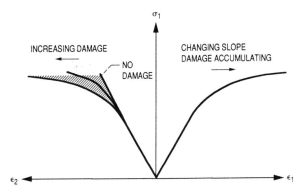

Fig. 11.8 Uniaxial stress–strain curve and possible transverse stress–strain curves for a material undergoing a damage process.

material. Once again, assuming the existence of an elastic potential energy function that admits a damage state variable, then

$$\Omega^e = (1/2)\,\varepsilon_{ij}\,\eta_{ijkl}(\omega_{mnop})\,C^0_{klrs}\,\varepsilon_{rs} \tag{55}$$

where $\eta_{ijkl}(\omega_{mnop})$ is a fourth-order tensor valued function that is dependent on a fourth-order damage state variable. Now, the Cauchy stress is given by the expression

$$\sigma_{ij} = \eta_{ijkl}(\omega_{mnop})\,C^0_{klrs}\,\varepsilon_{rs} \tag{56}$$

Thus, the effective stress and the Cauchy stress are related as follows

$$\sigma_{ij} = \eta_{ijrs}(\omega_{mnop})\,\overline{\sigma}_{rs} \tag{57}$$

where $\eta_{ijrs}(\omega_{mnop})$ is a transformation tensor. The simplest form of this transformation tensor is

$$\eta_{ijrs} = I_{ijrs} - \omega_{ijrs} \tag{58}$$

where I_{ijrs} is the fourth-order identity tensor. The effective stress tensor now becomes

$$\overline{\sigma}_{ij} = (I_{ijrs} - \omega_{ijrs})^{-1}\,\sigma_{rs} \tag{59}$$

where isotropic damage implies ω_{ijrs} is an isotropic fourth-order tensor, as opposed to stipulating that the damage state variable is a scalar which leads to unwarranted restrictions, as mentioned above. However, as long as uniaxial stress states are being considered it is not inappropriate to use a scalar damage state variable to illustrate how Weibull analysis can accommodate continuum damage mechanics to produce a time-dependent reliability model of practical utility. As a final comment, it would not be unreasonable to assume composite materials would undergo anisotropic damage, since these materials are anisotropic in the undamaged (or reference) state. Using a fourth-order tensor also allows modeling anisotropic damage in the most general fashion.

11.5.3 Damage Evolution

For time-dependent analysis, the rate of change of continuity $\dot{\eta}$ (or the damage rate $\dot{\omega}$) must be specified. This rate is functionally dependent on stress and the current state of continuity, that is

$$\dot{\eta} = \dot{\eta}(\sigma, \eta) \tag{60}$$

and is monotonically decreasing ($\dot{\eta} < 0$). For a uniaxial specimen, the dependence of $\dot{\eta}$ on stress is taken through a net stress defined as

$$\bar{\sigma} = P/A = \sigma_0/\eta \tag{61}$$

where P is the applied tensile load, and $\sigma_0 = P/A_0$. A power-law form of the kinetic equation is adopted, that is

$$\dot{\eta} = -B(\bar{\sigma})^n = -B(\sigma_0/\eta)^n \tag{62}$$

where $B > 0$ and $n \geqslant 1$ are material constants determined from creep rupture data, as discussed below. The authors recognize this form of evolutionary law is simplistic, stipulated *a priori*, and that experimental data may indicate some inconsistencies and/or inadequacies. Modification would be guided by experiment, and material science models for creep damage outlined in Duffy and Gyekenyesi.[76] As an example, physical processes that involve void growth mechanisms along grain boundaries typically exhibit threshold behavior. This is illustrated in a schematic plot of log of stress as a function of log of time-to-failure in Fig. 11.9. Marion *et al.*[79] suggest that along grain when stress levels are below a threshold value, liquid-phase sintered ceramics deform by a solution–precipitation mechanism without damage accumulation. Experimental data generated by Wiederhorn *et al.*[80] support the existence of this threshold for silicon nitride. Tsai and Raj[81] suggest methods of estimating values of a threshold stress for ceramics, and the above form of the kinetic equation could easily accommodate a threshold, that is

$$\dot{\eta} \begin{cases} = 0 & \sigma_0 \leqslant \sigma_{th} \\ = -B(\sigma_0/\eta)^n & \sigma_0 > \sigma_{th} \end{cases} \tag{63}$$

Dalgleish *et al.*[70] have presented experimental data that suggest the existence of a second threshold that delineates regions where subcritical crack growth and creep rupture failure mechanisms are operative. Chuang *et al.*[82] predict the value of this threshold stress by using principles of irreversible thermodynamics within the framework of several well-accepted models for crack growth. If this threshold (σ_{th}^*) exists, one can construct a composite reliability model such that

$$R \begin{cases} = R(\text{subcritical crack growth}) & \sigma_0 > \sigma_{th}^* \\ = R(\text{creep rupture}) & \sigma_{th} < \sigma_0 \leqslant \sigma_{th}^* \end{cases} \tag{64}$$

where R is the reliability of a component. Here, thresholds similar to those

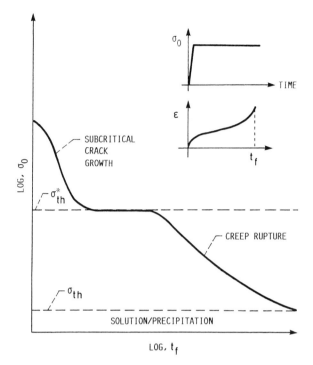

Fig. 11.9 Schematic plot of log of stress as a function of log of time to failure delineating threshold stresses, and outlining distinct regions where expected failure mechanisms would be operative.

outlined above for monolithic ceramics are recognized as a possibility for ceramic matrix composites. These thresholds would take on different values depending on the material orientation considered. However, a lack of quality experimental data leaves the authors unsure as to whether or not these thresholds are a universal phenomenon, and therefore ignoring the thresholds is expedient at this time.

It is postulated that during a creep rupture experiment σ_0 is abruptly applied and held at a fixed value (see the inset of Fig. 11.9). With $\eta = 1$ at $t = 0$, Eqn. (60) can be integrated to yield an expression for η as a function of time, stress, and model parameters, as follows:

$$\eta = [1 - B(\sigma_0)^n (n + 1) t]^{1/(n+1)} \tag{65}$$

An expression for the time to failure (t_f) can be obtained from Eqn. (65) by noting that $t = t_f$ when $\eta = 0$. Hence

$$t_f = 1/B(\sigma_0)^n (n + 1) \tag{66}$$

and the equation for η is simplified to

$$\eta = [1 - (t/t_f)]^{1/(n+1)} \tag{67}$$

This is consistent with the strong dependence of failure times on stress. However, the distribution of failure times for a given stress level σ_0 may be probabilistic or deterministic. Currently, the data are insufficient to postulate either case. Here, t_f is treated in a deterministic fashion noting that a probability distribution function for t_f could be introduced to the analysis in a manner similar to that suggested by Bolotin.[83]

To generate meaningful data, great care must be taken to determine the operative failure mechanism (i.e., subcritical crack growth or creep rupture). Dalgleish et al.[70] proposed using the Monkman–Grant constant to separate experimental rupture life data. However, the creep-damage tolerance parameter, defined by Leckie[84] as the total creep strain divided by the Monkman–Grant constant, may prove more suitable. After the data have been carefully screened, the model parameters n and B could be easily determined from creep rupture data. Taking the natural log of Eqn. (66) yields

$$\ln(t_f) + n\ln(\sigma_0) = -\ln[B(n+1)] \tag{68}$$

The value $1/n$ corresponds to the slope of $\ln(\sigma_0)$ plotted against $\ln(t_f)$, and B would be computed from the intercept.

11.5.4 A Reliability Theory Incorporating a Damage State Variable

Now consider that the uniaxial test specimen is fabricated from a ceramic material system with an inherent large scatter in strength. The variation in strength can be suitably characterized by the weakest link theory using Weibull's[85] statistical distribution function. Adopting Weibull's analysis, the reliability of a uniaxial specimen is given by the expression

$$R = \exp[-V(\sigma/\beta)^\alpha] \tag{69}$$

This assumes that the stress state is homogeneous and that the two-parameter Weibull distribution sufficiently characterizes the specimen in the failure probability range of interest. Taking σ equal to the net stress defined previously and, for simplicity, assuming a unit volume, yields the following expression for reliability

$$R = \exp[-(\sigma_0/\eta\beta)^\alpha] \tag{70}$$

Substituting for η by using Eqn. (65) gives

$$R = \exp\left\{-\left[\frac{\sigma_0}{[1 - B(\sigma_0)^n(n+1)t]^{1/(n+1)}\beta}\right]^\alpha\right\} \tag{71}$$

Alternatively, substituting for η by using Eqn. (67) yields

$$R = \exp\left\{-\left(\frac{\sigma_0}{\beta}\right)^\alpha\left[1 - \left(\frac{t}{t_f}\right)\right]^{-\alpha/(n+1)}\right\} \tag{72}$$

Here, it is clearly evident that in the limit as t approaches t_f, R approaches zero. Examples of reliability curves and their dependence upon time and model parameters are presented and discussed in Duffy and Gyekenyesi.[76]

Next, the hazard rate function is considered. By definition, the hazard rate (or mortality rate) is the instantaneous probability of failure of a component in the time interval $(t, t + \Delta t)$, given that the component has survived to time t. In more general terms, this function yields the failure rate normalized to the number of components left in the surviving population. This function can be expressed in terms of R or the probability of failure P_f as

$$h(t) = -\frac{dR}{dt}\left(\frac{1}{R}\right) = \frac{dP_f}{dt}\left(\frac{1}{1 - P_f}\right) \tag{73}$$

With Eqn. (72) used to define R, the hazard rate becomes

$$h(t) = \left(\frac{\alpha}{n+1}\right)\left(\frac{1}{t_f}\right)\left(\frac{\sigma_0}{\beta}\right)^{\alpha}\left[1 - \left(\frac{t}{t_f}\right)\right]^{-(\alpha+n+1)/(n+1)} \tag{74}$$

The hazard rate function can be utilized from a modeling standpoint in one of two ways. First, it can be used graphically as a goodness-of-fit test. If any of the underlying assumptions or distributions used to construct Eqn. (72) are invalid, one would obtain a poor correlation between model prediction of the hazard rate and experimental data. On the other hand, experimental data can be used to construct the functional form of the hazard rate and R can be determined from Eqn. (73). In a sense, this latter approach represents an oversimplified curve-fitting technique. Since it was assumed that the creep rupture failure mechanism can be modeled by continuum damage mechanics, the first approach is adopted. In this spirit, the hazard rate function would be used to assess the accuracy of the model in comparison to experimental results.

As a final note, the relative magnitude of the hazard rate is interpreted as follows:

(1) A decreasing hazard rate indicates component failure has been caused by defective processing.
(2) A constant hazard rate indicates failure is caused by random factors.
(3) An increasing hazard rate denotes wear-out of the component.

Here, negative values of α and n are physically absurd, hence

$$-(\alpha + n + 1)/(n + 1) < 0 \tag{75}$$

and Eqn. (74) yields an increasing hazard rate. This is compatible with the underlying assumption that creep rupture is the operative failure mechanism if creep rupture is recognized as a wear-out mechanism.

11.6 Future Directions

Ceramic material systems will play a significant role in future elevated temperature applications. To this end, there are a number of issues that must be addressed by the structural mechanics research community. We begin by pointing out that recent progress in processing ceramic material systems has not been matched by mechanical testing efforts. There is a definite need for experiments that support the development of reliability models. Initially, this effort should include experiments that test the fundamental concepts embedded in the framework of current stochastic models. As an example, probing experiments could be conducted along various biaxial load paths to establish level surfaces of reliability in a particular two-dimensional stress space (similar to probing yield surfaces in metals). One could then verify such concepts as the maximum stress response which is often assumed in the multiaxial reliability models proposed for these materials. After establishing a theoretical framework, characterization tests would then be conducted to provide the functional dependence of model parameters with respect to temperature and environment. Finally, data from structural tests that are multiaxial in nature (and possibly nonisothermal) would be used to challenge the predictive capabilities of models through comparison to prototypical response data. These tests involve inhomogeneous fields of stress, deformation, and temperature, and would include two-bar tests as well as plate and shell structures. Results from structural testing provide feedback for subsequent modification. *Ad hoc* models result in the absence of structured interaction between the experimentalist and the theoretician. The validity of these models is then forever open to question. Furthermore, we cannot overemphasize that this genre of testing supports the development of methods for designing components, and not the material. Currently, this effort is hampered by the quality and scarcity of data. Ceramic properties pertinent to structural design, which include stochastic parameters, vary with test method. The mechanics research community is beginning to realize this, and a consensus is beginning to form regarding the adoption of standards. However, the authors wish to underscore the fundamental need for experimental programs that are relevant to structural mechanics issues.

The authors note that the discussion in previous sections for the most part focused on time-independent analyses. It should be apparent to the reader that the utilization of ceramics as structural components in harsh service environments requires thoughtful consideration of reliability degradation due to time-dependent phenomena. Mechanisms such as subcritical crack growth, creep rupture, and stress corrosion must be dealt with. Computational strategies are needed that extend current methods of analysis involving subcritical crack growth and creep rupture, to multiaxial states of stress. Furthermore, the bulk of current literature dealing with stress corrosion highlights experimental observations, with very little attention given to failure analysis. Several authors have suggested dealing with this mechanism analytically through the use of chemical reaction rate theory. Indeed, this approach would be a good starting point, since it deals with the chemistry of failure at the microstructural level. In addition, if ceramic materials mimic ductile failure locally, cyclic fatigue will become a design issue. Under cyclic loads, the process zone advances as the crack tip extends and brittle fracture mechanics may need to be modified to account for pseudo-ductile fracture.[86] Hence, application of modified metallic fatigue analyses may be a distinct possibility. However, the development of current predictive methods for service life of ceramics is hampered by the lack of data (supporting the development of design protocols) on the behavior in various environments of interest.

Large strides have been made in understanding crack growth behavior in monolithic and whisker-toughened ceramics. However, one important aspect that has not been addressed in detail is the effect of rising *R*-curve behavior. Clearly, brittle materials need to be toughened, and this is often accomplished by creating a process zone around the crack tip. Within this zone, localized energy dissipation takes place which results in the development of increased damage tolerance through an increasing resistance to crack growth with crack extension. Under these conditions, fracture toughness becomes functionally dependent upon crack size. Failure of materials exhibiting rising *R*-curve behavior would not be dependent upon the initial distribution and orientation of flaw sizes, but on the rate at which resistance increases with crack growth. Several authors have discussed in theoretical terms the effect that *R*-curve behavior has on the stochastic parameters that are necessary for short-term failure as well as life prediction,[87] though again, very little data exist that correlate strength distribution to this behavior.

Current analytical research initiatives have focused on the analysis of first matrix cracking and ultimate strength (in the fiber direction) of an individual lamina. These analyses must be extended to multiple-ply failure in an angle-ply laminate. Failure in an individual ply causes local redistribution of the load to adjacent layers. In addition, delamination between laminae will relax the constraining effects among layers, which allows in-plane strains to vary in steps within a laminate. These effects require the development of rational load redistribution schemes. Failure analysis of woven ceramics must initially deal with quantifying damage induced by service loads in the absence of a discernible macrocrack. Furthermore, issues germane to component life, such as cyclic fatigue and creep behavior, must also be addressed analytically and experimentally.

Since ceramic composites, in general, and woven ceramic composites, in particular, have been identified by many government-supported technology demonstration projects (e.g., the EPM and HITEMP programs at NASA, the CFCC program at the DOE, and the IHPTET/NASP program sponsored by the Air Force) as high risk (but also high payoff) materials, industry has demonstrated a reluctance to invest the resources necessary to cover the research and development costs. For woven ceramic composites, current research efforts underway have been conducted at universities and in industry, but not usually in conjunction with one another. Funded research at universities is mainly focused on composite mechanics to design the material, and in some cases on weaving technology needed to produce preforms. Industry is developing the high temperature processing technologies, which require substantial investments to bring this type of fabrication equipment on line. It strikes the authors that to minimize CMC product development cycles, more integrated efforts are needed that would develop transition to practice technologies such as structural analysis, testing, and nondestructive evaluation. This also points to the need for more concurrent engineering. Analysts may specify certain optimized architectures for strength, stiffness and toughness, but processing of these tailored composites may not be possible. Hence, designing components will require compromises between material scientists and design engineers. However, research in most CMC processing, testing and modeling is still in its infancy, especially when one compares them to the progress made in polymer matrix composites.

In closing, we recognize that when failure is less sensitive to imperfections in the material, stochastic methods may not be absolutely essential. Currently, imperfections can be found in abundance in ceramic matrix composites since the high processing temperatures required to fabricate components generate flaws and residual stresses as components cool down during fabrication. These materials exhibit a sensitivity to interface conditions, reaction among constituents, residual stresses, fiber bunching, misalignment, and also sensitivity to complex fiber architectures. However, a synergy effect is obtained since an individual matrix crack or fiber break will not cause macroscopic failure due to the structural redundancy of the microstructure. Trends in design protocols are moving in the direction of probabilistic analyses (even for metals) and away from the simplistic safety factor approach. In this sense, brittle ceramics will serve as prototypical materials in the study and development of reliability models that will act as the basis of future design codes.

References

1. B. W. Rosen, "Analysis of Material Properties," in *Engineered Materials Handbook Vol. 1—Composites*, ed. C. A. Dostal, ASM International, Materials Park, OH, 1987, pp. 185–205.
2. R. L. McCullough, "Micro-Models for Composite Materials—Particulate and Discontinuous Fiber Composites," in *Micromechanical Materials Modeling—Vol. 2*, eds. J. M. Whitney and R. L. McCullough, Technomic, Lancaster, PA, 1990, pp. 49–90.
3. F. F. Lange, "The Interaction of a Crack Front with a Second-Phase Dispersion," *Phil. Mag.*, **22**, 983–992 (1970).
4. A. G. Evans, "The Strength of Brittle Materials Containing Second Phase Dispersions," *Phil. Mag.*, **26**, 1327–1344 (1972).
5. D. J. Green, "Fracture Toughness Predictions for Crack Bowing in Brittle Particulate Composites," *J. Am. Ceram. Soc.*, **66**, C4–C5 (1983).
6. K. T. Faber and A. G. Evans, "Crack Deflection Processes—I. Theory," *Acta Metall.*, **31**[4], 565–576 (1983).

7. P. Angelini and P. F. Becher, "In Situ Fracture of Silicon Carbide Whisker Reinforced Alumina," *Ann. Meet. Electron Microsc.*, **45**, 148–149 (1987).

8. P. F. Becher, C. H. Hsueh, and P. Angelini, "Toughening Behavior in Whisker Reinforced Ceramic Matrix Composites," *J. Am. Ceram. Soc.*, **71**, 1050–1061 (1988).

9. R. C. Wetherhold, "Fracture Energy for Short Brittle Fiber/Brittle Matrix Composites with Three-Dimensional Fiber Orientation," *J. Eng. Gas Turbines Power*, **112**[4], 502–506 (1990).

10. S. B. Batdorf and J. G. Crose, "A Statistical Theory for the Fracture of Brittle Structures Subjected to Nonuniform Polyaxial Stresses," *J. Appl. Mech.*, **41**[2], 459–464 (1974).

11. T. T. Shih, "An Evaluation of the Probabilistic Approach to Brittle Design," *Engng. Fract. Mech.*, **13**, 257–271 (1980).

12. S. F. Duffy, J. M. Manderscheid, and J. L. Palko, "Analysis of Whisker-Toughened Ceramic Components—A Design Engineer's Viewpoint," *Ceram. Bull.*, **68**[12], 2078–2083 (1989).

13. S. F. Duffy and S. M. Arnold, "Noninteractive Macroscopic Statistical Failure Theory for Whisker Reinforced Ceramic Composites," *J. Comp. Mater.*, **24**[3], 293–308 (1990).

14. B. N. Cassenti, "Probabilistic Static Failure of Composite Material," *AIAA J.*, **22**, 103–110 (1984).

15. M. Reiner, "A Mathematical Theory of Dilatancy," *Amer. J. Math.*, **67**, 350–362 (1945).

16. R. S. Rivlin and G. F. Smith, "Orthogonal Integrity Basis for N Symmetric Matrices," in *Contributions to Mechanics*, ed. D. Abir, Pergamon Press, Oxford, 1969, pp. 121–141.

17. A. J. M. Spencer, "Theory of Invariants," in *Continuum Physics—Volume I*, ed. A. C. Eringen, Academic Press, New York, NY, 1971, pp. 239–255.

18. A. J. M. Spencer, "Constitutive Theory for Strongly Anisotropic Solids," in *Continuum Theory of the Mechanics of Fibre-Reinforced Composites*, ed. A. J. M. Spencer, Springer-Verlag, New York, NY, 1984, pp. 1–32.

19. R. C. Wetherhold, "Statistics of Fracture of Composite Materials," Ph.D. Dissertation, University of Delaware, Newark, DE, 1983.

20. S. F. Duffy and J. M. Manderscheid, "Noninteractive Macroscopic Reliability Model for Ceramic Matrix Composites with Orthotropic Material Symmetry," *J. Eng. Gas Turbines Power*, **112**[4], 507–511 (1990).

21. M. Miki, Y. Murotsu, T. Tanaka, and S. Shao, "Reliability of the Strength of Unidirectional Fibrous Composites," *AIAA J.*, **28**[11], 1980–1986 (1991).

22. P. de Roo and B. Paluch, "Application of a Multiaxial Probabilistic Failure Criterion to a Unidirectional Composite," in *Developments in the Science and Technology of Composite Materials*, eds. A. R. Bunsell, P. Lamicq, and A. Massiah, Association Européene des Materiaux Composites, Bordeaux, France, 1985, pp. 328–334.

23. K. J. Willam and E. P. Warnke, "Constitutive Models for the Triaxial Behavior of Concrete," *Int. Assoc. Bridge Struct. Eng. Proc.*, **19**, 1–30 (1975).

24. N. S. Ottosen, "A Failure Criterion for Concrete," *J. Engng. Mech.*, **103**[EM4], 527–535 (1977).

25. J. L. Palko, "An Interactive Reliability Model for Whisker-Toughened Ceramics," Masters Thesis, Cleveland State University, OH, 1992.

26. S. F. Duffy, A. Chulya, and J. P. Gyekenyesi, "Structural Design Methodologies for Ceramic Based Material Systems," in *Ceramics and Ceramic–Matrix Composites*, ed. S. R. Levine, ASME, New York, 1992, pp. 265–285.

27. C. T. Sun and S. E. Yamada, "Strength Distribution of a Unidirectional Fiber Composite," *J. Comp. Mater.*, **12**[2], 169–176 (1978).

28. K. Prewo and J. J. Brennan, "High-Strength Silicon Carbide Fiber-Reinforced Glass-Matrix Composites," *J. Mater. Sci.*, **15**[2], 463–468 (1980).
29. K. Prewo and J. J. Brennan, "Silicon Carbide-Yarn-Reinforced Glass-Matrix Composites," *J. Mater. Sci.*, **17**[4], 1201–1206 (1982).
30. J. J. Brennan and K. Prewo, "Silicon Carbide-Fiber-Reinforced Glass-Ceramic Matrix Composites Exhibiting High Strength and Toughness," *J. Mater. Sci.*, **17**[8], 2371–2383 (1982).
31. J. J. Blass and M. B. Ruggles, "Design Methodology Needs for Fiber-Reinforced Ceramic Heat Exchangers," ORNL/TM-11012, Oak Ridge National Laboratory, TN, 1990.
32. K. Friedrich, *Application of Fracture Mechanics to Composite Materials*, Elsevier Science Publishing Co., New York, NY, 1989.
33. B. Budiansky, J. W. Hutchinson, and A. G. Evans, "Matrix Fracture in Fiber-Reinforced Ceramics," *J. Mech. Phys. Solids*, **34**[2], 167–189 (1986).
34. D. B. Marshall, B. N. Cox, and A. G. Evans, "The Mechanics of Matrix Cracking in Brittle Matrix Fiber Composites," *Acta Metall.*, **33**[11], 2013–2021 (1985).
35. M. D. Thouless and A. G. Evans, "Effects of Pull-Out on the Mechanical Properties of Ceramic-Matrix Composites," *Acta Metall.*, **36**[3], 317–522 (1988).
36. M. Sutcu, "Weibull Statistics Applied to Fiber Failure in Ceramic Composites and Work of Fracture," *Acta Metall.*, **37**[2], 651–661 (1989).
37. H. E. Daniels, "The Statistical Theory of the Strength of Bundles of Threads," *Proc. Roy. Soc. Lond., Ser. A*, **183**[995], 405–435 (1945).
38. B. D. Coleman, "On the Strength of Classical Fibers and Fiber Bundles," *J. Mech. Phys. Solids*, **7**[1], 66–70 (1958).
39. B. W. Rosen, "Tensile Failure of Fibrous Composites," *AIAA J.*, **2**[11], 1985–1991 (1964).
40. C. Zweben, "Tensile Failure of Fiber Composites," *AIAA J.*, **6**[12], 2325–2331 (1968).
41. D. G. Harlow and S. L. Phoenix, "The Chain of Bundles Probability Model for the Strength of Fibrous Materials—1. Analysis and Conjectures," *J. Comp. Mater.*, **12**[2], 195–214 (1978).
42. S. B. Batdorf, "Tensile Strength of Unidirectionally Reinforced Composites—I," *J. Reinf. Plastics Comp.*, **1**[2], 153–164 (1982).
43. D. J. Thomas and R. C. Wetherhold, "Reliability Analysis of Continuous Fiber Composite Laminates," NASA CR-185265, National Aeronautics and Space Administration, Cleveland, OH, 1990.
44. H. F. Wu, "Statistical Analysis of Tensile Strength of ARALL Laminates," *J. Comp. Mater.*, **23**[10], 1065–1080 (1989).
45. T. G. Hu and D. B. Goetschel, "The Application of the Weibull Strength Theory to Advanced Composite Materials," in *Tomorrow's Materials: Today*, Vol. 1 (Proceedings of the 34th International SAMPE Symposium and Exhibition), eds. G. A. Zakrewski *et al.*, SAMPE, Covina, CA, 1989, pp. 585–599.
46. S. J. McKernan, "Anisotropic Tensile Probabilistic Failure Criterion for Composites," Masters Thesis, Naval Postgraduate School, 1990.
47. S. B. Batdorf and H. L. Heinisch, "Weakest Link Theory Reformulation for Arbitrary Fracture Criterion," *J. Am. Ceram. Soc.*, **61**, 355–358 (1978).
48. S. F. Duffy, J. L. Palko, and J. P. Gyekenyesi, "Structural Reliability Analysis of Laminated CMC Components," ASME 91-GT-210, Presented at the 36th International Gas Turbine and Aeroengine Congress and Exposition, Orlando, FL, 1991.

49. J. Margetson and N. R. Cooper, "Brittle Material Design using Three Parameter Weibull Distributions," in *Probabilistic Methods in the Mechanics of Solids and Structures*, eds. S. Eggwertz and N. C. Lind, Springer-Verlag, Berlin, Germany, 1984, pp. 253–262.

50. S. F. Duffy, L. M. Powers, and A. Starlinger, "Reliability Analysis of Structural Ceramic Components using a Three-Parameter Weibull Distribution," Paper presented at the 37th International Gas Turbine and Aeroengine Congress and Exposition, Cologne, Germany, 1992.

51. A. S. Fareed, M. J. Koczak, F. Ko, and G. Layden, "Fracture of SiC/LAS Ceramic Composites," in *Advances in Ceramics, Vol. 22: Fractography of Glasses and Ceramics*, eds. V. D. Frechette and J. R. Varner, American Ceramic Society, Westerville, OH, 1988, pp. 261–278.

52. F. Ko, M. Koczak, and G. Layden, "Structural Toughening of Glass Matrix Composites by 3-D Fiber Architecture," *Ceram. Eng. Sci. Proc.*, **8**, 822–831 (1987).

53. F. K. Ko, "Preform Fiber Architecture for Ceramic-Matrix Composites," *Ceram. Bull.*, **68**[2], 401–414 (1989).

54. T. W. Chou and J. M. Yang, "Structure-Performance Maps of Polymeric, Metal, and Ceramic Matrix Composites," *Metall. Trans.*, **17A**, 1547–1559 (1986).

55. T. W. Chou and T. Ishikawa, "Analysis and Modeling of Two-Dimensional Fabric Composites," in *Textile Structural Composites*, eds. T. W. Chou and F. K. Ko, Elsevier, New York, NY, 1989, pp. 209–264.

56. C. L. Ma, J. M. Yang, and T. W. Chou, "Elastic Stiffness of Three-Dimensional Braided Textile Structural Composites," in *Composite Materials: Testing and Design*, American Society for Testing and Materials, Philadelphia, PA, 1986, pp. 404–421.

57. J. M. Yang, C. L. Ma, and T. W. Chou, "Fiber Inclination Model of Three-Dimensional Textile Structural Composites," *J. Comp. Mater.*, **20**, 472–484 (1986).

58. F. K. Ko, C. M. Pastore, C. Lei, and D. W. Whyte, "A Fabric Geometry Model for 3-D Braid Reinforced Composites," Paper presented at Competitive Advances in Metals and Metals Processes, 1st International SAMPE Metals and Metals Processing Conference, Cherry Hill, NJ, August 1987.

59. R. M. Crane and E. T. Camponeschi, Jr., "Experimental and Analytical Characterization of Multidimensionally Braided Graphite/Epoxy Composites," *Exp. Mech.*, **26**[3], 259–266 (1986).

60. Z. G. Wang, C. Laird, Z. Hashin, W. Rosen, and C. F. Yen, "Mechanical Behaviour of a Cross-Weave Ceramic Matrix Composite," *J. Mater. Sci.*, **26**, 4751–4758 (1991).

61. A. Chulya, J. Z. Gyekenyesi, and J. P. Gyekenyesi, "Damage Mechanisms in Three-Dimensional Woven Ceramic Matrix Composites Under Tensile and Flexural Loading at Room and Elevated Temperatures," in *HITEMP Review 1991*, eds. H. Gray and C. Ginty, NASA CP-10082, National Aeronautics and Space Administration, Cleveland, OH, 1991, pp. 53.3–53.11.

62. C-F. Yen, Z. Hashin, C. Laird, B. W. Rosen, and Z. Wang, "Micromechanical Evaluation of Ceramic Matrix Composites," Final technical report submitted to Air Force Office of Scientific Research, Washington, DC, 1991.

63. N. K. Naik, P. S. Shembekar, and M. V. Hosur, "Failure Behavior of Woven Fabric Composites," *J. Comp. Tech. Res.*, **13**[1], 107–116 (1991).

64. J. M. Yang, "Modeling and Characterization of Two-Dimensional and Three-Dimensional Textile Structural Composites," Ph.D. Dissertation, University of Delaware, Newark, DE, 1986.

65. K. P. Walker, E. H. Jordan, and A. D. Freed, "Nonlinear Mesomechanics of Composites with Periodic Microstructure: First Report," NASA TM-102081, National Aeronautics and Space Administration, Cleveland, OH, 1989.

66. S. M. Wiederhorn and E. R. Fuller, "Structural Reliability of Ceramic Materials," *Mater. Sci. Eng.*, **71**, 169–186 (1985).

67. A. A. Griffith, "The Phenomena of Rupture and Flow in Solids," *Phil. Trans. Roy. Soc. Lond. Ser. A*, **221**, 163–198 (1921).

68. G. D. Quinn, "Delayed Failure of a Commercial Vitreous Bonded Alumina," *J. Mater. Sci.*, **22**, 2309–2318 (1987).

69. J. R. Ritter, Jr., S. M. Wiederhorn, N. J. Tighe, and E. R. Fuller, Jr., "Application of Fracture Mechanics in Assuring Against Fatigue Failure of Ceramic Components," NBSIR 80-2047, National Bureau of Standards, Washington, DC, 1980.

70. B. J. Dalgleish, E. B. Slamovich, and A. G. Evans, "Duality in the Creep Rupture of a Polycrystalline Alumina," *J. Am. Ceram. Soc.*, **68**, 575–581 (1985).

71. L. M. Kachanov, "Time of the Rupture Process Under Creep Conditions," *Izv. Akad. Nauk. SSR, Otd Tekh. Nauk*, **8**, 26 (1958).

72. D. Krajcinovic, "Distributed Damage Theory of Beams in Pure Bending," *J. Appl. Mech.*, **46**, 592–596 (1979).

73. B. N. Cassenti, "Time Dependent Probabilistic Failure of Coated Components," *AIAA J.*, **29**[1], 127–134 (1991).

74. J. W. Ju, "Damage Mechanics of Composite Materials: Constitutive Modeling and Computational Algorithms," Final technical report prepared for the Air Force Office of Scientific Research, Washington, DC, 1991.

75. F. A. Leckie, "Advances in Creep Mechanics," in *Creep in Structures*, eds. A. R. S. Ponter and D. R. Hayhurst, Springer-Verlag, Berlin, Germany, 1981, pp. 13–47.

76. S. F. Duffy and J. P. Gyekenyesi, "Time Dependent Reliability Model Incorporating Continuum Damage Mechanics for High-Temperature Ceramics," NASA TM-102046, National Aeronautics and Space Administration, Cleveland, OH, 1989.

77. D. Krajcinovic, "Damage Mechanics," *Mech. Mat.*, **8**, 117–197 (1989).

78. C. L. Chow and F. Yang, "On One-Parameter Description of Damage State for Brittle Material," *Engng. Frac. Mech.*, **40**[2], 335–343 (1991).

79. J. E. Marion, A. G. Evans, M. D. Drory, and D. R. Clarke, "High Temperature Failure Initiation in Liquid Phase Sintered Materials," *Acta Metall.*, **31**[10], 1445–1457 (1983).

80. S. M. Wiederhorn, D. E. Roberts, T.-J. Chuang, and L. Chuck, "Damage-Enhanced Creep in a Siliconized Silicon Carbide: Phenomenology," *J. Am. Ceram. Soc.*, **68**, 602–608 (1988).

81. R. L. Tsai and R. Raj, "Creep Fracture in Ceramics Containing Small Amounts of a Liquid Phase," *Acta Metall.*, **30**, 1043–1058 (1982).

82. T.-J. Chuang, R. E. Tressler, and E. J. Minford, "On the Static Fatigue Limit at Elevated Temperatures," *Mater. Sci. Eng.*, **82**, 187–195 (1986).

83. V. V. Bolotin, "Verification and Estimation of Stochastic Models of Fracture," in *Proceedings of the 1st USA-USSR Symposium on Fracture of Composite Materials*, Sijthoff and Noordhoff International Publishers, Alphen aan den Rijn, The Netherlands, 1979, pp. 45–54.

84. F. A. Leckie, "The Micro- and Macromechanics of Creep Rupture," *Engng. Frac. Mech.*, **25**, 505–521 (1986).

85. W. Weibull, "A Statistical Theory for the Strength of Materials," *Ingenoirs Vetenskaps Akadamien Handlinger No. 151*, The Royal Swedish Institute for Engineering Research, Stockholm, 1939.

86. R. O. Ritchie, "Mechanisms of Fatigue Crack Propagation in Metals, Ceramics and Composites: Role of Crack Tip Shielding," *Mater. Sci. Eng.*, **A103**, 15–28 (1988).

87. K. Kendall, N. McN. Alford, and J. D. Birchall, "Weibull Modulus of Toughened Ceramics," in *Advanced Structural Ceramics*, Vol. 78 (Materials Research Symposia Proceedings), eds. P. F. Becher, M. V. Swain, and S. Somiya, MRS, Pittsburgh, PA, 1987, pp. 189–197.

Elevated Temperature Mechanical Testing

CHAPTER 12

Critical Issues in Elevated Temperature Testing of Ceramic Matrix Composites

D. C. Cranmer

12.1 Introduction

Ceramic composites are of interest to a variety of industries including those in aerospace, automotive, medical, chemical, defense, electric power generation and conservation, electronics, and environmental technologies. Applications include armor, bearings, combustors, cutting tools, automotive engine components such as valves, tappets, cam rollers and followers, exhaust port liners, seals and wear parts, heat exchangers, rocket nozzles, electronic packaging, bioceramics, filters and membranes. Not all of these applications require knowledge and understanding of high temperature properties, but many do. Proper design of systems and components incorporating ceramic composites will require reliable information about a number of properties and reliable databases from which to extract design information. Temperatures of use for "high temperature components" (those exposed to elevated temperatures during use) can range from below room temperature to temperatures in excess of 3000°C during service. The advantage of composites is that they can fail noncatastrophically, if designed and manufactured properly.

In this chapter, composites will include particulate-, whisker-, and fiber-reinforced ceramics, although the bulk of the coverage will be focused on fiber-reinforced materials. As pointed out earlier in this book, fiber-reinforced ceramic matrix composites have been in existence since the 1970s, beginning with the British research on carbon fiber-reinforced glass.[1-3] Since that time, significant advances in materials selection (fibers, matrices, and interphases), composite design, processing, and properties have been made. The term "ceramic" refers to glass, glass-ceramic, and ceramic materials. Also included is information on carbon/carbon composites because of their unique features. The purpose of this chapter is to review and discuss elevated temperature mechanical property measurement techniques for ceramic composites and, where necessary, their constituents such as the fiber reinforcement.

Testing of materials provides a number of different types of measurement which serve different purposes: to obtain design properties, to derive properties which are not easily measured or observed, and to provide the information to verify models and theories of behavior. There are, thus, three different communities to satisfy: system and component designers, developers and producers of materials, and theoreticians and analysts. Acquisition of the differing types of data and materials required by each of these communities leads to very different testing requirements. Testing for design data is usually labor intensive because of the range of tests required and the need for repetitions to ensure reliability. Testing to understand the mechanisms by which the composite fails, and testing to verify theories and models can also be very demanding due to the range of material structures, constitutive properties (parameters), and compositions which may be required.

There are several significant issues to be considered in designing a test procedure and methodology, including specimen design, specimen gripping arrangements, load-train design, heating method, temperature measurement, and strain measurement. All the issues must be addressed properly to obtain a valid test result. Failure to correctly implement them in the test will result in property data which cannot be readily used. Each of these test design issues will be discussed in detail below, but, first, a brief review of the state of ceramic composites is in order.

In examining the properties and testing methodologies for ceramic and carbon/carbon composites, it is helpful to remember that there are two distinct fiber/matrix systems: those in which the modulus of the fibers is greater than that of the matrix, and those in which the modulus of the matrix is greater than the fibers. These different fiber/matrix systems result in different composite behaviors, and may require changes in the test configuration. In the first instance, where $E_f > E_m$, which is typical of polymer and metal matrix composites, strength increases via load transfer requires strong bonding between the fiber and matrix in order to transfer the load from the low modulus matrix to the high modulus fibers. The second case, where $E_f < E_m$, is more typical of ceramic composites, and weaker fiber/matrix bonding is needed in order to achieve the desired graceful failure.

12.1.1 Historical Perspective

Testing of ceramic composites has been around since the earliest fabrication of these materials. For particulate- and whisker-reinforced composites, testing methods which are suitable for monolithic ceramics are generally used. These methods include three- and four-point flexure, uniaxial tension and compression, and many others. For fiber-reinforced ceramic composites, flexural testing was also used initially. However, as was recognized in the polymer composites area, flexural testing alone could not provide the type of property data needed to design with, and qualify, composites for use in application, nor for scientific understanding of their structure/properties/performance relations. Testing of polymer and metal matrix composites has a (comparatively) long history compared to ceramic matrix composites. Let us begin by reviewing the properties of some ceramic composites, followed by some observations on carbon/carbon composites, and a brief review of the state of testing for fiber-reinforced polymer and metal matrix composites.

12.1.1.1 Ceramic Matrix Composites

The first fiber-reinforced ceramic matrix composites were actually the carbon fiber-reinforced borosilicate glass matrix materials developed at AERE, Harwell, U.K., in the early 1970s.[1–3] Evaluation of these, and other early composite materials, at room and elevated temperatures (up to about 1000°C) was performed using three-point flexure tests.[4,5] A typical stress–strain curve showing the graceful failure of the composite is shown in Fig. 12.1. This type of stress–strain curve exhibits three regimes showing the different failure mechanisms controlling the behavior of the composite. The first regime is a linear region, corresponding to elastic deformation of both fibers and matrix. This response continues until the formation of the first matrix microcracks, shown as a slight bend in the curve. Until this point, the composite is generally regarded as undamaged. The second region corresponds to the nonlinear part of stress–strain curve and is associated with multiple cracking of the matrix. As the composite undergoes further loading, the fibers break and are pulled out of the matrix, corresponding to the third region of the curve. In this final region, the load-carrying capability of the composite is compromised and it fails.

Since the development of the first glass matrix composites, a large number of other systems have been developed incorporating a wide variety of reinforcements, interfaces, and matrices. As the proposed temperature of application has increased, the need to identify compatible high temperature systems (fibers, matrices, and interfaces) has also increased. Candidates for these high temperature applications include SiC/SiC, SiC/Si$_3$N$_4$, SiC/MoSi$_2$, and Al$_2$O$_3$ with other oxide matrices. It has also been recognized that flexural testing alone is inadequate to completely characterize fiber-reinforced composites and identify failure mechanisms in a meaningful manner.[6] Consequently, tensile, compressive, shear, and other tests have been developed both to enhance our understanding of ceramic matrix composites, and to provide property data suitable for materials development and design.

12.1.1.2 Carbon/Carbon Composites

For the purposes of this chapter, carbon/carbon (C/C) composites consist of carbon fiber preforms which are surrounded by a carbon matrix. Development of these materials began in the late 1950s, and was continued under a variety of U.S. Air Force, NASA, and other federal government programs.[7] They currently find use in structures such as the space shuttle nose cone and leading edges, and brake materials, and have the potential to be used in many

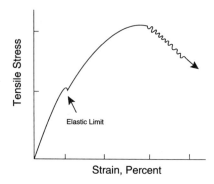

Fig. 12.1 Typical stress–strain curve for a ceramic composite showing graceful failure.

other applications.[8] Carbon/carbon composites satisfy a number of operating requirements in application including reproducible strength to 1650°C, excellent stiffness, low coefficient of thermal expansion, tolerance to impact damage, and sufficient oxidation resistance for short-term service. These composites can be manufactured in a wide variety of shapes and sizes, and have mechanical properties which can be extensively tailored due to the anisotropic nature of the fiber reinforcement. The properties of the underlying carbon fiber can be widely varied depending on the fiber precursor (pitch or polyacrylonitrile (PAN)) and the heat treatment the fiber receives. In general, increased heat treatment time and temperature, yield higher fiber graphite content, and result in higher modulus and strength.

The property which makes them unique from a testing perspective is that they do not lose strength with temperature (assuming no oxidation), unlike other materials, as shown in Fig. 12.2. This feature makes test fixture, specimen, and furnace design more difficult than is the case for materials where the highest temperature region is also expected to be the weakest strength region. Testing of carbon/carbon composites is also made more difficult because of the need for environmental controls to prevent oxidation for longer service times. Under proper conditions, these materials can be used at temperatures in excess of 3000°C. Testing methods, therefore, must be more extensively developed for this temperature range.

12.1.1.3 Polymer and Metal Matrix Composites Testing Methods

Numerous standardized test methods have been developed for evaluating the mechanical behavior of polymer and metal matrix composites. Mechanical properties of interest are ultimate strength, yield strength, elastic moduli, strain-to-failure, fatigue life, and creep. There are ASTM standard test methods[9] for composite tensile strength and modulus (D-638 and D-3039), composite compressive strength and modulus (D-695 and D3410), flexural

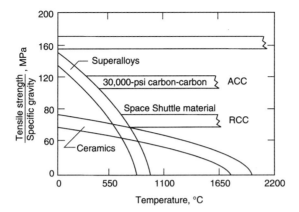

Fig. 12.2 Specific strength–temperature curve for a carbon/carbon composite showing stability of strength with temperature.

strength and modulus (D-790), composite shear strength and modulus (D-2344, D-3518, D-4255, and E-143), and composite tensile fatigue (D-3479). In addition, there are commonly used tests for shear strength and modulus such as the Arcan[10] and Iosipescu[11] methods. There are also ASTM standard tests for fiber density (D-276, D-792, D1505 and D3800), fiber tensile properties (D-3379, D-3544, and D-4018), dielectric constant and dissipation factor (D-150), and dielectric strength (D-149). The above list of tests is only a partial one and many lessons learned from these tests can be applied to ceramic matrix composites. However, the unique nature of continuous fiber ceramic composites (CFCCs) (a brittle matrix with a stiffness similar to that of the reinforcing brittle fibers) prevents the direct adoption of these test methods. As an example, matrix cracking in CFCCs must be considered, rather than the viscoelastic behavior of polymer matrix composites. This difference will result in differences in the details of items like load trains and gripping systems.

There is significant government, university, and industrial testing experience with nonstandard methods for determining fracture of ceramics and ceramic composites. For fiber-reinforced composites, this experience has been obtained predominantly at room temperature, although significant strides have been made in the past five years in elevated temperature testing.

Fiber alignment and fiber volume fraction affect the tests which can be performed, and may dictate other significant test design parameters such as specimen size and shape, and gripping scheme. The large number of fiber architectures (combination of alignments and volume fractions) which can be used has been demonstrated by experience with carbon fibers.[12] Ultimately, similar kinds of structures are expected to be found in ceramic composites as well, and each will lead to specific testing design issues. What then are the basic principles upon which elevated temperature testing of composites should be carried out?

12.2 Testing Design Issues

Evaluating composite performance requires test methods for mechanical properties of the constituent materials (fibers, tows, filaments, and matrices) as well as the composite materials themselves. The types and quantities of tests to be performed, and the selection of testing parameters, depends on the information desired. For material development, the tests may be much simpler and less numerous than those that would be chosen for design qualification, but may encompass a greater range of test parameters than would be expected in service. In the case of materials development, it is the trends in the data and the mechanisms by which failures occur which are most important, and it is crucial to examine the extremes of behavior. For component design and qualification, it is more important to know the reliability and reproducibility of the material under conditions which resemble the expected service conditions.

The complexity of the failure process in composites is very different from that of monolithic ceramics. As an example, in a four-point bend test of a SiC/glass-ceramic composite to determine strength, failure occurs either in compression when the thickness of the test specimen is 2 mm, or in shear when the thickness exceeds 3.5 mm, but never in tension. The calculation of a tensile strength from this configuration is therefore invalid. For a SiC monolith, tested in the same way, compression and shear are not normal modes of failure, and a tensile strength can be calculated from the strength result. A second example is in the fracture toughness test using a notched configuration. Failure in the composite occurs by delamination parallel to the fibers and normal to the notch. Calculation of the fracture toughness using a fracture mechanics relation based on single crack extension from the notch, as would be done for monolithics, is not valid. These examples reinforce the concept that the test configuration must be selected carefully to ensure that meaningful measurements are being made.

As an example of these differences, let us examine flexure testing. One common perception is that flexure testing is not an appropriate method for acquiring design data for fiber-reinforced composites, and may not be appropriate for materials development purposes. The major reason for this is that the results of the flexure test can not be interpreted in a consistent manner, due to the complex stress state and the possibility of compressive buckling and/or shear failure mechanisms in every test specimen. The tensile test, on the other hand, permits measurement of the stress–strain behavior in a manner that can readily be used by component and systems designers, and can allow us to investigate the more fundamental mechanics and mechanisms of damage accumulation and failure. In service, though, the component may experience flexural stresses, and it is thus important that bending tests be included in the qualification test suite. Additional advantages of the flexural test are that the specimen geometry is simple and inexpensive to produce, and that the test itself is simple to perform. Given those advantages, flexure testing has a proper place in evaluation of ceramic composites, but should not be used to the exclusion of all other test methods.

A specific example of this can be found in the evaluation of composites for heat exchanger applications.[13] In this case, the heat exchanger design calls for a tubular construction which will be pressurized. Under these conditions, a flexural stress will be present in service, and consequently, a C-ring test specimen configuration provides a reasonable way to examine the properties of the composite.

As noted earlier, the testing design issues which must be addressed include specimen design, specimen gripping arrangements, load-train design, specimen preparation, heating method, and temperature and strain measurements. Furnace design and selection of heating and insulation materials also need to be considered in designing test equipment. An additional complication is the length of time required for each test or test series. Each of these issues will be considered for tensile testing, then several testing configurations for tensile testing of fiber-reinforced composites will be illustrated. Similar considerations will apply to the testing required for other mechanical properties.

12.2.1 Specimen Design

For tensile testing, the specimen geometry can be one of a variety of types (straight-sided or dogbone, cylindrical or flat), and can be tabbed or untabbed. The types of specimens which have been utilized are shown in Fig. 12.3. The specific dimensions will depend on the test geometry, test conditions, and material availability. The straight-sided specimen type is the only one suitable for use with a unidirectionally reinforced composite. The others require the use of a cross-ply material to prevent failures in the grips. The cylindrical geometry is suitable for use with particulate- and whisker-reinforced composites, and, depending on the need for machining, may be appropriate for fiber-reinforced composites. This raises another issue associated with specimen design, namely the method of specimen preparation. Specimen preparation is a critical part of specimen design. Machining can have a significant impact on the behavior of the material, particularly where fibers are not parallel to the tensile axis, or when a transformation toughening additive such as ZrO_2 is used. In the case of fiber-reinforced materials, improper grinding can lead to broken, debonded, and/or improperly aligned fibers during the test, resulting in inaccurate or incorrect numbers for the tensile strength. For transformation-toughened materials, improper grinding or machining can cause transformation to occur in a surface layer of the material, which may lead to reduced strength when the composite is tested.

Design of specific test specimens can take advantage of finite element methods to ensure stresses are applied where needed, not in tab sections or transition regions. The utility of this has been demonstrated in design of a flat

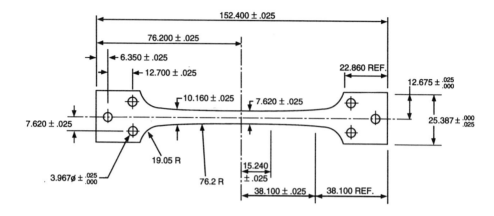

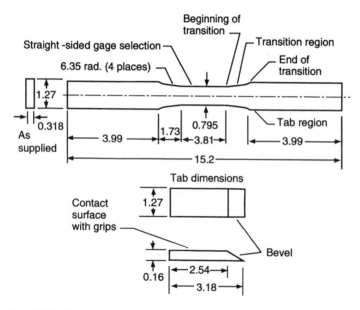

Fig. 12.3 Schematics of types of specimens used for tensile testing.

tensile specimen.[14,15] What cannot yet be accounted for in finite element analysis are the effects of fiber architectures. Changes in fiber architecture can result in very localized changes in stresses and strains which are not readily incorporated into the analysis.

Unidirectionally reinforced materials are more of a problem to test than multidirectional reinforcements because of the anisotropy of the properties. This anisotropy leads to differences in moduli, strain-to-failure, and other properties of the composite which allow failure of the material to occur in a mode other than that which the researcher may wish to investigate.

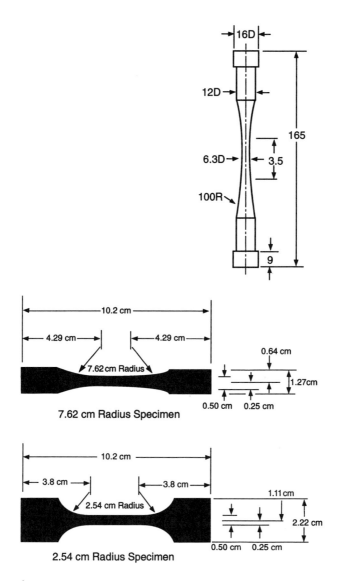

Fig. 12.3–*contd.*

12.2.2 Gripping Schemes

The gripping arrangements for tensile testing depend on several factors, and can result in grips which are either rigid or flexible, and may be cold, warm, or hot. One system which has been shown to work utilizes a rigid grip system as shown in Fig. 12.4. These rigid grips can be of either a clamping or

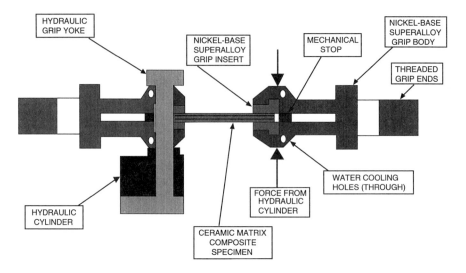

Fig. 12.4 Schematic of rigid grip system for composite testing developed for use by the U.S. Air Force Wright Laboratory Materials Directorate.

wedge type. The advantages of these grips are as follows. First, once the grips have been aligned, they tend to stay aligned during the test; therefore, only periodic checks of the alignment are necessary. Second, there is no "preload" required to remove any slack in the load train. Third, the test specimens can be consistently and correctly mounted using simple, standard tools such as calipers and depth gauges. Fourth, the rigidity of the entire load train provides a stable platform for the addition of required auxiliary instrumentation such as extensometer cables and thermocouples. Finally, the degree of bending introduced into the specimen remains nearly constant throughout the complete load range. The disadvantage is that while bending is nearly constant, it is frequently greater than that obtained with a flexible gripping system.

There are some commercially available gripping systems which may, or may not, meet the needs of composites testing. They include self-aligning grips capable of operating at temperatures up to 1600°C and fully articulated fixtures fabricated from silicon carbide which are good at temperatures up to 1500°C.

12.2.3 Load-Train Design

The load train itself can be of either a rigid or flexible type. When using a rigid grip system, it is best to use a rigid load train. If other gripping systems are used, a flexible load train should be used. Figure 12.5 shows a flexible load train consisting of pull rods, precision adjusters, chains, and universal joints.

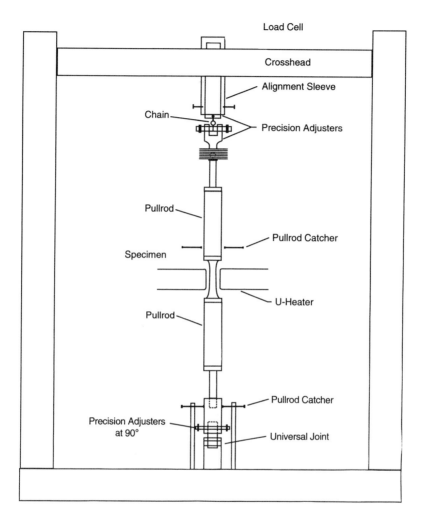

Fig. 12.5 Schematic of flexible load train developed at Southern Research Institute for high temperature composite testing.

12.2.4 *Heating Methods*

Heating of the specimen can also be accomplished in several ways including resistance heating elements and direct radiation methods such as quartz lamps. Each of these methods will provide the necessary temperature, but care must be exercised to ensure that thermal gradients and time variations are minimized. The absolute temperature must be measured to provide the necessary control of the test temperature. Temperature measurement, discussed in more detail later, can be accomplished by a number of methods. The method of choice depends on the need for temperature measurement versus temperature control. It is a good idea to use more than one method, either during setup or during the test, to provide checks for consistency.

12.2.5 Furnace Materials Requirements for Testing

Testing up to 1000°C can be accomplished using conventional furnace elements, insulation materials, and other components with gripping and load-train designs implemented with superalloys or other suitable materials as necessary. Above 1000°C, changes in gripping materials, and furnace designs and materials, are required. Different insulation materials must be used, different heating elements may be required, and more care must be taken to minimize gradients and ensure that the materials are compatible with the environment in the furnace. Heating elements for use in air environments typically are silicon carbide for temperatures up to around 1600°C, while molybdenum disilicide elements can be used in air up to 1800°C. For other environments, the maximum use temperatures can change significantly. Where long-term testing is required, as may happen in creep testing, the maximum use temperatures provided by the manufacturer may need to be reduced somewhat to ensure that the elements or insulation survive for the entire test duration. There are few things worse than getting two-thirds of the way through a test, only to have the heating elements fail, especially when the expected test duration is in the thousands of hours as happens in some creep testing.

Testing conditions (stresses, temperatures, environments) will ultimately have to match up with service conditions. If flexural stresses are expected in service, then a flexural test is one that should be performed as part of the test suite. If thermal or stress cycles are expected, then simulation of these conditions should also be included in the evaluation of the material.

12.2.6 Measurement Techniques

12.2.6.1 Temperature
Temperature can be measured using any of a number of techniques but the choice of a particular technique depends on the temperature to be measured and the environment in which it will be measured. The major methods are thermocouples and pyrometers. Each of these methods is straightforward, but the experimenter must pay attention to detail by performing periodic checks of the equipment. Thermocouples are simple, easy to use, and can be customized for range of use and length of service (i.e., made bigger or smaller). They are also relatively inexpensive. They also have some disadvantages, including an upper temperature limit of about 1650°C, and are subject to contamination and vaporization which can affect their accuracy, stability, and length of service. Pyrometers are non-contacting instruments and are therefore not subject to contamination, have no upper temperature limit (at least as far as composite testing is concerned), and can be used for a variety

of environments which would adversely affect thermocouples. They are subject to changes in emissivity of the material being tested, and do not work well at only mildly elevated temperatures. There are ASTM standard practices and guides for calibrating temperature measuring equipment (E207, E220, E230, E452, E633, and E988) which specify the procedures and time intervals at which calibrations should occur.

12.2.6.2 Strain

Strain can be measured using a variety of tools including strain gauges, laser dimension sensors, clip-on extensometers, and quartz-rod (or other high temperature rod) extensometers. The choice of strain measurement tool depends on the specimen geometry, and the test geometry and conditions, as well as the degree of precision desired. For room temperature measurements, glued-on strain gauges provide an adequate method for determining strain, although there is still an art to their attachment to the specimen, and the selection of appropriate adhesives with which to attach them. At mildly elevated temperatures, however, glued-on strain gauges will not work. Time can also be a significant factor in deciding on the precision required in the strain measurement. In practice, for strength measurements, because of the low strains-to-failure and high elastic moduli of these materials, clip-on extensometers and laser dimension sensors do not have the required precision, and attention should be focused on high temperature rod extensometers. The contact point on the specimen for the rod extensometers is a critical area. There are three methods which can be used to ensure that the specimen deformation is transferred to the contact rods: pin holes, grooves, and grooved paint. All three have been shown to work satisfactorily under some conditions, but the choice will depend on the specimen (whether it is coated) and the temperature of the test (the paint can creep).

Commercially available extensometers and their thermal limits are: water-cooled extensometers up to 500°C, quartz-rod extensometers up to 1000°C, and capacitance extensometers up to 1600°C. The latter extensometers can have either SiC or Al_2O_3 knife edges, and are therefore suited to different environments and test materials.

Since introduction of bending moment is a concern, strain should be measured on more than one side of the specimen. This is necessitated by the fact that the specimens are heterogeneous and may be bent or warped as a result of the manufacturing or fabrication process, and cannot be corrected by machining or grinding due to other testing constraints.

12.2.7 Control Variables

What are the variables which must be controlled in order to perform a useful test? The main ones are temperature, heating rate, environment, load, and loading rate. Control of the temperature is one of the keys to a valid test. Failure to adequately monitor and control the temperature may result in significant overheating, which, in turn, can change the chemistry and properties of the material. Failure to control the heating rate, whether too fast or too slow, can also result in undesirable changes. Too fast a rate may lead to overheating with the same result as the failure to adequately control the temperature. Too slow a rate may also expose the material to conditions under which significant changes in chemistry and/or structure can occur. This is particularly true if there is a glass at the interface. From nucleation and crystallization theories and experiments,[16–20] we know that crystallization of a glass can occur under many different conditions. Too long an exposure to temperatures below a critical temperature can allow the glass to form nuclei and/or grow crystals. Changes such as this can have a significant impact on properties such as creep rate[21] and strain-to-failure.

The test environment selected depends on the information desired, and should reflect the service environment. For example, if design data for a combustor application is the purpose of the test series, then a test environment reflective of that application should be used. The difficulties which can be encountered are illustrated in an analysis of the combustor environment for a high speed civil transport.[22] In this type of application, the combustor environment contains N_2, O_2, CO/CO_2, and H_2O, and may contain significant amounts of contaminants such as sodium sulfate (Na_2SO_4). Testing in complicated environments like those of the combustor require well-sealed systems to ensure that the correct environment is present in the testing chamber and for protection of those conducting the tests, mass flow controllers, and may require the use of scrubbers or other gas cleaning technology to meet environmental regulations.

12.2.8 Tensile Test Configurations

A number of viable test systems are currently being used to evaluate the tensile properties of ceramic composites. One such configuration (Fig. 12.3) uses a dogbone-shaped specimen containing three holes in the tab end. The three holes used to pin the specimen in the grips are the means by which the load is applied to the gauge section. The remainder of the load train consists of a load cell, pull rods, universal joints and chain to control alignment and prevent introduction of bending moments into the system, as was shown in Fig. 12.5. Additional details of this configuration and its use can be found in Ref. 15.

Another tensile test configuration, which works well, utilizes the rigid grip system.[23] This configuration is shown in Fig. 12.6. As noted above, rigid grips have several advantages. The specimen geometry can be straight-sided or dogbone-shaped, and can be tabbed or untabbed. Specific specimen dimensions depend on the test geometry, test conditions, and material availability. The straight-sided specimen is the only one suitable for use with a uni-

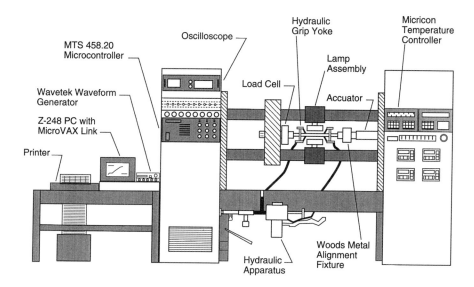

Fig. 12.6 Schematic of loading system used with rigid grip system shown in Fig. 12.4.

directionally reinforced composite with this configuration. Other specimens must use cross-plied material.

A third configuration[24] applies the load to the shoulders of the tab section. This configuration uses dogbone-shaped specimens, and requires very tight tolerances on both the specimen and the loading fixture. Depending on the testing requirements, this configuration can be operated with hot, warm, or cold grips, and can be used to measure strength, creep, or tension–tension fatigue. It has been used for tensile creep and fatigue[25,26] for ceramic matrix composites. The apparatus is similar to that used for the shoulder-loaded tensile strength test. The grips can be either ceramic (e.g., SiC) for high temperature use, or superalloy for intermediate temperatures. A schematic of the gripping arrangement is shown in Fig. 12.7. Front-to-back alignment in the fixture is controlled by precision-machined inserts which are fastened to the body of the fixture after the specimen is in place. This configuration has been used to evaluate tensile strength, tensile creep, and tension–tension fatigue of fiber-reinforced ceramic composites.

An additional test geometry for measuring tensile creep, fatigue, and strength has been developed at Oak Ridge National Laboratory, TN.[27] The specimen has a cylindrical geometry (see Fig. 12.3) with a machined button-head end for gripping. The shape is simple but requires very precise tolerances which are achievable with computer numerical controlled (CNC) grinding equipment. The technique has several advantages: symmetrical loading on the specimen, a simple gripping arrangement, relatively uniform load transfer which minimizes the bending moment on the specimen, and the ability to

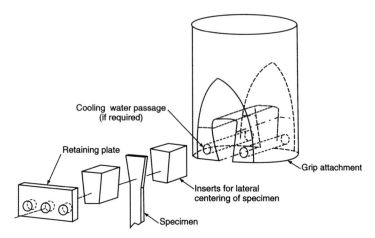

Fig. 12.7 Schematic of edge-loaded gripping system developed at the University of Michigan for high temperature composite testing.

achieve large volume to surface area ratios. There are, however, at least two difficulties with this specimen: the cost associated with machining of the material, and the large amount of material required for each specimen. This test can be used to obtain design data, but, given the machining requirements, it may not be appropriate for fiber-reinforced materials because of fiber breakage and misalignment. It is, however, suited for the testing of particulate- and whisker-reinforced materials.

12.2.9 Other Properties and Test Methods

12.2.9.1 Flexural Testing

Where flexural testing is appropriate, a four-point bend fixture of the type shown in Fig. 12.8 can be used, as can a C-ring configuration. The bend fixture shown can be implemented using steel, superalloy, silicon carbide, or other materials suitable to the temperature, loads, and compliances desired for the test. Although there is an ASTM standard for flexural testing of monolithic ceramic specimens, for fiber-reinforced composites, the length to depth ratio should be at least 20:1 in order to minimize unwanted failure mechanisms such as compressive buckling or shear. Such test methods should only be used where flexure will be a primary stress mode in service. It is otherwise considered unsuitable, even for materials development purposes, because of the difficulties in unequivocally determining the mode of failure (tension, compression, buckling, shear, etc.).

12.2.9.2 Compression

Compressive strength can also be an important property needed to characterize composite behavior, and qualify materials for design, since many

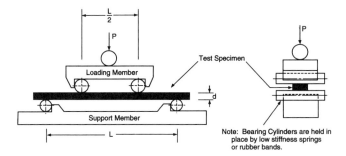

Fig. 12.8 Schematic of four-point bend test fixture as per ASTM C1161.

applications for these materials require the composite to be exposed to compressive stresses. A typical test configuration is shown in Fig. 12.9. The number of such tests applicable to fiber-reinforced ceramic composites is limited, and needs to be expanded. An alternative configuration relies on the Hopkinson pressure bar.[28,29] A typical arrangement for this geometry is shown in Fig. 12.10.

12.2.9.3 Creep

High temperature deformation is also an extremely important property of ceramic composites for many applications. The techniques required for measuring creep are very similar to testing techniques for determining tensile strength and flexural strength. These techniques, as well as creep properties and mechanisms, are covered in Chapter 4.

12.2.9.4 Impact

Many applications of ceramic composites will expose the material to impact damage. Here again, the amount of work which has been done in this arena is limited, and needs to be increased. Impact tests can be performed using drop towers or Charpy impact testing apparatus. Such measurements have been done at room temperature[30] and the apparatus can be adapted for use at elevated temperatures.

12.2.9.5 Fracture Toughness

The final property to be discussed is that of fracture toughness. While this property is not used in design analysis, it can provide an indication of the material's ability to withstand damage. A number of techniques can be used to determine fracture toughness in simple and complex systems including compact tension,[31] double cantilever beam,[32] double torsion,[33] chevron notch,[34] double cleavage drilled compression (DCDC),[35] and indentation techniques.[36,37] The direct crack measurement indentation techniques are generally not suitable for fiber-reinforced composites because the size of the indentation and the resulting cracks are much smaller than the spacing between the fibers, thus

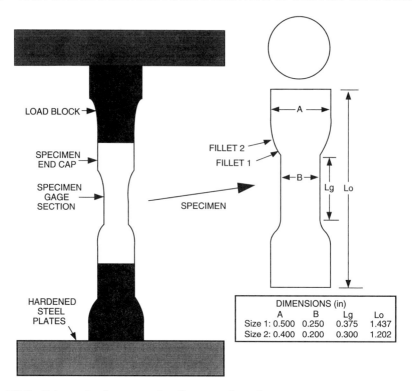

Fig. 12.9 Schematic of compression fixture and specimen.

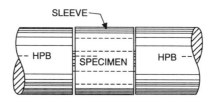

Fig. 12.10 Schematic of Hopkinson pressure bar for compression testing.

leading to crack propagation behavior that is not representative of the actual composite behavior, which includes fiber bridging and other mechanisms that substantially enhance the toughness of the composite over the monolithic material. The indentation strength method is more representative of what occurs in the composite, as has been shown for a SiC/Si_3N_4 composite,[38] because it interrogates all of the available bridging mechanisms, but the observed behavior is indentation load, and, therefore, crack size dependent. Indentations in the size range expected from surface flaws (approximately 50–100 μm), such as those expected from machining damage, may actually be the most realistic in simulating the failure mechanisms in composites. In the case of an Al_2O_3/SiC composite, catastrophic failure of the composite occurs when the crack becomes unstable before encountering the fibers; non-catastrophic failure occurs when a crack of the same length is bridged by a fiber. In addition to this dynamic effect of crack location, there is a clear distinction in behavior between long and short cracks, and frequently small crack behavior dominates the fracture process. The indentation strength method also lends itself readily to elevated temperature measurements. A more detailed discussion of the composite toughening mechanisms and models for them can be found in Chapter 2.

Changes in crack resistance behavior as the material configuration (grain structure, fiber alignment, and volume fraction) changes can readily be seen using the compact tension specimen.[39] The advantage of this technique is that the force–displacement (p–u) relation can be experimentally determined and extrapolated to small crack lengths. The p–u function can be regarded as an engineering material property, and can be determined from crack profile measurements. If p–u is known, then fracture under monotonic loading can be predicted for any shape and load change. The force–displacement function cannot be determined from indentation techniques because of unknown residual stress fields and unknown geometric factors associated with the indentations. The specimen geometry for a compact tension test is shown in Fig. 12.11. Whisker bridging in composites has been observed using this technique[40] and corresponding measurements have been made of the changes in toughness as the crack approaches and passes the whisker. A more extensive review of the application of this technique to alumina-based ceramics has recently been published,[41] which discusses the short crack regime. A disadvantage of this technique is that it is difficult to work in the short-crack regime because of the need to notch and pop-in a crack.

Using model materials consisting of small numbers of fibers in a representative matrix, the compact tension technique can also be used to determine the crack opening displacement and toughness of fiber-reinforced composites.[42] The information which can be collected using this configuration is shown in Fig. 12.12. From such measurements, the crack closure stresses can be determined, and, in turn, used to better understand, design, and incorporate toughening mechanisms in ceramic matrix composites.

The compact tension technique could be readily adapted to elevated temperature use with a little creative engineering, but may not be as informative, or as robust, as the indentation strength methods.

12.2.10 *Experimental Design Requirements*

The number and types of tests that are performed depend very much on the end use for the data, and, in part, whether or not any testing of the material has ever been performed. The tests required to develop trends in behavior suitable for guiding materials development are different from the test suites needed to qualify a material for design. One question which needs to be

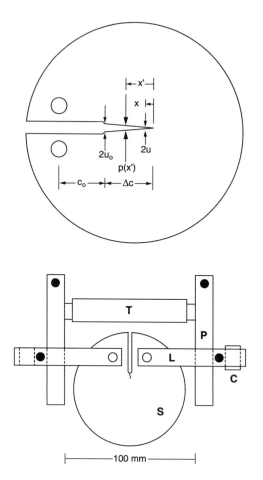

Fig. 12.11 Schematic of a compact tension specimen and *in situ* observation apparatus.

answered is: How can we, as simply and easily as possible, relate processing, microstructure and properties, to performance for a given application? Materials formulation and processing is still more art than science. Generally, there are a large number of processing variables for any given materials system, and the interactions among these variables are either complicated or unknown. Many processes are unrepeatable, which means that they are sensitive to unknown and/or uncontrollable factors. Formulation and processing also takes a long time because usually only one variable is chosen and changed at a time until the best result is obtained, then the process is repeated with other chosen variables. The result is very slow progress on developing new and improved materials.

However, a more scientific approach is available that can greatly speed

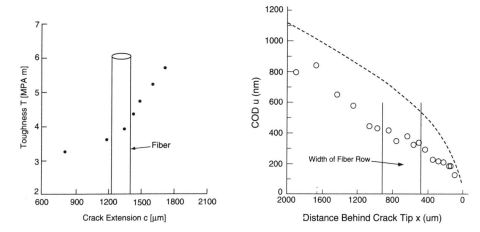

Fig. 12.12 Results of compact tension specimen showing increase in toughness as crack goes past fiber, and crack opening displacement from which closure stresses can be determined.

up the process of development. Since most materials know-how comes from physical experiments, the key lies in an approach which lets us obtain as much information as possible from as few experiments as possible. Such an approach is available by using statistically planned experiments. In general, statistical design and analysis can be used to determine the limits of a material's behavior and minimize sample and testing requirements. An example of the use of statistically planned experimentation and analysis can be found for SiC whisker-reinforced aluminum oxide.[43] In the particular case of the SiC_w/Al_2O_3 composite, a one-half fractional factorial experiment was set up to examine the processing–microstructure–properties–performance relationships for a potential cutting tool application. The term "one-half fractional factorial" means that of all the possible combinations of processing conditions, 50% of them were actually used in the experiment. The details of how to set up such an experiment and analyze the resulting data can be found in more detailed treatises on statistical analysis. A set of processing variables (whisker type, whisker volume fraction, hot-pressing temperature, hot-pressing pressure, and hot-pressing time), and their upper and lower limits were selected. Primary characteristics of the microstructure (which result directly from the processing) were selected for examination: homogeneity (uniformity of distribution of the whiskers), density, and matrix grain size; as were secondary characteristics: mechanical properties—flexural strength, reliability (Weibull parameter), fracture toughness, and hardness; and tertiary characteristics (reflective of the performance of the materials in service). Use of these statistically based experiments allowed establishment of a suitable window within which to consider optimizing processing conditions to achieve the desired performance of the material. This technique also minimizes the number of tests which must be carried out in order to obtain working relationships between processing conditions and in-service performance. Such schemes are also applicable to testing of fiber-reinforced composites.

12.3 Industrial Requirements for Composites Testing

In an industrial setting, as well as some others, there are a number of additional considerations, including the need to have low cost, simple, reliable test procedures for determining composite properties. Another problem related to the fiber–matrix interface is that of oxygen embrittlement. An example of this behavior is the reaction of carbon interfaces with environmental oxygen after first matrix microcracking has occurred. A number of currently available ceramic composites rely on a carbon or carbon-rich interface to obtain the debonding and sliding behavior necessary to achieve the damage tolerance behavior shown in Fig. 12.1. As shown there, a change in slope representative of first matrix microcracking is evident. At this point, the composite interface and fibers become exposed to the environment and can chemically react with it, thus, adversely changing the properties of the interface and/or fiber. In particular, the carbon layer can form CO or CO_2, which is then removed from the interface. The underlying fiber (such as SiC) can then react to form an oxide which bonds tightly to the matrix, thus destroying the desired fiber sliding behavior. A simple test has been developed for materials development purposes whereby the behavior of a representative composite can be characterized, and which forms a complement to indentation tests used to characterize the fiber–matrix interface (described later).

The test method mentioned above uses a simple rectangular parallelepiped (similar to a flexure bar), which is mounted in hydraulic wedge grips (as described in the tensile strength tests above) with metal foil shims. The mounted specimen is loaded until matrix microcracking is observed (precracking), then removed from the grips, exposed to the desired environmental conditions (temperature, and atmosphere). The post-exposure specimen is reloaded in the grips and pulled to failure.

This method has been applied to materials such as a calcium aluminosilicate (CAS) reinforced with SiC fiber, and having a carbon interface, with the results showing a significant reduction in strength after exposure. When the interface is changed to a micaceous material, the post-exposure strength is much nearer to the pre-exposure strength, indicating an enhancement in properties due to the interface. The test has some limits, at present, especially in that it has not been adapted for elevated temperature use, but for purposes of material development and the indication of trends in behavior, it is an acceptable test. Another limitation exists in the determination of the point of initial microcracking of the matrix, which at present is detected using the deviation from linearity of the stress–strain curve. A more sophisticated approach would use acoustic emission, real-time microscopy, or other more

Table 12.1 Industrial test suite requirements for design qualification

Test	−54°C	21°C	ET1	ET2	ET3	ET4	ET5
0° Tension	5	5	5	5	5	5	5
90° Tension	5	5	5	5	5	5	5
Cross-ply tension	5	5	5	5	5	5	5
0° Compression	5	5	5	5	5	5	5
90° Compression	5	5	5	5	5	5	5
Cross-ply compression	5	5	5	5	5	5	5
1–2 Shear	5	5	5	5	5	5	5
1–3 Shear	5	5	5	5	5	5	5
2–3 Shear	5	5	5	5	5	5	5
Creep-rupture					15	15	15
$R = -1$ Fatigue		15				15	15

sensitive techniques to determine the onset of microcracking. In addition, the chemistry of the interface should be determined, both before and after exposure in order to properly guide the changes required for improvements. It is important to note the need both to raise the elastic limit of the composite (the point at which microcracking occurs) and to control the interface chemistry and properties. This latter need can be addressed by the use of interface coatings and/or matrix dopants.[44]

For an end user such as an aircraft engine manufacturer, the kind of information obtained on ceramic matrix composites can be dictated by systems considerations. These considerations have resulted in the collection of wisdom related to design practice such as MIL-HDBK17.[45] This handbook essentially mandates the requirements for a structure which must be met before the structure can be placed in service. The result is a series of design allowables. A typical requirement is that the strength of the material shall be such that it is sufficient to sustain the ultimate load without failure. A similar requirement for damage tolerance is that the structure shall be capable of performing its function in the presence of expected manufacturing- and service-induced damage.

Additional factors which must be taken into account are environmental effects (thermal as well as chemical), effects of defects, statistical variability of the material, long-term behavior, and cyclic versus static loading effects. Assessment of these effects requires the end user to conduct a large series of tests using multiple specimens. A typical series will examine a unidirectional material in tension in the 0, 90, and cross-ply directions; 0, 90, and cross-ply in compression; and 1–2, 1–3, and 2–3 shear at different temperatures ranging from −54°C to the expected service temperature; creep rupture at temperatures up to the expected service temperature; and fatigue at room and elevated temperature. This series of tests, shown in Table 12.1, may require over 400 specimens.

The generic test method requirements to meet the series needs are (1) that the test be appropriate for the materials of interest, (2) that it be uniform and repeatable, and have a stable hot section, (3) that the strain be measured accurately and precisely, (4) that the test method be easy and repeatable, (5) that it be cost effective and have a reasonable turnaround time, and (6) that it efficiently use the available material. In attempting to meet these generic requirements, experience with composites testing has resulted in the use of ASTM D-638 and D-3039 tensile tests in addition to those described earlier and Cortest; ASTM D-695 and D-3410 for compression; ASTM D-3518 and D-3518(C) as well as an asymmetric four-point bend (AFPB) test (a variation of the Iosipescu test mentioned earlier) and saddle geometries for in-plane shear; and ASTM D-3846 and D-2344 as well as the AFPB for interlaminar shear. The Cortest for tension relies on shoulder loading of a dogbone-type specimen, as contrasted to the tab-pinned geometries of the other tensile tests.

Industrial experience has also led to identification of a number of future testing needs to obtain basic properties and general test requirements. For basic properties, the desired information includes tension, compression, interlaminar shear modulus, long cycle fatigue, and Modes I and II interlaminar fracture toughness. The general test requirements include the need for interlaboratory standardization and vendor qualification, and notched tension/compression/long cycle fatigue test, a biaxial test method, a pin bearing test, characterization of joints, and residual properties as a result of impact, erosion, and wear.

12.4 Other Elevated Temperature Testing Issues

Testing of the composites themselves, or portions of them, are only one part of elevated temperature testing. Other areas requiring attention include failure analysis, damage accumulation, nondestructive evaluation, microstructural evaluation, and information needed to validate predictive models. Chapter 4 refers to modeling efforts which require the collection of data on the constituents of the composites, i.e., the starting fibers, the matrices, and the fiber–matrix interfacial materials, to make predictions for properties of real composites. To use the models described previously, one must also know about the properties of the individual components. These issues are addressed in this section.

12.4.1 Nondestructive Evaluation

Nondestructive evaluation (NDE) of composites will be an important area of future development. NDE refers to the use of inspection techniques which can characterize a material or component without damaging it during the inspection process. In general, the types of testing described above are destructive in nature. Currently, there are a number of techniques which can be used to evaluate the integrity of ceramic composites, in both pre- and post-test configurations[46–48] but the ability to nondestructively interrogate composites during high temperature testing is nonexistent, although some techniques may be adaptable to limited temperature exposures. Techniques of some utility for pre- and post-test evaluation include simple techniques such as optical inspection for obvious delaminations, cracks, spallations, and regions of pitting; X-radiography; and ultrasonic inspection. Optical inspection may be aided by use of dye penetrants. More complicated techniques include computed tomography and scanning ultrasonics.

Generally, for a first cut, the "flaws" observed will be in the order of 100–500 μm. If discovered in whisker- or particulate-reinforced composites, such flaws would indicate a part which is not suitable for further testing, although the presence of these flaws should be recorded to provide feedback which may be useful in processing. For fiber-reinforced composites, it is not clear that flaws of this size are detrimental. For smaller "flaws," i.e., those less than about 100 μm, more sophisticated techniques such as microfocus X-radiography and acoustic microscopy must be used, but the utility and reliability of them for routine examination is questionable, primarily because of the time required for a complete inspection. The questions arise because of the time required to inspect a specimen, and the inability to detect some types of flaws, e.g., agglomerates, tight cracks, and large grains, all of which may act as flaw origins. More information can be found in the review references cited above as well as some more recent ones.[49,50]

In principle, systems, components, and materials can be designed to make inspection easier, although there is a clear trade-off between function and inspection. Three immediate points to note with respect to NDE are (1) all materials contain flaws, (2) flaws in a material do not necessarily render it unfit for its intended purpose, and (3) detectability of a flaw generally increases with its size. A set of design principles primarily related to metals has been developed[46] but some clearly do not apply to composites. Additional work is required in this area to determine changes in the design principles which are better suited to composites, and in some cases, development of techniques for measuring properties which are not, at present, readily obtainable.

One of the key issues for both ceramics and ceramic composites is the lack of suitable standards and standard reference materials.[51] In principle, this issue can be rectified by development of materials containing known types and numbers of flaws. In practice, it is difficult because of our lack of knowledge about the numbers and types of flaws which are important. Techniques suitable for some monolithic ceramics have been developed which incorporate known internal and surface-connected defects. Using these types of specimens, our knowledge of aspects of NDE related to probability of detection of different kinds of flaws, and the procedures which must be followed to optimize detection, will be increased.

Which of these techniques are most likely, in my estimation, to be applicable to ceramic matrix composites? Ultrasonics or acoustic emission evaluation techniques are adaptable for high temperatures if high temperature coupling materials can be used. The other techniques do not appear to be immediately applicable, even to monolithic ceramics. Laser holography has been shown to be useful in determining displacements and deflections in turbine airfoils.[28] For experimental laboratory setups, the use of such equipment is relatively direct but the most likely drawback for in-service conditions is the size and placement of lasers and detectors compared to the available space and design. Some creative engineering will be required here in order to utilize these kinds of techniques.

12.4.2 Damage Accumulation

Creation and accumulation of damage will control and likely degrade the elevated temperature behavior of composites. As pointed out in Chapter 4, at elevated temperatures, damage mechanisms may change or be influenced by additional phenomena. Changes may take the form of blunting of cracks, resulting in localized stress reductions, or oxidation phenomena, resulting in changes in local chemistry and subsequent changes in properties. Our state of understanding of how damage accumulates is primitive at best, and could benefit from an increased emphasis.

12.4.3 Failure Analysis

Failure analysis will be crucial to an understanding of what happened during testing. The source of the critical flaw or damage accumulation needs to be understood, as does the propagation path. Microstructural characterization will also be an important component of the evaluation process. Changes in grain size and orientation in the fiber may be as important as changes in the matrix and may be expected to have an impact on properties such as fracture, fatigue, and creep. The significance in whisker-reinforced composites has been shown by Hockey et al.[52] For example, in creep studies of SiC_w/Si_3N_4, it was found that the initial transient was dominated by devitrification of a glassy interfacial phase, and that cavitation at the SiC_w–Si_3N_4 interface enhances the creep rate, thus reducing the composite's lifetime. Compared to an unreinforced Si_3N_4, there was no increase in creep resistance as a result of introducing the whiskers. This unexpected result is explained by cavitation which occurs at the SiC_w/Si_3N_4 interface, and may possibly be altered by changing the initial glassy phase composition or content, or by appropriate surface treatment of the whiskers to minimize cavitation. Examples of how failure mechanisms change with temperature can be found in Ref. 53.

12.4.4 Constituent Testing

A clear area where there is a need for testing is that of the composite components, namely, the starting fibers, the matrix material, and the fiber–matrix interface. Some test methods for determining the interface properties are discussed in the previous section, others will be described later in this section. For determining the properties of matrix materials, recourse can be had to a large number of techniques developed over several decades. Several comprehensive reviews of these techniques have been published in the past 10–15 years,[54,55] and the topic will not be discussed further here.

12.4.4.1 Single Fiber Testing

Determination of the properties of the starting fibers is a difficult task, particularly at elevated temperatures. Among the factors to be considered are testing of a single filament versus a multifilament yarn, test system compliance, gauge length, strain rate, grip material and pressure, strain measurement technique, and fiber diameter measurement. Each of these factors can significantly affect the measured result. As an example, when considering the measurement of a single fiber, it matters whether the fiber was indeed single (as would be some of the large diameter silicon carbide monofilaments), or removed from a larger tow (as would be the case if testing were done on a small diameter fiber, multifilament tow). For those removed from a multifilament tow, the fiber selection process itself may be expected to bias the results toward a higher strength, as fibers that break during removal from the tow would be discarded, not quantitatively tested. This will affect not only the average strength, but any measures of strength distribution such as standard deviation and the Weibull modulus. Reliable test techniques for measuring the strength of the tow itself also need to be developed.

Room temperature testing can be accomplished fairly readily using standard tensile tests such as ASTM D-3379, D-3544, and D-4018. Elevated temperature testing provides a host of new considerations including furnace design, vertical versus horizontal testing, hot versus cold grips, grip materials, and temperature measurement and uniformity. The furnace design must consider potential chimney effects and the resultant difficulties associated with temperature stability and thermal gradients. An additional consideration is the effect of time at temperature, and the length of time required to thermally equilibrate the fiber prior to testing. The small diameter fibers typically are composed of small grains, which can grow rapidly when exposed to elevated temperatures. If it takes too long to perform the test, the material tested may no longer be representative of the initial fiber.

Strain of small diameter fibers can be determined at room temperature using noncontact laser methods.[56] While the technique yields good results at room temperature, it is not clear that it is readily adaptable to higher temperatures without significant modifications in coupling fluid, and mounting methods and materials.

Creep of fibers also occurs much more rapidly than happens in bulk materials, in part because of the much smaller grain sizes present in the fibers. Creep and creep rupture test techniques for composites are described elsewhere in this volume (Chapter 4: Weiderhorn and Fuller) but performing such tests on small diameter fibers requires somewhat different apparatus.[57] To date, only a small number of fiber compositions have been characterized.[58]

12.4.4.2 Fiber–Matrix Interfacial Properties Testing

The strength of the fiber–matrix interface is one of the key parameters responsible for the stress–strain behavior and damage tolerance of ceramic composites. Two different types of tests are available to measure the fiber–matrix interfacial properties in fiber-reinforced ceramic composites. The first is based on an indentation technique to either push the individual fiber into or through the matrix. The second test method relies on pulling a single fiber out of a matrix. These methods have been compared[59] to one another for a glass matrix material, and yield similar results.

The indentation tests were performed using an instrumented indenter, allowing for independent determinations of force and displacement during the complete loading and unloading cycle. Indentation tests use a minimal amount of material and can be performed on samples containing either large monofilaments or small diameter multifilament tows but provide information on only $\tau_{friction}$ (push-in), or $\tau_{friction}$ or τ_{debond} (push-out), depending on indenter geometry and material characteristics. The push-out test can be performed at slower loading rates and with a different indenter geometry, thus allowing separation of the debonding strength from the interfacial friction stress in the (force-squared)-displacement curve. Preparation of push-out samples is more difficult than for push-in samples but the analysis is simpler and the results appear to be more reproducible. An additional potential advantage of the indentation method is that it may be adaptable for use as a quality assurance tool, since it can be used on small pieces of the as-fabricated composite. This quality assurance application may not be realized until a clearer relationship is established between the debond strength/frictional shear stress and the macroscopic properties of the composite, such as strength.

12.5 Summary

A great deal of progress has been made in the past 20 years since ceramic composites were first reported. There are now viable tensile test methods for evaluating ceramic composites. For particulate- and whisker-reinforced materials, the appropriate test techniques are generally the same as those used for monolithic ceramics. For continuous fiber-reinforced composites, however, more thought should be given to type and quality of information desired. Depending on the needs, specialized tensile test techniques may need to be used, either alone or in conjunction with flexure or other techniques. There is a need for different tests for materials development, and component design and qualification. Test parameters including temperature, temperature and stress gradients, and environments need to be carefully controlled. Specimens must be properly designed and prepared for testing. There is also a need to develop standard tests and report information in order to facilitate comparison of data from one laboratory to another. (This activity has just begun with the formation of ASTM Subcommittee C28.07 on Ceramic Matrix Composites.) Toward that end, an interlaboratory comparison of continuous fiber-reinforced composites is needed to discern testing reproducibility and limitations, as well as material behavior and its variability.

References

1. R. A. J. Sambell, D. H. Bowen, and D. C. Phillips, "Carbon Fibre Composites with Ceramic and Glass Matrices, Part 1: Discontinuous Fibers," *J. Mater. Sci.*, 7, 663–675 (1972).
2. R. A. J. Sambell, A. Briggs, D. H. Bowen, and D. C. Phillips, "Carbon Fibre Composites with Ceramic and Glass Matrices, Part 2: Continuous Fibers," *J. Mater. Sci.*, 7, 676–681 (1972).
3. D. C. Phillips, R. A. J. Sambell, and D. H. Bowen, "The Mechanical Properties of Carbon Fibre Reinforced Pyrex Glass," *J. Mater. Sci.*, 7, 1454–1464 (1972).
4. K. M. Prewo and J. J. Brennan, "High-Strength Silicon Carbide Fibre-Reinforced Glass-Matrix Composites," *J. Mater. Sci.*, 15, 463–468 (1980).
5. K. M. Prewo and J. J. Brennan, "Silicon Carbide Yarn Reinforced Glass Matrix Composites," *J. Mater. Sci.*, 17, 1201–1206 (1982).
6. D. C. Larsen, S. L. Stuchly, and J. W. Adams, "Evaluation of Ceramics and Ceramic Composites for Turbine Engine Applications", AFWAL-TR-88-4202, Final Report, Wright-Patterson AFB, OH, December 1988.
7. John D. Buckley, "Carbon-Carbon Overview," in *Carbon-Carbon Materials and Composites*, NASA Reference Publication 1254, National Aeronautics and Space Administration, Washington, DC, February 1992, pp. 1–17.
8. Louis Rubin, "Applications of Carbon-Carbon," in *Carbon-Carbon Materials and Composites*, NASA Reference Publication 1254, National Aeronautics and Space Administration, Washington, DC, February 1992, pp. 267–281.
9. For ASTM standards, refer to the appropriate Annual Book of ASTM Standards, ASTM, Philadelphia, PA.
10. M. Arcan, Z. Hashin, and A. Voloshin, "A Method to Produce Uniform Plane-Stress States with Applications to Fiber-Reinforced Materials," *Exp. Mech.*, 18[4], 141–146 (1978).
11. D. E. Walrath and D. F. Adams, "Verification and Application of the Iosipescu Shear Test Method," NASA-CR-174346, National Aeronautics and Space Administration, Washington, DC, June 1984.
12. F. K. Ko, "Textile Preforms for Carbon-Carbon Composites," in *Carbon-Carbon Materials and Composites*, NASA Reference Publication 1254, National Aeronautics and Space Administration, Washington, DC, February 1992, pp. 71–104.
13. V. Parthasarathy, B. Harkins, W. Beyermann, J. Keiser, W. Elliot, Jr., and M. Ferber, "Evaluation of SiC/SiC Composites for Heat Exchanger Applications," *Cer. Eng. Sci. Proc.*, 13[7–8], 503–519 (1992).
14. D. W. Worthem, "Flat Tensile Specimen Design for Advanced Composites,"

NASA Contractor Report 185261, National Aeronautics and Space Administration, Washington, DC, November 1990.

15. H. S. Starrett, "A Test Method for Tensile Testing Coated Carbon-Carbon and Ceramic Matrix Composites at Elevated Temperatures in Air," *Cer. Eng. Sci. Proc.*, **11**[9–10], 1281–1294 (1990).

16. D. Cranmer, R. Salomaa, H. Yinnon, and D. R. Uhlmann, "Barrier to Crystal Nucleation in Anorthite," *J. Non-Cryst. Sol.*, **45**, 127–136 (1981).

17. P. I. K. Onorato, D. R. Uhlmann, and R. W. Hopper, "A Kinetic Treatment of Glass Formation: IV. Crystallization on Reheating a Glass," *J. Non-Cryst. Sol.*, **41**, 189–200 (1980).

18. D. R. Uhlmann, "Nucleation, Crystallization, and Glass Formation," *J. Non-Cryst. Sol.*, **38–39**, 693–698 (1980).

19. D. R. Uhlmann, P. I. K. Onorato, and G. W. Scherer, "A Simplified Model of Glass Formation," *Proc. Lunar Planet. Sci. Conf. 10th*, 375–381 (1979).

20. D. R. Uhlmann, "Glass Formation," *J. Non-Cryst. Sol.*, **25**, 42–85 (1977).

21. S. M. Wiederhorn and B. J. Hockey, "High Temperature Degradation of Structural Composites," *Ceramics International*, **17**, 243–252 (1991).

22. N. S. Jacobson, "High-Temperature Durability Considerations for HSCT Combustor," NASA Technical Paper 3162, National Aeronautics and Space Administration, Washington, DC, January, 1992.

23. L. P. Zawada, L. M. Butkus, and G. A. Hartman, "Room Temperature Tensile and Fatigue Properties of Silicon Carbide Fiber-Reinforced Aluminosilicate Glass," *Cer. Eng. Sci. Proc.*, **11**[9–10], 1592–1606 (1990).

24. J. W. Holmes, "A Technique for Tensile Fatigue and Creep Testing of Fiber-Reinforced Ceramics," *J. Comp. Mat.*, **26**[6], 916–933 (1992).

25. J. W. Holmes, "Tensile Creep Behaviour of a Fibre-Reinforced SiC-Si_3N_4 Composite," *J. Mater. Sci.*, **26**[7], 1808–1814 (1991).

26. J. W. Holmes, "Influence of Stress Ratio on the Elevated Temperature Fatigue of a Silicon Carbide Fiber-Reinforced Silicon Nitride Composite," *J. Am. Ceram. Soc.*, **74**[7], 1639–1645 (1991).

27. M. G. Jenkins, M. K. Ferber, and R. L. Martin, "Evaluation of the Stress State in a Buttonhead, Tensile Specimen for Ceramics," *Cer. Eng. Sci. Proc.*, **11**[9–10], 1346–1363 (1990).

28. J. Lankford, "Compressive Strength and Damage Mechanisms in a SiC-Fiber Reinforced Glass-Ceramic Matrix Composite," in *Proceedings of the Fifth International Conference on Composite Materials (ICCMV)*, eds. W. C. Harrigan, J. Strife, and A. K. Dhingra, TMS-AIME, Warrendale, PA, 1985, pp. 587–602.

29. J. Lankford, "Strength of Monolithic and Fiber-Reinforced Glass Ceramics at High Rates of Loading and Elevated Temperatures," *Cer. Eng. Sci. Proc.*, **9**[7–8], 843–852 (1988).

30. D. F. Hasson and S. G. Fishman, "Impact Behavior of Fiber Reinforced Glass Matrix Composites," in *High Temperature/High Performance Composites*, Materials Research Society Symposium Proceedings, Vol. 120, eds. F. D. Lemkey, S. G. Fishman, A. G. Evans, and J. R. Strife, Materials Research Society, Pittsburgh, PA, 1988, pp. 285–290.

31. S. V. Nair and Y.-L. Wang, "Failure Behavior of a 2-D Woven SiC Fiber/SiC Matrix Composite at Ambient and Elevated Temperatures," *Cer. Eng. Sci. Proc.*, **13**[7–8], 843–852 (1992).

32. S. W. Freiman, D. R. Mulville, and P. W. Mast, "Crack Propagation Studies in Brittle Materials," *J. Mater. Sci.*, **8**, 1527–1533 (1973).

33. S. W. Freiman, "A Critical Evaluation of Fracture Mechanics Techniques for Brittle Materials," in *Fracture Mechanics of Ceramics 6*, eds. R. C. Bradt, A.

G. Evans, D. P. H. Hasselman, and F. F. Lange, Plenum Press, New York, NY, 1983, pp. 27–45.

34. J. H. Underwood, S. W. Freiman, and F. I. Barrata (eds), *Chevron-Notched Specimens, Testing and Stress Analysis*, ASTM STP 855, ASTM, Philadelphia, PA, 1984.

35. E. P. Butler, E. R. Fuller, Jr., and H. Cai, "Interactions of Matrix Cracks with Inclined Fibers," *Cer. Eng. Sci. Proc.*, **13**[7–8], 475–482 (1992).

36. G. R. Antsis, P. Chantikul, B. R. Lawn, and D. B. Marshall, "A Critical Evaluation of Indentation Techniques for Measuring Fracture Toughness: I, Direct Crack Measurements," *J. Am. Ceram. Soc.*, **64**[9], 533–538 (1981).

37. P. Chantikul, G. R. Antsis, B. R. Lawn, and D. B. Marshall, "A Critical Evaluation of Indentation Techniques for Measuring Fracture Toughness: II, Strength Method," *J. Am. Ceram. Soc.*, **64**[9], 539–543 (1981).

38. H. H. K. Xu, C. P. Ostertag, L. M. Braun, and I. K. Lloyd, "Effects of Fiber Volume Fraction on Mechanical Properties of SiC Fiber/Si$_3$N$_4$ Matrix Composites," *J. Am. Ceram. Soc.*, **77**[7], 1897–1900 (1994).

39. J. Rödel, J. F. Kelly, and B. R. Lawn, "In Situ Measurements of Bridged Crack Interfaces in the Scanning Electron Microscope," *J. Am. Ceram. Soc.*, **73**[11], 3313–3318 (1990).

40. J. Rödel, E. R. Fuller, Jr., and B. R. Lawn, "In Situ Observations of Toughening Processes in Alumina Reinforced with Silicon Carbide Whiskers," *J. Am. Ceram. Soc.*, **74**[12], 3154–3157 (1991).

41. L. M. Braun, S. J. Bennison, and B. R. Lawn, "Objective Evaluation of Short-Crack Toughness Curves Using Indentation Flaws: Case Study on Alumina-Based Ceramics," *J. Am. Ceram. Soc.*, **75**[11], 3049–3057 (1992).

42. C. P. Ostertag, H. Xu, L. Braun, and J. Rödel, "Pressureless Sintered Fiber Reinforced Composites: Part II. In Situ Crack Propagation," *J. Am. Ceram. Soc.*, in press.

43. E. R. Fuller, Jr., R. F. Krause, Jr., J. Kelly, R. N. Kacker, E. S. Lagergren, P. S. Wang, J. Barta, P. F. Jahn, T. Y. Tien, and L. Wang, "Microstructure, Mechanical Properties, and Machining Performance of Silicon Carbide Whisker-Reinforced Alumina," *J. Research NIST*, in press.

44. D. C. Cranmer, "Fiber Coating and Characterization," *Bull. Amer. Ceram. Soc.*, **68**,[2], 415–419 (1989).

45. Military Standardization Handbook 17, "Polymer Matrix Composites Guidelines," MIL-HDBK17B, Vol. 1, 1988; "Plastics for Aerospace Vehicles, Part I. Reinforced Plastics," MIL HDBK17A, 1971, US Department of Defense, Washington, DC.

46. L. Mordfin, "Nondestructive Evaluation," in *Materials and Processes, Part B: Processes*, 3rd Edn, eds. J. F. Young and R. S. Shane, Marcel Dekker, New York, NY, 1985, pp. 1495–1519.

47. G. Birnbaum and G. S. White, "Laser Techniques in NDE," in *Nondestructive Testing*, Vol. 7, Academic Press, London, U.K., 1984, pp. 259–365.

48. R. J. Pryputniewicz, "Laser Holography," Worcester Polytechnic Institute, MA, January 1979.

49. A. Vary and S. J. Klima, "NDE of Ceramics and Ceramic Composites," NASA Technical Memorandum 104520, National Aeronautics and Space Administration, Washington, DC, July 1990.

50. H. E. Kautz and R. T. Bhatt, "Ultrasonic Velocity Technique for Monitoring Property Changes in Fiber-Reinforced Ceramic Matrix Composites," *Cer. Eng. Sci. Proc.*, **12**[7–8], 1139–1151 (1991).

51. A. Vary, "NDE Standards for High Temperature Materials," NASA Technical Memorandum 103761, National Aeronautics and Space Administration, Washington, DC, April 1991.

52. B. J. Hockey, S. M. Wiederhorn, W. Liu, J. G. Baldoni, and S.-T. Buljan, "Tensile Creep of Whisker-Reinforced Silicon Nitride," *J. Mater. Sci.*, **26**, 3931–3939 (1991).

53. A. Chulya, J. Z. Gyekenyesi, and J. P. Gyekenyesi, "Failure Mechanisms of 3-D Woven SiC/SiC Composites under Tensile and Flexural Loading at Room and Elevated Temperature," *Cer. Eng. Sci. Proc.*, **13**[7–8], 420–432 (1992).

54. S. W. Freiman and C. M. Hudson, "Methods for Assessing the Structural Reliability of Brittle Materials," ASTM STP 844, Philadelphia, PA, 1984.

55. S. W. Freiman, "Fracture Mechanics Applied to Brittle Materials," ASTM STP 678, ASTM, Philadelphia, PA, 1979.

56. R. M. Kent and A. Vary, "Tensile Strain Measurements of Ceramic Fibers Using Scanning Laser Acoustic Microscopy," *Cer. Eng. Sci. Proc.*, **13**[7–8], 271–278 (1992).

57. D. J. Pysher and R. E. Tressler, "Creep Rupture Studies of Two Alumina-Based Ceramic Fibers," *J. Mater. Sci.*, **27**, 423–428 (1992).

58. D. J. Pysher and R. E. Tressler, "Tensile Creep Rupture Behavior of Alumina-Based Polycrystalline Oxide Fibers," *Cer. Eng. Sci. Proc.*, **13**[7–8], 218–226 (1992).

59. D. C. Cranmer, U. V. Deshmukh, and T. W. Coyle, in "Thermomechanical Properties of Metal Matrix and Ceramic Matrix Composites," ASTM STP 1080, ASTM, Philadelphia, PA, 1990, p. 124.

Index